# THE ARMALITE AR-10

# THE ARMALITE AR-10
## World's Finest Battle Rifle

Joseph Putnam Evans

produced and edited by R. Blake Stevens

Collector Grade Publications
INCORPORATED

2016

# The Collector Grade Library, 1979 - 2016

**● = In Print and Available**

© 2016 Collector Grade Publications Incorporated

**ISBN 0-88935-583-5**

**Published by Collector Grade Publications Incorporated**
**PO Box 1046, Cobourg, Ontario K9A 4W5 Canada**
**Printed and Bound in Canada**

# *Acknowledgements*

This work is a continuation of the tradition of scholarship that already exists about the AR-10. I feel that I am but a novice compared to past experts, many of whom have been studying this topic and enjoying, shooting, and examining these rifles since long before my birth.

First and foremost, I wish to extend my deepest thanks and appreciation to my intrepid editor, publisher, and friend, R. Blake Stevens. To be considered a peer to this legend of firearms scholarship is truly an honor. The renowned author of the definitive compendia on the FAL, M14, AR-15, and many other modern arms, Blake stands clearly at the top of our historical fraternity, and working with him was a great privilege. His encyclopedic knowledge of the history of military small arms, vast worldwide connections, and most of all his enthusiasm for the topic, were critical in making this book what it is.

I must also pay tribute to Eugene Stoner, George Sullivan, Charles Dorchester, Richard Boutelle, L. James Sullivan, and Arthur Miller. Were it not for the contributions of these men, neither the AR-10, nor any of its numerous and successful modern offspring, would have ever existed. The Dutch opposite numbers to the American designers and engineers, Friedhelm G. Jüngeling, J. E. Manus Van Der Jagt, F. W. Spanjersberg, and M. A. Bakker, were crucial in developing and modifying Stoner's original design, allowing a fortunate (unfortunate for those on the receiving end) few a glimpse of its blinding prowess on the field of battle. Similarly, without the imaginative and dedicated representation, marketing, and promotion of the AR-10 by Samuel Cummings, Jacques Michault, and Robert "Bobby" MacDonald, the AR-10 might never have been purchased or adopted by even the few élite forces that did make use of it.

With regard to the actual combat use of the AR-10, I wish to express my most sincere gratitude to Lieutenant General Kuol Deim Kuol of the Sudan People's Liberation Army for taking the time to speak with me. Of the many brave Portuguese servicemen who carry on the proud tradition of their country's airborne armed forces, I would like to single out Chief Sergeant Carlos Alberto De Sà Canas and Sergeant Major Alfredo Serrano Rosa for their extraordinary contributions to my research. Much of what I learned about the harrowing combat service of the AR-10 in Africa is thanks to these men.

I had a rare and wonderful opportunity to visit Springfield Armory—now a museum and historic site—and personally disassemble and study the Dutch-made Second Transitional AR-10, serial no. 003769, which resides in their collection (fig. 300); one of the vanishingly few such weapons set up to accept an infra-red night-sighting device. Accordingly, I would like to offer my thanks to Alexander MacKenzie, curator of firearms, who arranged time for my visit and the private study of this rifle. His work is critical to the preservation of a vast collection of historic firearms, and he deserves the gratitude of our entire community.

On the collector front, *The AR-10'er* newsletter, published during the 1980s, served as the starting point for my inquiries into the AR-10's history, and its editor and publisher, Louis T. Carabillo, Jr., also deserves my heartfelt thanks. The extant body of knowledge about the AR-10 featured therein and elsewhere is due to the collective efforts of a comparatively small and insular community of enthusiasts, the names of many of whom I will never know, and thus I am unable to thank every one of these dedicated researchers by name. However, for being such good friends and adding immeasurably to my own understanding of this rifle, I would like to personally thank "Huckleberry Hollow", Dan Shannon, Rob Allison, Ralph Cobb, Bruno "Armeiro" Santos, and Vic Tuff. Without you, gentlemen, this book would not have been possible.

I also want to extend my compliments to Major Sam Pikula, U.S.A.R., for his scholarly work titled *The ArmaLite AR-10 - The Saga of the First Modern Combat Rifle*, which he published privately in 1998. Although I made a diligent effort to contact Major Pikula and acquire a copy of his book, I was unsuccessful in both attempts, and so none of what may be learned from it is reflected in this study. I was, however, able to locate and gain much inspiration from an article written by Major Pikula titled "First Modern Combat Rifle: The Unknown Legend of the AR-10", which was published in the Summer, 1996 issue of the Soldier of Fortune *Fighting Firearms* periodical.

Finally, my heartfelt appreciation goes out to Colonel (ret.) Mark Westrom, who founded the modern-day company that carries on the name ArmaLite and the AR-10 design, and Mr. C. Reed Knight, Jr., the owner of Knight's Armament Company and the proprietor of the SR-25. These men generously shared their vast expertise on matters concerning both the early AR-10 and its modern resurrections at their hands. Mr. Knight was also so kind as to make available the Institute of Military Technology collection, containing virtually every model of the AR-10 ever produced, for me to study and photograph. He is an expert without equal, and he guards vigilantly the heritage of the AR-10 for posterity.

In addition, the Editor also wishes to thank the following, who have contributed generously and materially to enhance the scope and veracity of this study:

- Christopher R. Bartocci
- Marcel Braak <marcel.braak@xs4all.nl>
- Terry Edwards
- the late William B. Edwards
- The [Edward C.] Ezell Archives, Defence Academy of the United Kingdom, Shrivenham, Swindon
- Dolf L. Goldsmith
- The Dutch Nationaal Militair Museum, Soesterberg <www.nmm.nl>
  - Matthieu Willemsen, Conservator
- Dr. Elmar Heinz
- Jean Huon
- Frank Iannamico
- The Institute of Military Technology, Titusville, Florida <www.instmiltech.com>
  - Corey Wardrop, Curator
- Edward R. Johnson
- Richard Jones, Editor, *Janes Infantry Weapons*
- James D. Julia Auctioneers
- Eric Kincel, Director of Research and Development, Bravo Company USA Inc.
- the late R. H. G. ("Kick") Koster
- Charles Kramer <chauchat1@aol.com>
- Law Enforcement International Ltd., St. Albans, Herts, U.K.
  - Greg Felton, Director
- Peter Marcuse, official photographer for the Artillerie-Inrichtingen AR-10 program
- John Miller

- Marc Miller
- Jeff Moeller
- Thomas B. Nelson
- Michael J. Parker
- Tim Rooker, Chief Technician, ArmaLite, Inc.
- Robert Segel, Editor Emeritus, *Small Arms Review*
- Dan Shea, Publisher, *Small Arms Review*
- Mike Spradlin
- L. James Sullivan <jim.armwest@commspeed.net>
- Dave Terbrueggen
- Masami Tokoi
- U.S. Bureau of Alcohol, Tobacco, Firearms & Explosives (BATFE)
  - Earl L. Griffith, Chief, Firearms & Ammunition Technology Division, Martinsburg, WV
  - Ed Owen, former Chief, Firearms & Ammunition Technology Division
- Eric de Wilde <eric.jan.dewilde@gmail.com>

# *Dedication*

I dedicate this work to my loving and supportive wife Huan, who is so kind and wants me to be happy so much that she condones and even encourages my interest and study of strange and arcane topics like this one.

# *Table of Contents*

# *Foreword*

**B**efore I begin what I hope will be the definitive history of the finest, most beautiful, and most advanced-for-its-time battle rifle ever designed, I would feel remiss if I did not briefly explain my infatuation with this weapon. Although I use what must seem to be a vast number of superlatives in the following pages, I have tried my best to dampen my enthusiasm at being able to chronicle the development of this singularity of firearms achievement. The use of such forceful rhetoric to describe this rifle is deliberate, because I firmly believe it would be quite difficult to overstate the true precociousness of its design, and the unsung impact it has had on the successful military and civilian rifles of today. Even so, my own feelings about the AR-10 pale in comparison to the high opinions held by the veteran AR-10 users whom I interviewed during the course of my research.

I have been a firearms enthusiast since turning 18, mostly as an extension of my love of history, but also due to my interest in the outdoors. Although not an engineer by profession, the inner workings of firearms have always fascinated me. When acquiring any new firearm I would always strip it down to its smallest parts and study their interaction. It is from this curiosity about what makes firearms 'tick' that I came to deeply respect an advanced design when I saw one. Throughout college I had acquired different surplus rifles (on an undergraduate budget), before finally getting my hands on my first semi-automatic rifle, a Norinco SKS.

After I graduated and headed back to China to work for a year, I had decided that two years of trying unsuccessfully to make an SKS work reliably and fire accurately was enough, and that when I returned to the U.S. for law school, I would acquire a truly excellent rifle, with no compromises made for cost. After no small amount of research, I decided on a pair of weapons, both filling different roles, but each a modern-representation of the AR-10. One was a modern AR-10 made by ArmaLite, and the other was a Knight's Armament SR-25.

I was so thoroughly impressed by every possible aspect of these rifles that I immediately fell in love with the design. I could not get over the fact that for once in my life, I could take a new rifle right from the box, study it, and just use it without worrying that each magazine might give out on me, that it wouldn't hit the target at which I was aiming, or that a blown primer would fuse my firing pin in the forward position, where a poorly-machined bolt would result in a slamfire that would cause the reciprocating charging handle to break my thumb. All of these horrors and more had been visited upon me by my SKS, and so the contrast could not have been sharper.

It seemed that the AR-10 was designed to address every complaint that had ever flitted through my consciousness about a firearm, and that it solved every one of them perfectly, as well as ones of which I had never thought. Just to name a few of its advantages: it was lighter; fired a superb cartridge with pinpoint accuracy; recoiled less; and never malfunctioned, no matter how dirty or hot it was.

As I worked on my pair of rifles and shot them more and more, I became such a convert to their design that I began to read about their history and origins. Little did I know that this would launch me on one of the most interesting, exciting, and intellectually stimulating quests of my life; the discovery of more than half a century of history which had led to the perfect amalgamation of ingenuity and functionality I held in my hands; and the realization that sacred stuff indeed resides in those polymer stocks and anodized aluminum metalwork.

In the following pages I propose to share with the reader a story that has enthralled me for countless hours. I hope that you are able to enjoy the same connection with this marvel of technology.

Joseph Putnam Evans
Arlington, Virginia
December 9, 2015

# Editor's Preface

## A Thumbnail History of the AR-10

Compared to the other military shoulder arms of its era, the ArmaLite AR-10 was in several ways a decided departure from the norm.

First, as the firm name itself makes clear, the paramount characteristic sought in every arm developed by ArmaLite, irrespective of type or action, was light weight.

True to form, the 7.62mm NATO caliber AR-10 was much lighter than any of its contemporaries, at least in the beginning. It was described in the 1957 ArmaLite brochure as weighing 6.85 lbs., without magazine. An ArmaLite document titled "Comparison of Basic Infantry Rifles" (fig. 66) shows that this was indeed substantially less than the weight of the standard M1 rifle (9.56 lbs.), the developmental T44 (8.45 lbs.), and the FN T48 (9.47 lbs.).

The light weight of the AR-10 was accomplished mainly by constructing the upper and lower receivers of aluminum alloy, the use of which was in turn made possible by the front-locking, multi-lugged bolt, which locked directly into the barrel extension, in emulation of the locking system previously employed in the Johnson rifle. In such a system the receiver elements are not required to withstand the violent stresses incurred in firing the powerful 7.62mm NATO cartridge, and function largely as mere guides for buffering and returning the recoiling parts.

Hand in hand with its light weight, the original ArmaLite AR-10 as tested at Springfield Armory in 1956 and 1957 produced remarkably light recoil, which is most unusual in a firearm firing a powerful military cartridge, where light weight and low felt recoil are generally considered to be mutually exclusive. This light recoil was largely the result of incorporating a large, cleverly-designed muzzle brake, the effect of which was described succinctly by Fairchild consultant Melvin M. Johnson Jr. in his report titled "Preliminary Tests of F E & A [Fairchild Engine and Airplane Corp.] AR-10 ARMALITE Light Automatic Rifle at Springfield Armory" dated December 2 - 7, 1956, as follows:

*A recoil reduction of 40 per cent was noted from the muzzle brake.*

In addition to light weight and low recoil, the AR-10 was remarkably controllable, especially in fully-automatic fire. This additional characteristic was due to a further salient design feature whereby the butt of the stock and the bore were in the same plane. As Mel Johnson described it in a later report dated September 14, 1957,

*. . a straight stock is superior to the ancient drop stock, especially for a fast-firing military automatic. Some reasons include:*

- *Reduced climb.*
- *Better shooter control.*
- *Better target area spotting visibility with high sights.*
- *Reduced effect of heat mirage on sight picture.*

- *Preferable bore-axis distribution of moving parts with large diameter mainspring in action line.*
- *Shorter body due to straight-line action partly housing in butt section.*

For various reasons, during the series of events which followed the 1957 agreement that licensed the mass-production of the ArmaLite AR-10 to the state-owned Dutch arms manufactory Artillerie-Inrichtingen, most of these advantages were traded away. Of the three original salient features (light weight, low recoil and controllability), low recoil became the first casualty when the distinctive can-shaped muzzle brake was abandoned in favor of adding versatility to the AR-10 to allow for bayonet mounting, plus the ability to launch grenades from the rifle's muzzle.

With the demise of the highly effective muzzle brake, the inexorable laws of physics turned the original low weight of 6.85 lbs. from a unique advantage into an instant liability, with the result that over time, the weight of the AR-10 slowly crept up. The May, 1958 Interarmco sales brochure describing the first (Cuban) model AR-10 manufactured by Artillerie-Inrichtingen listed it as weighing 3.25kg (7.16 lbs.) without magazine. In the first A-I handbook, dated September, 1959, titled "The ARMALITE AR-10 INFANTRY-RIFLE Caliber 7.62 (NATO)," the weight of the then-current model was given as 3.75kg (8.27 lbs.) without magazine. In the final A-I publication, "Handbook on the ArmaLite AR-10 Infantry Rifle Caliber 7.62mm NATO," dated June, 1961, the weight of the finalized NATO (Portuguese) Model AR-10 was listed as 4.05kg (8.93 lbs.).

This left only the third original advantage, controllability. As the Canadian writer and researcher Terry Edwards commented in his January, 1978 *Soldier of Fortune* article titled "Great Expectations: AR-10," concerning the Sudanese Model AR-10,

> . . *The kick is violent, and the noise when fired from the shoulder irritating. From the hip, the blast is deafening, but, control is excellent. It doesn't twist or climb; it just locks on target driving out 7.62mm bullets at 700 rounds per minute.*

Edwards further summed up his firing test of the Sudanese Model AR-10 as follows:

> . . *Recoil exceeds that of the FN/FAL or the M14, but is not unmanageable for anyone of average North American build. The rifle's remarkable feature is its combination of light weight (nine pounds empty), pointability and uncanny control under rapid or full-auto fire. The in-line stock configuration is designed to eliminate muzzle climb by placing the axis of the bore in line with the shooter's shoulder so the gun will kick straight back instead of pivoting around the shoulder. It works. Laying a three-shot burst on a silhouette target at 100 yards is no problem. After that, the kick begins to knock the shooter off balance, but when shooting 7.62 NATO, three shots should be enough . .*

Further events were to follow. The M14 rifle, adopted on May 1, 1957, became the shortest-lived U.S. service rifle of all time as it and the 7.62mm NATO cartridge itself were swiftly superseded by the 5.56mm AR-15 (M16). The AR-15, also an ArmaLite development, was essentially nothing more than the familiar straight-stocked AR-10 reduced in size, weight and caliber to become, in the words of General Willard Wyman, Chief of the U.S. Continental Army Command (CONARC), "the ideal SCHV platform."

As for the AR-10, powerful personality clashes within the Dutch military establishment culminated in the announcement on May 25, 1961 that the Dutch Ministry of Defense had prohibited Artillerie-Inrichtingen from making any further expenditures on arms. Faced with this order, A-I closed down all AR-10 research, development and production on July 6, 1961 after manufacturing a total of less than 6,000 examples, and effectively orphaning the relatively few AR-10s which were still seeing military service.

Meanwhile, during the latter half of 1956, representatives of the using branches of the U.S. military at Fort Benning, Fort Campbell and Fort Monroe had been very positively impressed with the U.S.-made AR-10 in the demonstrations they had witnessed, to the point where in November, 1956 General Maxwell Taylor ordered that the decision on adopting a new rifle be delayed indefinitely in favor of the AR-10.

Even before that, in late March, 1956, overtures had been made by the Ordnance Department through retired General Devers, who was then acting as a consultant for ArmaLite's parent company, Fairchild, to acquire the rights to the AR-10 and continue its development at Springfield Armory, with a royalty to be paid to Fairchild on any and all subsequent production. In a watershed decision that was to profoundly affect the entire future history of the AR-10 and the later AR-15, Fairchild rejected the government's offer, even though they had already denied ArmaLite's entreaties for more funding to expand their own facility. With the deal with the Dutch still in the future, no capability to undertake a program of intensive production engineering development or large-scale manufacture of the AR-10 thus existed.

## The Road Not Taken

Consideration of what might have been the future course for the AR-10 had the government's offer been accepted makes for some fascinating conjecture. At the same time as the rights offer had come in, in a show of good will the government had shared the results of the SCHV (Small Caliber, High Velocity) portion of the ongoing Project SALVO with ArmaLite, with the thought that they would be the perfect group to develop what was to become the AR-15, meaning that the seeds of demise of the 7.62mm NATO round as the standard U.S. military rifle cartridge had already been sown.

But what if the 7.62mm rifle chosen for adoption in 1957 had been a government-perfected AR-10 instead of the M14? The first ArmaLite AR-15s had already appeared that spring, and one wonders whether the comparison between the "apples and oranges" of the AR-15 and the M14 would have been so stark had the two rifles resembled one another as closely as did the AR-15 and the AR-10. This might have allowed time for a re-evaluation of the 7.62mm NATO cartridge for shoulder arm use where, in the hands of one or more members of every squad, its enhanced range, penetration and power—the 7.62mm round produces approximately twice the striking energy of the 5.56mm—fired from a truly controllable selective-fire rifle such as the AR-10, might have proven very useful indeed in any number of combat scenarios.

Also, as happened in later years when the AR-10 design was resurrected as the Knight's Armament SR-25, a certain degree of commonality might well have been engineered into both the AR-10 and AR-15 before mass production was begun, and the belt-fed light machine gun version of the AR-10 might also have been perfected, with the result that a really versatile weapons family sharing many common logistical features and components might have become a reality.

## The Truth - and the Legacy

This was not to be, of course, and the story of what actually happened to the AR-10 is told in the following pages in a fuller and more complete rendering than has ever before appeared in print.

Remarkably, for a rifle that has been out of production in its original form for over half a century, the legacy of the 7.62mm NATO caliber AR-10 remains strong. AR-10-

pattern rifles are still in production today, with the Knight's Armament SR-25 being adopted in 2000 by the U.S. Navy SEALs as the "Rifle 7.62mm Mk 11 Mod 0," and a version of Mark Westrom's resurrected ArmaLite AR-10 adopted by the Canadian Army Special Forces as the AR-10(T) in 2002. In addition, a modern modular 7.62mm NATO caliber sniper rifle designed by the Lewis Machine & Tool Co., based on the original direct gas system ArmaLite AR-10 and initially named the LM7, has been further developed by Law Enforcement International Ltd. and adopted in 2010 in a semi-automatic-only configuration by the British Army as the L129A1 "Sharpshooter." According to Richard Jones, the editor of *Jane's Infantry Weapons*, "Although it was purchased under a UOR [Urgent Operational Requirement], the operational success of the L129A1 has been such that in 2013 it was announced that it would, post-Afghanistan deployment, be taken into 'core', becoming an established element of the British Armed Forces infantry weapons fleet in the future."

## The Foot in the Door

Undoubtedly, the AR-10's greatest and longest-lasting legacy is as the forerunner of the 5.56mm AR-15/M16, which has gone on to become the longest-serving military shoulder arm in U.S. history. With no replacement in sight, the M16/M4 remains the benchmark against which all new military rifles, domestic and foreign, must be compared.

R. Blake Stevens
Gore's Landing, Ontario
December 28, 2015

# Part I: The Hollywood Years

*Chapter One*

# In the Beginning

So often in the history of human endeavor the institutions that seem set in marble to those who live in the world they have already shaped, have in reality come about by chance encounters and the vagaries of fate, and were in their infancy every bit as fragile and tenuous as a seedling clinging to the sheer face of a cliff. This was the case with the ArmaLite AR-10, which would seismically shift the small arms world, and be the genesis of the most pervasive and successful U.S. rifle family of the modern age.

This revolution in firearms design did not start within the design bureau of a well-established company, or the ordnance board of a first-rate military power; it came about by the chance meetings of a few remarkable individuals who saw the potential in one another to collaborate on the creation of something truly great.

## Introducing a Family of Innovators

Our story begins in California just two years after the end of WWII, when brothers-in-law George Sullivan and Charles Dorchester began utilizing their mechanical expertise and access to machine shops to indulge their interest in creating new types of firearms. This dream started rather traditionally, with the design of a bolt-action precision sporting rifle, but their methods of production and the materials they used were anything but conventional.

George Sullivan, the chief patent counsel for the Lockheed Corporation, the famous military and civil aircraft firm, had been fascinated for much of his professional life with the exotic "wonder materials" pioneered by the aircraft industry. With his interest in applying these innovations to the world of firearms, and his brother-in-law Dorchester's engineering expertise, the two took the first step on the journey that would culminate in the introduction of the AR-10.

As could be observed by fellow hunters in the country around Santa Monica, California in the late 1940s and early 1950s, there was something strange about the rifles carried into the field by George Sullivan in pursuit of game. Little did they know that the different experimental models using strange plastic stocks and aluminum parts, which presented themselves in often-garish colors, were the development stages of what would become the "Para-Sniper" rifle.

In 1952 Sullivan and Dorchester, the two pioneers of the Para-Sniper, decided to commit resources to the venture of developing both it and other new designs in the brave new world of aircraft-material firearms. They began as an unincorporated general partnership, with one clear goal being to explore the new frontiers of lightweight firearms design by incorporating the alloys and plastics of the aerospace industry. The concern took as its first name "S-F Projects," a title bearing no resemblance to the later well-known company name "ArmaLite," but from such humble beginnings, great things would emerge.

The entire premises and resources of S-F Projects consisted of a modest machine shop in George Sullivan's garage at 3085 Lake Hollywood Drive in what is today considered Los Angeles, and the fertile and inventive minds of Sullivan and Dorchester. A more modest beginning could hardly be imagined, but from this small factory of innovation, the first faint glow of the AR-10's rising sun, still far below the horizon, could be seen.

# The S-F Projects "Para-Sniper" Rifle

Various pre-standardization examples of Sullivan-Dorchester lightweight bolt-action rifles were chambered in .30-'06 Springfield, the standard U.S. military cartridge at the time. Some had muzzle brakes, and a different type of commercial telescopic sight sat atop almost every example.

The first design actually intended for commercial production by the Sullivan-Dorchester alliance, initially known briefly as "S-F Projects" of Hollywood, California, was the bolt-action "Para-Sniper" rifle, later designated the ArmaLite AR-1. The Para-Sniper sported many radical and previously unheard-of features.

While ostensibly describing one rifle, the Para-Sniper was a work in progress from its inception, and since it never entered a production stage, almost every example differed in some parameter or other from the others.

## The .257 Roberts Caliber "Paratrooper"

The Para-Sniper rifle first saw the light of day as the S-F Projects "Paratrooper," chambered for the flat-shooting .257 Roberts cartridge; a round that, while not as potent as the .30-'06, combined excellent range, acceptable power, and low recoil.

## Moving Up to the Developmental "T65" (.308 Winchester) Cartridge

On October 28, 1953, it was proposed to chamber the rifle for the finalized version of the developmental U.S. .30 caliber "T65" round, soon to be adopted as the "7.62x51mm NATO," and already available in a commercial version known as the .308 Winchester, which had been introduced in 1952 and had already become a favorite for use in short-action rifles.

The rifle so chambered was the first to be christened the "Para-Sniper," due to its intended use as both a fine sporting rifle and as a military sniper weapon.

On its surface, the Para-Sniper looked much like most popular hunting rifles of the time, and indeed of today. It was a bolt-action rifle based on the front-locking Mauser 98 bolt system (the basis for almost all precision bolt-action rifles today), with a plain-looking barrel, and a stock that was contoured the same as that of most sporting rifles. On closer inspection, however, the Para-Sniper was anything but conventional.

Originally built on a standard steel Remington Model 722A receiver, the .308 caliber Para-Sniper rifle was fitted with a brown-colored lightweight "Shellfoam" stock, made of a high-strength molded shell filled with a plastic foam core. The barrel, which appeared at first glance to be blued ordnance steel, was actually plain 7075 aircraft aluminum with an anodized finish. Sullivan referred to this material as "Sullaloy" (Sullivan's alloy). Later, the barrels were strengthened by swaging the aluminum alloy around a rifled liner made of stainless steel.

The initial steel-receivered Para-Sniper rifle weighed 6 lbs., complete with scope and sling. Unheard-of at the time, and still extremely rare on bolt-action precision rifles today, the Para-Sniper's barrel was threaded at the muzzle and fitted with an all-things-to-all-men muzzle brake, which according to the S-F Projects 1954 brochure "reduces recoil, flash, muzzle blast and barrel jump."

A second model, boasting a receiver made of aluminum, a fiberglass stock, and a composite aluminum/stainless steel barrel, looked and felt modern and, at 5¾ lbs. complete with scope and sling, impossibly light.

# The 1954 S-F Projects Para-Sniper Brochure

These pages from early 1954 represent the very first publication by S-F Projects, the organization that would become known as ArmaLite, which would give birth to the AR-10. This advertisement for the Para-Sniper before it gained its "AR-1" designation, as well as for the innovative design techniques championed by the new organization, also stands as the only known public use of "S-F Projects," the original name of the Sullivan-Dorchester partnership. As it was never incorporated or listed in a telephone directory, the name by which this earth-shakingly influential concern was first known would be lost to history without this important document.

While the Para-Sniper never proved successful in either military or civilian applications, its advanced design would be the catalyst for the formation

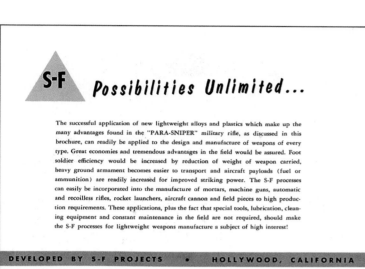

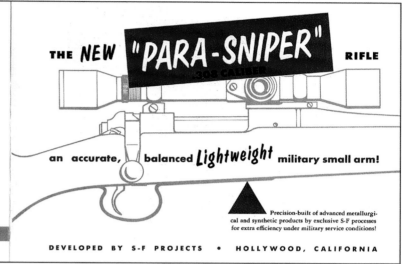

1. The outside pages of the fold-open "Para-Sniper" rifle brochure, copyrighted in 1954 by S-F Projects of Hollywood, California.

At this point the .308 caliber Para-Sniper was still fitted with a steel receiver, and even so the weight is listed as "approx. 5 ¾ lbs."     Smithsonian Institution collection

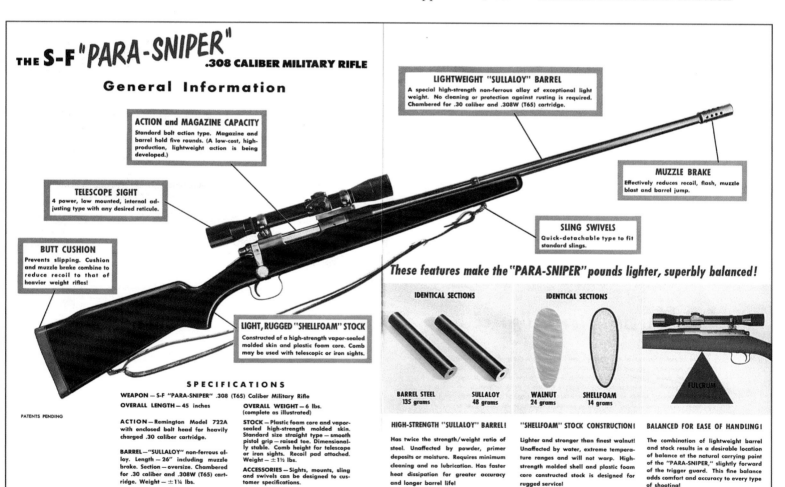

2. Inside the 1954 S-F Projects' brochure for the .308 caliber "Para-Sniper" military rifle, describing some of its innovative features.

The weight is here listed as 6 lbs. complete with scope.

Note under "Action and Magazine Capacity" that the steel receiver was soon to be replaced by a "low-cost, high-production, lightweight action" which was being developed.

Smithsonian Institution collection

of, and critical early investment in, ArmaLite, as well as the inspiration for many features found on later rifles like the AR-10.

# George Sullivan Meets Jacques Michault

Just months later, the AR-10 inched just a bit closer to reality as a result of a fortuitous meeting between Jacques Michault, the founder and president of SIDEM International, a Belgian arms sales firm which had been contracted by the U.S. occupation forces to supply West Germany's newly-reformed military with weaponry, and George Sullivan, acting in his then-professional capacity as representative of the Lockheed Corporation. The two became acquainted over lunch, and subsequently sat down together in Lockheed's California headquarters, where they shared ideas which, while gained from experience in very disparate fields, proved to be eminently complimentary. Michault, delighted to learn of his American contact's interest in firearms, shared with Sullivan his observations on German research towards the end of WWII, which had concentrated on producing lighter, more manageable infantry weapons.

Michault himself had already played a very important role in the development of the modern battle rifle, though he scarcely could have known it at the time. At his firm SIDEM International, Michault had been experimenting with different new cartridges since the end of WWII, and had been in close contact with the U.S. military's developers in the field. This was a time of great flux when the powers of western Europe and the United States, already formulating the alliance that would become NATO, had in their minds the goal of standardizing the equipment of all friendly forces. Needless to say, the new cartridge that would emerge successful in this grand game of a selection process would enjoy a degree of omnipresence and hegemony never before seen in military history.

Although the dream of a single infantry rifle for all Allied forces was quickly abandoned, due to the great stake and pride many nations had in their own weapons designs (some more ingenious and deserving than others), it was decided that all should at least fire the same ammunition. The field during the late 1940s and early 1950s was inundated with new cartridge designs, with some powers such as the British, Belgians, and Germans preferring an intermediate-power assault rifle load, while others, like the United States, France, and the Scandinavian member states, were searching for a shorter, more efficient cartridge that would provide essentially the same excellent ballistic performance as had the previous full-power rifle rounds.

Michault became sold rather quickly on the developmental "T65" .30 caliber (7.62mm) cartridge design championed by his contact and friend Colonel René Studler, Chief of Small Arms R&D of the U.S. Army Ordnance Department, and was instrumental in what essentially amounted to the strong-arming of the West Germans into adopting this cartridge.

This type of ammunition utilized a slightly shorter case, loaded with modern, more efficient powder which would essentially propel the same size-and-weight bullet as the previous full-power cartridges, to the same velocities. This shorter length was critical to the advent of the modern battle rifle because a shorter cartridge meant that the receiver, one of the heaviest parts of most rifles, could be shorter, thus saving weight and increasing efficiency.

American small arms doctrine at the time was opposed to the reduced-power theories championed by the British, Belgians, and Germans (until these countries were "convinced" by the U.S. to change their minds), and was informed by the excellent effectiveness of the M1 Garand rifle. It was believed that rapid fire controllability, although that could be achieved to a certain extent, was less important than long-range stopping power, and that keeping a full-power round as the main infantry issue rifle cartridge was necessary to this end.

It was due in no small part to Michault's advocacy that the T65 .30 Light Rifle cartridge found its first traction outside of the United States. After several improvements and increases in case length from the initial 47mm through 49mm to the final length of 51mm, this round would eventually become adopted in December, 1953 as the 7.62x51mm NATO, which would not only be the standard chambering for all the modern military battle rifles used by the NATO alliance, but would also be the round fired by the AR-10. However, the great power of the new NATO round meant that its heavy recoil would constantly dog all attempts to field controllable selective-fire rifles weighing less than the benchmark M1.

As for Sullivan and Michault, the special innovations that would comprise the AR-10, which were at once ahead of their time and never rivaled before or since, would combine to make the AR-10 the most manageable rifle in this caliber ever developed.

In their first informal meeting, as well as in subsequent conferences between the two, Sullivan introduced Michault to the idea that aircraft materials might be the answer to the question of controllability in a lightweight rifle firing the NATO cartridge. The two soon came to the conclusion that the ideal rifle of the future would be "aluminum  .  . using a stock of fiberglass constructed like a helicopter rotor blade, and with a straight in-line design which would require a higher sight." They even hypothesized the possibility of using the high rear sight as a carrying handle and scope base. Little did they know the paradigm-shifting effect this simple conversation would have on the future of small arms, and how it would lead both of their professional lives in new and fateful directions.

## Serendipity Strikes Again: Sullivan Meets Richard Boutelle

Meanwhile, in June, 1953, another serendipitous if unexpected meeting was to bring the fledgling S-F Projects partnership the funding and resources it needed to take off. Just one year after the formation of the company, George Sullivan happened to be attending an aircraft industry trade show and conference, where he struck up a conversation with Paul Stetson Cleaveland, secretary and chief counsel of the Fairchild Engine and Airplane Corporation. At this lunchtime meeting Sullivan, not likely aware that he was doing anything more than shooting the breeze with a like-minded enthusiast, described at some length to Cleaveland his ideas for advanced small arms, including the plan for a lightweight, in-line infantry rifle. Little did he know that the gist of this conversation would reach the ear of the president of Fairchild, Richard Boutelle.

Mr. Boutelle, a veteran of the U.S. Army Air Corps and an executive for whom the products of his company were more than just business, also happened to be an avid hunter and firearms enthusiast. At the time, Fairchild was in the process of diversifying, and its board of directors were looking to expand into new fields, wherein the materials and manufacturing techniques, as well as the industrial capacity of their firm, could be of benefit.

Through Paul Cleaveland, Boutelle and Sullivan quickly struck up a friendship, with Sullivan paying a visit to Boutelle in Washington that summer, and showing him some of the prototypes on which his small startup had been working. Almost on the spot, Boutelle, possessed of considerable personal financial resources, began not only cooperating with and testing the designs of S-F Projects, but by December, 1953 was actually underwriting these developments out of his own pocket.

Over the ensuing year, the two kept in close touch and made regular visits to one another's home regions, usually with Sullivan making his way east, which he already had to do often in his work for Lockheed. In addition to the development of the Para-Sniper from its original .257-caliber Paratrooper format, the two explored other ideas, and in the process formulated some interesting and eclectic prototype designs. One of the strangest was a pump-action shotgun firing .38 Special revolver cartridges specially loaded with shot instead of single bullets. Both Sullivan and Boutelle were avid handloaders of their own ammunition, so this project was one they were uniquely suited to undertake. Sullivan also fabricated a standard 12-gauge shotgun based on a receiver from the famous Ithaca company (fig. 14), using polymer and aluminum components for the remaining parts of the weapon.

Neither of these early shotgun designs was pursued into the future, but their strange and esoteric principles give a good example of the extreme creativity and innovation that was constantly simmering in the minds and workshops of Sullivan and Boutelle.

## Fairchild Acquires S-F Projects: ArmaLite is Born

After over a year of personal cooperation and collaboration, as well as serving as a private benefactor, Boutelle was finally able to get all of the necessary approvals, and on October 1, 1954, the firm we know today as ArmaLite became a division of the Fairchild Engine and Airplane Corporation. ArmaLite thus acquired the funding and support needed to capitalize on the enterprising initiative showed by Sullivan and Dorchester in their early work with Michault at S-F Projects.

## "Sullivan's Backyard": ArmaLite's First Real Workshop

So taken were Sullivan and Michault with one another's similar thinking and innovative plans at this chance meeting that they became partners in business, acquiring together what would become the operating office of ArmaLite; a small, unremarkable machine shop at 6567 Santa Monica Boulevard in Hollywood, California, located in the center of what is today Hollywood's art district. Few of those performing at the Comedy Central Stage (a standup comedy venue run by the Comedy Central television network) would know that one of the most profoundly influential events in firearms history took place just next door to them, so many years ago.

As Sullivan and Dorchester began to hire employees for their business, the grounds came to be known as "Sullivan's Backyard." It was here that the Para-Sniper was standardized for its submission to the military for testing, and new principles of firearms design and production were refined. It would take a bit of time, though, as clearing the title of the property was not an easy task, even for an experienced attorney like Sullivan. The facilities were in a constant state of upgrade, even as work continued in earnest, until all the machinery and equipment was in place by November 9, 1954.

Although Michault and Sullivan had formulated possibly the most prophetically-clear vision of the infantry rifle of the future, the original plan of action for their new venture was to develop firearms for the civilian market, which would employ new and innovative materials and designs which combined light weight with phenomenal accuracy and controllability, to achieve previously-impossible levels of performance. The American consumer firearms industry was then a field growing by leaps and bounds, and these innovations promised to be very profitable.

Fairchild had been interested in diversifying its assets for some time following the postwar boom, and ArmaLite fit the bill perfectly. This did not mean that the new division was given free reign, however. Although it was capitalized quite well now, the board of directors of its parent company were unwilling, and the firm was indeed financially unable, to establish a full-scale, full-service firearms manufacturing facility. Instead, ArmaLite was chartered as a design house, the business of which was to formulate new firearms designs using aircraft-industry components

3. The unassuming entrance to "Sullivan's Backyard", the cinder-block building located at 6567 Santa Monica Boulevard in Hollywood, which housed the ArmaLite machine shop.
      This building still exists today.
                              Smithsonian Institution collection

such as non-ferrous metals and polymers, and then license those designs for manufacture by other entities.

As noted, the original goal of ArmaLite was to develop these firearms for the civilian market, but Boutelle, Sullivan, Dorchester and Michault always kept it in mind that a military supply contract would be the ultimate goal, and nothing in the division's charter prevented them from devoting resources to that end. This course was known to be the ticket to large, sustainable sales, and each of the new firearm designs being developed was thoroughly and eagerly examined for imminently-useful military applications.

# Eugene Stoner, the Father of the AR-10

The inexorable turn of the world of small arms towards the greatest battle rifle ever to grace it was not due to the labor of Sullivan, Dorchester, Boutelle, and Michault alone, for as these fledgling titans of industry built a base of innovation and design, a prodigy among firearms designers named Eugene Morrison Stoner was hard at work, only a few miles away from the epicenter of ArmaLite's machine shop.

Laboring in his own free time to achieve a new era in infantry rifles, Stoner, the true fountainhead of the AR-10, was brimming with ideas of his own, many of which were so similar in doctrine and design to those hypothesized by the nearby S-F Projects team that they would appear to be products of the same mind. Even more fantastical, though, is that initially there was no actual contact between these two sides.

## Stoner's Early Life

Born in the small town of Gosport, Indiana on November 22, 1922, Eugene Stoner had begun his interest in small arms even as a child, experimenting with machine designs of every kind. His interest in all things mechanical took him to technical high school in Long Beach, California. Although desiring to enroll in a university engineering program upon graduation in 1939, he instead entered the workforce to help his parents who, like many farming families of those days, had suffered seriously during the Great Depression.

Stoner's first job was with the Vega Aircraft Corporation. Despite not having any formal engineering education, he proved to be an adept designer of aircraft components. It was at this organization—the military aircraft production subsidiary of the Lockheed Corporation—that Stoner got his first taste of the new "wonder materials" being used in the aircraft industry, and had already begun thinking about their possible utilization in firearms.

With the entry of the United States into WWII following the Japanese attack on Pearl Harbor on December 7, 1941, Stoner signed on with the U.S. Marine Corps, becoming an ordnance officer in the 1st Marine Division. Here he serviced all sorts of weapons, from infantry rifles to aircraft-mounted machine guns, as he accompanied the island-hopping Marines in their campaign across the Pacific. He later described how formative his time in the Marines had been, which had allowed him to work on and study various types of machine guns. He even built some prototypes of his own designs with the equipment and parts he had on hand in the armories.

At the end of the war, Stoner became one of the fabled "China Marines." This title refers to those members of the 1st and 6th Marine Divisions (less the 4th Marine Regiment) who were scrambled by Admiral Nimitz, concurrently with the Japanese general surrender, to Northern China. Their mission was to accept the surrender of the Japanese units in the areas around Beiping (later Beijing) and Tianjin Cities, Hebei, and Shandong Provinces on behalf of the National Revolutionary Army, the armed forces of the Republic of China. The urgency of this deployment stemmed from the Soviet entry into the war against Japan just three days after the first atomic bomb was dropped. The Soviet forces accepted surrenders of Japanese units and territorial administrations for the Communist Party of China, which was at war with the Republic of China's government for control of the nation. The U.S. saw it as of vital importance in the prevention of the spread of communism in the Far East that the Republic of China's forces be in control of as much of the country's heartland as possible.

## Johnson Weapons in the U.S.M.C.

Utilizing the patented concept of a short-barrel-recoil-operated rotating bolt locking into a barrel extension, the Johnson LMG weighed 13 lbs., somewhat less than most other contemporary designs. It fired fully-automatically only. Recoil was modulated by its straight-line design, which brought the shooter's shoulder in line with the bore axis.

The Johnson's detachable magazine was a single-stack design, meaning that rounds were not in a staggered column as in most other magazines, such as on the M1918. This meant that to achieve the same magazine capacity, the Johnson's had to be considerably longer. This could cause problems for both carrying and balance, as the magazine was mounted on the left side, making the LMG, particularly when fully loaded, tend to tilt to the left. As shown in fig. 6, a later postwar experimental prototype was belt-fed.

Although it was never adopted as the standard light machine gun of any military, the Johnson was still bought and used in numbers by the U.S. Marines, Canada, Britain, and the Philippines. Its light weight and rapid-fire controllability made it particularly in-demand by Allied special forces and infiltration units fighting behind the lines in Europe and the Pacific.

## Marine Armorer Stoner Studies the Johnson Rifle and LMG

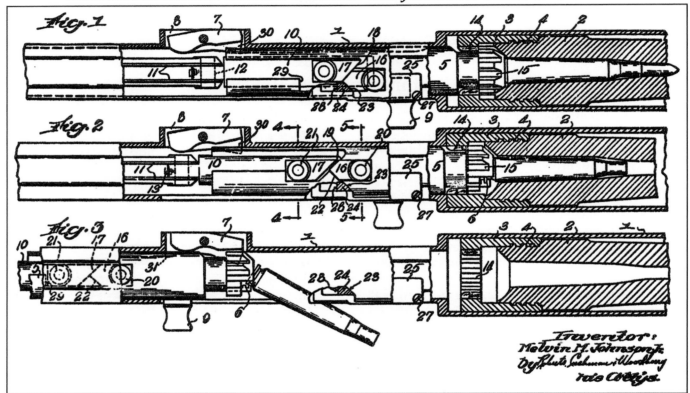

4. Figs. 1, 2 and 3 from Melvin M. Johnson's second U.S. patent, no. 2,146,743, granted on February 14, 1939, showing a number of features including the barrel extension (4) and multi-lugged bolt (15) which locks into it, both of which features were copied by Stoner in the ArmaLite AR-10.                                    U.S. Patent Office

As a Marine armorer during the harrowing days of WWII, Stoner had become familiarly acquainted with the M1941 Johnson rifle and M1941 Johnson light machine gun, a family of weapons which would later become a major inspiration for the AR-10. An interesting feature of both the Johnson rifle and LMG was its front-locking rotating bolt, which locked into an extension on the barrel of the weapon. This differed radically from all previous successful battle rifle designs, in which the bolt locked, either by rotating or tilting, into the receiver itself.

5 (right). Side and face views of the much more compact bolt assembly shown in Melvin M. Johnson's third U.S. patent, no. 2,181,131 from November, 1939, entitled "Breech Mechanism."                                    U.S. Patent Office

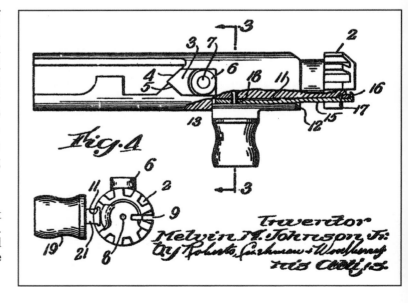

As no major military force had yet adopted a rifle containing this feature as a main weapon, the impact it could have on a very critical area of infantry weapon utility was not yet realized by the mainstream firearms design world. Stoner, however, in a flash of the inestimable precociousness and innovation for which he was to become known, realized the genius of isolating the bolt locking forces within the barrel, instead of the receiver. From here, he made the next leap by combining this innovation with the materials with which he had worked in aircraft parts design to achieve the greatest advantage of all: the key to reducing an automatic rifle's weight.

It was no secret that the U.S. military had been searching for means to lighten the naturally-heavy automatic rifles since before the adoption of the M1 Garand in 1935. However, unbeknownst to Stoner during his military service, the chief way that this endeavor was pursued before the Garand, and would be pursued after the war in search of its replacement, was by shortening the cartridge it fired. This decrease in cartridge length meant that the receiver could be shortened, and would therefore need less material to support the pressures of firing, even if the new round generated the same chamber pressure as its longer predecessor. This also slightly decreased weight by shortening the chamber within the barrel, which is the thickest and consequently the heaviest part of the barrel.

Stoner's innovation, however, took this lightweight concept further, and allowed the receiver to be made of lighter and weaker materials. The key to this was that when a firearm fires, whether it be bolt-action or automatic, the two main sources of mechanical stress occur from the high pressure in the barrel pushing the bullet out, and the bolt locking and holding the shell casing within the chamber so that the pressure does not destroy the firearm. This means that the barrel and the area where the bolt locks the action need to be the strongest, and thus the heaviest, parts of the firearm's action. Until the advent of the Johnson action, these two functions were performed by two distinct and separate parts; the barrel handled the pressure within it and the power of the bullet being forced down it, while the receiver provided the beefy, strong, and heavy locking areas for the bolt.

By designing a bolt that locked directly into the barrel, Johnson had eliminated the need for this high-stress load-bearing function of the receiver. Although it is not clear whether Johnson grasped the potential advantage of weight-saving from this innovation, the powerful U.S. .30 caliber (.30-'06) cartridge for which the Johnsons were chambered precluded it from being realized in his final designs.

It is a fair comparison to hold the M1941 Johnson rifle up against the M1 Garand, its direct competitor for mainstream military adoption, as not only did the two have identical empty (unloaded) weights, but the thicker gauge of the M1 Garand's barrel meant that the receiver of the Johnson was actually heavier than that of its traditional, receiver-locking adversary.

## Stoner's Own Initial Four Wartime Designs

Nevertheless, this innovation observed in the Johnson series of small arms inspired Stoner in his very first prototype design and creation. He named his prototypes similarly to U.S.-issue weapons, using an uppercase "M" followed by a number indicating the sequence of their invention.

Stoner's first four designs were drawn up while he was still serving in the Marine Corps, so none of these was ever completed. The first three of them were partially built; with only the steel receiver and action being machined by Stoner on armory equipment, while the fourth prototype only ever existed on paper. Despite the fact that they never made it off the drawing board, and were little more than the fanciful weekend hobby projects of a genius, they were fascinating, and hinted at the greatness which Stoner would achieve.

His first two prototypes, the M-1 and M-2, both from 1943, were submachine guns chambered for the .30 Carbine cartridge as used by the M1 Carbine, the shorter, lighter counterpart to the M1 Garand designed for issue to rear echelon troops, and commonly used in the Marines.

The M-3, his third design, appeared in 1944. Although no plans or the prototype receiver and action for it were ever found or studied, apparently the M-3 was very similar if not identical to the M1941 Johnson rifle. Stoner related during his lifetime that the M-3 was intended to be an infantry rifle, chambered in the standard .30 M2 (.30-'06) caliber, which made use of a rotating bolt locking into a barrel extension, powered by a short recoil action.

Stoner's M-4 design of 1945, which only reached the drawing board stage and never existed even as a prototype receiver and action, was another idea for a .30-'06 infantry rifle with a rotating bolt locking into the barrel, but which utilized the primer-actuated system, using an oversized primer which

6. Right side view of the postwar T48 Light Machine Gun,
a version of the Model 1944 Johnson LMG experimentally
reworked to belt feed.
Springfield Armory photo no. 6738-SA
dated November 18, 1948

was partially pushed out of the cartridge by the pressure of the expanding propellant gases (normally a dangerous sign that something is wrong with the ammunition), to force the firing pin back, unlocking the bolt and allowing the action to cycle. This operating system had been examined and tested early on in the design process of what became the M1 Garand, but while it was found to be reliable, it would not have allowed the military to proceed with a primer crimping operation which had been found necessary to obviate "popped" primers in machine guns, and Garand had perforce switched to gas operation.

With the end of the war and of his military service, Stoner experienced a period of time during which he did not have access to machine-shop facilities, and one might have expected him to lose some momentum for firearms invention. Stoner persevered, however, and continued to dream, plan, and draw, awaiting the time when he could experiment with production once again.

Following his honorable discharge from the Marine Corps, Stoner went to work for the Whittaker Aircraft Controls Company, back in the Hollywood area. Even though he did not possess a university education, Stoner proved to be quite talented in the design of new aircraft fuel gauges and valves, and by 1951 he had risen to the position of design engineer.

In control of the machine shop of the company where he worked, and as a part owner of the firm, Stoner set about once again designing innovative new firearms. This time, he was determined to take his inventions beyond the drawing board stage to their actual realization in the form of working prototypes. This endeavor was aided by his newfound access in the prospering postwar American economy to manufactured goods and materials. Now, with his financial resources and the ability to work with materials he acquired himself, Stoner found himself able to go beyond producing experimental receivers and actions, and actually build and test fully-functioning guns.

# From Imagination to Reality: the Stoner M-5

By early 1952 the M-5, the long-awaited first of the Stoner postwar designs, was completed and ready for the range. The first of Stoner's designs to reach the firing-prototype stage, the M-5 demonstrated its inventor's unique prowess in weapons design and fabrication. It looked quite conventional to the casual observer, and was, in fact, similar in some respects

to the M1 Garand. It sported a conventionally-contoured (as opposed to the straight-line design to be developed later) drop-heel wood stock, and fired the then-standard .30 M2 (.30-'06) cartridge. Instead of using the internal *en bloc* clip-fed magazine of the Garand or the rotary magazine of the Johnson, the

7. Right and left side views of the M-5, the first in the series of postwar Stoner self-loading rifle prototypes. The M-5 was a conventional long-stroke gas-operated .30-'06 caliber rifle utilizing an aluminum receiver with a rotating bolt which locked directly into the barrel extension. Weight: 6 lbs., 11 oz.

Note the commercial sights: the Lyman micrometer rear, and the hooded ramp front sight, which then sold for $8.25 complete from the C. A. Dahl Manufacturing Co. of Chicago.                Institute of Military Technology

M-5 fired from a detachable box magazine, like the Soviet SVT-40.

The M-5 utilized a rotating bolt that locked directly into an extension on the barrel. The barrel itself was of traditional ordnance steel construction with a relatively thin profile, further contributing to the low overall weight of the rifle. What made the M-5 so special, however, is that unlike the Johnson rifle, it capitalized on the greatly reduced load-bearing requirement of the receiver, and so this part was constructed out of thin aircraft aluminum. This cut the weight of the rifle drastically.

Although fitted with what would become known as Eugene Stoner's trademark rotating, barrel-locking bolt and an innovative aluminum receiver, the light weight and advanced nature of the M-5 were well hidden by its conventional lines and stock.

The use of detachable box magazines seems like a no-brainer to today's firearms enthusiast, but there was actually quite a bit of resistance to this idea in Stoner's day. The argument against the detachable magazine was the same as the early rationale for opposition to the machine gun: it would allow the soldier to waste too much ammunition.

Being in the then-standard caliber of the U.S. armed forces, the choice of magazines for the M-5 was simple. The magazine Stoner chose on which to base his new rifle's feed system was actually one that had existed since before his birth. Although the military establishment was steadfast in their stand against detachable box magazines in infantry rifles, the concept was accepted in the field of the light machine gun, and it just so happened that the U.S. M1918 Browning Automatic Rifle (BAR) fired from a steel box magazine, with a capacity of 20 rounds. Although low for an LMG, the contemporaries of which usually held around 30 rounds, this capacity was considered scandalously high for an infantry rifle at the time. Stoner found that producing lower-capacity examples for testing from surplus 20-round BAR magazines, large quantities of which were then available cheaply, was easy.

The M-5 design incorporated a commercial receiver-mounted aperture rear sight, which, like the M1 Garand, makes for a very intuitive sighting system through which the eye naturally centers the front post with a minimum of training. It also allows for a type of optical illusion, in which the target appears to the marksman's eye to be relatively sharp and clear, even while the eye focuses on the front sight (the correct method of aiming a rifle).

## Range-Testing the M-5

As his work began to bear fruit, Stoner and his space-age rifles became a regular sight at shooting ranges around the Hollywood area. Getting time on the range behind the stock of his new creation allowed Stoner to learn and evaluate his ideas in actual testing, giving him new insights never afforded him on his non-firing Marine Corps prototype designs. The need for a few changes quickly became apparent, and, after strengthening the bolt locking lugs and the

connection between the barrel extension and receiver of the M-5, Stoner's next prototype model was born.

Although the M-5 was never designated an ArmaLite production design, it led directly to the AR-3, and its vindication of so many of Stoner's innovations played no small role in bringing about the AR-10.

# The Stoner M-6

8. Right and left side views of the Stoner M-6, which was almost identical to the M-5 except for the addition of a cheekpiece on the walnut stock and a commercial hunting telescope. The magazine was cut down from a standard military 20-shot BAR magazine. Weight: 6 lbs. 15 oz. (with scope). Institute of Military Technology

Both cosmetically and mechanically, the gas-operated, rotating-bolt, semi-automatic M-6 was nearly identical with its predecessor, representing merely a product-improved version of the M-5.

A few modifications were made to the earlier M-5 design, which increased the reliability and proved that Stoner's simple and ingenious combination of a lightweight receiver and forward-locking rotating bolt could be perfected, even by a single man in his machine shop.

Like the M-5, only a single example of the M-6 was ever manufactured, and it differs from the only M-5 in that Stoner fitted it with a commercial telescopic sight, both to test the limits of the design's accuracy, and to satisfy himself that the aluminum

receiver was robust and stable enough for scope-mounting.

Stoner also fitted a stock with a higher comb (cheek rest) in order to bring the eye higher above the bore, so the shooter could peer through the scope comfortably. This aiming method allowed Stoner to test the inherent accuracy of his home-made rifle, which apparently was quite good.

It also incorporated a reliability improvement over its predecessor, which added greatly to the longevity of the working parts. The operating rod on the M-5/M-6 was connected to the gas piston, and presented the charging handle to the shooter. This assembly slid freely back and forth along the flat surface of the rear right side of the receiver with every shot. At the end of its travel it would slam hard into

9. Left side view of the modified .30-'06 BAR magazine cut down by Stoner for use in his M-5 and M-6 prototypes.
Institute of Military Technology

10. Top and underside views of a reduced-capacity BAR magazine as used in the Stoner M-5 and M-6.
Institute of Military Technology

the ridge that was designed to stop its movement, but the fact that it could move vertically meant that it struck at a slightly different angle with each shot, battering itself against the ridge and doing so in a non-uniform fashion, causing preliminary wear, especially on the lighter and less durable aluminum receiver. On the M-6, Stoner designed a channel, making the rear portion of the operating rod captive, enclosed on both sides by flanges which kept it uniform in its movement. This long area of surface

contact also worked to spread out the friction and wear on the moving parts, and slowed the operating rod, making it come to a much lighter rest against the back ridge on the receiver. This less violent action also reduced the felt recoil of the rifle, making for quicker recovery time between shots.

The M-6 proved to be quite durable, and did not give Stoner any more articulable problems in the testing to which he put it.

# Introducing the .308 Caliber Stoner M-7

Emboldened by his success in finally taking one of his ideas all the way from theory and drawing board models to a functioning, firing rifle, Stoner decided in the early days of 1953 to take his rejection of conventional firearm orthodoxy further, and added two new innovations to his M-5/6 design.

The first, and most easily-overlooked of these changes, but one which would be shown by history to be exceedingly forward-looking, was Stoner's decision to design his next prototype model in a new and portentous chambering. In 1952, the year that Stoner first began cutting metal in pursuit of his M-5 rifle, a new and innovative sporting cartridge was released to the American shooting public, called the .308 Winchester.

This new round, while boasting essentially identical ballistics to the famous and venerable .30-'06 Springfield of U.S. military and sporting fame, did so in a cartridge case roughly ½" shorter (51mm case length as opposed to 63mm for the .30-'06). Although the .308 was a bit wider at the shoulder than its ballistic ancestor, it still was able to attain the same

velocities with the same sizes and weights of bullets. This was possible because in 1906 (hence the '06' in .30-'06) when the Springfield round was designed, smokeless powders were relatively new, and consequently not very efficient. Over four decades later, when the .308 hit the market, powder formulation had come a long way, and a smaller amount of newer powder types could do what had required a larger amount of the .30-'06's powder to accomplish, allowing the shorter cartridge to still pack a serious punch.

Winchester, which was one of the commercial collaborators with the military on the T65 project, had seen the potential in the civilian market for a round that would do what the .30-'06 could, but in a shorter package and consequently in a lighter firearm. In 1952, the Army had finally given Winchester, which had been champing at the bit, permission to use the dimensions of the finalized T65 prototype cartridge in its commercial line. Just as Sullivan and Dorchester had done with their bolt-action "Para-Sniper" rifle, Stoner chambered his M-7 in .308 Winchester.

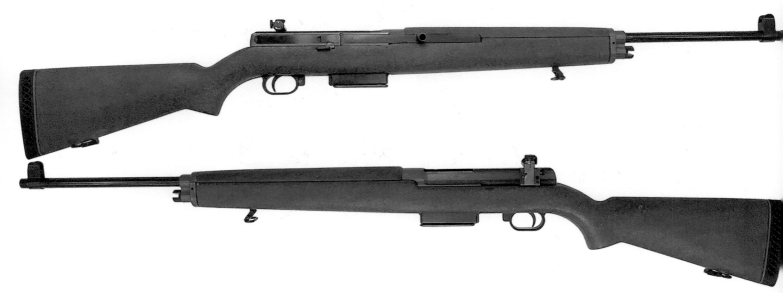

11. Right and left side views of the Stoner M-7 prototype, later renamed the ArmaLite AR-3.

Built upon the foundation of Stoner's M-5/M-6 developments, the M-7 would take the design idea further, both with the reliability improvements found on the M-6, and with the addition of a lightweight fiberglass stock and a new chambering.

Firing the then-new .308 caliber cartridge, the M-7 was extremely forward-thinking, but posed a technical challenge, as data and materials for this loading were not yet readily available.

Although the new chambering was dimensionally different than the .30-'06 Springfield used in his earlier designs, Stoner was still able to achieve reliability feeding for his new creation by using only minimally-reworked BAR magazines.　Institute of Military Technology

## Obsoleting the .30-'06

The introduction of the .308 Winchester cartridge had an unfortunate corollary for the venerable .30-'06. While its greater case capacity could hold more of the newer types of powder and thus produce higher velocities still, this was not allowed under the standards of cartridge specification. When a new type of ammunition is developed—an occurrence that happens with regular frequency down to the present day—the maximum chamber pressure generated by that round is set by the manufacturer, and firearms subsequently produced for it are designed to be able to safely withstand that pressure. Packing newer, more efficient powders into an older, larger case like the .30-'06 could easily be done, but this would increase the pressure level above what was set originally and could thus be unsafe for use in any existing firearm chambered for that cartridge. In this way the new .308 Winchester superseded the old .30-'06, although that venerable workhorse still remains a popular cartridge for sportsmen, both in the United States and abroad.

## Stoner's Audacity; Col. Studler's Compromise

The real sublimity of Stoner's picking this particular cartridge, out of the many then-new offerings on the commercial market, is that it could not have been clear to Stoner at the time he began work on his M-7 prototype that it would become the standard rifle round of the entire western world. For one thing, the T65 project was conducted within the military itself, and had been of a strictly classified nature. Additionally, even if Stoner were singularly well informed about the inner workings of the U.S. Army's cartridge selection and development project, of which there is no evidence, there were still many other competing designs with excellent potential from other prospective NATO member states.

The British, like the Americans, continued work after WWII on an "Ideal" cartridge development project that they had begun earlier, arriving at their own intermediate rifle cartridge designs. These different designs showed great promise, and the .280 British chambering in particular is considered by many military historians today to have been a better choice for NATO adoption than the U.S. design.

Nevertheless, the U.S. military establishment aggressively pushed the T65 round for adoption by NATO and, being the mightiest power in the new alliance, the other nations essentially fell into line. This was not easy, however, and it took very persuasive—not to say coercive—measures that went all the way up to the personal advocacy of Presidents Truman and Eisenhower to make holdouts like the British "see the light" regarding the T65.

The T65 was in itself a compromise, the result of Colonel Studler's attempt to both accommodate the obvious need for progress in the direction of a lighter rifle, and at the same time appease the traditionalists in the U.S. Army who insisted on not losing any of the power of the .30-'06, which they still viewed as the ideal infantry cartridge. Had he been left to his own devices, Studler may well have been inclined to go in the "assault rifle" direction together with the British, Belgians, and Germans, but ironically, as events actually transpired, he was responsible for pushing cartridge and light rifle requirements that were essentially incompatible, at least for all rifles except the AR-10.

# More on the Stoner M-7

These innovations came together to form what Stoner christened his M-7 prototype rifle. This weapon bore an almost identical cosmetic appearance to the M-5, as it was fitted with a traditional front sight and aperture rear sight, instead of the telescopic sighting device of the M-6. The only divergence in appearance from its two immediate predecessors was an extended magazine well, in which the detachable magazine could be inserted and held tightly to make for reliable feeding of ammunition.

The new, revolutionary cartridge for which the M-7 was chambered, while shorter and slightly fatter, did not differ enough in its basic dimensions from the workhorse .30-'06 to necessitate a completely new magazine design. Therefore, Stoner took a very minimalist approach to the M-7's feeding device, simply putting together a magazine based on that of the M1918 BAR, only shortened somewhat.

Like the M-5 and M-6, the M-7 was a marvel of not only Stoner's ingenuity and inventive spirit, but also his pragmatism and resourcefulness. While he designed the entire rifle and invented and produced from scratch the parts that made it unique and revolutionary, Stoner did not shy away from utilizing "off-the-shelf" parts produced by established manufacturers when he got the chance. To allow for efficient use of the superior receiver sight, he fitted as original equipment on both the M-7 and M-5 (as well as theoretically the M-6, if he had produced one with iron sights instead of a mounted scope) the commercially-available Lyman 48 WJS elevation and windage-adjustable aperture sight with micrometer adjustment, working it into the design of the rifles.

For the barrels of these prototype rifles, he started with blanks—pre-manufactured rifled tubes of barrel steel that could be turned on a lathe to produce the external contour and profile desired by the final producer of the rifle. This is a production method no different than that used by many major firearms manufacturers of both his and the present day.

Once received, Stoner would mill down the outside of the barrel blank to give it the contour he wanted. In the case of his early designs, which called for a slim barrel profile, this meant that a considerable amount of the barrel blank was cut away, so that the overall weapon weight would be kept as low as possible.

## Some Pros and Cons of Light Barrels

The barrel is one of the areas with the most potential for weapon weight reduction, and a thin barrel profile also results in faster handling and aiming of the rifle. The downsides of using a light barrel profile are the tendency for it to heat up quickly under rapid fire. A barrel that heats up to high temperatures and is continually fired at those temperatures burns out its rifling more quickly, with a resulting degradation of accuracy. Firing when a thin barrel is very hot also can cause it to crack or even rupture. That is why, compared to their rifle counterparts, machine guns as a rule have heavier, and usually quick-changeable barrels, because the sustained firing which they must withstand would quickly burn out or burst a lighter barrel.

In addition, a thin barrel is more sensitive to foreign objects making contact with it during firing. The physical principle that makes a barrel suffer in accuracy from contact with foreign objects is the barrel's natural tendency to flex. If viewed in ultra-slow-motion as the rifle is fired, the barrel actually bends and flexes as the bullet moves through it. The key to accuracy in this counter-intuitively malleable material, then, is that a good barrel tends to flex in exactly the same way with every shot, putting each bullet on the exact same path. A foreign object making contact with the barrel affects the direction in, and degree to, which it flexes, which can send the bullet off in a different direction. As a thin barrel flexes more than a heavy one, such contact affects it more.

On an interesting note, a thin barrel is not inherently less accurate than a thicker barrel, within reasonable parameters. However, the fact that it heats up more quickly and is springier along its length means that under normal operating conditions, a thin barrel is less forgiving to heat and contact with objects, making it less accurate than a thicker barrel, all other factors being equal.

In addition to producing military-ready 20-round magazines, Stoner also fabricated several smaller, 5-round flush-fitting magazines that were more suited to the field testing of the M-7 as a sporting piece.

# A Legendary Collaboration Begins

In early April, 1954, several months before S-F Projects had joined Fairchild as the ArmaLite Division of that corporation in October, Stoner attracted some attention while test-firing one of his creations at his usual stomping grounds, the Topanga Canyon Shooting Range. While walking the range, taking a break from sitting behind a shooting bench and running an accuracy test on his own Para-Sniper rifle, George Sullivan himself looked over the shoulder of a young man at the conventional-but-unfamiliar rifle he was firing, and was at once taken with interest. In that moment when Stoner turned around at Sullivan's greeting, all the key players in the brotherhood that would become ArmaLite were finally in touch, marking the establishment of the creative and technically collaborative spirit of genius which would make the AR-10 a reality.

Conflicting reports exist as to exactly which rifle Stoner had to his shoulder at this seminal moment, whether it was the M-6, M-7, or a far more radical prototype on which he had recently been working. Stoner himself was silent on the topic of exactly which of his rifles was noticed by Sullivan, only recalling that he was testing some of his designs that day.

In any case, it seems logical that Sullivan would have been drawn as a moth to a flame by a rifle that appeared to be the long-lost sibling of his own Para-Sniper. The striking similarities, including the use of aluminum receivers in both rifles, exemplifies the scientific concept of multiples, in which great innovations are developed by different people with no contact with one another at roughly the same time. Examples from history are Lipperhey, Janssen, and Metius, who without knowing one another simultaneously developed the telescope; or Leibniz and Newton, both of whom, unaware of the other's work, built the foundations of calculus. In this case Stoner and Sullivan, who lived just a few miles from one another, met by chance while each was working on and testing his own designs.

Regardless of which of Stoner's creations was on view at this chance meeting, the result was that S-F Projects now had an opportunity for great advances into a new market sector, and Sullivan knew it. On the spot he offered Stoner first a part-time consulting position, with the promise of the post of chief design engineer to follow. Stoner accepted, sold his interest in Whittaker, and almost immediately was hard at work in Sullivan's machine shop on Santa Monica Boulevard.

The unique military promise of their new employee's prototype was not lost on either Sullivan or Dorchester. Correspondence picked up in the following weeks from George Sullivan about this "new man" he had hired who would streamline and invigorate the firm's development projects, and he began to show a sudden interest in the military requirements for a new weapon under Col. Studler's Lightweight Rifle Program.

## Sullivan Dubs the Stoner M-7 the "Autorifle"

Around this time Sullivan began referring to Stoner's M-7 as the "Autorifle", which he regarded as one of S-F Projects' more promising developments. The Autorifle fanned into flame Sullivan's barely dormant passion for pursuing military projects. He even took it hunting in the high Sierras in September of 1954, after having Stoner outfit it with a scope. This was not particularly difficult, as Stoner had already done so with his M-5 prototype to create the M-6.

Meanwhile on July 16, 1954, Sullivan, who had previously been introduced, contacted retired General Jacob Loucks Devers, and requested him to get in touch with U.S. Army Ordnance to ascertain the exact requirements of the Army's Lightweight Rifle Program, which had been ongoing for almost a decade by that time, with its specifics still a closely-guarded secret. Having been in the employ of Fairchild since his retirement in 1949, serving as a consultant personally to Richard Boutelle, General Devers had not only been extremely helpful in Fairchild's quest for military aircraft contracts, but was particularly well-suited to advising on arms developments. He had, after all, reached the pinnacle of his career as Chief of the U.S. Continental Army Command and, as a recently-retired four-star general, was in a perfect position to pry the desired details from his former colleagues. He was also asked to speak with personnel of various élite infantry formations to deduce which features they desired in a new rifle.

# A Preview of the Revolutionary Stoner M-8

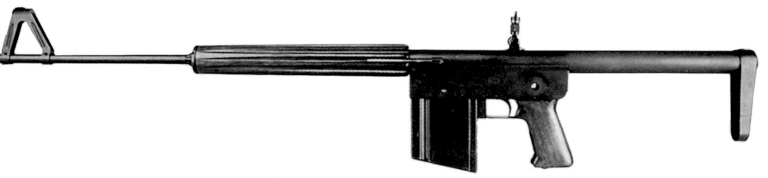

12. Left side view of the revolutionary Stoner M-8, the "first AR-10", as depicted in *Small Arms Review* Volume I No. 4. The M-8 is described below, and further discussed in Chapter Two.

Note the change lever is missing, and was possibly recycled for use in the following model, the X-02 (fig. 40).

courtesy Dan Shea

The rifle, if indeed it was the M-7, that had caught Sullivan's eye at the Topanga Canyon Shooting Range in the spring of 1954, which served as the catalyst for one of the most propitious collaborations in firearms history, was not, however, the only prototype Stoner then had under development.

It is not known whether Stoner's mysterious fourth prototype was with him at the firing range on that fateful day, or whether he had left it hanging up in his machine shop. However, it is known that, concurrently with the development of the M-7 (or possibly even the M-5 or M-6), Stoner completed and produced a design that radically diverged in both cosmetics and adherence to convention from all his previous work, and the norms of the firearms industry in general. As further discussed in Chapter Two, Stoner's final solo work, designated the M-8, appeared different from anything seen before on a firing range. Although never divulging, even to his closest friends and associates, the exact time at which he first conceived of this rifle, Stoner had certainly drawn up the first blueprints by early 1952. He subsequently went so far as to cut the date notation off the original design drawings, for fear that his previous company might claim an interest in his gas system patent.

Lacking any semblance of what would be considered a buttstock, a steel tube extended straight backward out of the M-8's steel, manganese-phosphated receiver, at the end of which hung a rudimentary metal buttplate. A wooden pistol grip was attached to the lower rear of the box-like receiver. The barrel extended forward from the front of the receiver, perfectly in line with the tube jutting back from the opposite end, and on it was a simple, tubular wooden forend.

No semblance of a stock encapsulated the receiver, and forward of the pistol grip and trigger assembly was a hollow cutout serving as a well for the insertion of an M1918 BAR magazine. To allow for the raised line of sight, both the front post and rear aperture sights were raised high over the bore axis, standing up on towers above the receiver and barrel.

Even more peculiar to the experienced observer of automatic rifles, was the fact that the barrel appeared to stand alone, with no piston or operating rod apparatus apparent. In fact, besides the barrel, the only other component, visible only at the juncture of the handguard and barrel, was a small-diameter stainless steel tube that ran back from a forward point on the side of the barrel into the left side of the receiver. Unheard-of in the world of American fire-arms design at the time, this was a stationary gas tube, which fed propellant gases directly into the receiver to power the action, thus making this strange and wonderful rifle the most modern iteration of a very small but honorable family of rifles powered by the direct-impingement gas operating system.

## Chapter Two

# The ArmaLite "AR" Models

## The Para-Sniper Becomes the "AR-1"

13. Right side view of the newly-renamed ArmaLite AR-1, formerly the "Para-Sniper" rifle, in standard coloration and finish (if one can really be said to exist), presented to General Curtis LeMay of the Strategic Air Command.

Inset, far right: an enlargement of the legend engraved on the right side of the barrel, visible in the main view under the front of the scope.

Institute of Military Technology

The ArmaLite division of the Fairchild Engine and Airplane Corporation, now backed by an influx of capital from its parent company, consisted of a design and machine shop, a number of worthy prototypes, and the greatest firearms designer since John Moses Browning on board. ArmaLite's long-term strategy was to develop their designs to a level of perfection that would allow them to be licensed for mass-production by arms manufacturers. However, immediately upon Eugene Stoner's accession to his post of chief design engineer, the AR-10, and indeed military rifles in general, were not the main focus of the firm.

Despite Stoner's designs all being of a military nature, the management of the company was still in the hands of Sullivan and Dorchester, with close oversight from and involvement by Richard Boutelle, the firearms-fanatic president of Fairchild. Seeing as how Sullivan and Dorchester had already spent close to a decade developing the Para-Sniper, which was primarily a sporting rifle (although with a name and features that were intended to appeal to the military), and had formed ArmaLite to further pursue this end, the natural focus of the firm's earliest efforts were rifles to be offered to the burgeoning American civilian market. Therefore much early effort was put into pursuing the Para-Sniper, which upon the acquisition of ArmaLite by Fairchild in 1954, was designated the AR-1.

## Stoner Objects to the Composite Barrel

Stoner quickly found that he had to split his focus somewhat from working exclusively on his earlier autoloading military designs, as he was soon put to work on the bolt-action side of ArmaLite's developments. Although his design experience was eminently applicable to work on the AR-1, one modernistic facet of it with which Stoner took exception right from the start—an objection from which he never wavered—was the use of the composite barrel developed and championed by Sullivan and Dorchester, which was constructed of an aluminum jacket swaged around a stainless steel liner. His unique experience as a military armorer had taught him the importance of simplicity and reliability in designs, and he considered the two-part barrel as representing an unnecessary risk of parts failure from the heat and vibration that would be generated in firing, which could cause the two components to separate.

# Plans for Other "AR" Designs

This early drive did not, however, put blinders on management, as Stoner had been brought into ArmaLite expressly because of his genius in creating rifle designs with military potential. What was at least a plan to leave the military avenue open, and was more likely a signal of more concrete plans to explore in the future, the AR-3, Stoner's conventionally-stocked M-7, was actually only the second weapon to be given an "AR" designation. The gaps in numbering designs bearing ArmaLite's signature designation (AR-1 and AR-3, but no AR-2) was due to the fact that ArmaLite had several other models, both in the prototype stage and still on the drawing board, about which no decision had yet been made whether to pursue for commercial or military applications. Some chronological designations were purposely left open for future designs that figured in a later-defunct ArmaLite plan of action.

## ArmaLite Formally Adopts the M-7 ("AR-3") and the M-8 ("AR-10")

The gas-operated, long stroke, rotating-bolt M-7 (fig. 11) was the first of Stoner's prototypes to be formally adopted as a development project by ArmaLite, where it was slotted into the "AR-" prefix heirarchy as the AR-3, where its conventionally-shaped fiberglass stock, much like that found on the Para-Sniper (AR-1), gave it a further weight-saving advantage, weighing in at 6 lbs. 8 oz.

For its part, the strange and futuristic creation of parkerized steel, known at its birth in 1953 as the Stoner M-8 (fig. 12 and below), would be christened anew upon its acceptance for development by ArmaLite as the "AR-10."

It was once said by the satirist Machiavelli that "it is not titles that honor men, but men that honor titles", and the same indeed holds true for rifles. Despite the bad luck which was time and time again to dog the development of the AR-10, the "AR" designation was to turn an unremarkable combination of letters and numbers into a respected and universally-known symbol, and the pinnacle of combat rifle design.

## An Early Pump-Action Shotgun

14. Right side view of what was envisaged as the prototype of a line of lightweight sporting shotguns, this aluminum-barreled pump gun, built on a traditional steel receiver sourced from the Ithaca Gun Company, was never assigned an official "AR-" designation.
Smithsonian Institution collection

One design that was never assigned an official "AR-" designation was an experimental pump-action shotgun, intended to be the starting point for a line of hunting shotguns for the commercial market. It was built of Sullivan-designed parts and fitted with an aluminum barrel on a traditional steel receiver, bought from the Ithaca Gun Company of New York.

## The Autoloading 12-Gauge ArmaLite AR-9

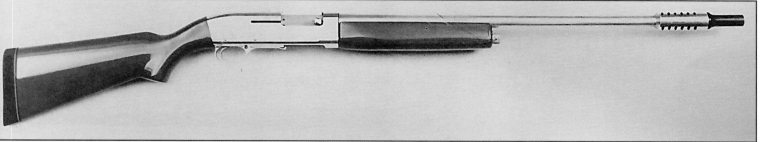

15. Right side view of the ArmaLite AR-9 autoloading shotgun, as depicted in an early hand-lettered prospectus prepared by the ArmaLite Division for Fairchild's corporate board. Its trademark aluminum receiver and barrel and plastic stock and handguard are easy to discern, and would have looked even more radical in their time.

Smithsonian Institution collection

Another early design of ArmaLite's, in line with Sullivan and Dorchester's original goal of marketing their innovative arms to the American shooting public, was the AR-9. This was a bold semi-automatic 12-gauge shotgun, featuring an aluminum receiver and polymer stock, which was produced in both standard brown as well as pink. Its distinctive and bulbous muzzle brake also served as the housing for the detachable choke, which could be changed out to alter the pattern of the shot as it left the barrel. Though it never reached the production stage, the prototype AR-9 showcased the versatility of the young company.

## Setting Aside Other AR-Numbered Designations

By May, 1955, ArmaLite had purposely set aside the AR-2, -4, -6, -7, and -8 designations for use on future projects, which included the very peculiar .38 Special-chambered shotgun, concerning which Sullivan and Boutelle had corresponded prior to the acquisition of ArmaLite by Fairchild; a pump-action, 12- or 20-gauge shotgun with which Sullivan had been experimenting; and the "Paratrooper", the AR-1 Para-Sniper's predecessor, chambered in .257 Roberts.

The AR-8 designation was expressly left open in anticipation of a line of lightweight hunting rifles in .270 Winchester caliber, which would bear a strong resemblance to the AR-1. Discussions with Hugh Drane, the owner of a firearms firm in Texas, were ongoing for the development and production of the AR-8 during this period, although this project never materialized.

# Time Out for an Unexpected Opportunity

Richard Boutelle's connections came in handy in an unexpected way during the winter of 1954, to the decided benefit of the new company, and not for the only time. Having served as a major in the U.S. Army Air Corps during WWII, and maintaining close and cordial ties with his parent armed services branch (the U.S. Air Force after 1947) since the early postwar years, Boutelle had developed close personal friendships with several high-ranking figures in military aviation, particularly General Curtis LeMay, head of the newly-formed Strategic Air Command. LeMay and Boutelle were good friends, both being fanatical firearms enthusiasts who often shot together at Boutelle's Maryland estate.

General LeMay, having been the architect of the successful strategic bombing campaign against the Japanese mainland late in WWII, as well as a veteran of many dangerous missions in enemy skies, knew intimately the countless facets of offensive military aviation, and one area about which he felt particularly passionate was that of aircrew survivability.

During the early development of military aviation and throughout the war, aircrews were, to this end, issued survival weapons. These differed significantly from infantry rifles. They had special requirements, resulting from their having to be stowed but remain easily accessible in the severely space- and weight-restricted cabins of combat aircraft, and needed to be easily attached to parachutes and bail-out bags.

# A Short Review of USAF Survival Rifles
## The Rudimentary H&R Bolt-Action M4

16. Right side view of the .22 Hornet caliber H&R M4 Survival Rifle, with wire stock extended.

Jeff Moeller collection

17. The H&R M4 Survival Rifle, shown disassembled for storage in the seat pack of a fighter aircraft.
The 5-shot magazine is the same as that furnished with the commercial Savage 220 rifle.    Jeff Moeller collection

The M4 bolt-action Aircrew Survival Weapon, the culmination of prewar development in this area, consisted of little more than what is known as a "barreled action", that is a rudimentary receiver, bolt, magazine, barrel, and trigger assembly. It was not equipped with a handguard of any kind; the barrel simply protruding naked from the front of the tubular receiver. The collapsible wire buttstock had a bent-

18. Left side closeup of the USAF M4 survival rifle, showing markings.
     The tubular receiver is marked "HARRINGTON AND
RICHARDSON ARMS CO./WORCESTER, MASS. U.S.A."
and the trigger housing is marked "RIFLE, SURVIVAL,
CAL. 22. M4 (HORNET CARTRIDGE/U.S. PROPERTY".
     Note the commercial Lyman 33M receiver sight.
                                         Jeff Moeller collection

down loop at the end to be placed against the shoulder. The M4 was chambered for the commercial .22

Hornet cartridge, suitable for small game but largely
ineffective against enemy personnel.

## The Over/Under Ithaca M6

In the immediate postwar period, the newly-consti-
tuted United States Air Force had adopted the M6
Aircrew Survival Weapon as a replacement for the
less-versatile M4. This exceedingly strange weapon
was capable of being folded up, making it only 15"
in total length when stowed as part of survival gear,
and boasted a combination of innovative and stunningly anachronistic features. The M6 had two 14"
barrels, one above the other. The upper barrel was
rifled and chambered for the same .22 Hornet round
as used in the M4, while the bottom barrel was
smoothbored and chambered for the M35 3" .410
shotgun shell. The idea behind this combination was
to allow an airman downed in the wilds to be able to
hunt game like rabbits, deer, and squirrels using the
M6 as a rifle, while also having the ability to down
game birds with the shotgun barrel.

The glaring downside of this weapon was that
each individual round had to be manually loaded
directly into the chamber by the shooter. To this end
a hefty top-mounted latch connecting the barrel
monobloc to two "horns" riveted onto the top of the
stamped receiver was lifted up, allowing the barrels
to break open around a hinge located on the bottom
front of the receiver. A single spring-loaded extractor
pulled the spent shell casing(s) part way from both
chamber(s) at once, allowing them to be fully extracted by hand. Fresh rounds were then inserted into
either or both of the barrels by hand, then the action
was closed. Fire selection was accomplished by pulling up a round-headed switch on top of the hammer
to align the firing pin with the .22 Hornet cartridge,
or lowering the switch to address the .410 shotgun
shell.

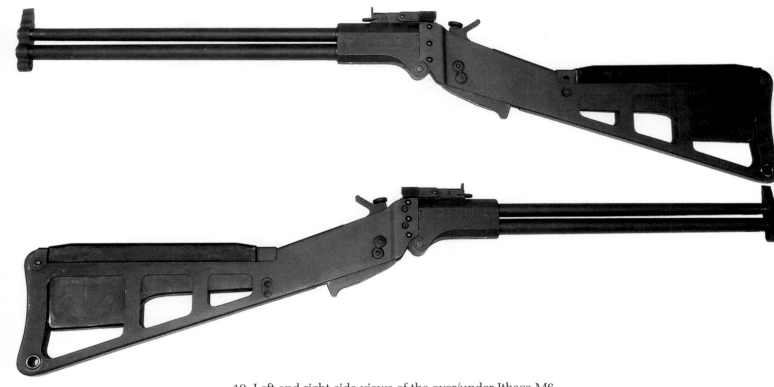

19. Left and right side views of the over/under Ithaca M6 Survival Rifle, with upper barrel chambered for the commercial .22 Hornet round and the lower barrel chambered for the M35 .410 ⅕-oz. Rifled Slugs.

Jeff Moeller collection

20. Left side view of the Ithaca M6 over/under Survival Rifle, with action broken to take up as little room as possible in the storage position.

Note the rubber-covered cheekpiece, which is hinged over the ammunition storage area in the stock.

Jeff Moeller collection

21. Closeup of the ammunition storage area in the butt of the M6 with the cover raised (right), showing spaces for nine .22 Hornet cartridges and four M35 3" .410 shotgun shells.                    Jeff Moeller collection

The top of the framework stock had a compartment with spaces for nine .22 Hornet cartridges and four M35 3" .410 shotgun shells. The cartridges were covered by a rubber-covered cheekpiece, hinged at the butt end and held closed by a push-in latch at the front.

The trigger of the M6 was a simple bar, with no trigger guard, which was simply gripped with the entire hand and squeezed to fire the rifle. This appears to have been the only component of the M6 to take into consideration the fact that the reloading procedure described above might have to be performed in conditions of extreme cold weather, when the operator would presumably be wearing thick gloves or mittens.

The break-open action, although as far outside the scope of military rifle design at the time of ArmaLite's formation as any could be, would nonetheless show up later in the evolution of the AR-10, illustrating the incomparable ingenuity of Stoner and the other members of the ArmaLite creative team in adapting a totally unrelated principle to improve an already-unconventional system.

General LeMay, being a strong proponent of expedience and light weight in all matters, including protection for his air crews, was not satisfied with either the M4 or M6 survival weapons. For one thing, they were both quite heavy for what they did; the WWII M1 Carbine, developed as a personal defense weapon for rear echelon troops and used also as a main combat rifle by many airborne and amphibious units, weighed only about one pound more than either of these totally combat-ineffective survival weapons.

Still adhering to the basic philosophy of adopting a weapon that would dissuade airmen from engaging in needless combat while attempting to avoid detection, LeMay decided not to go the route

22. Left side closeup of the M6 survival rifle showing markings reading "U.S.A.F. PROPERTY/[serial no.]/RI-FLE-SHOTGUN SURVIVAL/CAL. .22 / .410 M6".
Jeff Moeller collection

of simply issuing a lightweight combat arm like the M1 Carbine to flight crews, but felt equally strongly that a more desirable survival weapon would be lighter, more modular, and have better ergonomics than the M4 or M6.

In one of his frequent contacts with Boutelle, LeMay was intrigued to learn that Fairchild, with whom he already had dealt favorably on many occasions, had expanded into the firearms business. Of particular interest to the flamboyant and controversial general, was the potential for extreme light weight and modularity offered by ArmaLite's designs. Sensing the imminent possibility of pursuing a military contract—a firearms designer's dream—Boutelle had the foresight to order the ArmaLite team to prioritize the development of a survival rifle design of their own.

# Soft-Point Cartridges <u>NOT</u> for Use Against Enemy Personnel

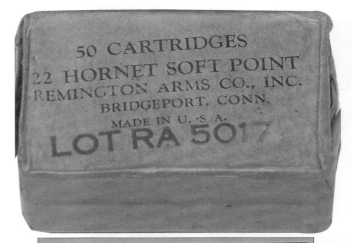

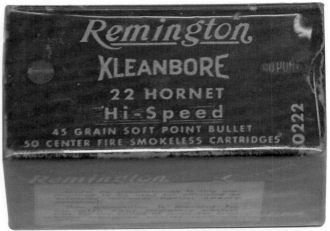

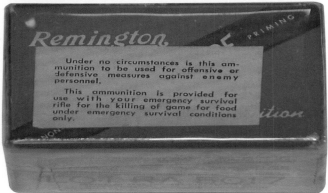

23. Top and bottom of a plain cardboard box of 50 .22 Hornet soft-point cartridges, manufactured by Remington for a USAF contract.

Note the stern warning on the underside label.

Jeff Moeller collection

24. A commercial box of 50 rounds of soft-point .22 Hornet ammunition, also made by Remington for a government contract.

Below: the warning label pasted onto the front of the box.

Jeff Moeller collection

The intended use of survival rifles was also drastically different from the use of arms by soldiers, with their purpose being foraging and hunting for food by downed airmen. In fact, the manual for the M4 contained clear instructions that the weapon was not to be used against enemy personnel. This strange requirement was intended to keep U.S. forces in compliance with the Hague Conventions, which banned the use of commercial ammunition issued with these pieces, in war.

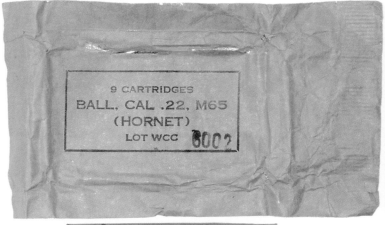

25 (right). A sample round of military full-patch M65 .22 Hornet ammunition produced under a Winchester contract, shown along with its nine-round plain cardboard box (devoid of warnings) and waterproof foil-lined envelope. The M65 was developed and issued specifically for use in the M6 Survival Rifle, which as shown in fig. 21 was designed to hold nine of these cartridges in the storage area of the butt.

Dave Terbrueggen collection

Generally used for hunting, soft-point ammunition had been ruled inhumane by the international body on the rules of war, memorialized in the Hague Conventions, as causing more severe wounds. The requirement for military ammunition was and still is that it uses a non-expanding, full-metal-jacket bullet, which in theory produces cleaner wounds, increas-

ing the chances of soldier survivability, while still providing the desired incapacitation of an enemy combatant.

As shown in fig. 25, a full-patch version of the .22 Hornet cartridge, designated the "Ball, Cal. .22, M65 (Hornet)", was issued later, in nine-round packages, specifically for use in the M6 Survival Rifle.

# The Design and Adoption of the ArmaLite AR-5

26. Left side view of an early ArmaLite AR-5 .22 Hornet caliber Survival Rifle. The easily-disassembled metal components were dimensioned to fit inside the hollow fiberglass stock, shown here with a gray spatter-painted finish, which would keep the rifle afloat in water.

Simply on its weight and ergonomics alone, the bolt-action AR-5 represented a substantial leap in the design of an aircrew survival weapon.

Note the early magazine, the same as the commercial Savage 220 magazine, which was also used in the M4.

Institute of Military Technology

With direct input from the Air Force generally and LeMay in particular, Sullivan and his team quickly formulated a drawing-board design. The requirements were settled between the parties on November 9, 1954, just a month and a few days after the formation of ArmaLite. Work commenced immediately on the production and testing of the first prototype of the new rifle, which was designated the AR-5.

The basic design was that of a manually-operated bolt-action, similar in function to the M4 Survival Rifle, but this is where the similarity stopped. Instead of the conventional steel receiver of the M4, the AR-5 made use of ArmaLite's trademark aluminum receiver, with the barrel liner and bolt being the only steel parts. In place of what passed for a stock, as well as the complete lack of a handguard on the M4, the AR-5 sported a buttstock of large but traditional appearance, but of course constructed as a foam-filled fiberglass shell. The magazine well, an extension on the aluminum receiver, housed a detachable four-round magazine, originally the same as that used with the M4, and provided a convenient handhold for the support hand, making it no longer necessary to either grip the barrel directly, or rest the

rifle on a support for effective use. The barrel was constructed similarly to that of the Para-Sniper, with an aluminum jacket swaged around a thin, stainless steel liner.

The finishing touch that made this design a true *tour de force*, was the previously-unheard-of ability of the entire rifle to be stored inside its own hollow buttstock. Without the use of tools, an airman could detach the barrel, receiver, and bolt, insert them into specifically-designed storage compartments in the buttstock (along with space for two magazines), and snap the plastic buttplate back on over them. This allowed the rifle to be carried in a tiny package, and what is more, it was light enough to float, which would be useful to a crew member downed in water. Even when fully assembled and loaded with a full magazine of ammunition, the rifle would still float.

The special ability of the AR-5 to be completely stowed in its watertight polymer buttstock added greatly to its utility. The buttstock also featured compartments which contained fishing equipment, a compass, matches, and other survival gear.

The AR-5 retained the same .22 Hornet caliber as previous survival weapons. Its smooth bolt action and detachable magazine provided the shooter with the capability for rapid follow-up shots to ensure hunting success on small- to medium-sized game, but without going so far as to make it an effective combat arm, thus discouraging the downed airman from fighting it out with enemy soldiers.

The incredible light weight of only 2 lbs. 12 oz., made possible by the combination of what was fast becoming ArmaLite's signature characteristics of aluminum parts and polymer furniture, was taken to new heights (or literally, lows) in the AR-5. Even loaded with a full magazine, it weighed slightly less than an M1911A1 service pistol.

Another interesting feature that not only significantly decreased reloading time, but was also directly inspired by a similar feature Stoner developed on his first AR-10 prototype, was a push-button magazine release.

Traditionally in rifles with detachable magazines, the user must apply pressure to a lever and pull the empty magazine out of the rifle before a fresh magazine can be retrieved and inserted, bringing the rifle back into action. The installation of a magazine release inside the trigger guard allowed the shooter to move his trigger finger forward and push the lever, automatically releasing and dropping the empty magazine, while his support hand simultaneously acquired a new magazine and brought it up to insert it into the magazine well. Although this feature existed on various contemporary pistols, one of the first times it was ever adapted to a rifle was on Stoner's marvelous M-8 and, as Stoner was heavily involved in the design of the AR-5, it appears almost certain that he was the direct source of this innovation.

There was also substantial design input on the AR-5 project from the Air Force Survival School's Major Burton Miller and Colonel Robert Enewold. The former actually became employed by ArmaLite after the completion of his military service.

By the beginning of January, 1955, the first prototype of the AR-5 was ready for testing, and the first one off the line (if ArmaLite's machine shop could really be called a "line") was presented to General LeMay as a gift, with an inscribed brass plate on the receiver dedicating it to him. This first prototype sported a reddish-brown, wood-toned plastic stock, giving it a somewhat traditional look.

The Air Force evaluated this first series of AR-5s, and came back with a few recommendations, such as strengthening the bolt, modifying the rear sight, and simplifying the magazine. The requirement cited by the Air Force for "barrel life," which is the number of rounds a weapon's barrel can usefully

fire before substantially losing accuracy and velocity, was 400. The composite barrels of the AR-5s were tested to 6,000 rounds, and no significant loss in accuracy was recorded when the tests were concluded.

The AR-5 was an excellent example of "framing," wherein one can clearly see which facets of the design came from which of the principal innovators, Stoner and Sullivan. The bolt was clearly representative of Sullivan's work. Stoner had never concerned himself with manual-action firearms, often expressing that automatic weapons in .30 caliber were his only real interest. Also, Sullivan's and Dorchester's early project, the one that led to the formation of ArmaLite, was the development of the bolt-action Para-Sniper rifle. The origin of the bolt is also made quite clear in that one of the recommended changes from the Air Force was adding an additional locking lug to the bolt head, bringing the total number of lugs to two. Stoner's signature was a 7-lug (eight balanced lugs with one as part of the extractor) bolt face, which would have been adaptable to a bolt-action rifle as well. However the two locking lugs made the bolt simpler to machine, and with the action strength deemed satisfactory by the evaluators, ArmaLite settled on this bolt design.

Another feature of the rifle that can be easily credited to Sullivan is the steel-and-aluminum composite barrel.

The variety of colors in which the stocks of the production model were available served as a nostalgic throwback to the early days of development of the Para-Sniper, when Sullivan and Dorchester worked on polymer rifle stocks in their garage. The colors ran the gamut from brown in the first prototype, pea-green in the second, to both of these and blue in the production version.

On the other hand, the detachable magazine, push-button magazine release, and barrel extension-locking action are easily attributable to Stoner. The polymer stock, aluminum parts, and ability of the AR-5 to be disassembled and stowed inside a floating buttstock could have sprung from either of their minds, but are more likely the result of productive collaboration between these two design pioneers.

The M4 rear sight consisted of a commercial Lyman 33M adjustable aperture sight which was mounted by means of a small bracket on the left side of the receiver and extended over the top, with the aperture held in place only by the rigidity of the extension arm. This was considered too flimsy, and so on the AR-5 this was changed to a fixed sight, made as an integral extension of the receiver. This far more

27. Left side view of an updated AR-5A Survival Rifle, with fiberglass stock finished in a brown color.

Thanks to the work of Tom Tellefson and George Sul-

livan, the furniture of the AR-5A could be made available in a dazzling array of colors.

Institute of Military Technology

28. Right side view of a later AR-5A with updated plastic stock colored blue.

Together with the rifle shown in fig. 27, these examples of the updated AR-5A show that in addition to the standard

light gray color of the original model (fig. 26), deep blue and reddish brown were also produced for Air Force testing.

Note here and above the redesigned "Ronson Lighter" magazine.                     Institute of Military Technology

robust design eschewed any means for adjustment, but as the rifle was only issued with one type of flat-shooting ammunition, there was no perceived need for an in-field sight adjustment. The sights were to be set and tested at the factory to make sure that their "point of aim" (POA) was the same as the rifle's "point of impact" (POI) at the distance desired by the buyer. This fixed sight was also intended to improve reliability and dependability, as an adjustable sight is more prone to being knocked out of alignment by the type of abuse to which a survival weapon could be subjected. In addition, trying to recalibrate the adjustable sights of a rifle in the wild, using up one's small and precious supply of ammunition while evading enemy patrols, is not a desirable situation.

True to form for ArmaLite, the magazine for the AR-5 was both innovative and lightweight. In an era where the detachable magazines utilized by rifles were made out of steel, an ammunition feeding device constructed of lightweight, corrugated aluminum stampings was revolutionary to say the least. The first prototype utilized the M4 magazine, the manufacture of which required a number of stamping operations. This was changed at the Air Force evaluation board's request to a simpler design, which coincidentally resembled a cigarette lighter that was ubiquitous at the time. This new "Ronson Lighter" magazine, while not able to be disassembled for servicing the way the original was, proved reliable, simpler, and easier to produce, so its design was given the go-ahead by the Air Force.

# The (Unproduced) U.S.A.F. MA-1 Survival Rifle

29. A photograph taken while the Air Force was testing the AR-5A. Here two NCOs look on as other Air Force personnel put some sample AR-5As through their paces in the Nevada desert. The man at right, sitting on a crate of Remington .22 Hornet ammunition, is checking the barrel of one rifle by examining it through a bore scope.

Smithsonian Institution collection

With the second prototype approved on March 28, 1955, ArmaLite standardized the production design as the AR-5A, which was subsequently adopted as the Air Force MA-1 Aircrew Survival Rifle. At a stroke, and with less than six months from drawing board to military acceptance, the AR-5 represented, at least theoretically, a major coup for this small, fledgling firearms design house.

However, the AR-5 was destined never to become the major profit-generating product for which Sullivan and his team had hoped. In the end, LeMay's enthusiasm was not enough to convince either the Joint Chiefs or Congress to fund the project, meaning that no actual orders were ever placed. The grand total of Air Force AR-5s in inventory never exceeded twelve, with the test examples of the AR-5A being put into service.

The establishment institutional thinking that would come to plague the AR-10 was actually quite different from the considerations which caused the curtailment of AR-5 production. There were still large stocks of M4s and M6s in inventory, which the Air Force pragmatically deemed to be sufficient to fill the role of a survival weapon. At the time, the U.S. Air Force was at the forefront of the nuclear arms race with the Soviets, and so in the grand game of strategic arms dominance, a new aircrew survival weapon was a low priority indeed.

Hope was not lost for the AR-5, however, as there was general interest in it as well. In 1957, none other than the famed polar explorer Bernt Balchen inquired as to the possibility of acquiring a quantity for equipping wilderness expeditions. He also requested a feasibility study of firing a shotshell cartridge from the small survival rifle.

George Sullivan himself wrote an article about the AR-5 titled "New Survival Weapon", which appeared in the January, 1957 issue of *The American Rifleman*.

# The ArmaLite AR-7 "Explorer" - a Popular Consolation Prize

30. A portion of the front page of an ArmaLite brochure describing the civilian .22LR caliber AR-7 "Explorer". Here the plastic butt cap has been removed to reveal to an appreciative admirer the metal components stored inside the hollow fiberglass stock.

The unique and distinctive AR-7 continues to this day to be a strong tribute to its designer, L. James Sullivan.

Michael Parker collection

The civilian follow-on to the AR-5, called the AR-7 "Explorer", was destined to become the sole member of the AR lineup to reach the large-scale production stage, but not as an incarnation of any of the concepts for which these designations were originally intended.

The short barrel of the Air Force AR-5 survival rifle made it illegal for regular civilian ownership, and so the AR-7, sporting a civilian-legal 16" barrel, chambered for the popular and affordable .22 Long Rifle rimfire cartridge and firing semi-automatically from detachable 8-shot magazines, actually became the most- and-longest produced weapon design to which ArmaLite would retain rights under a licensing agreement, with production commencing in the 1960s and continuing today under different ownership, with no sign of stopping.

Weighing a scant 2 ½ lbs., the AR-7 retained the revolutionary survival-minded design features of its military cousin, including the ability to stow all the metal components within the floatable plastic stock.

An excerpt from the text of the above brochure reads as follows:

*The ArmaLite Model AR-7 Explorer is the civilian version of the AR-5 U.S. Air Force survival rifle. Military exhibits of the AR-5 throughout the country brought over 3,000 letters asking if this unique new rifle was available for public sale. The AR-7 is now in production . .*

The AR-7 has remained the most successful and popular lightweight survival weapon in the world shooting community, with all rights later being sold to the Charter Arms Corporation of Stratford, Connecticut.

# More on the M-8 (Model X-01) - the First AR-10

This pyrrhic victory of the military adoption of the AR-5, even though unaccompanied by any appreciable monetary return, greatly boosted morale, and caused ArmaLite to become of one mind in pursuing military contracts first and foremost. Just the interest shown by the Air Force in late 1954 was enough for ArmaLite to re-evaluate its "civilian first" strategy.

Concurrent with the very early stages of the AR-5 development, considerably before it was submitted for evaluation, ArmaLite took another look at the M-8 (fig. 12 and below), the strange steel contraption Stoner had brought to the company a few short months earlier, and moved it to the forefront for development. With a serial number/model number of X-01 already assigned by Stoner before his meeting with Sullivan, ArmaLite hailed this prototype as the AR-10, and fast-tracked it for rapid testing and development.

The Model X-01, given this unique name because of its status as one of many unique variants in the evolution of the AR-10's design, bore almost no direct resemblance to the later, perfected form. Since this model was produced by Stoner and was only regarded by ArmaLite as a model on which to improve, it was never marked "AR-10" by the new company. It remained almost unknown, but luckily, thanks to the dedication of C. Reed Knight, who purchased the Model X-01 for the Institute of Military Technology collection, it has not been forgotten.

## Further Describing the One (and Only) Model X-01

32. Right side view of the Stoner M-8 prototype, designated the ArmaLite AR-10 Model X-01. This was the pinnacle of Stoner's pre-ArmaLite solo work. The one-off M-8 prototype rifle holds the most honorable distinction of being the very first AR-10. A left side view is shown in fig. 12.

Note the positioning plate, welded onto the side of the BAR magazine.                    Institute of Military Technology

This exact version was never produced beyond a single prototype, and so serial (or model) number X-01 was the only one ever made. The term "AR-10" was actually first applied to ArmaLite's first redesign of the M-8, but this model was later referenced in the files of ArmaLite and Fairchild as the "Prototype AR-10."

The minimalist furniture that adorned the M-8's humble features was made of wood, and not fiberglass, as Stoner and Sullivan had adapted to the M-7. In an additional throwback, the M-8 was chambered in .30-'06, and not for the .308 Winchester/T65 Light Rifle cartridge. It utilized standard 20-round M1918 BAR magazines, positioned by means of steel plates crudely welded to their right sides.

The heart of the selective-fire M-8, and what made it the first AR-10, was the rotating, front-locking bolt and the direct-impingement gas system. This, combined with the straight-line stock and raised sights, encapsulated the principles of light weight and controllability that would go on to make the AR-10 the spectacular weapon it became.

Weighing 7 lbs. 8 oz. overall, the X-01's receiver was constructed not of forged aircraft aluminum, but from stamped-and-welded sheets of ordnance steel. As a result, it was quite heavy, and this, combined with the ingenious and advanced features found at this time in the world only in this one solitary rifle, helped to further control recoil.

31 (preceding page). A famous photo, originally published in the March, 1957 issue of *Guns* magazine, shows a proud Eugene Stoner standing beside the four seminal design steps to the perfected ArmaLite AR-10.

From top: the M-8 (Model X-01); the scope-mounted Model X-02; the X-03 (AR-10A), and the AR-10B. The first three of these rifles were produced as single prototypes only.                                    editor's collection

One striking feature of this first AR-10 was the fact that its action and barrel were all in a straight line, and what would be the buttpad on a standard rifle stock actually extended down, with only its top connected to the steel tube that housed the return spring. This tube also housed the buffer, at this point integral with the bolt carrier but later a separate assembly, and so it became known as the "buffer tube". The buffer portion of the bolt carrier, encompassed by the action spring (also called the recoil spring), moves backward with the bolt carrier until it reaches the plate at the back of the buffer tube. Then, as the gas pressure subsides but before the return spring begins to decompress, it rebounds and pushes the carrier forward again. The purpose of having the buttplate extend below the buffer tube was to allow the shooter's head to be positioned as low as possible on what was already a design with a high sight plane.

The purpose, little understood in its time, of this strange stock configuration, was to bring all of the recoil forces of the AR-10 perfectly in-line with the shooter's shoulder. On a traditionally-stocked rifle, the stock drops down behind the receiver, with the buttplate, and the shooter's shoulder, positioned well below the bore axis, so that recoil causes the rifle to rotate upwards, with the buttplate acting as a pivot point. This rotation is the main problem in recoil control, as the muzzle must be brought down manually before another shot is fired, lest the second and third shots drive the muzzle higher and higher, until the rifle is shooting far over its original target.

In the AR-10's straight-line configuration, the recoil travels straight back into the shooter's shoulder, and thus there is no pivot point to make the muzzle rise and take the rifle off target. This means that the only effect of the recoil is to push the rifleman backward. The deleterious effects of rearward recoil, unlike those of muzzle rise, are something which proper training and experience can almost completely eliminate, allowing the shooter to keep his rifle on target during rapid fire, providing more accurate and effective shots on target. This function, although important in all calibers and sizes of firearm, was particularly valuable in controlling the heavy recoil from powerful rounds like the 7.62x51mm NATO, especially in the exceedingly light package which would come to be one of the hallmarks of the AR-10's revolutionary design.

Another effect of the in-line action's raising the shooter's head relative to the bore axis, is that the sights must be positioned higher above the barrel than on traditional designs. A notable combat accuracy advantage is realized by this high sight radius as well, because the heat mirage from the barrel under rapid fire causes far less visual distortion two inches above it. Anyone who has shot skeet with a shotgun in hot weather can attest to the fact that once the barrel heats up, the shimmering heat rising from it makes the sight picture directly along the top of the barrel quite blurry, and although a scattergun can still function with this indistinct sight picture, such a condition would render a precision rifle, mechanically capable of pinpoint accuracy, almost useless.

In achieving these breakthroughs in the AR-10, Stoner took a page from what was already a great source of inspiration: the Model 1941 Johnson Light Machine Gun.

As discussed in Chapter One, Eugene Stoner had become closely acquainted with the Johnson LMG during his time as a Marine armorer, so it is perhaps no surprise that his X-01 AR-10 prototype bore a striking resemblance to it, both cosmetically and in function. In addition to the bolt locking system, which served as the inspiration for all of his post-war designs, the Johnson's in-line action and raised sights were the models around which the AR-10's handling and sighting ergonomics were formulated. In fact, for his first prototype, the X-01, Stoner actually used a Johnson LMG rear sight he had acquired on the surplus market. Although the front sight was fabricated by Stoner, it appears almost identical to the front sight of the M1941 as well, and was clearly a faithful copy.

Stoner was not content merely to utilize the innovations embodied in the Johnson LMG, but improved greatly upon them. Where the Johnson design featured a traditional wooden buttstock surrounding the action spring tube that extended back from the receiver, Stoner did away with the weight and expense of the conventional stock altogether, simply attaching a buttplate to the rear end of the receiver/buffer tube.

The inner workings of this first prototype were revolutionary, encompassing some new and original ideas of his own, along with improvements on the ideas of others.

In essentially all automatic rifles and machine guns up to that point, the receiver, which contained the entirety of the bolt and carrier movement, was a closed box, and the buttstock was a solid piece of wood or plastic attached to the back of it, functioning merely as a place on which to rest the shooter's shoulder and cheek.

The Johnson's receiver tube only extended part of the way to the buttpad of the stock, whereas Stoner's buffer tube *was* the stock. In incorporating an actual function of the receiver into the buffer tube, Stoner greatly improved the efficiency of the automatic rifle concept, and greatly reduced the overall length required for its operation. Whereas in a conventional design the rearward movement of the bolt and carrier had to stop short of the beginning of the buttstock, in the AR-10 the recoiling parts moved into and occupied the entire length of the buffer tube, meaning that this part was no longer wasted space, and that the rifle's forward receiver could be correspondingly shorter.

A shorter weapon (especially one that achieves this goal by redesigning the receiver, as opposed to merely chopping off a portion of the barrel), is a boon to the rifleman for many reasons, not the least of which being maneuverability in close quarters. Additionally, having the rifle's muzzle closer to the shoulder fatigues the shooter far less than a longer rifle of identical weight, which will tire the user more quickly because of the mechanical multiplication of the barrel weight caused by its being further from the support hand. The reader will likely be familiar with this concept, in that it is far easier to hold a heavy object at shoulder level close to one's body than at the end of an outstretched arm.

Finally, the weight of a shorter rifle is reduced by its not being composed of as much material as a longer one, particularly in the heavy working parts. This was one of the driving forces behind the quest for a shorter standard round, which resulted in the 7.62x51mm NATO, and was furthered to a much greater extent by the shortened overall action made possible in the AR-10.

Adding an additional boon to the straight-line stock system was that the bolt, carrier/buffer, and action spring could be easily disassembled for cleaning and servicing by simply turning and removing the buttplate and pulling the action components out of the rear of the buffer tube. This was another feature which the X-01 borrowed from the Johnson LMG, and on which Stoner and his team at ArmaLite would improve.

Furthering the reduction in weapon weight to an even greater degree was Stoner's revolutionary direct impingement gas system, discussed below.

# Genesis of the Direct Gas Impingement System

## The French Rossignol B1 Rifle (1901)

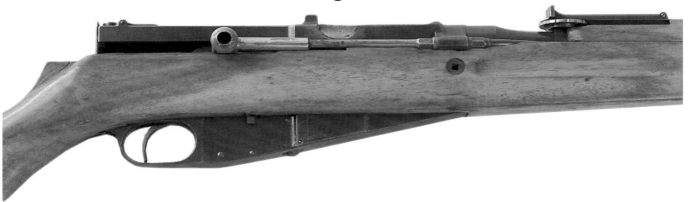

33. Right side closeup of the flap-locked Rossignol B1 rifle, invented in 1896.
Note the rear portion of the *adducteur* (gas tube), running along the side of the action and nesting inside the forward cocking handle extension.    courtesy Jean Huon

As far as is known, the direct gas impingement system of autoloading rifle operation was first used in the little-known Rossignol B1 rifle, invented in 1896 by a French weapons inspector of that name and perfected by the *École Normale de Tir* (ENT; the French Army Musketry School) in 1901.

## Rossignol B1 Machine Rifles, Some with Dual Gas Feed (1900)

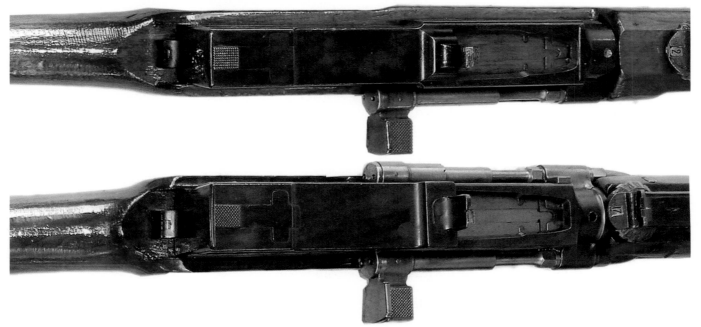

34. Top closeups of two versions of the Rossignol B1 machine rifle.
Above: with single gas tube.
Below: with dual gas tubes.          courtesy Jean Huon

The year before, in 1900, two versions of the flap-locked Rossignol design were also produced as selective-fire machine rifles, one of which, as illustrated here, was fitted with *two* symmetrical gas tubes. As noted in the 1995 Collector Grade title *Proud Promise*, by French weapons expert Jean Huon, with this design and the B1 rifle of the following year, the ENT School had produced the first real weapons *system* in military history.

All of the Rossignol designs were chambered for the French developmental 6x60mm ENT cartridge, and featured detachable box magazines which could be replenished through the open action by means of cartridges loaded on stripper clips.

As further noted in *Proud Promise*, the 6x60mm round was just one of the "remarkable experimental smokeless French cartridges produced for the 'First Wave' of autoloading rifle trials. These astoundingly modern, micro-bullet designs produced excellent ballistics, but were far in advance of their time."

## Further Applications of the *Gas* System

Following WWI the pistonless, direct-gas system featured in the long series of military rifle prototypes produced by the French State Arsenal *Manufacture Nationale d'Armes de St-Etienne* (MAS), beginning with the MAS *Modèle* 1922 and culminating in the excellent MAS *Modèle* 1949 and 49/56 battle rifles.

The direct gas impingement system was also used in the Swedish Ljungmann AG m/42b and its Egyptian clone, the Hakim rifle.

In all of these early applications, some of the propellant gas was simply piped to the rear from a port in the barrel to impinge directly on the outside of the bolt or bolt carrier, driving it to the rear and providing the impetus to unlock the bolt.

# Stoner's Ingenious Application of the Direct Gas System

Stoner had likely read about this operating system only in the most general terms, and had probably never actually examined or seen such a rifle in action. Nevertheless his 'take' on the use of this gas system introduced a new aspect which was at once novel and ingenious, yet sublimely simple.

In the French MAS and Swedish Ljungmann versions of the direct gas impingement system, a jet of gas was blown onto either a flat or dimpled surface on the face of the bolt carrier, driving it back by brute force, in exactly the same manner as a gas piston would. This method of operation was quite inefficient, however, since to equal the same pushing power generated by a piston, an excessive amount of propellant gas had to be vented into the action.

An additional issue caused by this rather indiscriminate blasting of hot and dirty propellant gases into the receiver was the rapid heating and fouling of the working parts. Although not nearly as large a cause of overheating as the friction between the bullet's jacket and the bore, the propellant gases generated by the burning powder which push the bullet down the bore are, needless to say, extremely hot. In all operating systems except for direct impingement, these gases are confined to either just the barrel, or the barrel and expansion chamber ahead of the gas piston. When the gases are contained forward of the action, this heat source does not directly impact the receiver and fire control parts of the rifle. Its introduction into the receiver, however, adds this variable into the equation of rifle performance, for which at the time there were no data on effects.

The facet of this issue that had at least the potential for an even more deleterious effect on the weapon was the inherent existence of fouling particles carried in the expanding gases, which are the result of the chemical reaction of the gunpowder when exposed to the flame from the impacted primer. Although the advent of smokeless powder had made firearms propellants exponentially cleaner, as well as capable of generating much higher and more consistent pressures, there is still an inevitable byproduct, consisting chiefly of carbon, in the form of soot. These residues build up on the parts exposed to it and, if not intermittently scraped free (like the bore being scoured by the passing bullet), can wreak havoc on areas requiring tight tolerances to function properly.

Needless to say, therefore, the direct gas impingement system was not without its problems. Even having no direct experience with this operating principle, the aforementioned downsides, which had previously either been ignored or reluctantly accepted, were immediately obvious to Stoner, who arrived at a way to minimize all of them with a simple mechanical solution.

Instead of following the hide-bound doctrine of simply substituting a fixed gas tube for a short-stoke piston, as the earlier French and Swedish designers had done, Stoner noted the inefficiency and host of problems which attended the propellant gases essentially blasting the face of the bolt carrier, and found a way around it.

Although the starting point for his bolt design on the AR-10 was the Johnson barrel-locking rotating bolt, which he copied almost identically (the exact same 22.5° turn for bolt unlocking, and the same multi-lugged bolt face), Stoner vastly improved, and added functionality to, this excellent concept. The bolt carrier and bolt in the X-01 prototype interacted in a way not previously seen in any firearm, going far beyond their original purpose to create a new type of synergy, the result of which was softer recoil, further weight saving, and a much cleaner mechanism.

In conventional automatic rifle operation, the bolt carrier's sole job is to receive the impulse either from a tappet, operating rod, or inertia; unlock the bolt; pull it to the rear; and relock it once it had been driven forward again by the action spring. In his X-01 AR-10 prototype, Stoner called on the bolt and carrier to work together in an entirely new way, necessitated by the elimination of a piston in the gas system.

The gas port on the X-01 was located on the left side of the barrel under the very front of the handguard, from which a stationary gas tube assembly ran back alongside the barrel. Instead of having the gas tube stop flush with the receiver and blow in a wide and indiscriminate arc onto the face of the bolt carrier, Stoner devised a dog-legged gas transfer valve with an integral inner gas tube extension, which took in the gas where the main tube ended on the outside of the receiver and relayed it into an internal gas tube, which extended almost two horizontal inches further into the left side of the receiver, covering more than half of the internal length of the portion of the receiver exposed when the ejection port on the right side was open. This rear end of this internal gas tube nested inside a precisely-cut hole on the left face of the bolt carrier. At that point, the ingenious design of the bolt and carrier came into play.

The inside of the carrier was hollow and cylindrical, exactly the same width as the body of the bolt, holding the bolt rigidly in-line, but also allowing it to rotate inside the space. The carrier's cylindrical interior came to an abrupt end about halfway down its length, where the interior wall descended at a 90° angle into a smaller-diameter hole. The tail of the bolt fit snugly into this hole, making an air-tight seal at the rear end of the carrier chamber.

Upon firing, the propellant gases flowed down the gas tube, through the gas transfer bracket and its integral internal tube, and into the air-tight expansion chamber between the tail of the bolt and the bolt carrier. The rising pressure in this sealed environment drove the carrier to the rear, camming the bolt head 22.5° and unlocking it from the barrel extension. The entire mechanism was then forced backward against the action spring, cycling the weapon.

The absolute surgical precision of the gas delivery in the AR-10's operating system was orders of magnitude more effective and energy-efficient than the previous method of spraying a torrent of burning propellant against a receiving plate or cup on the bolt carrier, meaning that a far smaller and more pre-

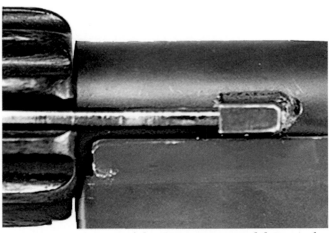

35. Left side closeup of the Stoner M-8 (Model X-01), the first AR-10, as depicted in *Small Arms Review* Volume I No. 4, showing the gas tube leading into the gas transfer bracket, the tubular extension of which channeled the gas into an opening in the left side of the bolt carrier.

courtesy Dan Shea

cisely-modulated amount of gas—and hence fouling—was sufficient to unlock the action.

# The Stoner Gas System Patent

Stoner applied for a patent on his system of direct gas impingement on August 14, 1956, for which U.S. Patent no. 2,951,424, titled "Gas Operated Bolt and Carrier System", was granted four years later on September 6, 1960.

## The AR-14: Guarding the Secret of the AR-10

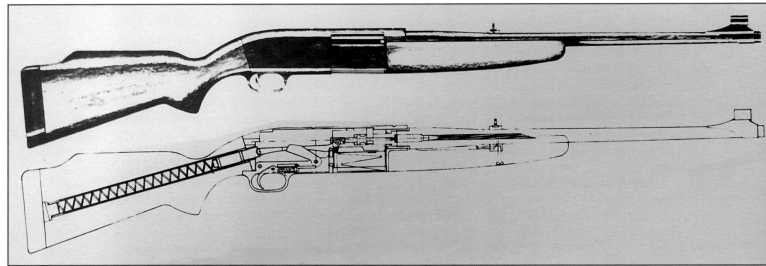

36. Right side views of the exterior and a vertical section of the ArmaLite AR-14, a conventional rifle with a two-piece cheekpiece stock, utilizing the patented Stoner gas system design and a forward-locking rotating bolt locking into a barrel extension.

As can be seen in figs. 1 and 2 from U.S. patent no. 2,951,424 (fig. 37), this was the design chosen by ArmaLite to illustrate the Stoner gas system rather than divulging the AR-10 at this time. courtesy Charles Kramer

As an interesting aside, one can see in the patent diagrams that the gas system is illustrated as part of a rifle of a conventional design, much like Stoner's M-7, which was named the AR-14. ArmaLite was playing the AR-10 very close to the chest at this time, and in fact it was not until April 23, 1955 that Sullivan informed Fairchild president Boutelle of its existence. Considering that Stoner was an employee of the company, Sullivan initially claimed responsibility for its creation, and it was not until much later

that it came out publicly that it was, in fact, Stoner's design.

The earliest newspaper and magazine articles about the AR-10 referred to Sullivan as the inventor. One such, which accompanied a clip of the famous photo of Richard Boutelle firing the AR-10B at his farm in Hagerstown, Maryland (fig. 57), stated, "[The AR-10] was invented by George Sullivan, Lockheed Aircraft Corp. patent attorney whose hobby is guns."

## Illustrating the Stoner Gas System Patent

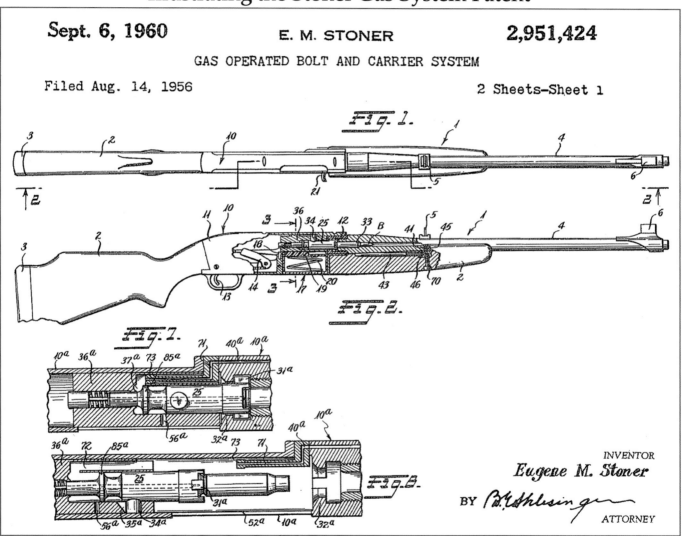

37. Figs, 1, 2, 7 and 8 from U.S. patent no. 2, 951,424, titled "Gas Operated Bolt and Carrier System", applied for on August 14, 1956, granted to Eugene M. Stoner on September 6, 1960, and assigned to the Fairchild Engine and Airplane Corporation.

As noted, the rifle depicted is the conventional ArmaLite AR-14.                                    U.S. Patent Office

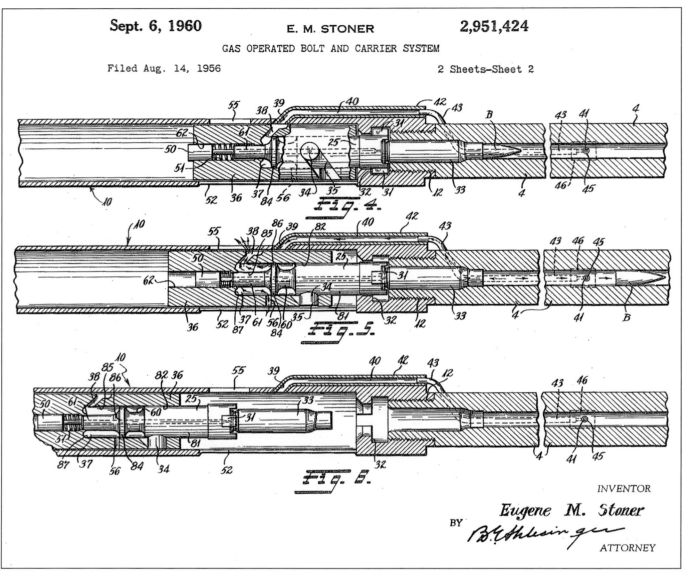

Sept. 6, 1960          E. M. STONER          2,951,424

GAS OPERATED BOLT AND CARRIER SYSTEM

Filed Aug. 14, 1956                          2 Sheets—Sheet 2

*Fig. 4.*

*Fig. 5.*

*Fig. 6.*

INVENTOR

Eugene M. Stoner

BY

ATTORNEY

38. Figs. 4, 5 and 6 from the Stoner gas system patent, illustrating how his innovation works.

In this early version the gas transfer bracket and integral rear gas tube are shown located on the outside of the receiver.

Although the illustrations appear quite complex, one need only look at a few parts to understand how Stoner's revolutionary gas system operated. The steps are as follows:

1. After the bullet (B) has passed (fig. 5), gas enters the gas port in the barrel (45).

2. It flows down the gas tube (43), into the gas transfer bracket (40 and 39), and into a hole in the side of the bolt carrier (38).

3. It blows into the hollow "piston" space (37) in the bolt carrier's interior tube (36), between the tight fitting bolt tail (61), and the wide bolt piston rings (84).

4. This forces the bolt carrier back, and causes the cam pin (34) to ride inside the camming cut in the bolt carrier (35), rotating and unlocking the bolt.

5. The bolt and carrier then move backward, extracting and ejecting the spent shell casing, and then return forward, stripping a new round from the magazine into the chamber, and locking the bolt behind it.

U.S. Patent Office

To understand the effectiveness of Stoner's innovation, consider the case of the paper wrapper on a straw at a restaurant. Just about everyone exposed to these straws as a kid would, at some point, discover that they could remove one end of the wrapper and then blow on that end of the straw, turning the wrapper into a missile. This author learned this on the receiving end of a volley from his dad and great grandfather over a milkshake, at the age of five. The air-tight expansion chamber made by the bolt and carrier in the AR-10 can be thought of like the straw wrapper, still wrapped around and forming a more-

or-less airtight seal with the gas tube, which serves very nicely and literally as the straw. The fact that the paper wrapper encompasses and essentially forms a seal around the end of the straw, means that not much of a breath is required to send it flying in your target's direction.

At a stroke, Eugene Stoner had vastly improved the efficiency of the direct gas impingement concept, while simultaneously remedying to a large degree the disadvantages considered inherent to it. The bolt and carrier now moved in a far more controlled and concentric fashion, and the much smaller amount of gas needed to power it greatly curtailed the heating and carbon fouling issues caused by the direct use of propellant gas.

## Controlling the Bolt on the Model X-01

The X-01 prototype has stood as somewhat of a mystery to contemporary students of the AR-10, as it would seem from looking at the outside that there was no means of keeping the bolt unlocked during the cycling of the action. This is extremely important, because without some mechanical means of keeping the bolt in the open or "unlocked" position within the carrier, it would move forward into the locked position as it contacts the next round in the magazine and feeds it up into the chamber. If this happened, the bolt locking lugs would no longer be aligned with the open slots in the barrel extension, and would instead slam into the solid portions of the extension and cause a malfunction. Guiding the bolt was generally achieved in firearms with rotating bolts by having a recessed cut called a "cam track" inside the receiver, in which a pin sticking out of the bolt would ride, unlocking and locking the bolt as the cut angled upwards or downwards.

The solution in Stoner's early bolt operation was to incorporate a thin ridge on the inside of the receiver body at the top, which would slide between two of the bolt's lugs when the bolt was unlocked and began to move backward. The bolt would ride on this ridge to the end of its rearward travel, and then would slide along it again forward until the ridge ended, whereupon the unlocked bolt would drop into the locking recesses in the barrel extension. The continued forward movement of the bolt carrier, acting through the cam pin riding in a curved cam cut in the carrier's body, would then cause the bolt to rotate and lock.

The reason Stoner opted to use a raised ridge inside the receiver body instead of the more tradi-

39. Right side closeup of the action of ArmaLite AR-10 Model X-02 (fig. 40) with dust cover open, showing the metal ridge descending from the top of the receiver which guided the bolt on both the X-01 and X-02 prototypes.

This ridge kept the bolt in the unlocked position throughout its reciprocation until the bolt had returned to its place inside the barrel extension.

*Institute of Military Technology*

tional groove in which the cam pin, already present in the X-01's bolt assembly, would ride, was that of modularity, and ease of disassembly. The method of field stripping Stoner chose, in which the buffer, carrier, and bolt could be withdrawn from the rear end of the buffer tube, meant that a cam pin that protruded from the carrier and rode inside a camming groove would be too high to pull back through the buffer tube and remove. Therefore, this alternative means of keeping the bolt unlocked during cycling, and allowing it the freedom to rotate and lock only when safely all the way forward, was devised.

## The Lack of Substance in the Criticisms of Direct Gas Impingement

As it turned out, many of the concerns about the theoretical pitfalls of the direct impingement system, which Stoner worked to alleviate, never materialized. The Swedish AG m/42b never received complaints of unreliability or increased parts breakage due to the huge blast of gas used to operate it. The French MAS-49, in addition, was actually used extensively in combat, from the arid deserts of Algeria to the steamy and corrosive coastal wetlands of Indochina, and despite being fielded in these harshest of combat environments, attained a reputation for extremely robust reliability. This is all despite the fact that these two rifles employed a much cruder version of the direct impingement system, which used far more gas and introduced a much greater amount of heat and fouling into action than Stoner's simple-but-ingenious carrier-bolt-gas tube mechanism.

In fact, Stoner's means of minimizing the amount of gas used to power the action down to a trickle, compared to the designs that went before it, meant that very little additional fouling was introduced into the action, when compared to that which is inherent in any automatic action. The spent shell casing is the epicenter of carbon and fouling residue from firing, containing a much higher concentration than that of the expanded propellant gas at the gas port, and when it is automatically ejected, just a fraction of a second after it is fired, its open mouth releases a far greater amount of carbon, soot, and other fouling particles into the interior of the receiver than a gas tube (at least on Stoner's design) blows in.

Furthermore, many of the perceived issues with heat, in particular, would end up not being borne out by testing. After firing a large number of rounds of ammunition by the same manufacturer in quick succession (100 in one minute) the bolt and carrier of the author's personal AR-10, which he immediately disassembled, were only slightly warm to the touch and comfortable to hold, although the forward portions of the bolt and carrier were relatively dirty, just as the bolt of any automatic rifle or shotgun would be. The same test performed with the author's Kalashnikov, which sports a long stroke operating system in which the piston head, operating rod, and bolt carrier are parts of the same component and thus always in physical contact, ended up with at least as much fouling on the bolt and carrier as the AR-10, while its bolt and portions of its carrier were too hot to touch, and had to be cooled for five minutes before they could be handled.

The debate about the downside of direct gas impingement has flared up once again during the last decade, and there has been a trend toward "fixing" the M16, the most successful U.S. combat rifle in history which, as the direct descendant of the AR-10, uses Stoner's original gas system, essentially unchanged from that found on the X-01 prototype, by adding the complexity and expense of one or other design of a mechanical piston system. To this author, this is merely a passing novelty, and the admirably concentric Stoner direct impingement operating system, with its many advantages, continues to stand the test of time.

## ArmaLite's New Military Focus

ArmaLite's shift at this time towards the prioritization of military weapons over civilian models meant that Stoner's star was rising, while interest in the primarily manual-action sporting rifles and shotguns developed by Sullivan and Dorchester was on the wane.

Although plans had already been drawn up for the development of a civilian version of the AR-5 survival rifle as a lightweight semi-auto .22 rimfire, already designated the AR-7, and the AR-9 trap-and-skeet shotgun, these were put on hold for the immediate future, as resources were channeled into the improvement and production-engineering of Stoner's M-7 (AR-3) and M-8 (AR-10) prototypes.

Of these two, Stoner foresaw (correctly) that the M-8 was the superior design in every way, with the potential to usher in a new era of infantry rifle effectiveness. However he acquiesced to Fairchild president Boutelle's request that attention also be paid to preparing the conventional M-7 to face military trials, both men recognizing that U.S. Army Ordnance, with its conservative and cautious mindset, might well look askance at a rifle of such revolutionary design and strange appearance as the newly-crafted AR-10, and that by concurrently submitting the AR-3 for evaluation, a comparison between the two might prove to be beneficial and profitable.

This plan ended up serving a twofold function. In addition to offering the Army something of both the conventional and the modern, ArmaLite was hedging their bets that at least one of their designs would gain favor. Also, however, was the added benefit of making the AR-10 look good by comparison. Stoner was confident that when contrasted with a weapon utilizing a conventional stock that caused the rifle to pivot on the shooter's shoulder, as did the AR-3, the great recoil control benefits offered by the AR-10's straight-line configuration would become obvious and attractive. Perceived recoil is one of the most subjective aspects in the overall impression a firearm can make on its user, and the experienced sporting and military firearms enthusiasts that comprised ArmaLite's staff knew that this was one of the hardest attributes to convey on paper. Therefore, direct comparison by those who were making the selection decisions, they deduced, would be bound to favor the AR-10, and increase the chances of its acceptance as a military service weapon.

# The Model X-02: the Second AR-10

40. Left side view of the ArmaLite AR-10 Model X-02 prototype as depicted in *Small Arms Review* Volume I No. 4.

Only this one single example of the X-02 was ever constructed, which today resides in the Institute of Military Technology collection. However, no examples of the special Model X-02 magazine, which was not interchangeable with later production, remain in existence.

courtesy Dan Shea

Although having to split their resources between two different designs which were clearly (at least to those familiar with them) not equal in usefulness, the good news for the AR-10 was that Stoner had essentially worked out all of the issues with the AR-3 when it was still the M-7 in his machine shop, meaning that the AR-3 was already basically ready to face military evaluation with just minimal technical testing, polishing, and the streamlining of production methods. This meant that although both prototypes were being pursued aggressively, the majority of the attention and resources could be focussed on the AR-10.

Though ArmaLite was a design firm and not a full-fledged manufacturer, it was considered acceptable under their corporate charter to prepare a weapon for speedy and economical production, and to experiment with methods which would make that process more appealing to prospective licensees. This was considered by the Fairchild board to be a legitimate and integral part of marketing weapons designs, and with the backing of Boutelle, it was approved. This role of adapting designs to production and streamlining that process would prove later to be one of Stoner's most important roles in the story of the finest battle rifle ever devised, although it is an unsung and virtually-unknown achievement today.

## Describing the X-02

ArmaLite's first refinement of Stoner's original X-01 prototype (the first AR-10) incorporated several new, revolutionary design concepts. Foam-cored polymer furniture was added to this model, replacing the wood found on the M-8. The X-02 also marked the first time an AR-10 was chambered for the 7.62x51mm NATO cartridge.

The X-02 had a receiver constructed of steel, and weighed 8 lbs. 4 oz., but finalized plans were drawn up for the construction of a production model using aluminum forgings, which was projected to weigh only 6 lbs. 4 oz. The X-02 was thus intended to only be the prototype of a production model, which was cancelled when it became apparent that further major changes were needed before the design was finalized.

The new magazine and push-button release helped both in weight reduction and reloading speed.

Just as Stoner had started what would become the AR-3 with his M-5 design and improved it in

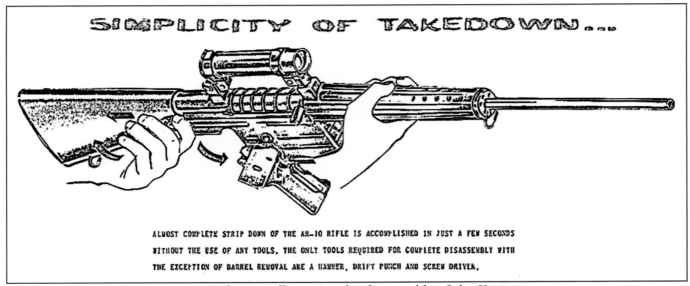

41. A drawing illustrating the disassembly of the X-02, from a promotional brochure produced by ArmaLite in early 1955.                    Institute of Military Technology

several steps, arriving at a useful and relatively bug-free battle rifle, fitted with polymer furniture and chambered for what would become the new NATO round, he was likely planning the same course of development on his own for the AR-10. However, the fact that it was a newer idea when he was discovered by S-F Projects meant that the improvements made to this more advanced design would happen with the help of a team, backed by the resources of a successful company. One of the first design changes, and one which Stoner likely had on the docket for the next AR-10 prototype, was adapting lightweight aircraft aluminum into the design. This was drawn up in the design bureau for future production of this model. However, as mentioned, in the X-02, the first working prototype, the receiver was fabricated from sheet steel, stamped out and welded at specifically-marked connection points, which nevertheless was a step up in production efficiency and weight saving from the partly-machined steel receiver of the X-01.

This use of steel instead of aluminum in the first prototype of what was planned to be a series was considered by the technical staff to only be a stopgap measure. The ArmaLite literature produced as a marketing tool for this design to the military announced that while "the weight of the prototype unit with a sniper scope is approximately eight and one half pounds with a twenty round magazine[, t]he weight of the production model will be approximately six and one half pounds."

Within the first two months of 1955, the receiver group of Stoner's M-8 X-02 prototype was

42. Left side closeup of the X-02 prototype as depicted in *Small Arms Review* Volume I No. 4, showing the selector and its markings.

Note the spring-loaded takedown button for removing the spatter-painted fiberglass-and-foam buttstock, at upper right.                    courtesy Dan Shea

adapted (theoretically) to the much lighter material, and along with the proposed aluminum housing a new means of rifle disassembly was also developed. The part of the lower receiver that contained the fire control group and pistol grip was now made a separate unit, hinged at the front and locked into the receiver by the overhanging buttstock and removal lever at the rear. When taking the rifle apart, the shooter only had to pull back on the spring-loaded takedown button, shown in fig. 42, which would release the buttstock and trigger group. One would then twist the buttstock counterclockwise while pulling it back. This freed up the rear of the detachable fire control housing, which could then be swung down and forward against the hinge, and removed.

The method of disassembly can be seen above in fig. 41.

From there, the bolt, carrier, buffer and action spring could be removed as before from the back of the buffer tube. Further disassembly, taking the rifle down to its very basic components, was accomplished with only a few standard tools. This was a big step forward over the X-01, in which the fire control group could not easily be removed in the field.

The next improvement to be incorporated in the X-02 prototype, one which would seem to be inevitable for any rifle with which Stoner or Sullivan were involved, was the use of polymer furniture. A new pistol grip, buttstock, and handguard were crafted using the same fiberglass shell and plastic foam interior as found on the AR-1 and AR-3. These lightweight components were modeled so exactly after Stoner's wood furniture found on the X-01, that even the ridged contours on the handguard were preserved in the new material.

A novel reliability feature added to the X-02, which was not only absent from the X-01 but was not present on any automatic rifle of its time, was a "dust cover", intended to keep the bolt assembly covered except during the actual working of the action. Such a cover had been a rather cumbersome and heavy feature of the Japanese Type 38 and 99 bolt-action infantry rifles of WWII. The X-02's dust cover was much simpler, and consisted of a ribbed shell that covered the top of the receiver under the scope. This piece was hinged on the left, and when the charging handle moved, either by being pulled back by the rifleman or from the automatic action of firing, it would pop open, opening the ejection port and allowing the bolt to eject spent shell casings. Once firing was completed, it could be easily reclosed manually, keeping dust and dirt out of the action when the rifle was not in use. A view of the dust cover in the open position can be seen in fig. 39, which also shows the camming ridge Stoner devised to guide the bolts of the X-01 and X-02.

Furthering what were already two coups in weight reduction from the original M-8 prototype design, the next task, accomplished with the X-02, was adapting the AR-10 to fire the new 7.62mm NATO round.

This shortened the receiver length over that of the .30-'06-chambered X-01, and required less material. It also gave the design team an opportunity for some innovative and novel thinking, and led to the first patent Eugene Stoner acquired for his developmental AR-10, discussed below, covering a new lightweight box magazine, constructed from aluminum.

## Scoping the X-02 - No Iron Sights

As shown in fig. 40, another interesting change seen on the X-02 was the entire elimination of iron sights, and their replacement with an integrally-mounted scope. For the sole X-02 prototype, the ArmaLite designers decided to tap once again into the rich postwar surplus market, and utilize a scope that had already been specifically designed for the special rugged use of a battle rifle. During WWII, the German *Luftwaffe* (Air Force) had produced and fielded a battle rifle known as the *Fallschirmjägergewehr* 42 (Paratrooper Rifle Model 1942, or FG42), which was not only ahead of its time in many ways, but bore a striking resemblance to its contemporary, the M1941 Johnson LMG. A modular type of scope was developed for this rifle which was capable of being mounted on virtually all German rifles, although this plan never came to full fruition. The German thinking was that iron sights could be replaced completely with a new generation of rugged, low-power (usually 2 - 5x) telescopic sights, which would greatly increase the fighting effectiveness of the combat rifleman.

The 9-lb. 3-oz. FG42 utilized many of the same features as the Johnson light machine gun, such as feeding from a magazine mounted on the left-hand side, firing in the full-automatic mode from the open bolt and semi-automatically from the closed bolt, and sporting a straight-line stock. Both the FG42 and the Johnson were developed secretively on opposite sides of the war, and these similarities can thus be chalked up to convergent evolution. Intended for use by paratroopers, the FG42 was, compared to other weapons of the era, notable for its massive firepower and above all its controllability.

The telescopic sight used was called the ZFG-42, later simplified as the ZF-4. Made in different variations, it was capable of being attached via a quick-detachable dovetail mount to first the FG42, and later the *Gewehr* 43 battle rifle and the StG44 assault rifle. It provided 4x magnification, and was narrow with a thin exterior and long eye relief (the distance from the eye to the scope lens offering the best resolution), allowing the user a clear field of vision of the combat area, while still being able to look through the scope. It was also filled with nitrogen to prevent fogging, and very strongly constructed, sacrificing a bit of optical clarity for durability.

In a purchase that makes the mouth of the military collector of today water, Stoner and Sullivan picked up a mint-condition ZF-4 scope for a few dollars on the surplus market (priced in the tens of thousands today), and as a part of the X-02's receiver, designed and produced ring mounts to hold the front and rear of the scope, mounting it solidly on top of the receiver.

While the FG42 allowed for the mounting of the scope right over the folded rear iron sight, the ArmaLite X-02 prototype completely eschewed the use of iron sights. It is not clear whether this was temporary and the design team meant to incorporate backup iron sights later if the prototype was shown

interest, or if they had in mind to have the scope replace completely the need for iron sights. Such a move would have been revolutionary and unprecedented at the time, but few characteristics of the AR-10 could not be so described. It should be noted that the Steyr AUG, a very successful assault rifle in use today as the standard service weapon of Australia, Austria, and Ireland, as well as by élite units of more than thirty other nations, uses exactly this principle, having only an integrally-mounted scope over the receiver, with no provision for main (iron) sights.

Interestingly, neither the barrel of the X-02, nor of the finished AR-3, was threaded for or fitted with a muzzle device of any kind, although as discussed below, such a device would be incorporated on later models of the AR-10.

# Notes on the Lightweight Rifle Program

Around the same time that the X-02 AR-10 was being prepared for its unveiling, the AR-3 was on its way to a destination that would become a common haunt for ArmaLite weapons over the coming three years: the Weapons Evaluation Board of the Infantry School, sometimes called the "Infantry Board," located at Fort Benning, Georgia.

For almost a decade prior to ArmaLite's first full year as a division of Fairchild, the U.S. Army had been conducting what was then called then the "Lightweight Rifle Program". Under this program the search for a shorter cartridge and a lighter rifle went hand-in-hand. The various arms and cartridges developed during the course of this program are discussed in detail in several books, notably the Collector Grade titles *U.S. Rifle M14* and *The FAL Rifle*, as well as the classic *History of U.S. Military Small Arms Ammuniton, Volume II* by F. W. Hackley, W. H. Woodin, and E. L. Scranton.

The .308 caliber AR-3 made its appearance before the Board during the first few months of 1955. The élite infantrymen who comprised the Weapons Evaluation Board were blown away by the light weight of the AR-3. Even when fitted with the heavy and laughably-oversized M1918 BAR magazines,

which looked so comical on such a sleek aluminum rifle, the troops who informally evaluated the AR-3 on maneuvers found Stoner's pre-ArmaLite design to be a real winner, and it was immediately recommended for official evaluation in the program.

The Lightweight Rifle Program had begun in earnest in 1945, when the requirements for the Garand's successor were decided and formalized. These criteria, which would have such a strong effect on the evolution of the AR-10 from its X-01 beginnings, declared that the rifle to eventually win the contest and be accepted would:

1. fire a shortened, full-power .30 caliber rifle cartridge;
2. feed from a detachable box magazine, which by 1955 was set at an ideal capacity of 20 rounds;
3. weigh seven pounds when loaded (although this was later changed to when empty).

Although Col. Studler was resigned to the idea that the originally specified 7-lb. weight limit was unattainable, the AR-3 actually achieved it.

## The Short-Lived T25

Designed at Springfield Armory by Earle Harvey, Col. Studler's protégé, the T25 was as shown here originally fitted with an unorthodox-looking straight stock which necessitated a high sight plane. Ironi-

cally, the advantages of such a configuration were studiously ignored, and the T25 was superseded by a redesign fitted with a standard drop-heel stock, and renamed the T47.

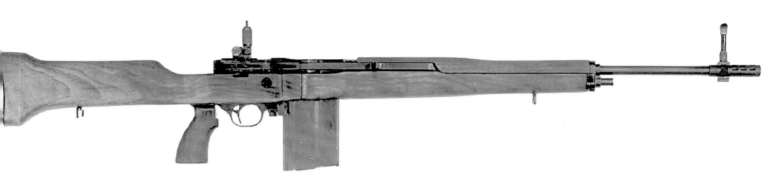

43. Right side view of the 7.62mm T25, designed by Earle Harvey at Springfield Armory, fitted with a straight stock and high sights. Both the rear-locking T25 and its successor, the conventionally-stocked T47, were soon rejected.
Springfield Armory collection

Both of the Harvey designs, to Col. Studler's chagrin, quickly lost out to the other two main contenders, the T44 and the FAL.

## The Two Top Light Rifle Contenders - the T44 and T48

44. Above: right side view of the last of the long-receiver T44 models, which featured in the "sudden death" Arctic trials of December, 1953. This was the rifle that turned the tide in favor of the T44, away from the T48.
Below: left side view of the FN T48, as manufactured for troop trials by the H&R Arms Co.    editor's collection, courtesy the late Ed Hoffschmidt

The two top contenders already in the running for adoption under the Lightweight Rifle Program were the T44, a modernized M1 Garand which was the subject of a long-running developmental program at Springfield Armory, and the T48, the U.S.-designated FAL rifle, developed and submitted by Belgium's Fabrique Nationale and already produced in some numbers by two U.S. arms firms, H&R and High Standard, for troop trials.

Both of these rifles weighed significantly more than the originally specified 7-lb. weight limit—the T44 at 8.45 lbs., and the T48 at 9.47 lbs.

# Springfield Tests the "ArmaLite, Caliber .30, Lightweight Rifle (T65E3)"

45. Right side view of the plastic-stocked, selective-fire 7.62mm Armalite AR-3, which featured in tests at Springfield Armory.

As Stoner had predicted, the conventionally-stocked, lightweight AR-3 proved absolutely uncontrollable in full-auto fire, paving the way for the straight-stocked AR-10 which was to come. *Springfield Armory photo*

In just three short weeks, the AR-3, which thanks to Stoner's previous intensive development of it from the M-5 to its final stage as the M-7, required almost no further improvement, was submitted to Springfield Armory for official testing as the "ArmaLite, Caliber .30, Lightweight Rifle (T65E3)", with the last parenthetical designation noting simply the 51mm caselength version of the T65 ".30 Light Rifle" cartridge for which it was chambered.

The AR-3 performed the standard function tests quite well, but it was quickly found to be utterly uncontrollable, even on rapid semi-automatic fire, let alone in the fully-automatic mode. It was rejected for this reason, and recommendations were made on reworking it for further consideration.

This decision, which had been foreseen by Stoner, played right into his plan for the ascendancy of the AR-10. Instead of trying to make the AR-3 do something outside of its design capabilities, ArmaLite, at Stoner's urging, elected instead to concentrate all of their resources on perfecting the AR-10. Stoner summed up the military's tacit realization that the seven-pound weight limit was far-fetched, commenting, "Like a lot of other things, once you had it in hand, they were all rather dubious about whether this is really what they wanted or not."

The remarks made about the AR-3 and its uncontrollable muzzle rise under sustained firing confirmed the hypothesis that Stoner had developed a decade earlier in the Marines, that a straight-line stock configuration was absolutely critical to the recoil control of any automatic rifle. Even though it helped to concentrate ArmaLite's attention onto the AR-10, the impracticability of the AR-3 when the specifications of the Lightweight Rifle Program were applied said as much about the practical impossibility of the demands as it did about any failings of the traditional stock design. As Stoner put it three decades after this intensive period of development and testing, "The United States Army was thrashing around, trying to come up with a seven-pound full automatic [7.62mm] NATO-caliber weapon. Well, we all know the practicality of such a thing because of its weight and recoil."

The AR-3 presented the Army with the most incontrovertible yet unwelcome proof that their requirements for a lightweight rifle firing such a powerful cartridge effectively in the automatic mode were mutually exclusive, and thus unattainable.

Although they knew the inherent risks involved in coupling these two criteria together, ArmaLite elected to achieve them both in their new AR-10 and thus gain an edge in the competition as the only rifle in the running for U.S. adoption to actually meet the original weight requirement of the Lightweight Rifle Program.

## The ArmaLite X-02 (AR-10) is Unveiled

The publicity surrounding the fledgling AR-10 began to pick up steam in mid-1955, as ArmaLite began to put out feelers to assess interest in their radical and revolutionary design. This newfound confidence was likely a byproduct of the Air Force's decision to adopt the AR-5 survival rifle.

In their new push for military contracts, their parent company, Fairchild, directed to ArmaLite the services of General Devers. Not only was Devers involved in the process of military arms review during his service, but his former command at Fort Monroe, Virginia was an important site for weapons evaluations. Fairchild's aggressive lobbying often made the ArmaLite staff, in their small facility in California, feel overexposed, as they were still just a small design operation, regardless of how innovative their developments were.

At ArmaLite, the necessity for improvement in controllability over the AR-3 was now clear to even the most skeptical, and the AR-10 was seen by all as their best hope to answer this problem. Stoner looked scornfully on the Army's weight requirements in particular, even though he preferred the heavy-hitting .308 caliber, and one can consider that the failure of the AR-3 to provide usefully accurate automatic fire served as further proof of the incompatability of the specifications. Regardless, however, he felt that thanks to the innovative straight-line stock, and the other what he called "crutches" to recoil control, discussed below, he could at least provide a level of recoil reduction in the AR-10 that would be considered sufficient, if not ideal; and would outclass even far heavier weapons of conventional design in the same caliber.

Upon its completion in the late fall of 1955, the first working prototype of the AR-10, the scope-mounted X-02, was immediately rushed to the Infantry School, the gatekeepers of U.S. rifle evaluation.

The X-02, now known simply as the AR-10, generated the same amount of excitement at its introduction as had the conventionally-stocked AR-3. On December 13 and 14, 1955, the X-02 amazed the experienced onlookers at Fort Benning. Stoner personally conducted the demonstrations, occasionally firing the rifle one-handed in order to demonstrate its ease of control and light weight. He even brought along the AR-3, both to demonstrate the variety of weapons types his company could conjure up, but more importantly, to give a frame of reference for how effective the AR-10 really was at taming the recoil of the NATO cartridge. In all of the firing, over two days and with round counts approaching one thousand, the X-02 only experienced one jam, and this was ruled not a fault of the weapon, but simply a defective cartridge. The AR-3, also firing a similarly large number of rounds, only malfunctioned once.

Within a few days it was reported back to George Sullivan and in turn to Richard Boutelle how interest was high for the AR-3, but nothing approached the enthusiasm shown for the AR-10. For his part, Boutelle was so happy at the glowing report of the X-02's performance in initial testing, which provided a resounding vindication of Stoner's beliefs, that he sent Sullivan a note congratulating him and telling him to buy a bottle of whiskey for his team as a congratulatory gift, and to write it off as a "conference expense."

Interestingly, there were those in the organization who pushed for the immediate submission of the X-02 to Springfield Armory for evaluation in the Lightweight Rifle Program, in a belief that it might stun them to such an extent that it would unseat the current frontrunners. Wiser counsel prevailed on this point, however, as one expert in the industry noted that Springfield Armory's being considered the final arbiter of infantry rifle acceptance was much like Winchester being allowed to judge a competition among sporting shotguns, when it had itself a model in contention for the title.

Encouraged, the ArmaLite technical staff stood ready to take notes on what the testers thought of the X-02. The Evaluation Board loved the straight-line stock and raised sight, the simplicity and ease of maintenance of its gas system, and the lightweight-but-strong (for its time) polymer furniture. More importantly, they also made several suggestions for updates that would be necessary for it to be a real contender to replace the Garand. First, although they found that the ZF-4 scope made long-range accuracy more attainable, they considered any telescopic sight to be too fragile to be the only means available for sighting the rifle. Therefore, they recommended that iron sights be standard equipment, but encouraged the modular mounting of a scope.

Next, as they had done with all entries in the Lightweight Rifle Program, especially those that most closely adhered to the original weight criterion, the Evaluation Board at Fort Benning expressed trepidation over the rapid-fire controllability of the X-02. They were particularly concerned by this, because the prototype that was actually demonstrated to them, the stamped-steel model, was two pounds heavier than the production model slated for formal testing in the final evaluations at Springfield Armory. ArmaLite got the message, and decided to develop another means of further increasing controllability; this time by reducing the recoil at its source.

# Melvin M. Johnson Comes On Board

With the keen enthusiasm and attention being paid to their new AR-10 rifle, ArmaLite realized that the time for aggressive marketing had arrived, and Fairchild decided to spare no expense in its attempt to garner attention for this promising design. This publicity campaign, which included advertising and lobbying, was run concurrently with ArmaLite's overhaul of the AR-10 to incorporate the Infantry School's recommendations, and brought on board none other than Melvin Maynard Johnson, the father of the M1941 Johnson Rifle and Light Machine Gun, another giant of American small arms design and a kindred spirit to Stoner and Sullivan.

## A Revealing Reminiscence

Mel Johnson's younger son Edward has kindly supplied the following interesting glimpse of his father's strong character as he remembered it from when he was growing up:

## *Camp Ogontz*

### *A Reminiscence by Edward Johnson*

*When I was very young, circa 1946, my father, Melvin M. Johnson, Jr., brought his family to Camp Ogontz, located in "Big W" Township on the west shore of Moosehead Lake, Maine. This camp had been built in the 1930s by my grandfather as a Masonic retreat. I would describe it as a large hunting and fishing camp consisting of a main lodge, separate dining room and kitchen, guides' camp, boathouse, and a long main pier for visiting boats. The only normal access from civilization at that time was by boat or seaplane, or by driving over the ice in winter.*

*Dad continued to bring the family there for two weeks every summer for "vacation," but he really utilized it as his favorite field combat "research facility" for his weapons. And the wildest 4th of July I ever observed was at a neighborhood party and cookout at Camp Ogontz, also circa 1946. Even the local Game Warden attended. Dad's version of the celebration and entertainment consisted of him standing on the beach, dressed in a USMC jump suit with a big Cuban cigar in his mouth, firing a Johnson .30-'06 light machine gun into the air and at rocks on the shore (for the ricochet effect). There were no boats out on the north bay of the lake - they were all moored at the Ogontz pier, where they were safe.*

*Dad fired at least six full magazines of ammo mixed with tracer rounds (one tracer, three standard rounds). As one barrel would overheat, he would dislodge it from the receiver (using just the point of a bullet) into the cold lake water, which made a loud hissing noise in the process. He would then quickly insert another fresh barrel and keep right on shooting.*

*For me, for noise and excitement, no other 4th of July will ever touch that event.*

*Nevertheless, despite his tall size and great strength, Dad was keenly aware that the modern soldier and law enforcement officer needed to carry as little weight as possible for mobility. Dad always liked the size and weight of the US M1 .30 Carbine, even though he felt that the .30 Carbine cartridge lacked knock-down power. He also felt that reduced recoil would assist in more accurate, rapid semi-automatic fire.*

*Later, circa 1954, when my older brother Melvin III and I were teenagers, Dad had us assist him with an experiment. He took us back into the deep woods behind Camp Ogontz and we set up a "combat course" where we would each be armed with a model M2 Carbine (selectable full and semi-automatic). We loaded each time with one round in the chamber, on safety, with two rounds in the magazine.*

*We would then each walk slowly forward and, upon a timed signal, would suddenly whirl to the right, release the safety and shoot up a hill at a black-and-white silhouette target mounted in the trees 50 yards away. We would each shoot three shots in fast semi-auto, and try again later on full-auto. We each performed this three times on each setting.*

*We found (not surprisingly) that all three of us did our best on the third trials, with Dad (also not surprisingly) performing the best of us. And all three of us hit in the black on SEMI-auto within about three seconds from the initial signal to shot number three.*

*However when we all shot on FULL-auto, Dad was the only one who was consistently able to land shot number one in the black and shot number two on the upper white area of the target, with the third shot off the target, in spite of his training and very strong arms. Brother Mel and I did OK with shot number one in the black, but shots number two and three went high off the target, even though Mel III also had strong arms.*

*So, although Dad felt machine guns definitely had their place, such as in aircraft, he did not feel that full-auto was so useful in jungle combat, and that it would waste ammunition. He always felt that rapid semi-auto was far more effective, and would soon utilize the above field experiment in his next project.*

*Indeed, not long after the above Ogontz experiment, Dad became involved with the AR-10 project. His patented short-rotation, multi-lugged bolt design was used in the AR-10, along with the actual style of the rifle itself, patterned after Dad's previous light machine gun. He also favored the .223 cartridge for its high-velocity performance and lighter weight, which allowed the shooter to carry more ammunition. He favored the 1 in 16" rifling twist, which created a "tumbling" effect when the bullet hit a fluid target. And I was once told later by a Colt engineer (in 1967) that had it not been for Dad's involvement and persistence, what eventually became the AR-15/M16 product might never have been developed.*

*Later, in 1963, Dad would return to the light rifle concept by designing and modifying a high-velocity 5.7mm cartridge for the M1 Carbine action, which was later marketed in small quantities as the 5.7mm Johnson "Spitfire" to the law enforcement and commercial varmint hunter market. He would deliberately limit production to semi-automatic carbines only.*

Johnson came to work as a consultant for Fairchild in the summer of 1956. The parallels of his life to those of the two principal figures in ArmaLite were startling: he was a lawyer who was interested in patents and machinery, like Sullivan; and was a Marine Corps Reserve officer, the same branch of the service in which Stoner had served during WWII. Johnson had developed his designs on his own time and working alone, occasionally making use of the machinery to which he had had access in the military. For his rifle, which would be used by his very own Marines, he developed the concept of the rotating bolt which locked into the barrel, and in his companion light machine gun, he instituted a straight-line stock design; both of which features Stoner would incorporate in the AR-10.

46. Melvin M. Johnson, in a photo taken later when he was promoting his 5.57mm MMJ caliber "Spitfire" carbine.

Despite an exhaustive search, it appears that very little if any photographic evidence exists regarding Johnson's time as a consultant for Fairchild.

Michael Parker collection

By the time ArmaLite was formed, Johnson had transferred to the Army Reserve, and had attained the rank of colonel, working at none other than Springfield Armory; where the finalists in the Lightweight Rifle Program all met their last tests. His deep connections in the Army's small arms research, development and analysis establishment went further, as was shown by his position as a consultant at Johns Hopkins University's Operations Research Office (ORO). This think-tank was sponsored by the Army, and had been set up three years after the cessation of hostilities in WWII to introduce the novel practice of statistical analysis into the way the Army made operational, weapons development, and procurement decisions.

By 1955, Johnson had become intensely intrigued when hearing the stories filtering up through the ranks about a rifle utilizing a rotating bolt that locked into the barrel and a straight-line stock. Due to its revolutionary design and performance, rumors of this strange and wondrous lightweight weapon being tested in Georgia were already an exciting topic of conversation in the small arms world, but for Johnson it was something more, because he had personally developed two of the major innovations of which it made use.

Johnson and ArmaLite first came into contact around the time the ArmaLite design team returned to California from the demonstrations of the soon-to-be-discussed next version of the AR-10 to the Continental Army Command and the Infantry School. On May 8, 1956, Johnson sent a handwritten note to Richard Boutelle, essentially asking to become involved with the ArmaLite project. Johnson was delighted to examine the AR-10 in person, and was retained on the spot as a new consultant for the company, with the purpose of updating the design for field trials, and of publicizing it within the military community. Having a retired Army Ordnance officer on staff, particularly one who had actually worked at the testing ground during the Lightweight Rifle Program, was extremely helpful to ArmaLite to begin with, but that officer being none other than the inventor of the rifle and light machine gun that had served as such a close inspiration for the AR-10 was seen as a dream come true for those engaged in the pursuit of the AR-10's adoption by the U.S. Army.

Recognizing that the AR-10 represented a real breakthrough in small arms design, a field that he felt had stagnated elsewhere in the U.S., Johnson decided to go "all in" with the new initiative. He consequently gave up his consulting position for Winchester Arms, on whose own entries into the Lightweight Rifle Program he had been working, and quit his position at Johns Hopkins University.

## William B. Edwards Beats the Drum for ArmaLite

Following the Fort Benning trials, the X-02 went on to be shown in demonstrations at a limited number of other U.S. military bases. No details of these events survive, but the fact of their occurrence is confirmed by secondary sources who were present, not all of whom were military personnel. The AR-10 had been developed by private industry, and thus was not subject to military classification, so members of the press and representatives from industry were also allowed to attend some of the demonstrations.

One notable representative of the world of firearms academia who attended at least one of these spectacles was William B. Edwards, the Technical Editor of the authoritative *Guns* magazine. As later recounted by associates and friends, Edwards was one of the first members of the press to take note of Fairchild's venture into the firearms business, an industry in which he was extremely well connected. The interest the X-02 kindled in him that year would be the catalyst for some of the most fantastical travels and stories in the AR-10's history.

# The Model X-03 (AR-10A)

## The Forerunner: the Futuristic AR-12, On Paper Only

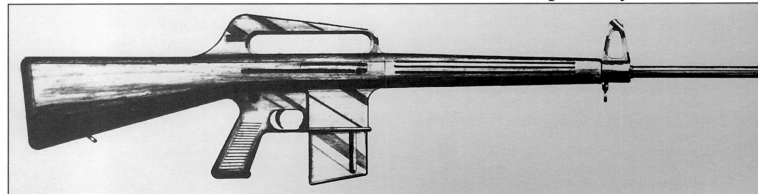

47. An ArmaLite rendering of the futuristic AR-12, which was never produced but which set the stage for the next iteration of the AR-10, the Model X-03, soon renamed the "AR-10A".                    Charles Kramer collection

## The AR-10A: One Step Closer to Military Evaluation

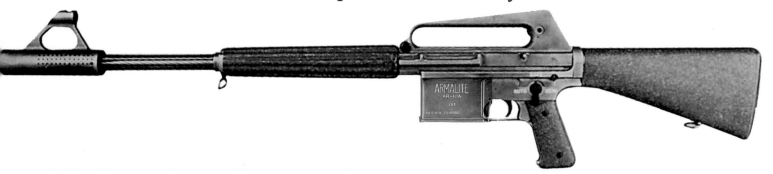

48. Left side view of the single AR-10A prototype, the ArmaLite model (and serial) no. X-03.

    This illustration was originally featured in *Small Arms* *Review* Vol. I No. 4, and the AR-10A is currently part of the Institute of Military Technology collection.

courtesy Dan Shea

With the help of their newfound and highly-placed ally Mel Johnson, Stoner, Sullivan, Dorchester, and John Peck (who according to the noted arms designer L. James Sullivan was an experienced and well-respected designer in his own right, already an ArmaLite employee before Stoner himself had joined the company), began in earnest to build a truly-lightweight, iron-sighted AR-10, with recoil control characteristics never before witnessed in a combat rifle. The X-03, the third version of "Tomorrow's Rifle Today", would be dubbed the "AR-10A", both to denote that it was a revision of the first ArmaLite AR-10, the X-02, and also to give the impression of a sophisticated design and modification process.

    It was decided at the outset that aluminum would be the material of choice for the receivers of even the first prototype, and so Dorchester, as chief of production, located a suitable source for aluminum receiver forgings in Harvey Aluminum, another California company later involved in supplying aluminum AR-15 receiver forgings to ArmaLite and Colt's.

    This manufacturing process, while more costly to set up, results in a stronger end product. Forging the metal by heating it and hammering it down actually compresses the material's crystalline structure, eliminating flaws and hollow pockets inside it, and producing occlusion-free metal structures.

    The added strength and consistency added by the forging process augmented the load-bearing portions of the aluminum receiver, allowing for tighter tolerances, and thus an even more Spartan use of resources, which further aided the quest to decrease the weight of the weapon. With a mind to keeping this third model AR-10 under the 7-pound weight requirement, no expense was spared in the production of working parts, with the thinnest and often-flimsiest pieces that could do the job being employed.

## Describing the X-03

49. Right side view of the AR-10A X-03 prototype.

    ArmaLite's second redesign of Stoner's X-01, the AR-10A incorporated many of the design changes that had been suggested by the Infantry Board, and shows a clear progression from the X-02 (fig. 40).

Institute of Military Technology

# Lightweight Upper and Lower Receivers, Made of 7075 Aluminum

50. Right side closeup of the X-03 (AR-10A). The spatter-painted furniture shows up clearly in this view.

The upper and lower receivers attached to one another by means of longitudinal flanges on the upper receiver mating with grooves in the lower, the two being held together and locked in place by a single takedown lever (shown rotated to the open position, above the rear of the pistol grip). Both receivers were constructed from aluminum forgings. Designer John Peck was put in charge of the changeover from steel to aluminum receivers.

This illustration was originally featured in *Small Arms Review* Vol. I No. 4.          courtesy Dan Shea

The selective-fire X-03, weighing 6 lbs. 11 oz., was the first model to boast both upper and lower receivers made of forged 7075 aircraft aluminum, with longitudinal flanges on the upper receiver mating with grooves in the lower, the two being held together and locked in place by a single takedown lever, located on the right side. For disassembly, the takedown lever was rotated, which freed the upper receiver to be guided forward off the lower against the pressure of the action spring, allowing quick access to the bolt and carrier/buffer for rapid cleaning.

## Contents of the Two Receiver Portions

At the front of the lower receiver was the magazine well, followed by the fire control group, the pistol grip, and the buffer tube in which the action spring sat, and into which the buffer and bolt carrier unit recoiled during the cycling of the action. The fiberglass buttstock mounted around the buffer tube was also connected to the lower receiver.

The upper receiver held the barrel at the front, the integral carrying handle with rear sight on the top, and the bolt and one-piece buffer/carrier assembly (fig. 51), fitted with a reciprocating charging handle mounted on the right side.

With the goal of reducing wear on the moving parts, the design team decided to abandon the camming ridge Stoner had included inside the upper receiver tube of the X-01 and X-02 prototypes, and replace it with a traditional groove in which rode the top portion of the cam pin, mounted in a curved track in the bolt carrier, which caused the bolt to lock and unlock. This was essentially the opposite of the X-01 and X-02's apparatus, in that instead of a ridge protruding downward to slide between two of the bolt lugs (which could cause premature wear on those lugs), the pin that rode inside the slot in the carrier and rotated the bolt to open and close it would be longer, and would slide in a cam track machined in the top of the upper receiver. The cut would hold the pin, and thus the bolt, in the unlocked position during the cycling of the action, so the bolt would not close before returning to the barrel extension into which it would lock.

## Bolt and Carrier/Buffer Assembly

51. Top view of the  buffer and bolt carrier group of the
ArmaLite AR-10 X-02 and X-03, with the bolt (far right)
shown in the unlocked position. The cam pin has been
removed to show its curved track in the bolt carrier.
Note the buffer assembly is an integral part of the bolt
carrier.                                      Institute of Military Technology

The bolt carrier on this third version of the AR-10 still incorporated the buffer assembly, which extended into the buffer tube even when the bolt was locked forward. The buffer assembly, which would become a separate component on subsequent versions of the AR-10, cushioned the carrier's stop as it reached the end plate of the buffer tube.

For the first time, in a feature that would become commonplace on the AR-10 and its descendants, vent holes were drilled in the right side of the bolt carrier. These allowed excess gas powering the action to be released harmlessly into the air, taking their heat and fouling with them, and at a stroke making the AR-10 a cleaner-operating weapon.

## Windage-Adjustable Iron Sights in the Carrying Handle

Iron sights were present on this version, just as on the X-01. For several reasons the design team diverged from Stoner's original application of having a thin, foldable rear tower sight stand alone on top of the receiver. This not only was seen as possibly too fragile (although the actual part had been used successfully on the Johnson LMG), but it also did not permit integral scope mounting. It was settled instead that the top of the receiver would now include what Johnson coined a "briefcase handle," with the rear sight incorporated within the rear portion of the handle, ahead of which a scope could be mounted. This carrying handle, forged integrally with the upper receiver, was rock-steady, and protected the rear sight aperture with sides that stood up on the left and right of the sight, extending forward the length of the handle.

This was a return to the original drawing in which Stoner first envisioned the M-8 back in 1952 (or earlier; he never revealed the exact date of his first blueprint for the M-8), as the drawing for the M-8 prototype showed a carrying handle just like that

found at last on the X-03. Due to his own lack of resources and the need to build the first prototype out of steel instead of forged aluminum, Stoner had decided to simply fabricate his X-01 without a handle, using the rear sight from a Johnson LMG, and work up to the later incorporation of the handle-mounted rear sight.

The rear sight was of the aperture type, and featured an adjustment for windage only. A knob on the rear right side of the handle, shown in fig. 50, could be turned to adjust a screw on which the aperture sat. The screw's threads would move the aperture left or right, allowing the rifle's lateral aim to be centered on the point of impact of the bullet.

No means of adjusting elevation for extended ranges were present on this third incarnation, but the plan was to get the original X-03 ready to demonstrate as quickly as possible, and then use the time it would take to set up and schedule the Ordnance Corps tests at Springfield Armory to incorporate a rear sight elevation adjustment in the model to be tested.

## A Short Treatise on Muzzle Flash, and the Function of a Flash Hider

Muzzle flash consists of the burning powder that follows the bullet out into the air while it is still combusting. Longer barrels produce less muzzle flash, as the powder has more time to burn before the bullet exits and releases the residual gas to the oxy-

gen-rich environment which causes it to burn so quickly and brightly. On a plain barrel, the natural shape of the muzzle flash is that of a globe, glowing briefly but brightly ahead of the muzzle.

A flash hider, contrary to popular belief, does not actually hide the flash from others watching, but rather it breaks up the shape of the flash so that it does not cross the eyeline of the marksman. This is obviously a concern for one who is staring intently along the barrel, as the sudden flash has the effect of ruining low-light vision.

The traditional flash hider has horizontal vents or slots angled away from the shooter's eyeline cut into it at intervals, which direct the flash out to the sides and minimize the formation of the fireball at the muzzle.

These devices actually concentrate the flash and make it more intense at the vents, but since these areas are not directly the focus of the shooter, the desired result is achieved. The effect on the shooter's vision of rapid firing, and in particular the consequent rapid, repeated and violent exposure to bright muzzle flashes, make the flash hider very important on infantry rifles capable of high rates of fire.

## The Sophisticated AR-10A Muzzle Device

52. Left side closeup of the recoil compensator, with integral high-mounted front sight, of the ArmaLite AR-10A model X-03.

The multifunctional compensator aided materially in reducing recoil and muzzle flash.

Note the fluted aluminum sleeve, described on the following page, fitted over the front portion of the barrel.

Institute of Military Technology

The final change made to the X-02 design to prepare it for military demonstrations was the addition of a novel muzzle device that combined the functions of a recoil compensator and a flash hider, introduced for the first time on the X-03.

Stoner called this device his new "crutch" for adding to the controllability of the rifle. It was an admirable design which actually countered and thus reduced the amount of recoil energy produced, thus adding to and strengthening the synergistic relationship between the direct impingement gas system's reduction of recoil and the straight-line stock channeling it straight rearward in the least disruptive direction.

The muzzle device was configured as a large, hollow cylinder, threaded onto and locked around the muzzle end of the barrel. In its face was a central orifice through which the bullet exited, and the cylindrical body had numerous small holes drilled in both sides of its circumference.

As discussed above, when a bullet leaves a barrel with a plain muzzle, the propellant gas follows straight out behind it, "blooming" into a fireball in front of the rifle. In addition to this flash of light, the burning gas also acts like a rocket engine which, in keeping with the physical principle of an equal and opposite reaction to the bullet's energy directed forwards, pushes the weapon backwards.

Since this device also reduced recoil, it was considered a "muzzle brake" or "recoil compensator" as well as a flash hider. Apparati of this kind had been put on firearms previous to the AR-10, and in fact one of the first prominent such devices was the Cutts Compensator, optionally available beginning in 1927 on the Model 1921AC Thompson submachine gun, and first marketed when Eugene Stoner was just five years old.

However, the AR-10A muzzle brake was novel in that compensators had never before been employed on mainstream military rifles, and its design was both more effective than the Cutts model and eliminated its major drawbacks.

The small number of large vents in the Cutts design meant that a correspondingly large volume of gas would be vented through each, creating a shockwave that went along with the rocket effect. This wave of compressed gas, normally directed straight out the front of a plain barrel's muzzle and away from anyone shooting or watching, now shot upwards, buffeting the shooter and others beside him. This phenomenon translates into a significant increase in the perceived volume and intensity of the muzzle blast. Muzzle brakes of this variety are still in use on various firearms today, and as many readers will attest, shooting next to one at the range is not a pleasant experience.

The large vents on the Cutts compensator also concentrate a large amount of the muzzle flash in a bright, glowing jet of flame that leaps upward, directly through the shooter's line of sight. This is far worse for the shooter's low-light vision than even the fireball at the end of the plain muzzle, since that at least expands in all directions, and is not intentionally blasted in front of his sight picture.

The high front sight, which lined up with the rear sight in the innovative carrying handle on the rifle's upper receiver, was made an integral part of the X-03's muzzle brake, so that none of the vent holes faced directly upwards. Instead, they only started in a position 45° down the surface of the cylinder on either side of the front sight, so that the muzzle flash exiting from them would form a horizontal "V" shape on each side, keeping the burning propellant out of the shooter's line of sight. The fact that much of the gas was still forced slightly upwards, however, meant that it was no less effective in taming muzzle rise. The use of many very small vent holes instead of a few large slots proved to be just as

effective at keeping the muzzle on target, and further benefits were realized by this arrangement.

Both the main drawbacks of the Cutts Compensator—the increase in blast and noise, and the blinding increase in perceived muzzle flash from the point of view of the shooter—were functions of the large venting slots. In addition to controlling muzzle rise just as well as the Cutts design, and directing the flash out of the shooter's eyeline, the large number of tiny vent holes in the AR-10's muzzle device only allowed a very small amount of gas out of each one, so the unburnt propellant present in it, the cause of muzzle flash, would consequently burn off far quicker, resulting in a jet of flame that was weaker, less bright, narrower, and did not extend nearly as far from the vent. Thus the entire muzzle brake produced what was essentially a diffuse, shallow glow along its entire length, instead of the proverbial dragon's belch that blasted up from the large venting slots of the Cutts model. This made the AR-10's recoil compensator into a *de facto* flash hider as well.

The small holes had the aggregate effect of keeping auditory disruption from the muzzle equivalent to what it would be on a plain muzzle; essentially gaining a recoil advantage with no increase in blast. The large expansion chamber inside the muzzle brake reduced noise even further, actually making the AR-10A quieter than its plain-muzzled predecessors.

The novel use of large numbers of small vent holes in the AR-10's muzzle device was inexplicably ignored for the next five decades, and has only recently been rediscovered by modern gunsmiths. New muzzle devices are finally being offered with this configuration, and the great benefits both in recoil control and flash suppression discovered more than half a century ago by the bright minds of ArmaLite are being realized by present-day enthusiasts of automatic rifles, including this writer.

## The Experimental Barrel Shroud

An aluminum sleeve or shroud with a longitudinally ridged pattern, resembling very closely the horizontal ridges and grooves on the handguard, was experimentally added to the front of the barrel of the AR-10A. This was intended to lend additional stability to the extremely thin barrel, which in turn had been necessitated by the light weight requirements under which ArmaLite labored, and to allow more rapid cooling. The corrugated surface greatly increased the surface area, which allowed the surrounding air to wick heat from the barrel more efficiently.

The shroud is seen installed on the X-03 in fig. 52 and in some photographs taken in the ArmaLite machine shop, but it is notably absent in other pictures taken of the finished rifle around the time it was presented and demonstrated to the military in early 1956. It seems likely that the team reasoned that they could do without the added weight the shroud represented, in order to make the rifle even more attractive to the Ordnance Corps.

# Stoner Patents the Aluminum AR-10 Magazine

This new innovation, first introduced in the X-02, the second prototype of the AR-10, was covered in U.S. patent no. 2,903,809, applied for on February 21, 1956 and granted on September 15, 1959. Titled "Cartridge Magazine of Aluminum or Magnesium", this took Stoner's success with light alloys and applied it to ammunition magazines. The BAR magazines that Stoner had used to feed his previous creations were made of light-gauge sheet steel. On rifles as lightweight as the AR-3, and the new aluminum-receivered AR-10, such a magazine added unnecessary extra weight, even when empty. In keeping with his characteristic style of coming up with new and ingenious designs, which were nevertheless very simple, reliable, and intuitive, Stoner simply adapted the standard method of producing magazines to the new materials in which he and his employer specialized.

The body of the detachable box magazine, in existence for over half a century, is normally fabricated from a box of stamped sheet metal, welded together. There appeared to Stoner no reason why in the jet age such devices could not be made the exact same way using aluminum (or possibly magnesium, with which he also experimented). The end result was a magazine that followed the basic principles and bore a similar shape and appearance to the M1918 BAR magazine, but which was less than one third the weight of its steel counterpart.

To make up for the relative fragility and malleability of aluminum, at least when compared to steel of the same thickness, Stoner increased the rigidity and strength of the side walls, the parts most vulnerable to denting or bending, by creasing them vertically and horizontally in a distinctive "waffle" pattern.

An excerpt from the patent disclosure reads as follows:

*. . Aluminum and magnesium sheet metals are not only lighter than similar grade steel sheeting but also are less expensive and much more durable under weather conditions than steel, tending to resist corrosion under the severest conditions. Prior to this invention, no satisfactory magazine of light metal was ever manufactured which would stand up under combat conditions . .*

The X-03 magazines were slightly modified over those used in the X-02. This included riveting a thin steel reinforcing strip behind the inside front

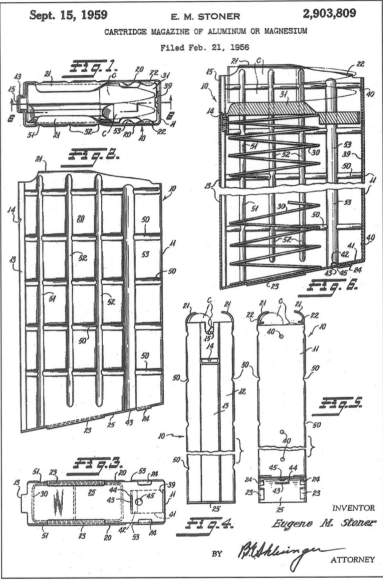

53. The single sheet of drawings forming part of U.S. patent no. 2,903,809 titled "Cartridge Magazine of Aluminum or Magnesium", applied for on February 21, 1956, granted to Eugene M. Stoner on September 15, 1959, and assigned to Fairchild Engine & Airplane Corporation.

U.S. Patent Office

face of the piece, increasing its strength in this vulnerable area. In a further excerpt from the patent disclosure (referencing fig. 53),

*. . The strip **39** is an important feature of this invention because it prevents the points of the cartridges **C** from denting or rupturing the front wall **11** of the cartridge case during firing operation. The shock is absorbed by this strip **39** and*

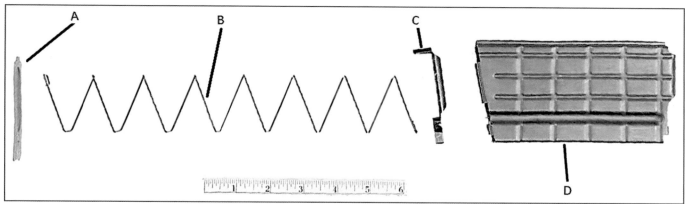

54. A standard aluminum AR-10 magazine, disassembled. The components, starting from the left, are the floorplate (A); the magazine spring (B); the follower (C); and the body (D). The rear extension of the follower actuated the bolt-hold-open device, discussed below.

Springfield Armory photo from Technical Report No. TS2-2015, dated February 4, 1957

*wear is, therefore, eliminated on the forward wall*
**11** . .

The X-03 magazines were also the first to be interchangeable with those of later standard-production AR-10s. Not only did the earlier magazine for the X-02 have slightly different dimensions, meaning that later units would not interchange with them, but as noted earlier there are no known X-02 magazines still in existence today.

55. Top view of an early X-03 magazine. Note the plain, rounded feed lips, and cast follower.

The floorplates were first bare, and later stamped "AR-MALITE."
author's collection

## Devising the "Throwaway Magazine" Concept

These "waffled" magazine side walls were still not as rigid or strong as steel, and so ArmaLite adopted as part of its marketing strategy the doctrine that the magazines were meant to be discarded after use in combat, and were not intended to be reloaded and reused, as detachable magazines generally were. This was justified in the company's promotional literature and by its sales representatives by the low cost-per-unit over steel magazines. Visually appealing as they were, these magazines were in short supply during the future combat deployments of the AR-10, and thus ended up being reloaded and reused numerous times, to the point where they would be literally the only aspect of the final version of the rifle about which any veterans interviewed for this work complained.

Aside from being lighter than traditional magazines, the design of the magazine well in the receiver meant that the magazine could simply be inserted straight up into the well, where a spring-loaded catch engaged a cut-out area in the side of the magazine body to hold it firmly in place. The other end of this spring-loaded catch was configured as a button on the right side of the receiver, which allowed the rifleman to simply move his right index finger from the trigger to the button and push, automatically ejecting the empty magazine, while his support hand retrieved a fresh magazine from his equipment. Traditionally, the reloading of a rifle or machine gun with a detachable magazine that had to be "rocked" into place was a somewhat more time-consuming process. The AR-10's push-button magazine release thus decreased the amount of time during which the rifle was out of action - a valuable consideration in combat.

# Tom Tellefson's Trademark Fiberglass Furniture

With the exception of the pistol grip, the furniture on the AR-10A was identical to, and even interchangeable with, that of the X-02, still utilizing the foam-filled fiberglass buttstock and handguard that were the trademarks of long-time ArmaLite employee Tom Edward Tellefson. A plastics engineer and expert machinist with a previous career at Northrop Aircraft, Tellefson is considered the father of the plastic furniture used on both the AR-10 and later AR-15.

Tellefson was a skilled and intelligent man who was a very early member of the ArmaLite team. L. James Sullivan remembers Tellefson setting up a large clear plastic goldfish bowl filled with water at the Fairchild booth at a trade show, wherein he floated an AR-7 to emphasize its handiness as a survival rifle.

The AR-10A pistol grip frame, unlike that found on the X-02, was actually an integral aluminum extension of the lower receiver, with thin polymer grip panels, also designed by Tom Tellefson, on either side held on by screws. The polymer panels extended up over the top of the trigger guard, providing a more comfortable and temperature-insensitive surface for the rifleman's forefinger when not positioned on the trigger. It also incorporated a newly designed winter trigger guard, hinged at the rear, so that by pressing in a spring-loaded plunger at the front with the point of a cartridge, the guard could be folded down flush with the front surface of the pistol grip, allowing the trigger to be easily accessed and manipulated by soldiers wearing thick gloves or mittens.

## Adding the Roxatone "Spatter" Finish

As shown in several illustrations, a unique finish was added to the furniture of a number of U.S.-made ArmaLite rifles by spraying them with Roxatone paint. This was very popular in those days, and was used to spray the interiors of automobile trunks and in other industrial applications. Roxatone consisted of a base layer in a neutral color, usually beige or tan, and contained a myriad of small droplets of different-colored paint, usually black, white, and grey, which due to their composition remained in suspension until sprayed on a surface, where they dried to produce an effective "spatter" pattern of random colors.

It is not known whose idea it first was to apply this finish on ArmaLite furniture, but Jim Sullivan recalls that George Sullivan liked it, while Gene Stoner did not.

# How Rifle Accuracy Works, and How the AR-10 Maximized It

The properties of polymer, as opposed to wood, lent themselves to improved weapon accuracy for two reasons. First, wood can warp, when rifles are moved to and used in different environments, if they get wet, or just from sustained firing. Warped wood can place pressure on the barrel in places where it wasn't before, degrading the accuracy potential by throwing off the original point of aim. The actual warpage that can affect accuracy is generally imperceptible to even the trained observer, and can be continuous when the rifle is exposed to different atmospheric conditions. This problem is known as the "wandering zero."

The other means by which accuracy is aided in the AR-10 is that the fiberglass shell of the handguard not only does not warp, it does not contact the barrel at all, and thus does not have any impact on the rifle's accuracy.

Just as warping can have an adverse impact on a rifle's accuracy, a foreign object contacting the barrel, or pressure either on the barrel or on something contacting the barrel can also move the point of impact away from where the sights are calibrated. This is especially true of light, thin barrels, but can also result from the shooter's hand gripping a traditional wood handguard very strongly, or a sling attached to the rifle's barrel being pulled too tightly to steady the weapon for shooting. In addition, clear experiments have shown that resting the actual barrel of a rifle on a branch, sandbag, or pile of dirt (instead of having only the handguard contact the rest) can throw off accuracy by several arc-minutes. The results and procedures of these tests can be seen on the popular firearms research website *Box O' Truth*. This is something to remember when one is out hunting, where the thought of bracing the barrel of one's rifle on a tree limb before making a long shot might seem like a good idea, but in fact often degrades accuracy more than firing "off-hand". Time and time again, this author has observed people at shooting ranges attempting to calibrate the scope on their new, expensive precision rifle by resting the actual barrel of the weapon, instead of the stock, on a sandbag. It is little wonder that these rifles do not hit where they are supposed to in the field, and are often incorrectly blamed for being inaccurate.

The ballistic principle behind this ability of light barrel pressure to affect accuracy is called "barrel harmonics." As also described in Chapter One, contrary to what one would perceive by watching a rifle fire with the naked eye, even a thick and heavy barrel made of very hard and strong steel actually flexes as the bullet travels down it. A meaningful illustration of this phenomenon is beyond the scope of the print medium, but videos abound, both in documentaries and on the Internet, of rifle barrels exhibiting this flex. A particularly good example is shown in the Discovery Channel documentary "Greatest Military Clashes: M16 v. AK47." The thin barrel of the Soviet Kalashnikov demonstrated in this documentary, combined with its heavy projectile and construction from weak, cheap steel, means that it exhibits an extreme amount of barrel flex, which is easily observed in slow motion.

The only point, in fact, where anything makes contact with the AR-10's barrel is the handguard cap, which is the small, round aluminum plate that fits snugly around the barrel, with recessed ridges that slide over the front end of the handguard, holding it in place. This is the only location at which pressure or pull on the handguard can have any effect on the barrel, and as such the AR-10 was more forgiving than any rifle before it to changes in grip, or to the use of a rest when placed against the handguard. Due to the thin material of the handguard, and its important function of shielding the gas tube, however, the ArmaLite design team decided to mount the front sling attachment point to the handguard cap, which meant that tension on the sling would be transferred directly to the barrel. The AR-10 was a military rifle, after all, and Stoner later recounted that the staff of ArmaLite made a conscious decision that the AR-10 was to exceed the accuracy requirements for the M1 Garand, and no more. Regardless of military requirements, the modular handguard system created by Stoner for the AR-10 was still the most conducive to accuracy of any yet devised, and would result in later versions as well as modern examples of the AR-10, and its direct descendant the AR-15, being the most accurate automatic rifles ever built.

# The AR-10A in the Spotlight

The X-03 AR-10A was the first AR-10 to be given a full-scale publicity demonstration tour, where it easily garnered the attention and enthusiasm that would guarantee it a spot in the military selection process.

On January 14, 1956, in an obvious show of the political capital ArmaLite was gaining within the Fairchild organization, Sherman Fairchild, founder of Fairchild Engine and Airplane Corporation and chairman of its board, toured the ArmaLite works. Just four days later, he was there to witness the completion of the very first, and only, AR-10A, model number X-03. All of the improvements were made and a working copy in firing condition had been fabricated, and it was time for the AR-10A to go on the road. Its first stop, just like that of the X-02, was at the Infantry School at Fort Benning, the very next month; preliminary firing tests at the Topanga Canyon Range having exposed a few minor bugs that had to be worked out first.

## America's Military Decisionmakers Praise the AR-10A

Finally presented with a design that combined the seemingly-impossible light weight of the AR-3 and recoil control characteristics of the X-02, as well as other welcome additions, the reviewers at Fort Benning were unabashedly enthusiastic about the AR-10A's performance. It was put through extensive firing demonstrations, where it received the rapt attention of all present. After the Infantry School presentation, the AR-10A was taken first to Fort Bragg, in North Carolina, and then Fort Campbell, in Kentucky, for similarly successful firing performances, where General Thomas Sherburne of the 101st Airborne Division at Fort Campbell was particularly impressed. Being the chief of one of America's most élite infantry formations, Sherburne knew a rifle with potential when he saw one. Also, being involved with air-mobile troops, for whom weight and modularity of weapons carry special importance, the light weight of the AR-10 was of especially great appeal.

The AR-10A continued its whirlwind tour of various bases, ending up in Fort Monroe, Virginia, in March, 1956, where it functioned magnificently in front of none other than the Headquarters staff of the Continental Army Command (CONARC). Combined with the successful lobbying by General Devers, who until just six years before had been the chief of that command, the AR-10A's *tour de force* before this most distinguished body of military decisionmakers ensured that Stoner's brainchild would receive favor and fame from the highest places in the United States defense community.

## An Unprecedented Offer from the Ordnance Corps

The buzz surrounding this new "wonder rifle" was such that it attracted the attention of the Ordnance Corps members responsible for the ongoing evaluation of entries in the Lightweight Rifle Program. In addition to their interest, Ordnance was subsequently ordered by the Continental Army Command to test and evaluate the AR-10 as a possible replacement for the M1 Garand. Melvin Johnson, who was an experienced insider at Springfield Armory, was assigned the task of liaising with the Ordnance Corps to establish parameters for a series of tests that would both be fair and would showcase the particular advantages which the AR-10 brought to the table.

Overtures were immediately made from Ordnance Small Arms R&D (ORDTS), now with Col. Studler's protégé, Dr. Frederick H. Carten, in charge, to Fairchild, by way of General Devers, for the Army to acquire the development and production rights to the AR-10, with ArmaLite being paid a royalty for every rifle produced, if it were adopted.

This offer of support, engineering expertise and funding, with a production royalty to be paid to a private rifle designer, was absolutely unprecedented. The fact that such an invitation was extended at all speaks to the fact that CONARC viewed it as superior to anything previously put forward, and that it would almost certainly be adopted if the offer were accepted.

```
ORDTS                                          30 March 1956


Fairchild Engine & Airplane Corp.
1317 F St., N. W.
Washington, D. C.

ATTENTION:  Gen J. L. Devers (Ret)

Gentlemen:

        This letter confirms agreements reached during a conference
held in this office on 29 March 1956, attended by Gen J. L. Devers
(Ret) and Mr. W. R. Smith of your company, with respect to the con-
tinued development of the Armalite 10 rifle.

        At this meeting the relative advantages and disadvantages of
the development of shoulder fired weapons under Army sponsorship
compared to commercial development of the item on a proprietary
basis were discussed.

        It is understood that this office will be advised by Fairchild
Engine & Airplane Corp., with respect to company policy regarding
continued development of the Armalite 10 rifle.

        FOR THE CHIEF OF ORDNANCE:

                                              Sincerely yours,


                                          F. H. CARTEN
                                          Assistant
                                          G. P. GRANT
                                          Lt Col, Ord Corps
                                          Assistant
```

56. The letter from the Office, Chief of Ordnance dated March 30, 1956, rubber-stamped by Dr.Fred Carten, the "father of the M14", confirming the "agreements reached" with Fairchild "with respect to the continued development of the ArmaLite 10 rifle."

Smithsonian Institution collection

## A Disastrous Decision

Unfortunately, however, ArmaLite took the position that this offer was an attempt by a hostile Ordnance Corps to gain the rights to, and then bury, the AR-10 once and for all. It was true that during the course of the original search for a self-loading rifle, which had resulted in the adoption of the M1 Garand on October 9, 1935, the Ordnance Department had purchased the designs of several inventors, but had found reason not to accept any of them. Fairchild therefore concluded that without their continued ownership of the design, it would be easy for those opposed to its progressive characteristics to simply decide not to pursue it, leaving ArmaLite without even the option of marketing it abroad. Fairchild therefore flatly rejected the Ordnance Corps' offer.

In hindsight, this decision could not have been more disastrous for the future of the AR-10. By disallowing the U.S. government the rights to its professional production engineering development, Fairchild was turning its back on its very likely adoption.

The repercussions of Fairchild's refusal added to the sour grapes Ordnance already felt at this upstart private concern which by all accounts had turned out a better rifle than they had, on a much lower budget. Their decision to offer to buy the AR-10 design was very much forced on them by the Infantry

Board and the Continental Army Command, and this embittered them further, increasing their feeling of isolation and insularity, in which it was them against the world.

## The SCHV Concept Rears Its Head

Interestingly, another result of the AR-10A's stunning performance, together with the behind-the-scenes efforts of General Devers, was the decision by the Army to open their books to ArmaLite on another classified project which was running concurrently with the .30 Lightweight Rifle Program. This was known as Project SALVO, which was investigating a number of possible avenues aimed at increasing military hit potential. Among these was the small-caliber high-velocity (SCHV) concept, which had been pioneered at Aberdeen Proving Ground. The extreme light weight and recoil control characteristics exhibited by the AR-10 gave the proponents of this program, which competed directly with Colonel Studler's "full-power" initiatives, the idea that ArmaLite might be the perfect firm to develop a weapon to meet their own, divergent requirements in the future.

The decision of the defense community to open their books to this relatively unknown division of Fairchild was, just as the offer to buy the design of the AR-10, completely without precedent and indica-

tive of the inestimably high regard in which those on the "sharp end" in the U.S. Army, particularly the Infantry Board, held the AR-10. The disclosure of fledgling defense project results to even much-more established private companies was simply not done in those days, as can be seen in Winchester's multi-year struggle for permission to market a commercial version of the T65 cartridge.

The SCHV portion of Project SALVO would lead to the birth of the AR-15 which, as the later military M16, has gone on to become the the longest-serving infantry rifle in the history of the U.S. armed forces.

Meanwhile, Springfield Armory announced that it would not be able to schedule any AR-10 tests for at least half a year, and the ArmaLite design team decided to make the best use of this grace period to upgrade the AR-10 even further, improving its design and adapting it for production. Thus begins the account of the version of the AR-10 which would meet its baptism of fire in the Springfield Armory Lightweight Rifle Trials.

# The AR-10A Becomes the AR-10B

Only two weeks after the AR-10A was unveiled to the world, work had already begun on the next stage in the AR-10's evolution. On April 11, 1956, George Sullivan contacted Warren Smith, an executive at Fairchild's Hagerstown, Maryland headquarters, with the announcement that specifications would be forthcoming on the next installment in the AR-10 saga. It was also at this time that the first drop in the flood of bad blood from Springfield Armory was felt, as General Devers was rebuked by Dr. Carten who, reversing the previous announcement that the AR-10

tests could not take place for at least six months, told General Devers that if the AR-10 was to receive any consideration at all, it must be ready for testing very soon.

Warren Smith speculated at the time that perhaps pressure from Lieutenant General James Gavin, a particularly brave and distinguished commanding officer from the WWII era, to wrap up the Lightweight Rifle Program as quickly as possible, was behind this peremptory change of plan.

## Redesigning the Receivers

In the new string of upgrades applied to the AR-10A that would result in the AR-10B (later redesignated simply as the "AR-10" to indicate that it had superseded the X-02 as ArmaLite's new standard rifle), the dual receiver system was modified further, into a perfect piece of simple ingenuity that not only made field-stripping and cleaning a breeze, but was never again changed on all the future versions of the AR-10,

or indeed in any of its modern descendants. The sliding action of assembly and disassembly was abandoned, and the new AR-10's upper and lower receivers were held together by two transverse pins, one at their rear juncture and one at their front. The front pin was not intended to be removed in field stripping, while the rear pin could be pushed out with the tip of a loaded round of ammunition, or even

57. An oft-used publicity photo showing Fairchild president Richard Boutelle demonstrating one of the first five AR-10Bs on his farm near Hagerstown, Maryland.

The first appearance of this photo may have been in *Army Times* on October 20, 1956, and of course it was later famously featured on the cover of the March, 1957 issue of *Guns* magazine, with the provocative heading, "Is This the Next G.I. Rifle?"                    editor's collection

by the thumb of the rifleman, allowing the two receivers to swing open, and allowing easy withdrawal of the bolt and carrier group from the rear of the upper receiver. In furtherance of this design concept the lower receiver now contained a separate buffer assembly held captive by a spring-loaded plunger ahead of the recoil spring in the buffer tube. The design staff reasoned that since the buffer and

spring were protected from the ingress of dirt or fouling during firing or even rough use, there was no need to remove them for normal cleaning, although doing so if desired was extremely simple and could be accomplished without the use of tools. Thus the operator could concentrate on the bolt and carrier as the only parts that needed cleaning and servicing.

## The Distinctive Non-Reciprocating Charging Handle

On this improved model the convenient carrying handle, aside from protecting the rear sight and providing a solid base for the optional mounting of a scope, now also housed a distinctive trigger-like charging handle.

Stoner's charging handle was a separate, hollow component, with a lower extension at the front that hooked onto a protrusion on the bolt carrier to draw the bolt back in order to clear the weapon, or feed the first round into the chamber, but which would not move with the bolt and carrier during firing. This non-reciprocating charging handle rode in a slot cut

in the top of the receiver that allowed it to be pulled to the rear, shielded by the overhang of the carrying handle. The cocking handle guideway in the upper receiver also served as the "cam track" for the new bolt locking/unlocking system introduced in the AR-10A, keeping the bolt in the unlocked position during the cycling of the mechanism.

Up until that point, virtually all charging handles had been attached directly to, and reciprocated with, the bolt or bolt carrier. Indeed, this sort of fixed charging handle had been present on the first three models of the AR-10 (counting the X-01), but Stoner

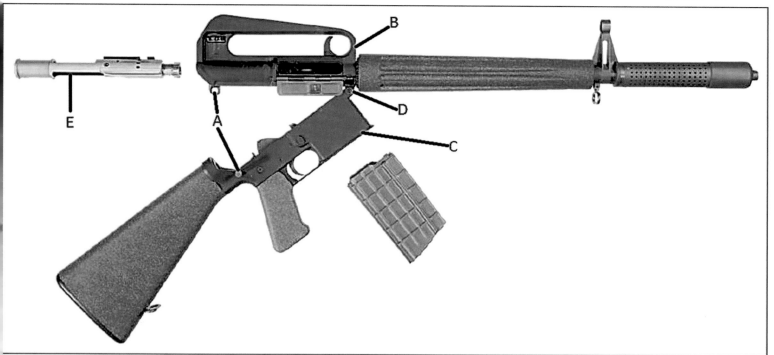

58. Right side view of the AR-10B, field stripped. Note the spatter-painted furniture.

 This photograph illustrates the ingenious receiver attachment principle of the AR-10B, and how with the simple retraction of the rear locking pin (A), the upper receiver (B) and the lower receiver (C) could be swung open, hinged at the forward locking pin (D), whereupon the insides of both receivers could be cleaned, the bolt and carrier assembly (E) could be removed for servicing, and the barrel could be easily accessed by a cleaning rod.

Springfield Armory photo from Technical Report<br>No. TS2-2015, dated February 4, 1957

realized that there were several potential advantages in using the new non-reciprocating type of charging handle.

 It had long been known that a charging handle that protruded from the side of the rifle and moved violently back and forth with the recoiling parts could catch on items in tight spaces during firing, not the least of which were the rifleman's fingers, clothing, or straps from load-bearing equipment. In addition to causing the weapon to jam if anything arrested the movement of the handle, this author can attest personally to how cleanly and forcefully it will snap a thumb like a twig.

 An additional result of adding the non-reciprocating charging handle as a separate component inside the carrying handle was that it decreased the weight of the recoiling parts, further reducing recoil because less mass was moving back and forth when the action cycled.

## Adding the Bolt Hold-Open

As an additional means of action control beyond the stand-alone charging handle, an entirely new pair of devices was added that significantly increased the efficiency and speed with which the AR-10B could be reloaded. On previous models, the bolt would close on an empty chamber after the last shot was

59 (right). Right side closeup of the AR-10B, illustrating the bolt hold-open mechanism.

 When the magazine is empty, the back of the follower (A) pushes up on the extended arm (B) of the bolt-hold-open (C), which causes its flat back to rise, blocking the forward path of the bolt face (D). The bolt stays locked back until the outside control lever of the bolt-hold-open (located on the left side of the receiver) is actuated by the rifleman. Institute of Military Technology

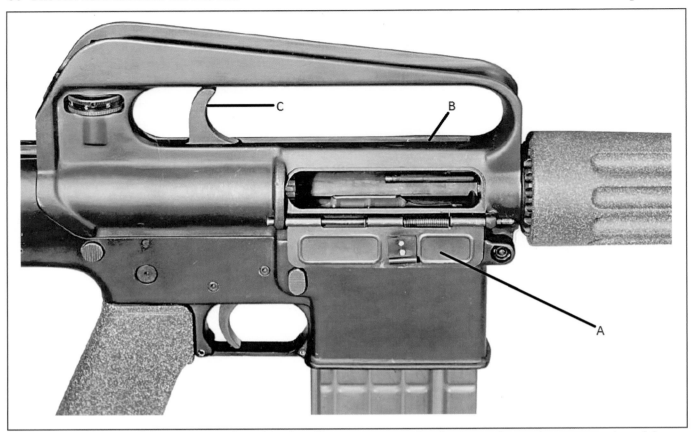

60. Right side closeup of the AR-10B with cocking handle (C) to rear, bolt held open, and ejection port cover (A) open.

Note the early form of the cover latch, formed from a simple curl of spring steel, riveted in place. Compare this with the later, more substantial latch assembly, shown in fig. 59, which was fitted with an internal spring-loaded plunger.

Note also the internal gas tube, an extension of the gas transfer bracket mounted in the left side of the receiver. As shown below, this mated with a longitudinal hole drilled in the bolt carrier to transfer gas to the chamber between the bolt and bolt carrier.

Springfield Armory photo from Technical Report No. TS2-2015, dated February 4, 1957

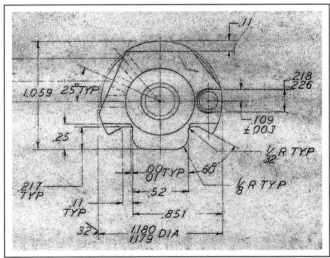

61. A portion of a dimensioned blueprint drawing titled "Carrier - Bolt" approved by E. M. Stoner on March 21, 1955, showing a face-on view of the bolt carrier.

Note the aperture, center right, in which nested the internal gas tube shown above in fig. 60.

L. James Sullivan collection

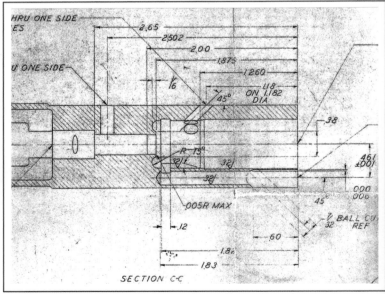

62. A further portion of the dimensioned blueprint for the original AR-10B bolt carrier, showing an underside view of the carrier with gas aperture leading into the central expansion chamber.                L. James Sullivan collection

fired, giving the shooter the false impression that the rifle was still loaded and ready to fire, but with nothing in the chamber.

To cure this, a "bolt hold-open" was added on this fourth model of the AR-10. It consisted of a lever that extended across the back of the magazine and upwards into the path of the bolt carrier, with an arm that overhung the magazine. When either there was no magazine in the rifle or a loaded magazine was inserted, this lever stayed down and out of the way of the bolt carrier, under pressure from an internal spring. However, when the last cartridge came out of the magazine, the spring-loaded follower caught the extending arm, and the far stronger magazine spring overcame the spring pressure holding the hold-open down, forcing the arm up into the way of the bolt face. The angled front of this lever allowed the bolt carrier to move backward and eject the last shell casing, but

when it tried to return to battery, the flat face of the bolt would run straight into the flat back of the bolt hold-open lever, now standing directly across its path, which would hold it securely suspended there.

On the outside of the rifle, a hinged "bolt release lever" was located on the rod which also served as the hinge on which the bolt hold-open rotated. This was used to manually engage and disengage the bolt hold-open, allowing the rifleman to simply push the top of the lever to reclose the bolt once a fresh magazine had been inserted. Just as had the new push-button magazine release system pioneered on the X-02, this further decreased the time required for a reload.

Pressing in the bottom of the lever while manually pulling back on the charging handle would lock the bolt open when no magazine was inserted.

## Improving the Ejection Port Cover

The action of the spring-loaded ejection port cover, another reliability feature which had debuted on the X-02, was made simpler by the advent of the new top-mounted, non-reciprocating charging handle. This cover could be manually rotated into the closed position, where it was held, first by a simple flat spring riveted onto the inside of the cover (fig. 60), and later by a spring-loaded plunger (fig. 59). When closed, it completely covered the ejection port, effec-

tively keeping almost any kind of mud, dirt, sand, or other debris out of the action.

No operator action was necessary to make the rifle ready when the cover was closed, as the bolt carrier, moving either forward or backward, would brush against the extension inside the cover, forcing it to snap open automatically under pressure from the relatively strong coil spring that kept it snug against the side of the lower receiver when open.

## The Aperture Rear Sight Now Adjustable for Elevation

While the aperture rear sight in the carrying handle was retained from the AR-10A, a new range elevation adjustment feature was introduced. As shown in fig. 60, a transverse wheel was located in a slot under the aperture, replacing the windage adjustment knob on the side of the AR-10A's carrying handle, with numbers to indicate the range for which the sight was set, so that the rifleman could dial in the range of his target and the sights would be automatically zeroed for that distance.

If needed, windage adjustment was performed by raising the rear sight to its highest position by means of the elevation wheel, which would allow access to a screw that clamped the rear aperture laterally in place. This screw could be loosened, allowing the rear sight to be shifted to the left or right by hand. Once the screw was retightened, the sights could be checked again to insure that the windage was correct.

The rationale for making windage adjustment more difficult was that the knob design found on the AR-10A could easily be turned accidentally, throwing the rifle's sights off target to the left or right. Because windage is generally not something that needs to be adjusted in the field, this intentional increase in difficulty in its adjustment was seen as an improvement. The plan was that windage adjustment would be set by test-firing at the factory, so the soldier to whom the rifle was actually issued would never have a practical need to adjust it. If this factory sight calibration was not done correctly, or if a unit was issued a type of ammunition that had a divergent lateral point of impact, the unit armorer could easily recalibrate the windage. Having served in the Marines, Stoner well knew that one less concern for the combat rifleman to worry about was always a good thing.

## Redesigning the Front Sight and Muzzle Device

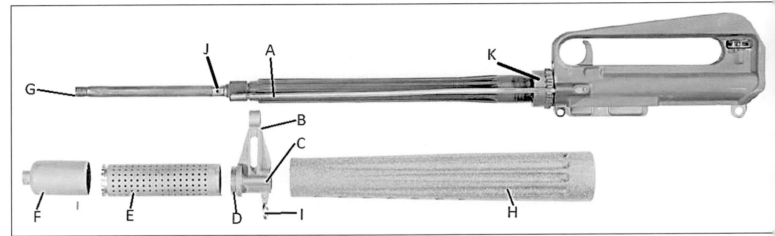

63. Left side view of the barrel and upper receiver of AR-10B serial no. 1004. The barrel group has been stripped and the components labelled, as follows:

From top left: the original stainless steel and aluminum composite barrel (G); the gas port (J); the elongated gas tube (A); the barrel nut (K).

From bottom left: the muzzle device cap (F); the external perforated compensator sleeve (E); the multifunction front sight tower (B) with positioning ring for the rear of the compensator (D); the tube-shaped portion (C), part of the nozzle following the internal 90° turn where the propellant gases are transferred to the gas tube; the front sling swivel (I); and the longer fiberglass handguard (H).

Springfield Armory photo from Technical Report No. TS2-2015, dated February 4, 1957

The configuration of the front sight was also altered from its position on the AR-10A. The designers found that while the revolutionary new recoil compensator invented for the previous AR-10A performed very well, it quickly became extremely dirty, as it served as a receptacle for almost all of the fouling produced by the burning propellant. Precise front sight alignment on the original AR-10A combination muzzle device/front sight unit required its very sturdy and durable attachment to the barrel. Its removal necessitated tools, which meant that it was practically impossible to clean the interior of the muzzle brake in the field. On the AR-10B, therefore, the muzzle device and front sight tower were made as two separate assemblies. The new muzzle brake could be removed simply by unscrewing its front cap from the threaded end of the barrel, which freed up the rear portion from its mounting bracket on the front sight tower, whereupon the entire assembly could be removed.

The new front sight base also served as the attachment point for the front sling swivel. Although this meant that sling tension would have a relatively strong effect on barrel harmonics and flex—and thus accuracy—it was considered more important that the reliability and ruggedness of the handguard be preserved, at the possible expense of a small amount of mechanical accuracy when using the sling to steady the rifle.

The great success realized in the minimization of muzzle flash from the original muzzle brake design prompted an expansion of the area covered by the small vent holes around the entire top hemisphere of the device. The combination of the very short jets of flame from the tiny vent holes, and the extreme height of the AR-10's sight plane above the bore, meant that vertically-directed gas would not interfere with the shooter's low-light vision, and the flash hiding effect was still achieved. The addition of more vent holes, particularly on the top of the muzzle brake, allowed for even more beneficial use of gas-venting to reduce recoil.

Another admirable feature of the new muzzle device was that it did not extend appreciably beyond the muzzle of the barrel, which provided the benefits of the compensator with almost no added length to the rifle.

In the promotional literature released by Fairchild to aid in marketing the AR-10 over the ensuing years, it was claimed that this revolutionary muzzle device achieved a 40% reduction in recoil energy, felt recoil, and muzzle climb, in addition to the other active and passive recoil-reducing and control-enhancing features found on the AR-10. Although felt recoil is a subjective concept and is perceived differently by every shooter, scientific testing did show that the claimed amounts of reduction in muzzle rise and actual recoil energy were accurate.

# The Gas Port Is Relocated Further Forward

On the first five serially-numbered AR-10Bs, nos. 1001 through 1005, the new stand-alone front sight base, also called "front sight assembly" or "front sight tower," was located by a series of splines on its interior mating with corresponding splines on the barrel, behind the gas port. These also aligned the gas port in the barrel with a corresponding hole in the front sight base, that directed the gas through a 90° turn and into a nozzle at the back of the assembly, where the new longer gas tube began. The lengthened gas tube allowed the propellant gases to enter at a slightly later point, after they had had more time to expand within the barrel and decrease in pressure, resulting in a smoother, less violent recoil cycle and a reduction in full-automatic rate-of-fire, both of which were a boon to rapid-fire controllability. The longer tube also gave the gas more distance to cover, allowing it to spread its heat over a longer area, and cooling it a bit more before reaching the working parts of the action, as well as allowing the rifle to dissipate heat better.

Moving the gas port closer to the muzzle also resulted in even less fouling being introduced into the bolt group, due to the shortening of the time from when gas first entered the tube to when the pressure began dropping back to normal after the bullet left the muzzle. The gas cycle was thus shorter, in addition to being smoother, so that by the time the bolt started to unlock, gas would have almost finished flowing into the carrier. This brief spurt of propellant gas meant that the time the bolt carrier took in moving back far enough so that the gas transfer tube on the inside of the receiver was no longer connected to the expansion chamber between the bolt and bolt carrier was shorter. This had the effect of concentrating all of the ejecta inside the specially-designed expansion chamber in the carrier, keeping the receiver and bolt face relatively free from carbon deposits. In fact, the main source of fouling elements observed on the actual bolt and inside the upper receiver of the AR-10 was found to be caused not by the direct impingement system at all, but by the powder residue emanating from the spent shell casing as it was being ejected.

In addition to providing the nozzle where the gas tube received the impulse of propellant gas, the

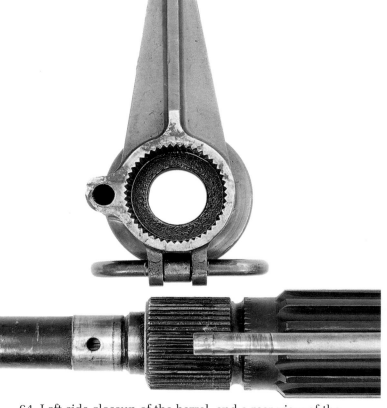

64. Left side closeup of the barrel, and a rear view of the front sight tower, from AR-10B serial no. 1003.

On the tower, note the nozzle to accept the front portion of the gas tube, and the internal mounting splines.

The barrel is the original composite design, made of a stainless steel liner and swaged-on fluted aluminum jacket. Note the gas port at center left, the splined portion onto which the front sight tower mates, and the gas tube.

Institute of Military Technology

rear of the new front sight base also served the function filled by the handguard cap on the X-02 and AR-10A. The new handguard, while still retaining its iconic horizontal ridges and grooves, was made longer, both to give the shooter more area to hold when manipulating the rifle in different shooting positions, but also to cover and protect the correspondingly lengthened gas tube. The front of the handguard was slotted, and these cuts dovetailed into ridges found on the rear of the front sight base, which locked it securely in place.

## More on the Controversial Composite Barrel

Finally, in the interest of further weight reduction, the two-part composite barrel design that Sullivan and Dorchester had championed was retained on the new prototype AR-10. Just as it had been on the AR-1 and AR-5, this was composed of a thin, stainless steel liner containing the actual rifling, encapsulated in an aluminum sheath or sleeve, swaged tightly around the liner. The concerns that Stoner repeated at this juncture were that the interaction between these two materials could degrade accuracy, and while their durability might be fine for the low rates of fire possible from bolt-action rifles, the intense heat produced by rapid fire could create separations at the swaging juncture, which could result in a catastrophic failure of the weapon. In short, the barrel might explode during firing.

It is important to note that the AR-10 was at this point essentially a combination of Stoner's individual innovations and those of the team headed by George Sullivan (not to be confused with designer L. James Sullivan, who, while no relation, was soon to start working for the firm). As a co-founder of the company, Sullivan wanted to see his own designs in the mix as well, and had already decided on incorporating the composite barrel, which he had developed himself, into the AR-10.

It so happened that around this time Sullivan received news that encouraged his selective bias in favor of the composite barrel. In the information about Project SALVO which the Army had entrusted to ArmaLite there were discussions of theories on using composite metal barrels. In subsequent discussions with the Ordnance Corps, Sullivan floated the idea of having barrels made completely from aluminum treated with modern hard coatings. The Army ruled these out as being inherently not strong enough, but did not reject out-of-hand Sullivan's proposal for an aluminum/stainless steel composite.

Stoner continued his opposition and warnings, however, stating that what had seemed to hold up in the very limited experimental testing to which the AR-1 had been subjected, was not a good idea for a military rifle. He was not swayed by the military exploration into the topic, and reiterated his formal opposition to submitting AR-10s for testing fitted with composite barrels of this kind. He further reasoned that with so many new features on the AR-10 which had not been tested and debugged the way new rifles developed by large companies and military institutions generally were, adding another unknown that would not necessarily decrease weight by very much was simply too risky.

As discussed below, Stoner's case against the composite barrel was given strong reinforcement by another military test of the AR-1 at Aberdeen Proving Ground. This test not only supported Stoner's concerns about the Sullivan-Dorchester composite barrel, but provided a prophetic warning for the soon-to-take-place tests of the AR-10B, one that should have been heeded.

The original AR-10B, with its composite aluminum/stainless steel barrel weighed an exact 7 lbs. As discussed below, this was later increased to 7 lbs. 2 oz. when the composite barrel was replaced with a fluted steel barrel.

## The First Five Numbered AR-10Bs on the Demonstration Circuit

As the time for the AR-10B to face its ultimate challenge at Springfield drew nearer, Melvin Johnson accomplished a major coup, and secured an appointment at Fort Campbell, Kentucky for the AR-10 to be "put through its paces" by none other than the élite 101st Airborne Division.

The initial five AR-10Bs earmarked for the Springfield Amory tests were issued to individual infantrymen on November 10, 1956, and taken on field trials. The reaction to their performance could not have been more positive. In all of the firing, not one rifle experienced a malfunction of any kind, and all displayed accuracy considerably outclassing the issue M1 rifles. The extreme light weight of the AR-10s was also particularly appreciated when carrying the rifles in rough terrain. This was seen by all involved as a universally good omen concerning the AR-10's performance in the upcoming Springfield Armory evaluations. In fact, Airborne General Maxwell Taylor ordered that his unit's selection of a new rifle be delayed indefinitely in favor of the AR-10.

On November 30, with the Airborne demonstration completed, Eugene Stoner took the same five rifles down to Fort Benning, so the Infantry Board could have a look at the newest incarnation of the AR-10. The outcome there was also very promising, and Stoner's next stop with the new AR-10Bs was Fort Monroe, Virginia, for an additional demonstration before the Continental Army Command. Only a few days after that, Stoner made the fateful trip up to Springfield, in order to be on hand as ArmaLite's

observer and representative as the Armory weapons trials began.

## ArmaLite's Diversified Publicity Initiative

65. A swimsuit model hired by Sullivan and Dorchester blithely demonstrates the floating capability of the ArmaLite AR-5A, even when fully assembled and ready to fire.                    Smithsonian Institution collection

Celebrating their success and still eyeing the civilian market, at the strong urging of George Sullivan, ArmaLite around this time staged a photo shoot in the Hollywood Hills, which preserves the fabulous hopes the division of Fairchild had for its creations.

These shots offer an alluring glimpse of the intense and optimistic spirit of innovation and progress that was so alive and well in the 1950s. It was in these heady days of the flourishing American Dream that the AR-10 was conceived, and no other

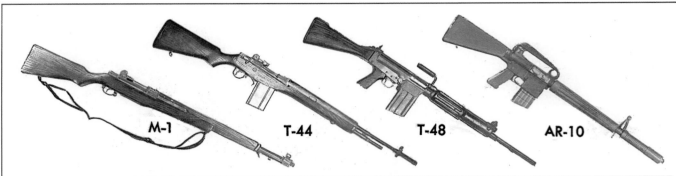

| | M-I | T-44 | T-48 | AR-10 |
|---|---|---|---|---|
| WEIGHT OF RIFLE | 9.56 | 8.45 | 9.47 | 6.85 |
| WEIGHT OF MAGAZINE | ——— | .53 | .57 | .25 |
| LENGTH OVERALL | 43.06 | 44.25 | 44.63 | 41.25 |
| OPERATING ROD | YES | YES | YES | NO |
| TYPE OF FIRE | SEMI AUTO | SEMI AUTO & AUTO | SEMI AUTO & AUTO | SEMI AUTO & AUTO |
| CORROSION RESISTANT METALS | NO | NO | NO | YES |
| STOCK MATERIAL | WOOD | WOOD | WOOD | PLASTIC |
| CARRYING HANDLE | NO | NO | YES | YES |
| QUICK BARREL CHANGE [ON LMG] | NO | NO | NO | YES |
| WINTER TRIGGER | NO | NO | YES | YES |
| BARREL RADIATOR | NO | NO | NO | YES |
| DUST COVER | NO | NO | NO | YES |
| EFFECTIVE MUZZLE BRAKE | NO | NO | NO | YES |
| FRONT LOCKING BOLT | YES | YES | NO | YES |

66. A handwritten mock-up of the later ArmaLite AR-10 brochure featured in Chapter Three included this "Comparison of Basic Infantry Rifles", which compared some salient characteristics of the M1 Garand, the T44, T48 and the AR-10B.

As shown, at 6.85 lbs. overall, the AR-10 was the only submission that met the stated goal of the Ordnance Department's Lightweight Rifle Program for a seven-pound rifle.                                     Charles Kramer collection

words or images could capture the *Élan Vital* that was every bit as necessary to the AR-10's creation as Eugene Stoner's genius or the lightweight metals and plastics used in its construction.

## An Ominous Rejection of the Composite Barrel on the AR-1

Almost forgotten by now in the intensity of the push to develop the AR-10 for military testing after the huge splash it had made, was the bolt-action AR-1 "Para-Sniper" rifle. Roughly concurrent with the incipient transition from the X-02 to the AR-10A, Army Ordnance had accepted the AR-1 for trial as a potential specialty sniper rifle for both élite forces and designated unit marksmen. Beginning on July 1, 1955, it had been evaluated at Springfield Armory, where both of the samples submitted experienced numerous issues with reliability and improper ejection of cases, as well as very poor and unpredictable accuracy.

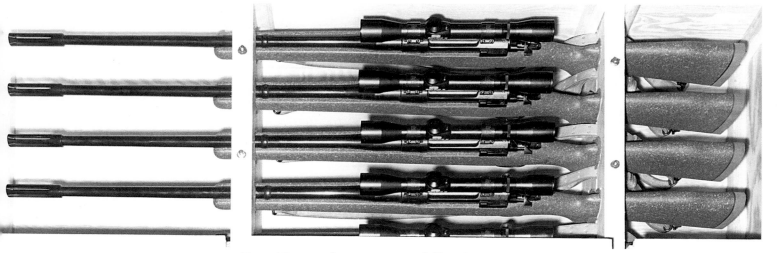

67. Four AR-1s, with spatter-painted fiberglass stocks, as crated up in Hollywood for shipment to Aberdeen Proving Ground for examination and limited firing trials.

It was this evaluation, discussed below, that should have brought home the point Stoner was trying to make concerning the weakness of the composite barrel.

Smithsonian Institution collection

ArmaLite made modifications to address the concerns initially expressed by Springfield, and the slightly more progressive Aberdeen Proving Ground in Maryland decided to give the AR-1 another chance. They ordered five additional examples of the rifle from ArmaLite at a total cost of $3,750, and in October, 1955, commenced a limited second evaluation.

Further firing and a metallurgical examination concluded in June, 1956 that the source of the poor accuracy was in fact the composite barrel. Not only was its two-part construction not conducive to increased accuracy by dampening barrel harmonics, as had been theorized by George Sullivan and Charles Dorchester, but examination of the individual components actually revealed signs of separation, as Stoner had predicted.

It was worrisome indeed that this dangerous issue should emerge even in a rifle with a very slow rate of fire like the AR-1, and this should have at least cast considerable doubt on the advisability of submitting the selective-fire AR-10Bs fitted with composite barrels.

Stoner, bolstered by these new tests results, renewed his vehement opposition to the composite barrel on the AR-10, but he was again outranked by Sullivan and Dorchester, whose decision was final, and so the issue was settled. A certain amount of quibbling took place within the leadership of ArmaLite at the time, which blamed the AR-1's poor performance on unfair testing criteria. Although these complaints may have had some merit, the firm's principals ought not to have let personal preferences cloud their judgment, and should have been content to delegate the making of technical decisions to those whose professional purview it was.

## Springfield Vindictively Awaits the New AR-10

While all of this modification and tweaking was being carried out on the AR-10B and the test parameters were being hashed out, the final evaluation and decision for adoption of a new rifle by the Army was put on hold in favor of the AR-10, by the strongest recommendations made by the Continental Army Command who, having seen firsthand the prowess of the AR-10, ordered a chagrined Springfield Armory to give it a chance.

Needless to say, this was a resounding endorsement of the AR-10, which only served to highlight the fact that, after spending over a decade of work at Springfield Armory on their own new rifle designs, the AR-10 was better.

As we shall see, the testing of the ArmaLite AR-10 at Springfield Armory would be the furthest thing from a fair shake. The same "NIH" (Not Invented Here) prejudice that had stymied the earlier hopefuls, the British EM-2 and the Belgian FAL, which was the only reason the U.S. had not adopted the FN T48 as early as 1951 or 1952, would now be turned full force on the AR-10. While both those foreign entries had sprung from powerful and well-funded design institutions, as well as having their respective governments behind them, the AR-10, developed by fewer than ten people and under a budget of less than one hundred thousand dollars, had no hope of advocacy at such high levels.

The depths to which Springfield would stoop to sabotage the AR-10 will forever stand as a black mark on what was and remains today, although in a very different form, a respected and storied institution.

*Chapter Three*

# Domestic Defeat

## The Watershed Springfield Tests of the AR-10B

As 1956 wore on, Charles Dorchester and Melvin Johnson made frequent trips to Springfield Armory to lay the foundations for the test procedures to be used for the AR-10. The fact that the T44 had been put on hold at the insistence of the Infantry Board, made vociferously enough to sway even the traditionalist General Dahlquist, chief of the Continental Army Command, was a serious sore spot for the Ordnance Corps in general and Springfield Armory in particular. With memories fresh of how badly the Ordnance representatives and personalities of the Lightweight Rifle Program had behaved, even towards their own servicemen, to the point of souring the members of the Infantry Board on them, Johnson and Dorchester fought hard with the Springfield staff to establish a series of tests that could be administered fairly. Not only was Springfield openly hostile to these "outsiders" (they even considered the infantrymen at Fort Benning outsiders), but since the tests would be conducted at the Armory, and not by the Infantry Board, which had been extremely impressed by the ArmaLite weapon, opportunities would abound for Springfield to derail the AR-10.

It is hard to overestimate the level of institutional resistance against which the AR-10 had to labor when it faced its first official trials at Springfield Armory late in 1956. The T44 and its predecessors, the T20, T25 and T47, had been under development for more than ten years by this point. The attitude at Springfield can be best summed up in the promise that Colonel Rayle made to Charles Dorchester upon learning that he had been ordered to halt the T44 adoption process and test the AR-10: "We will send you home in one day with your parts in a basket."

**SPRINGFIELD ARMORY**

RESEARCH AND DEVELOPMENT

TECHNICAL REPORT

PROJECT TITLE:  Rifle, Shoulder-fired, Infantry Type (Lightweight Development)

PROJECT NO.:  TS2-2015

REPORT TITLE:  Evaluation Test, Armalite Rifle, 7.62mm, AR-10

ITEM: 7.62mm AR-10  DATE: 4 February 1957  SA-TR11-1091

FOR OFFICIAL USE ONLY

FORM NO. ORDBD-719
10 AUG 51—E

**SPRINGFIELD, MASSACHUSETTS**

68. The cover sheet of the official Springfield Armory AR-10 test report dated February 4, 1957. This report is missing in the Armory archives, and from all other government sources.                    author's collection

# Round One of the Testing Begins

69. Left and right side views of one of the two ArmaLite AR-10B rifles submitted for tests at Springfield Armory in 1956. The report (fig. 68) stated that "Rifle No. 1002 was reported by company representatives to have fired approximately 100 rounds; Rifle No. 1004 approximately 500 rounds."  Springfield Armory photo dated December 4, 1956 from Technical Report No. TS2-2015, dated February 4, 1957

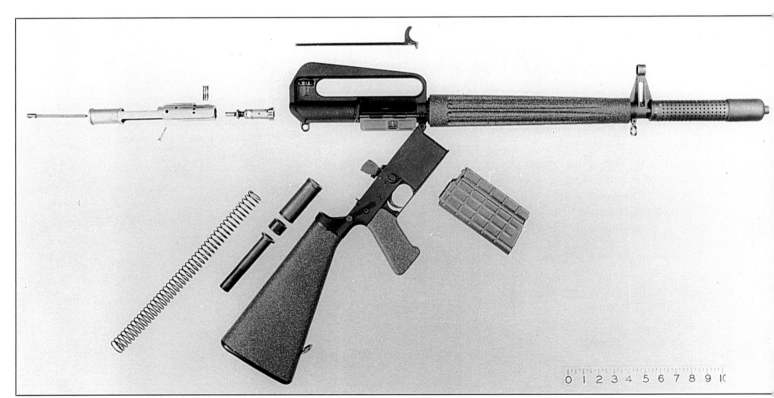

70. Right side view of one of the test AR-10B rifles, captioned "Stock and Receiver Group - Disassembly".  Springfield Armory photo dated December 4, 1956 from Technical Report No. TS2-2015, dated February 4, 1957

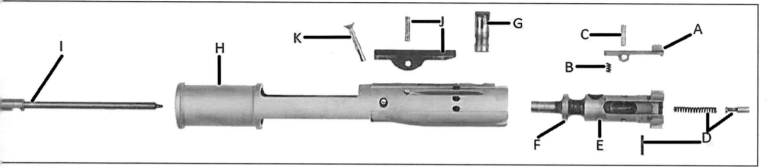

71. The AR-10B bolt and carrier group, completely disassembled with components labeled. These are:

    A. extractor
    B. extractor spring
    C. extractor pin
    D. ejector, spring and pin
    E. bolt (head)

F. piston ring - gas seal
G. cam pin (locking and unlocking)
H. bolt carrier
I. firing pin
J. bolt guide and pin
K. firing pin retaining pin

Springfield Armory photo
dated December 4, 1956 from Technical Report
No. TS2-2015, dated February 4, 1957

On December 3, 1956, the evaluations of the ArmaLite weapon began at the test range with two of the original five serially numbered AR-10s, nos. 1002 and 1004. There to observe the evaluations was, of course, Eugene Stoner, who was present for the duration. During these first days of testing, several other ArmaLite and Fairchild representatives were also on hand both to give technical assistance, and to keep a sharp eye out for any shenanigans. Warren Smith was on site for the 3rd and 4th of December, Melvin Johnson stayed from the 4th to the 7th, and George Sullivan himself was present on the 4th and 5th.

    The military personnel who conducted the actual testing at Springfield were not all from the hostile camp by any means. The first shooter was Colonel Fossum of the Continental Army Command. He was joined by two majors from the Fort Benning Infantry School Board Number 3. General Gavin sent Major Curtis from his personal staff, and Mr. Bonkemeyer, a civilian employee of the Armory, completed the roster of shooters.

    Rifle number 1002 was assigned to undergo an endurance test, while number 1004 was scheduled to perform the accuracy and functioning trials.

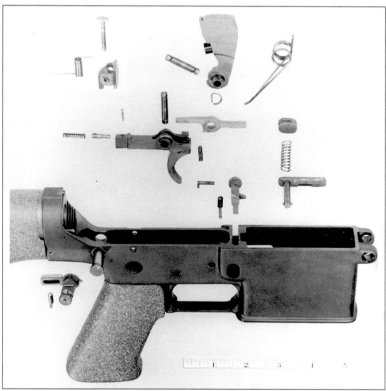

72. Right side closeup of the AR-10B lower receiver with all components disassembled (except the buffer assembly), and shown with their springs and pins.

    From left: the auto sear, trigger, hammer, sear, bolt holdopen, and magazine release.

Springfield Armory photo
dated December 4, 1956 from Technical Report
No. TS2-2015, dated February 4, 1957

## Issues Arise as Round Counts Mount Up

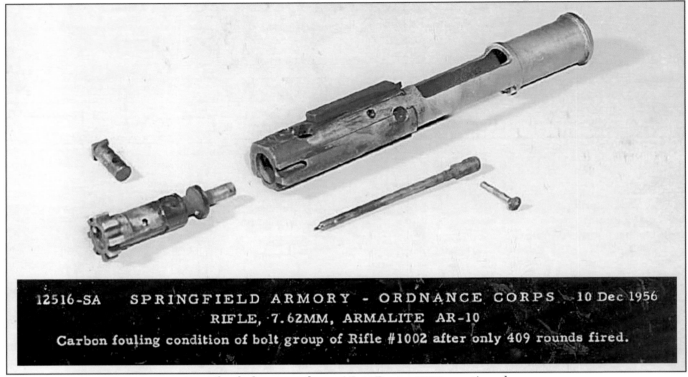

73. The bolt group from AR-10B no. 1002, captioned
"Carbon fouling condition of bolt group of Rifle # 1002
after only 409 rounds fired."
Springfield Armory photo no. 12516-SA
dated December 4, 1956 from Technical Report
No. TS2-2015, dated February 4, 1957

Initial firing trials reinforced the excellent impression made by the earlier prototype a year before, but as the round counts increased, issues began to emerge with the two AR-10s. There were no failures of the rifle to feed or extract rounds of ammunition, but the recoil compensators, constructed out of extremely lightweight aluminum, overheated and began to warp. It was worried that this would become so severe that the very small bullet exit hole would become misaligned, so that if a shot contacted it the entire device could be blown off the end of the barrel.

Sure enough, during the endurance test the interior sleeve of the rear venting portion of the compensator on rifle no. 1002 started to detach from its housing, and fragmented. A piece from it flew forward, tearing a hole in the cap which held the venting portion in place (fig. 74). It was clear that this device required serious attention and revamping before it could function reliably or even safely under the punishing conditions of the endurance test.

In addition, the extreme heat induced by the high round-counts in the endurance trial also exposed one more weak point of the finished "B"-pattern AR-10. The gas tubes, constructed of stainless steel, became flexible and began to deform (fig. 75). This did not actually impede functioning in any way, but it was decided that having them keep their shape under intensive firing was a must for the image of durability the military craved. Stoner quickly figured out that this material had a much lower melting point than did other types of steel, and that simple melting of the tubes, from the heat transfer of the hot gas piped through them, as well as the conduction from the extreme barrel temperatures, was causing the issue.

An extractor—the weak link in essentially every automatic rifle prototype, both successful and unsuccessful, tested since 1919—on one of the test rifles also broke, due to the heat and stress of sustained automatic firing.

One final complaint that came up during this stage, which was not considered a stoppage, but

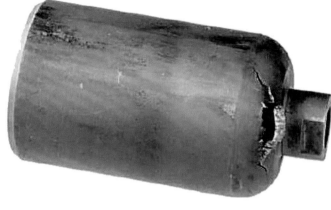

BAFFLE
FRAGMENTS

MUZZLE BRAKE

12504-SA SPRINGFIELD ARMORY - ORDNANCE CORPS 10 Dec 1956

RIFLE, 7.62MM, ARMALITE AR-10

74. This photograph, taken six days after testing began, shows the collected baffle fragments and fractured muzzle cap of the compensator of AR-10B no. 1002.
Springfield Armory photo
dated December 10, 1956 from Technical Report
No. TS2-2015, dated February 4, 1957

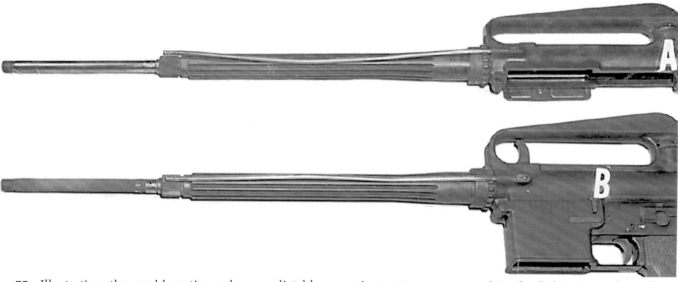

75. Illustrating the problematic and unpredictable warpage experienced by the stainless steel gas tubes.

Rifle "A", above, is no. 1004, after 681 rounds fired.

Rifle "B", below, is no. 1002, after 785 rounds fired.

The more extreme warpage seen on rifle no. 1004, which was used in the accuracy testing, was due to the use of a tight support sling and different heavy grips during the testing, as opposed to the light grip and sandbags employed for endurance testing on no. 1002. This showed that the heating made the tubes flexible, but it was pressure transferred through the handguard that actually caused most of the warpage.
Springfield Armory photograph,
Smithsonian Institution collection

which the Springfield Armory personnel still unfairly counted as a "malfunction," was that the bolt did not lock open reliably after the last round was fired. Although Johnson did not count this a real problem, Ordnance was searching for any reason to count points off the AR-10, and Stoner decided that he should remedy this issue as well. Luckily for the ArmaLite team, this was not a difficult goal to accomplish.

## Mel Johnson Calls Time Out

It came down to Melvin Johnson to decide what to do, and at 4:30 P.M. on December 6, at the end of that day's testing, he recommended that the tests be postponed so that modifications could be made on the test rifles. Colonel Rayle, who was overseeing the tests, conferred with Johnson and Stoner, and they all agreed that the testing would resume by January 8th of the coming year. This was an extremely short window to allow for much revamping, but since the AR-10s had turned in a stellar performance so far and seemed to require only a few minor changes, it seemed reasonable. It was also made perfectly clear to the ArmaLite representatives, both in their conference with Colonel Rayle, and earlier in backroom meetings with other current and former Ordnance officials, that if the AR-10 were to receive any consideration at all, the testing had to be completed extremely quickly.

Clearly expecting to be hearing the last of the AR-10, Rayle was taken aback when Stoner and Johnson seemed confident in their ability to correct the noted deficiencies and have the rifles back in one month's time.

In a way to reestablish the excellent performance of the AR-10, and thumb his nose at the hostile attitude of Springfield Armory, Johnson defiantly took the AR-10s out after the tests had been suspended and continued to fire them, after simply re-indexing the muzzle brake and replacing the broken part on the one that had been perforated. He determined that the problem had simply been that they had not been attached with the right alignment, and that the redesign to make them stronger and more heat-resistant should also include an allowance for easier indexing. He fired the rifle using the most diverse loadings of commercial .308 Winchester and military 7.62x51mm NATO ammunition available, firing one-handed, from the hip, and even at beer cans which he threw in the air. Through all of this "venting" of Johnson's, as well as during these initial days of testing, not one hiccup from the rifles was experienced.

# The Great Scramble: the AR-10B is Updated

## Rethinking the Gas Tube

With no time to spare, Stoner rushed the two rifles back to ArmaLite headquarters in Hollywood and took them apart. He had his work cut out for him: the gas tube, recoil compensator, and extractor all needed work. He started with the easiest of the tasks, drawing up plans for a simple material change in the gas tube. He substituted standard 4130-alloy ordnance carbon steel for the problematic stainless variety used on the previous examples, and specialized testing, both with sustained firing (once the other modifications had been incorporated), and heating of the individual part in an oven, showed conclusively that the warping and melting problems had been solved. The new gas tubes rigidly kept their shape, both internally and externally, and were reassembled into rifles nos. 1002 and 1004.

## Mel Johnson's List of Suggested Corrections

In addition to the obvious need to strengthen the gas tubes, Melvin Johnson prepared an extensive itemization of all the modifications he considered necessary. Several of his suggestions were overruled out-of-hand by Stoner, as simply not being in keeping with the overall nature of the AR-10 project, while others were taken seriously and the machine shop acted upon them. Finally, there were three suggestions that were not incorporated, mainly due to lack of time and resources for engineering, but these would prove quite important in the future, and be part of the finishing touches to the final incarnation of the AR-10.

Johnson's report touched only briefly on the gas tubes, as these were already slated to be fixed by Stoner. The areas where his suggestions would be followed had to do mostly with parts durability, and were able to be added to the test rifles in the time frame available. These were first to improve the durability of the muzzle device by both using different materials, and simplifying its interior design; and second to change the contours of the extractor. He also urged enlargement of the handguard, and the addition of more vent holes to aid in barrel cooling. He had particularly noticed how hot the receiver, cocking handle, and forearm got during his extended firing. This particular suggestion was taken into account, but changing the handguard would have to wait for the future, as ArmaLite did not have time for a handguard redesign during December, 1956.

The receiver heating (never to a dangerous or even uncomfortable temperature), was actually an intended part of Stoner's design, in that it helped to radiate and disperse heat from the barrel, with the aluminum acting as a heat sink, wicking energy from the barrel. The very low specific heat of the forged aluminum, as well as its considerable volume and mass, kept heat dissipating, and allowed the AR-10 to maintain lower barrel temperatures, with the same number of rounds fired in the same elapsed time, than any other contemporary design.

## Extractor Modification

The implementation of the extractor modification was rather detailed, and actually worked somewhat in a counter-intuitive way. Johnson described the procedure as "filleting" the extractor claw, meaning essentially the removal of some metal. He had noticed that on all of the spent shell casings, there was a very pronounced mark from the extractor, particularly where its body contacted the rim of the case. He realized that actually it was, in large part, the thickness of the extractor, intended to beef it up and make it last longer, that was its Achilles' heel. The thick extractor arm was making contact that was clearly enough to dent the hard brass cases, and if the metal were relieved there, the mechanical fatigue would be eliminated.

This simple milling process was performed on the existing extractors, and test firing confirmed that they were much easier on the ammunition's brass cases. Breakage and wear did not appear to be occurring, despite the smaller amount of steel present in the extractor.

## The New Titanium Flash Hider

The next update was a bit more involved. The warping problems seen on the recoil compensator/flash hider (these two distinct terms were used simultaneously, since the device performed the function of both) were clearly a sign that the thin, lightweight aluminum material used in their construction was not sufficiently rigid. Compounding the problems of the metal warpage was the tendency of the threaded muzzle cap to become loose. If the cap were also to become warped, the area where it threaded to the barrel could be damaged, making it either not lock sufficiently to the barrel, or break the seal, causing the compensator to fly off the muzzle at an inopportune time.

Stoner's rework first focused on identification of a new material from which to construct the baffle tubes of the muzzle device. He was resigned to the fact that the updated version would weigh a bit more, but he was not willing to fabricate the new compensator from steel. This would not only have massively increased weight, due to the large and complex composition of the device, but would concentrate this weight on the end of the rifle, making it front-heavy.

The solution was to replace the aluminum with titanium, and shorten the compensator a bit, preventing too much of a weight increase from the heavier material. A type of metal that was somewhat new to the American firearms industry, titanium was expensive and difficult to work, but it provided extreme strength, greater than that of steel, with weight more comparable to aluminum. Its relative scarcity and cost, as well as the reluctance with which it melted or allowed itself to be machined (a function of the strength that made it so desirable) were the only reasons why it was not the favored metal of manufacturing. At that time titanium was just starting to be used in revolver frames and cylinders, and today it is relatively common in firearms, but is less suited to some such jobs than steel. It is unsuitable for barrels, for example, because it has a tendency not to hold the precise and sharp edges of the rifling.

The most difficult part of the entire reworking process, which had to be accomplished in just 30 days, was the successful fabrication of the new muzzle devices, but with a lot of work using dedicated and expensive milling attachments on a lathe, Stoner turned out two polished examples which matched perfectly with his specifications, and subsequent testing done in early January proved them suitable. In the new design Stoner replaced the original aluminum sleeves with titanium tubes, pierced with small holes to vent gas.

Next, Stoner turned his attention to formulating a new form of attachment which would secure the two parts of the compensator together, making both the spontaneous unscrewing of the cap portion and the separation of the two parts no longer a concern. To this end, he designed a steel ring which fit around a slot he drew into his plans for the new flash hider cap, ending with a locking extension which went through holes in overlapping portions of both the cap and the venting sleeves of the compensator. This prevented the two parts from rotating in relation to one another, and with the rear venting portion now containing a metal ridge that locked into a slot that he machined into the front sight base, neither part could rotate relative to the weapon. This eliminated the issues of the front cap becoming unscrewed, and of the entire assembly losing its tight hold on the barrel and front sight base. Removal of the compensator was still simple, as the locking extension of the ring simply had to be pulled from the indexing holes first, and then the front cap unscrewed.

This portion of Stoner's rework of the two AR-10Bs was completed by December 15th, and news of its successful testing was announced to Col. Rayle at Springfield Armory on that day.

Concerning the recommendations made by Melvin Johnson that were either disregarded out of hand or not explored because of time and financial constraints, the main goal of most of them was easier extraction of the spent casings. Johnson was particularly avid about easing this process, partly because it had been accomplished so well on his military designs of the previous two decades.

A suggestion of his that was never taken seriously, or particularly needed for any AR-10 variant, was chamber fluting. This is a process of cutting longitudinal slits in the chamber wall, so that propellant gas could flow around the outside of the shell casing as well as inside it, equalizing the pressure, so that the cartridge case was easier to extract while pressures remained relatively high. This principle had been used to great effect in the Spanish CETME,

and indeed had been found necessary for all firearms employing the retarded-blowback "roller lock" operating system wherein the cartridge case was subjected to almost continual movement. Locked-breech, gas-powered actions, such as that of the AR-10, normally did not need this extra floating cushion to avoid case separations.

Several of Johnson's other proposed design changes were based on the principle of "primary extraction." Extraction, as the reader knows, is simply the bolt's pulling the shell casing out of the chamber after the round has been fired. The actual tossing of the spent case out of the ejection port is a separate function known as "ejection." Therefore, Johnson's theories were all focused on that crucial two milliseconds (the time extraction takes in an AR-10) when the spent shell case goes from sitting still, obturated against the chamber wall, to streaking backwards in the grip of the extractor claw.

In conventional automatic rifles, the bolt disengages from its locking surface, either by tilting or rotating, and then begins to move backward, commencing the extraction process. The concept of primary extraction is that the extraction process actually begins before the bolt is fully unlocked. Primary extraction was proposed to be accomplished in the AR-10 by milling the ends of the bolt and barrel extension's locking surfaces at a slight angle, at the ends of each that last make contact with each other.

In the original AR-10 bolt design, and indeed in Johnson's original patent for that very bolt locking mechanism, the bolt lugs and their attendant cuts in the barrel extension were squared off at a 90° angle, so that when the bolt lugs were turning to unlock, they and the locking surfaces in the barrel extension were in constant contact with one another at a perfect right angle until the bolt lugs had rotated completely clear of the recesses. Johnson proposed that for about the last one fifth of the contact area the locking surfaces and the bolt lugs should be cut at a 45° angle, so that during this last fifth of the contact distance the bolt would actually be moving back, riding on the correspondingly angled surfaces in the barrel extension, and beginning to extract the case before unlocking had been completed. The function of this was not only to allow more time for extracting the spent case, but to add this relatively slow component to the extraction cycle, allowing it to crack the seal between the cartridge case and the chamber wall before the high-velocity actions of extraction and ejection took place.

Johnson theorized that not only would this process be easier on the case, particularly the rim, but it would make the forces on the extractor, bolt, carrier, and barrel extension softer and more gradual, and lead to longer parts life, particularly of the vulnerable extractor. Finally, to aid in extraction, Johnson proposed increasing the chamber area by making the barrel's chamber wider at the base by one additional degree of angle. This would give the cartridge more headspace and room to expand, so that it would not adhere as tightly to the chamber walls during extraction. Like the proposed fluting of the chamber, this suggestion was rejected summarily, as not only having the potential to cause headspace problems and cartridge ruptures, but for having a

deleterious effect on accuracy. One key component of accuracy in a rifle is that the chamber hold the rounds of ammunition tightly in the same position, shot after shot, so that the bullet starts down the bore with precisely the same alignment every time.

Both of these latter suggestions were rejected. Stoner felt that the AR-10's chamber dimensions already struck the perfect balance between ease of extraction and tight headspace for accuracy. He reiterated in later remarks that the AR-10's gas system allowed for enough power in the extraction process so that there was never a need for primary extraction.

## Meanwhile, Face to Face with the Archenemies of Rifle Innovation

In a chilling event which would hint at the predetermined outcome Ordnance had in store for the AR-10, William Johnson (another ArmaLite consultant, no relation to Melvin Johnson) paid a visit on December 11, 1956, to the head offices of the Ordnance Committee of the United States Army in Washington. During this trip he initially met with everyone from the Assistant Secretary of Defense for Research, Frank Newbury, on down through the Air Force's procurement officers, and others.

The first portions of the exchange went cordially and well, as discussions with Assistant Secretary Newbury and the Air Force representatives were not only conducted with great professionalism, but led to genuine interest and future contacts for several aircraft innovations being developed by Fairchild's main research facility in Maryland. Having such important and high-level officials conduct themselves in a professional manner would hardly be noteworthy, until one considers the situation in which Johnson soon found himself.

When he finally reached the Office of the Chief of Ordnance, Small Arms Research and Development Branch, all the civility and good faith that had been experienced so far vanished. The ArmaLite representative was surprised indeed to see not only Dr. Frederick Carten, the Chief of Ordnance Research and Development, sitting behind the conference table, but William Bonkemeyer, an actual judge and representative at the ongoing Springfield evaluations, and none other than Dr. Carten's mentor; retired Colonel René Studler, who was by that time a consultant for Olin Mathieson, the parent company of Winchester Repeating Arms.

With no exchange of pleasantries or small talk of any kind, Carten and Studler, acting in concert and clearly in a pre-planned attack, launched into a del-

uge of criticism of Fairchild in general, the AR-10 and every aspect of its program in particular, and even Johnson personally. They first told him in no uncertain terms that the AR-10 would probably never be perfected to even approach acceptable reliability, and if it were done, it would take at least ten years. They continued that a rifle so light could simply never be sturdy enough for military use. Johnson retorted stoutly that Studler's own T25 was about the same weight, which clearly struck a nerve, as Studler was still quite bitter that this rifle, which he had ordered produced in large quantities for pre-adoption troop trials, had been cancelled by his superiors.

Carten and Studler went on to contend that a rifle weighing as little as the AR-10 "would be useless in combat due to its recoil," and "must weigh 8.75 to 9.25 pounds to be controllable." They were not swayed by the accurate counterpoint that it was Colonel Studler himself who had dreamed up the fantastical seven-pound weight limit, which none of his organization's designs had been able to meet. He had ramrodded this idea, combined with that of the .30 Light Rifle cartridge, down the throats of both the U.S. military and that of the NATO alliance, and now he not only scoffed at a rifle that actually achieved it, but openly treated it as an impossible and wholly unrealistic goal. As for the issue of controllability, both master and pupil sneered at the AR-10's straight-line construction (ironically also a design feature of the original T25), and dismissed the efficacy of the recoil compensator out of hand.

Dr. Carten then took the lead and informed Johnson that at the upcoming meeting of the Ordnance Technical Committee, a military ordnance trade show of sorts to be held in February at Eglin Air Force Base, the AR-10 would be banned from being demonstrated or even presented, claiming that doing so would "commercialize" the conference, since it had been developed by what was "just a private company" and as such was not, in their definition, a military rifle. When asked for the basis of this bizarre and arbitrary classification, they stated that any rifle that had not been evaluated by Springfield Armory was a "commercial rifle." Johnson came back immediately with the two-pronged rebuttal that not only had the AR-10 already been entered into a testing program at Springfield, but that the FN FAL T48 was also not a Springfield design. Dr. Carten's response to this was simply that Colonel Studler had "invited" the Belgians to come and demonstrate their rifle, while he had been ordered by his superiors to conduct an evaluation of the AR-10, clearly against his wishes.

Next on the docket for the *de facto* kangaroo court that was taking place in the conference room was that ArmaLite was involved in designing weapons for the commercial market, including what they derisively referred to as "shotguns in several colors." In another leap of the most convoluted logic, Carten and Studler accused Fairchild of impropriety by using money it earned from military contracts to invest in research on commercial rifles and shotguns, as well as to promote the AR-10. In support of this, Studler even had the audacity to claim that his current employer, Winchester, which had a stake in the Lightweight Rifle Program as well, did not use funds earned from military contracts for commercial applications, a point that was demonstrably false. In a final insult, Studler even went after William Johnson personally, accusing him to his face of un-

ethical self-promotion and impropriety in working for ArmaLite, when he had in the past been a consultant for Winchester. Needless to say, Johnson, a hardworking firearms expert and engineer, was taken aback and hurt by this insult.

Even before walking into this nest of treachery, William Martin, Director of Research for the Department of the Army, had expressed interest in the AR-10, and lamented to Johnson the fact that Ordnance was allowed to serve as a "judge in [their] own cases." Even not having been briefed on the AR-10, Mr. Martin was well aware that this was exactly what was afflicting ArmaLite at the time.

Licking his wounds and trying to remove the salt that Studler had rubbed into them, Johnson commiserated with none other than the former Secretary of Defense, Robert Lovett, who described his own experiences of dealing with Colonel Studler's Research and Development Department as "backing into a buzz saw." Johnson added that his take on it was more like "crawling through a snake pit."

The impropriety, unprofessionalism, and just-plain spite that was levelled at William Johnson that Tuesday afternoon was not only a blatant display of the very worst type of bureaucratic territorialism and intransigence towards what was clearly a better idea, but it actually permanently shook Johnson's faith in the future of the AR-10. He concluded his report of this meeting by recommending that the U.S. military application of the AR-10 be terminated, and it be turned into a semi-automatic sporting rifle, and marketed within the new NATO alliance instead for military adoption.

## Springfield's Arsenal of Dirty Tricks: Lying in Wait for the AR-10

Exactly one week after Mel Johnson had recommended postponing the remaining portion of the AR-10 evaluations in order to perform modifications on the test rifles, a very interesting document came across George Sullivan's desk; one which would add to the mounting pile of evidence that the branch of Army Ordnance that was testing the AR-10 had it out for the upstart ArmaLite design. This document detailed the standard testing procedures for rifles in contention for the Lightweight Rifle Program, which had been applied to the T48 and T44. On December 4, one day after the tests had begun, Eugene Stoner had been furnished with a copy of the testing itinerary. This outlined the stages through which the AR-10 would have to pass, and was titled "Standard Operating Procedure . . Evaluation Tests for Rifles." The problem was that this differed considerably

from Springfield's normal testing schedule, of which Sullivan had procured a copy, and he lost no time in comparing them to confirm in which direction of difficulty the "standard" procedures had been modified by Springfield expressly for the AR-10.

The deviations from the standard procedure affected all portions of the AR-10 test that involved firing, and began with the accuracy test. In this area, the rifle was supposed to be fired some to heat and foul the weapon before its accuracy checks were conducted. This was standard, as it was meant to simulate "hot bore accuracy," such as a soldier, probably having already fired his weapon considerably, would likely encounter in combat. The figure that was given to ArmaLite was that the rifle be fired 100 times rapidly before the accuracy evaluation was to begin. Interestingly, the standard test, which had been applied to the T44 and T48, was that it be fired 60 times, at a rate of only 15 to 20 rounds per minute. Not only would this lower round count foul the bore less and thus have a less detrimental impact on accuracy, but the slow rate of fire combined with the smaller number of rounds would heat the test weapon far less.

The double standard did not end there, however. The next phase, which tested general weapon performance, called for the AR-10 to be fired 600 times with no limit on the rate of fire, while the standard test called for a limited number of rounds per minute for this phase.

Perhaps the most egregious deviation was outlined in the endurance testing. Both the standard and the toughened procedure applied to the AR-10 in-

volved firing 6,000 rounds to determine whether the weapon would hold up to sustained combat use. The procedure for the AR-10 called for the rate of fire to be a minimum of 20 rounds per minute, while the standard minimum was 15. Each test allowed the rifle to be briefly cooled and lubricated every 100 rounds, but the standard allowed the weapon to be disassembled, cleaned, and fully lubricated every 600 rounds. This was in stark contrast to the AR-10's requirements, which only allowed it to be disassembled for cleaning and lubrication if the test staff deemed it "necessary." Such a standard was essentially an illusory promise, which the test moderators from Springfield, who had a vested interest in the AR-10's failure, could simply disallow.

# Adverse Condition Tests - Only for the AR-10

Next followed a litany, almost too numerous to catalog, of completely novel adverse condition tests which were not standard, and had not been applied to any previous contender for military adoption. First was a "muddy water" test, in which the rifle was submerged in muddy water while loaded before being fired, and also called for loaded magazines to be left in mud for 18 hours before being used to fire 100 rounds. The dry test came next, and called for the AR-10 to be completely degreased in solvent before being fired 100 times in rapid semi- or fully-automatic fire, the choice of which was left to the test moderators.

The low temperature evaluation added the requirement that the rifle and magazine be removed from Springfield's infamous cold chamber after being chilled to 65° below zero Fahrenheit, and allowed to form copious amounts of condensation before being put back into the chamber to allow the condensation to refreeze. It was plain that these modified standards and newly-concocted adverse condition tests were meant only to increase the malfunction rate of the AR-10 in comparison to that of the T44.

Next came the freshly-invented "high temperature test," in which the rifle, ammunition, and magazines were heated to 160°F and left at that temperature for three hours before being fired rapidly for 280 rounds, with the actual type of fire, fully- or semi-automatic, being left to the tender mercies of the personnel conducting the evaluation. The humidity test required for the rifle, its magazines, and 560 rounds (the number of rounds to be fired seemed to be ever-increasing) be exposed to a temperature in

excess of 100°F with 95% humidity before rapidly firing the ammunition. The subsequent "sustained firing test", in which 600 rounds would be fired at a relatively slow rate of 20 rounds per minute for half an hour, was also completely new.

The final two procedures were the "rough handling test" and "sand drag test." The first of these ominous-sounding trials, both without precedent in U.S. military rifle evaluations, and being more reminiscent of Soviet small arms adoption doctrine, called for the weapon to be dropped three feet onto a flat hardwood bench many times, so that the left side flat, right side flat, toe of the stock, and heel of the stock would all make direct first contact with the solid surface. Then the weapon was to be held aloft horizontally by its muzzle, and twisted and thrown at the same time so it would rotate 90° in the air before striking the corner of the bench, with ArmaLite's innovative fiberglass stock clearly the intended casualty.

The sand drag test, punishment of the utmost severity, required the AR-10 to be dragged through sand and fired periodically until malfunctions occurred. No criteria were given for a rate of fire or number of rounds to be fired, nor was type or grade of sand, depth of dragging, duration, or any other parameter specified, meaning that all were essentially left up to the discretion of the extremely biased Springfield staff.

In total, the battery of adverse condition tests to which the AR-10 was to be subjected would require it to fire 2,225 rounds, not including the unspecified number for the sand drag, which compares starkly to the 600 required of other rifles, for which the only adverse condition test was the "rain test," in which the weapon would be fired a set number of times over 75 minutes while periodically being sprayed with mist.

Impropriety, even at this early stage, however, was sadly not limited to the setting of prejudicially harsh test standards, but also involved what could amount to material and informational sabotage. First, the initial two days of testing, which involved stripping and disassembly of the weapon for photography, examination, and x-raying, were much more involved compared to standard operating procedures, all of which was done by personnel untrained in the use of the AR-10. George Sullivan remarked in

his own personal notes that it was indeed a wonder that the weapons would function at all after these excessive repetitions of disassembly and reassembly. Also, only two of the five furnished weapons were selected for testing, and were to be used with no modifications of any kind for the entire two-week battery of evaluations, while Springfield had the practice of sending a freshly modified and polished T44 for almost every day of testing when evaluating their own product. The "home team" also had access to a full machine shop for repairs and modifications on the Armory grounds, while ArmaLite, just like their Belgian counterparts who had presented the T48, were denied its use and services.

## "No Frills" for the AR-10

When delivering their heavily modified test schedule on the second day of testing, the Ordnance officials also made it clear that no daily data sheets would be provided to ArmaLite, no control weapon was to be used, no evaluation of performance would be furnished to ArmaLite, and no comparison would be made, officially or unofficially, with any other weapon under consideration; all of which were standard procedures which had been observed for every other entrant in the Lightweight Rifle Program.

## Specifying a "Bad Lot" of Non-Standard Test Ammunition

Finally, perhaps the most dastardly twist of all came with the ammunition assigned to the AR-10. During the entire development process of the rifle, Army Ordnance had, as a rule, banned the provision of any ammunition other than standard "ball" type rounds. This was the type of ammunition which would be normally issued to combat troops, while other types of rounds, if selected, would be reserved for special purposes. Instead of using regular-issue ammunition to evaluate the AR-10, as they had done with every other rifle submitted, and in conformity with their early requirements that private industry-developed weapons only be permitted to fire standard ammunition, the ammunition they required the AR-10 to use for the entire duration of the testing was a special armor-piercing round still under development. Not only were the bullet weight, shape, and length different, but the powder burn rate, pressure, and point-of-impact characteristics of this ammunition were all different from regular combat loads. Achievable accuracy with armor-piercing ammunition is in general also not up to the standard of normal ball rounds.

When a rifle is designed to function with a particular cartridge and bullet combination, changing the contours of the bullet is one sure way to increase the number of malfunctions. The armor-piercing ammunition, due to its heavier projectile,

already produced higher operating pressures than standard "ball" ammunition, which increased wear and strain on parts and made catastrophic malfunctions more likely. In sum, the experimental "AP" round specified for use in the AR-10 tests gave the perfect combination of strange bullet contour, to induce feeding malfunctions; substandard accuracy, which would artificially lower the AR-10's performance in an area where it was known to excel; and elevated pressures, that would be sure to break parts and possibly even blow a weapon apart.

Far worse than this, however, was the particular batch of ammunition assigned. Each lot of ammunition produced for military use in the U.S. is marked with a particular lot number. One of the majors from Fort Benning's Infantry Board Number Three, an organization that had always been friendly to ArmaLite because they, unlike Springfield Armory, were in direct touch with what the regular infantryman needed, immediately recognized the particular lot number on the ammunition earmarked for the AR-10's testing. He informed the ArmaLite representatives that it was a batch that had actually first come to his own organization at Fort Benning, and had been rejected by them as being a "bad batch." It had not only suffered from unusually poor accuracy, even for AP rounds, but it was found to generate even higher pressures than specified for its type, and demonstrated dangerously inconsistent pressures and velocities.

All of these additional strains that were to be put on the AR-10, coupled with the almost complete discretion of the testing staff, resulted in gross excesses that would plague the AR-10 during its evaluation. A friendlier officer in the Pentagon was able to get his hands on a copy of Springfield's preliminary report on the AR-10's performance on December 20, and sent a copy to George Sullivan. This act of clandestine kindness helped ArmaLite be better prepared for what to expect, and modify their test rifles accordingly. Surprisingly, however, even these harshest of torture tests would not end up being an insurmountable obstacle for Eugene Stoner's wonder rifle.

## The Final Stages of Testing: the (Missing) Official Springfield Armory Test Report

Details of how Springfield had altered the test requirements and conditions in their attempt to sabotage the AR-10 at every turn during these evaluations have been explored above, but something for which a historical view at the AR-10's development also calls is an empirical look at exactly what was required for the full tests of the rifle. The following is just that: a summary of the different types of trials to which AR-10s nos. 1002 and 1004 were subjected before the Evaluation Board at Springfield Armory.

Interestingly, the actual official test report, published on February 4, 1957, which is conspicuously missing not only from the Springfield Armory archives but all government sources, tends to paint a different picture from that relayed to the ArmaLite staff and the general military community. Suffice it to say that someone high up wanted the results of the AR-10's performance buried, but at least one copy has survived in an unlikely place, where it is well looked after to this day.

### A. Function Check

1. The rifle, lubricated, was put in a vice and fired by means of a pull-string for one full magazine in both semi-automatic and then fully automatic modes.
2. By an individual rifleman, the weapon was fired from the shoulder for a total of 100 rounds, with each of the five 20-round magazines being expended in a different position. These were:

    a. Semi-automatic from the shoulder;
    b. Rapid bursts from the shoulder;
    c. Full-automatic from the shoulder;
    d. Semi-automatic with a loose grip; and
    e. Full-automatic with a loose grip.

### B. Flash

The rifles were fired with first 1, and then 5 rounds in complete darkness, and the flash was recorded by a camera placed at a 90° angle to the side of the muzzle, at a distance of four and a half feet. This single exposure allowed for the aggregate effect of both single rounds and 5-round bursts to be visually recorded (fig. 78).

### C. Accuracy

1. 10 rounds were fired at each of 3 targets at a range of 100 yards. During this portion of firing, the rifle was either held in a mechanical vice to keep it on target, or fired from a bench rest by an expert shooter.
2. A range of combat-condition accuracy tests were then fired at 100 yard-targets by three separate riflemen on each rifle. These consisted of:

    a. 10 rounds with a hot and fouled bore from a bench rest. This means that the rifle was fired a few times beforehand so that its bore would be somewhat dirty and warm, simulating the state in which an infantryman's rifle would be during a combat engagement.
    b. After disassembly, cleaning, and lubrication, the same test was done, but with a clean and cold bore (unfired since cleaning).
    c. One target was then shot with 10 rounds from the prone (lying down) po-

sition, using the sling wrapped around the arm for support.

d. 100 rounds were fired as rapidly as the weapon could be reloaded in fully-automatic mode, to simulate the condition of a very hot rifle.

e. Immediately after the extreme heat exposure, the first 3-target accuracy test was again performed from a bench rest.

f. An additional 10-round target was then immediately fired from prone position with use of a sling to gauge this type of accuracy in a very hot weapon.

3. The next stage tested the useful close-range accuracy of burst fire. This was meant to simulate close combat, and to assess how effective the rifle would be at replacing the desired close-quarters functionality of the submachine gun. Each rifle would be fired by the three different shooters who conducted the 100-yard accuracy trials, and would consist of each firing ten 3-round fully-automatic bursts at man-sized targets at 25 yards, from a standing position. This was then repeated from a prone position. Although a test rifle (normally an M1 Garand) was generally used

as a control for comparison, it is claimed that no such example was used with the AR-10. One can surmise that this was so that the riflemen would not have a rifle with a conventional design to compare with the AR-10, which particularly excelled in this category.

4. Each rifleman would then complete a 100-yard rapid fire test three times with each of the two rifles. This consisted of firing as many shots as one could aim, fire, and reload in a one-minute period. This was done first in semi-automatic, and then in fully-automatic fire mode. The total number of hits on each target (one was used for each stage by each rifleman) was then recorded. The targets were the size and shape of a kneeling infantryman, simulating the likely size of enemies at various engagement distances.

5. Six different riflemen performed a standard military rifle qualification course (the test used to determine a soldier's readiness to engage in combat with a rifle) with each rifle. This course varies among branches of the armed services, and it is not clear whether the more rigorous Marine course was used, or whether the Army's more relaxed regime was applied.

## D. Endurance

This was the infamous 6,000-round test. The conditions called for the rifle to be allowed to cool for a short time after each 100 rounds, with each string of 100 rounds fired semi-automatically or fully-automatically, switching at each 100-round mark. Cleaning and lubrication was permitted only when the testing staff deemed it "necessary," which ended up being never. The average velocity of ten rounds fired from each test rifle was recorded before and after the test. This was intended to calculate barrel wear. As a barrel becomes more worn, its interior becomes slightly larger, and a barrel with looser tolerances will produce less velocity, because some of the gas, instead of using all its energy in propelling the projectile, will flow around the bullet and drive past it.

Several different positions and attitudes of the rifle were used, to test its mechanical functioning in any conceivable stance or grip of the shooter, or angle of fire. The following is an outline of how each 600-round string was performed:

1. 100 rounds were fired in semi-automatic mode from the shoulder.

2. 100 rounds were fired in rapid, semi-automatic bursts.

3. 100 rounds were fired in full-automatic mode.

4. 40 rounds were fired with a loose hold from the hip, in semi-automatic mode.

5. 40 rounds were fired with a loose hold from the hip, in full-automatic mode.

6. 40 rounds were fired with a loose hold from the hip, in semi-automatic mode, with the weapon rotated 90 degrees to the left, so that the left side of the rifle was parallel with the ground.

7. The same test as above was performed three more times; once with the rifle flipped onto its right side, then on its left again in full-automatic fire mode, and then finally on its right side in full-automatic.

8. Finally, 60 rounds were fired from the shoulder with a normal tight grip in full-automatic mode.

## E. Dust Test

The first of the adverse condition tests, the Dust Test started with the rifle being lubricated lightly, and then wiped for external oil. The muzzle was then taped shut, a dummy round was loaded into the chamber with the safety on, and a magazine was loaded to one round less than full (19 rounds).

> 1. The rifle was then placed in a dust box with dust in the bottom, and more blown in with a fan. It was set this way upright for one minute, and another minute with it upside down.

2. The rifle was then extracted from the dust, the shooter being permitted to use his hands to wipe away excess dust and blow on the congested areas. The magazine was then inserted, and the 19 rounds were fired in semi-automatic mode.
3. An additional full magazine, exposed to the dust, was then inserted and fired on full-automatic mode.

## F. Rain Test

The rifles were cleaned and lubricated, and then placed on their left side (ejection port up) underneath a soaker hose (a type of garden hose that contains many small holes which allows a single faucet to water a long row of plants), through which water was run. The rifle was exposed to this deluge of water for five minutes, with the bolt locked open and a magazine inserted. The bolt would then be shut, charging the rifle with a round in the chamber, and it was fired for 80 rounds in semi-automatic mode. Next, it was again exposed to the spray for an additional five minutes. After this, four more magazines would be fired through the rifles on full-automatic fire mode. These two tests were repeated until 640 rounds in total had been fired, which totaled 4 complete cycles of the test in both full and semi-automatic fire.

## G. Muddy Water Test

Rifles in this stage were again cleaned and lubricated, and tape was stretched across the muzzles, to prevent mud from entering the barrel. A bore that is obstructed by a foreign object will almost certainly cause a catastrophic failure when fired, likely injuring or killing the shooter. No firearm can be designed to defeat this hazard, and to protect the shooters and the rifles, this precaution was taken. It is more apt to call this a mud test, as the "water" was essentially the consistency of pudding, and the name of the test was changed in the next two years to simply the "mud test." This event was conducted in two parts:

> 1. The weapons were submerged completely in the thick, muddy water for 15 seconds. The shooter would retrieve the rifle, wipe and blow on the action if necessary, wipe the bore to make certain there was no barrel obstruction, charge the weapon, and fire it. A maga-

zine that had not been submerged in mud with the rifle was then inserted and fired in full-automatic mode.

2. The second step was slated to take place only if time allowed, and involved the same procedure as the first, except that in addition to the magazine in the weapon, four more loaded magazines were submerged in the muddy water with it. They were kept under the surface for five minutes, and then withdrawn. Water was allowed to drain out of the bore for 30 seconds, and then the rifle and magazines were left in a muzzle-down and feed lip down position respectively (so that obstructions of mud substrate would not form in the barrel). The rifle and magazines were to sit for 18 hours to allow the mud to harden, and then firing of all magazines with the muddy ammunition was attempted.

## H. Unlubricated Test

Both test rifles were cleaned with solvent, but not lubricated. They were then fired in normal temperature conditions for 100 rounds. With each change of a 20-round magazine, the fire mode was changed in the order similar to that observed during the endurance tests:

1. 20 rounds, semi-automatic;
2. 20 rounds, rapid bursts of semi-automatic;
3. 20 rounds, full-automatic;
4. 20 rounds, semi-automatic with a loose grip from the hip; and
5. 20 rounds, full-automatic with a loose grip from the hip.

## I. Extreme Cold Test

In this test the rifles were cleaned and lubricated, and then placed in Springfield's special cold chamber at a temperature of -65°F, with a loaded magazine inserted, and with the bolt closed and hammer down on an empty chamber. After 12 hours of exposure, the rifles were retrieved, charged by pulling the charging handle to cock the hammer and bring a round from the magazine into the chamber, and fired for the full capacity of the magazine in semi-automat-

ic fire mode. If this stage was passed satisfactorily, the bolt would be closed and the hammer dropped on an empty chamber, an additional full magazine would be inserted, and the rifle would be put back into the cold chamber at the same temperature for two more hours. It would then be retrieved and charged again, and fired, this time in fully-automatic mode.

## J. High Temperature Test

The rifles were cleaned and lubricated, and then placed with 400 rounds of ammunition in a climate-controlled chamber which was then heated to 160°F, and left there for three hours. The rifles were then withdrawn and fired according to the following procedure:

1. 40 rounds, semi-automatic;
2. 40 rounds, rapid bursts of semi-automatic;
3. 40 rounds, full-automatic;
4. 20 rounds, semi-automatic with a loose grip from the hip;

5. 20 rounds, full-automatic with a loose grip from the hip;
6. 20 rounds, semi-automatic with a loose grip, rotated 90 degrees to the left;
7. 20 rounds, semi-automatic with a loose grip, rotated 90 degrees to the right;
8. 20 rounds, full-automatic with a loose grip, rotated 90 degrees to the left;
9. 20 rounds, full-automatic with a loose grip, rotated 90 degrees to the right; and
10. 40 rounds, full-automatic.

## K. Humidity Test

The rifles were cleaned and lubricated, and placed with 600 rounds of ammunition in the same climate-controlled chamber as in the high temperature test, but this time with the temperature set to 100° and the humidity regulated at 95%. Time was not specified for this portion of the tests, but it was required to be sufficient for the "material to condition to the temperature and humidity." The rifles were then taken out of the chamber and fired using the same procedure as in the high temperature test outlined above. They were then allowed to cool without cleaning, and fired with the same number of rounds through the same firing procedure.

These two tests which involved heat—the high temperature test and the humidity test—were both concocted specifically for the AR-10, and were never used on any other tested rifle, before or since. It is important to note that heating of cartridges increases the pressure they generate when fired, and this, when added to the fact that higher-pressure ammunition had already been assigned for use during the tests, and that the lot of ammunition selected had been rejected by the Infantry School for, among other things, producing dangerously high pressures and temperatures, serves as strong circumstantial evidence of a serious attempt to cause the AR-10 rifles to fail their tests.

## L. Cook-Off Test

This test was to determine the minimum number of rounds fired in rapid succession it would take to heat the chamber sufficiently to make the rifle fire on its own, simply by the powder inside a chambered cartridge igniting from the conducted heat; a phenomenon called a "cook-off."

Many loaded magazines were prepared for the test, and the weapon was fired in fully automatic mode, until a chambered round fired on its own. At that point, an attempt was made by the observers and shooters to determine exactly which round it was that "cooked off."

## M. Sustained Fire "Emergency" Test

This battery of examinations was meant to assess the rifles' performance in a scenario in which an entrenched position was being overrun, and infantry-

men had to operate their rifles continually for long strings of fire without cooling, cleaning, or maintenance. The rifles were to be fired at a rate of 20 rounds per minute for half an hour, and temperature was to be noted at the end of firing. Several subjective criteria were also added to this *ad hoc* evaluation regimen, which was essentially thrown together for the AR-10. These required the shooter to note the round count at the times that the stock and handguard

1. became uncomfortably hot;
2. became too hot to hold;
3. began to emit disagreeable odors (something far more possible with a fiberglass handguard than one composed of wood);

4. displayed a change in physical appearance; and
5. smoke began to be visible.

This test had not been performed with any of the other then-current contenders in the Lightweight Rifle Program, but it was hinted by sympathetic members of the Springfield Amory staff that Colonel Studler's darling T25 had burst into flames during a similar test regimen back in 1952.

## N. Rough Handling Test

Next came a procedure that appeared only calculated to provide an outlet for anger, perhaps at the performance of an upstart, outside contender in an expensive and career-defining rifle evaluation program. The details of how it was done are outlined above, but suffice it to say it attempted to provide proof for the perceived fragility of the fiberglass furniture in which the AR-10s were clothed. One could say that it was

an honest attempt to assess the durability of a fiberglass stock compared to one of walnut, but because no control weapon was allowed in the AR-10 tests (a procedure that was always followed in all other evaluations), and no other entrant in testing was so subjected, it can be chalked up as one more attempt to smear the ArmaLite rifle.

## O. Sand Drag Test

Finally came the last adverse condition evaluation. The details of how it was to be done are not particularly important, but suffice it to say the rifle was

dragged for some distance through sand, before firing was attempted.

# Time Out for Armed Air Travel: Stoner's Strange Brush with History

Meanwhile, back in California, with the modifications to the test rifles completed, it was time to send them back to Springfield. History does not tell us which of the rifles was reassembled first, but Stoner performed the retrofitting one rifle at a time. The first to be completed was taken by Charles Dorchester when he made his way back across the country on January 2. Stoner spent the next few days incorporating the new parts into the second AR-10, and had to rush the rifle to the test grounds on his own on January 6, two days before testing was set to resume.

Flying was an entirely different experience in the 1950s than it is today, and both the convenience of the travel and the attitude with which it was then regarded would seem strange to us in the second decade of the twenty-first century. Flying was then a novelty, or an adventure. Men would wear suits and women would wear dresses, and people of all walks of life would all sit in the same class of seats on the day's airliners. Flights were short, and changing planes was required much more often than it is today.

Especially fortuitous for a young firearms engineer running late to catch his connecting flight to Massachusetts was that security checks and metal detectors were things of the future of which no one had yet heard. International terrorism and hijacking had not yet come about as political phenomena, and as such, there were no laws regulating the carrying of weapons aboard aircraft.

The final connecting flight was about to take off when Stoner arrived, out of breath, and so he was not able to check any luggage. He was ushered onto the plane swiftly and the doors were shut. Upon walking to his seat in the back, the star of ArmaLite noticed two passengers seated at the front who seemed somewhat out of the ordinary. They were both military-age men with regulation haircuts, but also dressed in suits with black ties. They seemed unusually alert, bending their gaze to each new passenger that boarded. However, Stoner settled into his seat, buckled up, and forgot about the unexpected glances he had received when boarding.

When the plane landed at its destination, however, the two men at the front of the plane stood up, and announced so that all the passengers could hear, that they would appreciate it if everyone would remain seated for a few minutes to allow Vice President Richard Nixon to disembark first. Watching all this from the rear of the passenger compartment, was a firearms engineer carrying a suitcase in his lap, which held a selective-fire AR-10, separated into its upper and lower halves. After opening the suitcase, it would have taken Stoner less than five seconds to lock the components together, insert a magazine, and commence firing. As it happened, of course, he had

had no thought whatsoever of doing anything illegal, and after Nixon had got off he and his unassuming luggage calmly deplaned and made their way to the Armory.

Although not directly related to the performance of the AR-10, this anecdote, related by one of the few remaining giants of the firearms industry who knew Stoner well and worked with him, serves to paint a picture of the heady days in which this precocious and marvelous rifle was developed. It also represents a bit of a frolic before the deadly serious work of the AR-10's final evaluation was to begin.

# The Final Round of Testing Gets Under Way

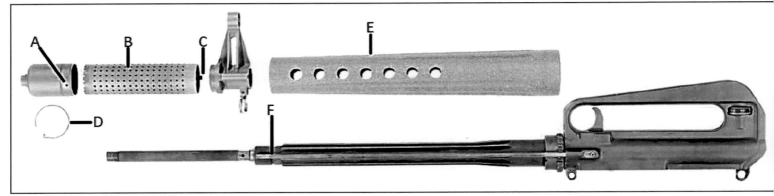

76. Left side view of the barrel and receiver group in a picture taken after the first day back at Springfield Armory on January 8, 1957, showing the revised parts added by Stoner over the space of a few weeks before testing was resumed.

The hole in the front cap of the recoil compensator (A) into which the locking arm of the ring (D) fits can easily be seen.

The cylindrical venting portion of the compensator (B)

and its extension (C) that locks into a cut in the front sight base are evident, as is the handguard (E). The holes that appear on the side of the handguard were always present on it, they simply were indexed at the 12-o'clock and 6-o'clock positions when installed on the rifle, to allow top and bottom venting.

The new gas tube (F), although discolored by the heat of firing, kept its shape and functionality perfectly.

Springfield Armory photo, from Technical Report
No. TS2-2015, dated February 4, 1957

Upon arriving back at Springfield, the tests began with a disassembly and inspection of the newly-modified AR-10s, facilitated by the ArmaLite staff. The two rifles were taken apart, examined, and weighed. The modifications made by Stoner, chiefly the redesign of the recoil compensator, brought the updated AR-10B's weight to seven pounds, two ounces. This was still stunningly low, given that as the T44 and T48 were continuously and serially modified to improve their reliability, they gained more and more weight. Even with a full 20-round magazine of ammunition inserted, the updated AR-10B weighed less than either the American or Belgian rifle did when empty with no magazine.

Firing was resumed the next day, January 9, 1957, and Stoner's hypotheses were validated. The new gas tubes stayed rigid both when hot and under sling and handguard pressure. Despite being made even hotter by round counts higher than those fired in December, they did not become clogged, and kept functioning normally. The titanium muzzle device experienced no warpage, and continued to provide excellent recoil control.

It was noted that after a very large number of rounds without cleaning or scraping the interior of the compensator, it would stop reducing flash as effectively. This did not affect its recoil control abilities, but was a function of a large amount of fouling,

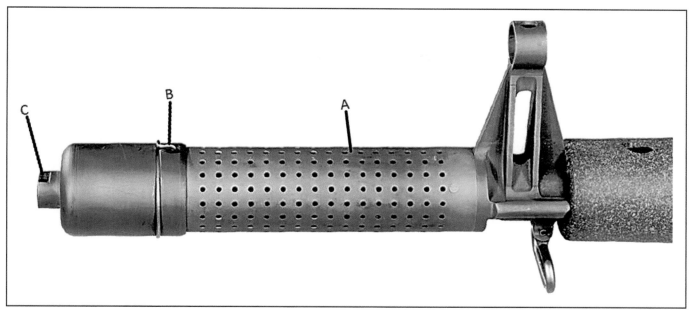

77. Left side view of the new AR-10B muzzle device, composed of two nested, perforated titanium cylinders (A), held onto the barrel by the muzzle cap, with a removal flat on its forward spigot (C).

The locking ring, set into a circumferential groove, is configured with a locking arm (B), which snaps into a hole in the back of the cap and downward to lock into one of the castellations in the front of the cylinder (A).

Springfield Armory photo from Technical Report No. TS2-2015, dated February 4, 1957

composed partly of unburnt powder building up on the inside surfaces, which would ignite every few dozen rounds, causing a completely safe, but very bright and large muzzle flash. None on the evaluation board considered this to be a fatal issue, just something to be improved in future development.

It should be noted that the current Russian infantry rifle, a version of the Kalashnikov, is issued standard with a muzzle device very similar to that used on the AR-10. Some authorities suggest that the similarity between the two is not the result of mere convergent evolution, but of deliberate Cold War industrial espionage. The Russian device also suffers from the same problem of erratic flash suppression at high round counts (in this author's personal experience, it performs poorly compared to modern incarnations of the AR-10 device), but this is considered acceptable by the capable Russian forces, due to its recoil control properties.

The firing tests continued for a number of days, during which the AR-10s demonstrated remarkable reliability, and accuracy that was peerless next to all of its competition. Some issues did occur, however. Freezing conditions caused a few reliability concerns, as they had on the T48, but not to such a degree that had been present in the infamous Arctic trials of late 1953. Icing and refreezing caused the tightly-toleranced working parts to lock up, making the weapon

difficult to charge, and short-stroking would occur in such environments. Of course, the added abuse of allowing the weapon, magazine, and ammunition to form condensation at room temperature, and then forcing that condensation to refreeze, a requirement unique to the AR-10, exacerbated this issue. However, it still did not rack up nearly as many stoppages in this course of fire as had the T48. About one week into testing, as shown in fig. 79, the gas transfer bracket on rifle 1004 suffered a failure.

It was surmised that this failure in the transfer bracket stemmed not only from its lightweight construction, but from the extreme temperature changes encountered during several of the tests inducing thermal shock. Cold weather tests were something of which an organization with the limited resources of ArmaLite was not capable, and so, just like the infinitely better-funded and supported Fabrique Nationale, they were more-or-less blindsided by the now-infamous cold chamber at Springfield Armory. Once the problem was diagnosed, the ArmaLite team pulled the problematic bracket from the accuracy and performance-testing rifle, and Eugene Stoner, thanks to sympathetic staff, used the Springfield machinery to fabricate a drop-in replacement from sturdier material.

This inopportune breakage actually had an upside and would, in the long-term, serve to obviate the

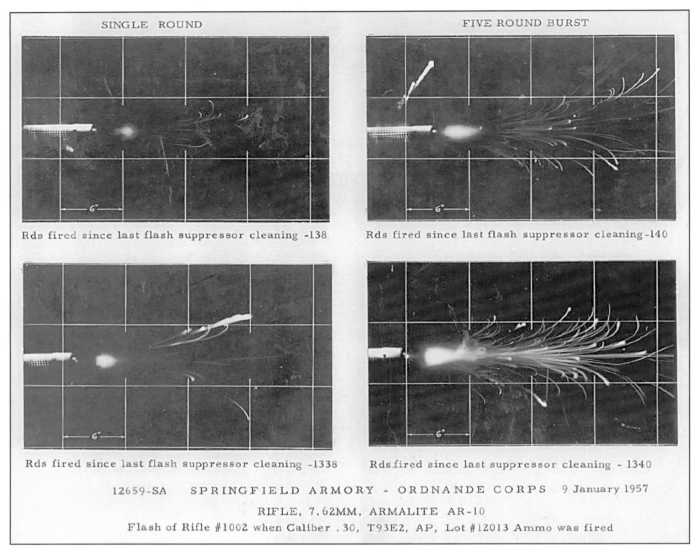

SINGLE ROUND

FIVE ROUND BURST

Rds fired since last flash suppressor cleaning -138

Rds fired since last flash suppressor cleaning -140

Rds fired since last flash suppressor cleaning -1338

Rds fired since last suppressor cleaning - 1340

12659-SA    SPRINGFIELD ARMORY - ORDNANDE CORPS    9 January 1957

RIFLE, 7.62MM, ARMALITE AR-10

Flash of Rifle #1002 when Caliber .30, T93E2, AP, Lot #12013 Ammo was fired

78. The performance of the new titanium muzzle device is illustrated in this series of light-accumulation photographs. In addition to staying completely rigid and holding up to serious abuse, the compensator did an excellent job of taming recoil while not measurably increasing sound and blast. Its flash suppression properties, however, are shown here to degrade once in excess of 1,000 rounds had been fired between cleanings.

Each picture was taken using an open aperture, so that all light present during the entire time of firing was exposed on the film. The photographs on the left are of the flash generated by a single round, and on the right, the aperture was left open for automatic bursts of five rounds, showing their cumulative flash.

Springfield Armory photo no. 12659-SA
dated January 9, 1957, from Technical Report
No. TS2-2015, dated February 4, 1957

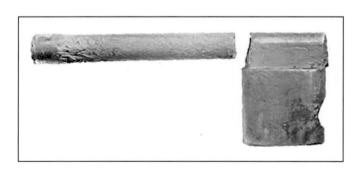

79 (left). The broken gas transfer bracket, showing the integral rear extension tube snapped off. This photo was taken after the failure of the aluminum gas transfer bracket was discovered. Note the warpage and disfigurement, particularly on the end of the extension tube.

The main gas tube fed into the bottom-right of the bracket, which remained intact and robust, but the extension tube proved to be a weak link.

Springfield Armory photo from Technical Report
No. TS2-2015, dated February 4, 1957

need for a gas transfer bracket in the subsequent AR-10 models, in which it was replaced by a better, simpler, and far more robust means of incorporating

Stoner's ingenious direct impingement gas system. At the time, however, new brackets were installed in both rifles, and testing continued.

# Catastrophe at Springfield: Stoner's Dire Prediction Comes True

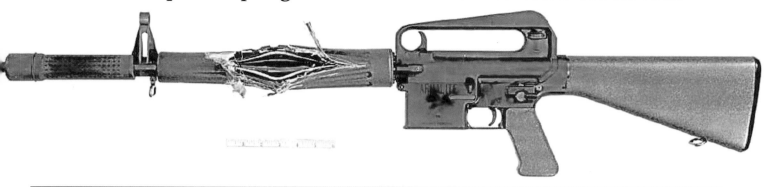

**12590-SA     SPRINGFIELD ARMORY – ORDNANCE CORPS     15 Jan 1957**
**RIFLE 7.62MM, ARMALITE AR-10**
**Barrel failure of Rifle #1002 after 5564 rounds were fired.**

80. Left side view of AR-10B no. 1002, showing the catastrophic barrel failure which occurred after 5,564 rounds of ammunition, including some identified as a "bad" lot of experimental AP rounds, had been fired with only the most minimal cleaning, cooling, or lubrication.

The evidence of the intensive firing can be seen from the extreme fouling on the recoil compensator and around the gas tube's connection to the gas transfer bracket.
Springfield Armory photo no. 12590-SA dated January 15, 1957, from Technical Report No. TS2-2015, dated February 4, 1957

With their new gas transfer brackets installed and gas tubes and muzzle brakes holding up well to heat and pressure, the two AR-10s chugged on into their second week of testing. However, disaster struck at an unexpected time. While rifle no. 1004 was being disassembled for a photographic record of the testing, no. 1002 was undergoing the most severe endurance trials, in which it was being fired in alternating rapid semi-automatic and fully automatic modes with no respite or inspection. This test consisted of a total of 6,000 rounds, with all stoppages or other issues recorded at their approximate round count.

Many loaded magazines were prepared for the test, so as soon as one was empty, another could be inserted and firing resumed. After each series of 100 rounds the rifle would be allowed to cool for an extremely brief time while five more magazines were brought up for the next hundred rounds. A field clean-only (the most cursory type of maintenance) and a spurt of lubrication was allowed to be applied only when the test staff considered it "necessary." The standard requirement for every other rifle at Springfield Armory had been thorough cleaning after each string of 600 rounds.

It was further reported by ArmaLite staff that even these meager allowances for cooling, cleaning, and lubrication were not observed by the Springfield test staff, which had a vested interest in the AR-10's failure.

Following an astounding 5,564 rounds fired with only the most minimal cleaning, cooling, or lubrication, the unthinkable happened. On January 15, 1957, the barrel of AR-10 no. 1002 ruptured and the bullet deviated in its course, exiting at the 10-o'clock position of the rifle (when viewed from behind), and forward at around a 30° deviation from the bore axis. In the process it split open the stainless steel barrel liner and swaged aluminum barrel cover, blew the sturdy gas tube aside, and tore a gaping hole in the handguard.

It is worth noting that if the standard battery of fire-for-endurance tests had been applied, by which the T44 and T48 had been judged, AR-10 no. 1002 would have finished this portion of the testing without incident, and almost free of any malfunctions at all.

By a miraculous stroke of luck, not only for the testing staff, but for the AR-10, the rifleman conducting the test at the time had his support hand positioned well back on the handguard, partly cradling the magazine. Aside from being extremely shaken, he sustained no injury from the accident, as the bullet and high-pressure gas were all vented forward and away from him. The damaged piece was immediately rushed into the machine shop for photography and diagnosis. After having pictures taken of the results of the failure, the offending barrel was removed, and sent for metallurgical analysis.

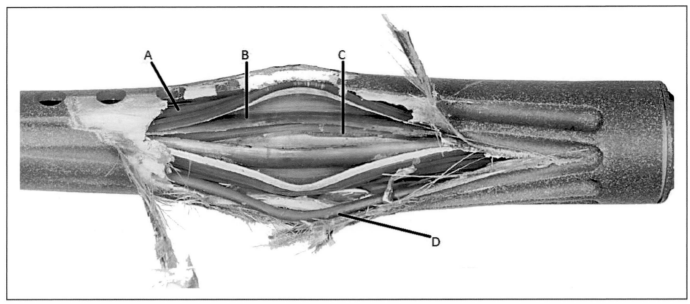

81. Left side closeup of the blown barrel on AR-10B no. 1002.

The fluted aluminum barrel jacket (A) is ballooned and folded up far from the ruptured stainless steel liner (B), from which it had separated. The rifled interior of the barrel liner (C) is clearly evident, as is the strong new gas tube (D), which was bent but unbroken by the blast.

Finally, the light-colored foam plastic of the handguard is exposed in a few places, and the woven fiberglass skin of the handguard shows some broken and frayed strands jutting out in all directions.

Springfield Armory photo from Technical Report No. TS2-2015, dated February 4, 1957

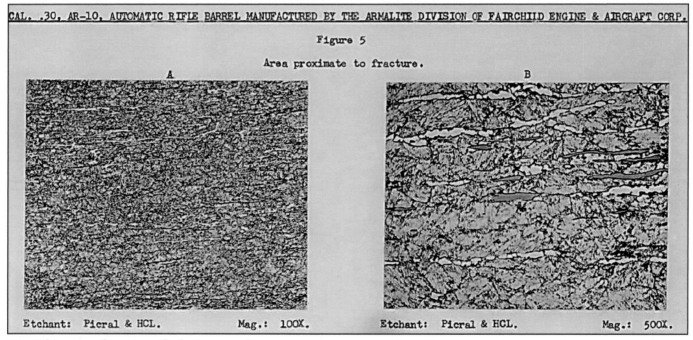

82. This greatly magnified image, known as a "macrograph," clearly displays the sulfur stringing that weakened the barrel steel and caused the catastrophic failure. The latticework of lines, present on both images, displays the strings of sulfur, and the lighter-colored hori-zontal striations (most obvious in the 500x magnification view at right) are the microscopic fractures that caused the steel to burst.

Springfield Armory photo from Technical Report No. TS2-2015, dated February 4, 1957

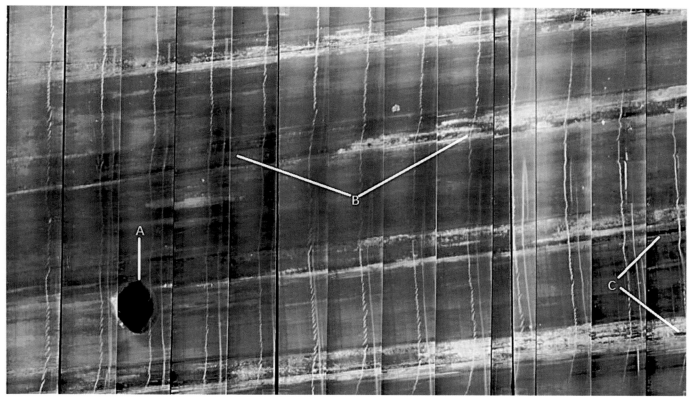

83. A borograph of a section of the blown barrel from AR-10B no. 1002, showing a composite of the interior of the barrel. This process involved inserting a tiny camera on a rod into the bore (hence "borograph"), and stopping it at every frame length, rotating it 360° for each exposure. What is produced is a single strip that shows as a flat image the entire circumference of the interior of the portion of the bore within the camera's frame. Many individual borographs are then added together to produce an image of the entire length of the bore.

The labeled areas are the gas port (A); the rifling (B), lighter-colored in some areas because of the leftover copper residue from fired bullets; and the dark beginnings of fractures (C).

Springfield Armory photo from Technical Report No. TS2-2015, dated February 4, 1957

It was quickly found that not only had the composite barrel been the actual cause of the accident, but it had happened exactly as Eugene Stoner had prophesized. Heating from the high round count caused warping in the aluminum, which was not reciprocated by its stainless steel liner, which caused the two to separate. A source of gyration was also introduced that had not been part of Sullivan's calculation, although Stoner had brought this to his attention as well. The recoil compensator, adding its considerable mass to the end of the barrel, caused barrel flex to be accentuated, and added much more strain to the two composite parts under flexion, due to the mechanical advantage it had being on the unsupported end of the barrel.

This separation of the two components of the barrel also expedited the weakening of each individual part. The Springfield engineers quickly identified the stainless steel used in the barrel liner (type 416) as a variety that was particularly unsuited to wide ranges of, and rapid changes in, temperature. It had also been heavily alloyed with sulfur, without the chief designer's knowledge, to allow for easier and less expensive machining. This sulfur was discovered to have condensed into strings in the steel, which served as flaws and weak points where the metal could split. No longer provided with even the pitiful support of the aluminum outer shell, and exposed to shock and heat for such an extended spell, the steel had finally given way.

The splitting of the stainless barrel liner was found to have been taking place for almost the entire length of the barrel, with images of the interior of the bore showing telltale stress fractures present from the gas port all the way back through the barrel throat (fig. 84; the area where the bullet of the chambered cartridge rests when the rifle is in battery and where the rifling begins).

It was therefore extremely fortuitous that the bullet exited the fractured barrel where it did, for had its yaw taken place any earlier, the marksman would doubtlessly have lost his hand. The ensuing bad press would have likely spelled the end for the AR-10 design, and the liability for injuries would have drained ArmaLite's already-meager coffers, in addition to making Fairchild lose confidence in its rogue division.

A few bright spots did emerge from this disaster, however. The other parts of the AR-10 were shown to be made and designed well, as they were not damaged by the rupture. The fiberglass handguard, although made at a time when such technologies were in their infancy, had performed well in shielding the shooter's hand from the extreme barrel and gas tube temperatures, and had not cracked or bent in any way. Indeed, it took a 7.62x51mm NATO bullet and its attendant propellant gases blasting through it to actually break the handguard.

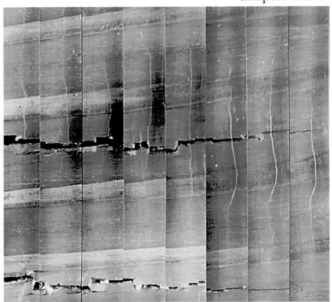

84. A second borograph, taken at the throat of the barrel, shows the cracks in the stainless steel liner in much better contrast. They extend in roughly the same direction as the rifling, showing the particularly lateral direction of the strain placed on the composite barrel.
Springfield Armory photo from Technical Report No. TS2-2015, dated February 4, 1957

## Replacing the Composite Barrel on AR-10 No. 1002

Needless to say, testing on both rifles was immediately and indefinitely suspended while an investigation was conducted on no. 1002. Once it had been determined that the composite barrel had been the sole source of the rupture, and that the rest of the rifle had held together well under this type of shock, for which it had not been designed, consent was given for the testing to be continued once a replacement barrel had been satisfactorily developed and installed.

Despite having exploded and almost grievously wounded a member of their staff, the AR-10's performance was gaining quick converts at Springfield. It was for this reason that some of the engineers—it is not known whether these were part of a progressive clique that never agreed with the .30 Light Rifle concept and Lightweight Rifle Program, or simply concerned employees who saw the ArmaLite design as the best answer to the requirements of the program—offered to help Stoner in fabricating a new barrel. Having gained considerable experience with T44 barrels that delivered good accuracy and stood up well to heat, they shared their data on a particular material. Later to be known as "chromoly", this 4100-series steel was heavily alloyed with chromium, molybdenum, and vanadium.

ArmaLite production manager Charles Dorchester and designer Eugene Stoner accompanied the sympathetic Springfield engineers to the nearby Mathewson Tool Company of New Haven, Connecticut, which had already done considerable work for the Armory, and had constructed several receiver prototypes for the T44. There they quickly worked out the tolerances and dimensions to have an existing T44 barrel blank machined to fit the AR-10 and its unique bolt-locking barrel extension.

In the ultimate vindication of Stoner's ideas of using an ordnance-grade steel barrel instead of George Sullivan's composite on the AR-10, the finished alloy steel barrel actually weighed less than its two-part aluminum-and-stainless steel predecessor. The original reason for the development of the composite barrel, and the only articulable rationale for its ill-advised use in the military trials, was its weight saving, and Stoner's ingenuity was even able to overcome this.

A good part of the weight reduction was achieved by continuing the process known as fluting, which consists of milling a series of parallel, horizontal (other configurations are sometimes used today, but at the time, they were horizontal) grooves in the thick portion of the outside of the barrel, behind the front sight tower. These serve not only to decrease

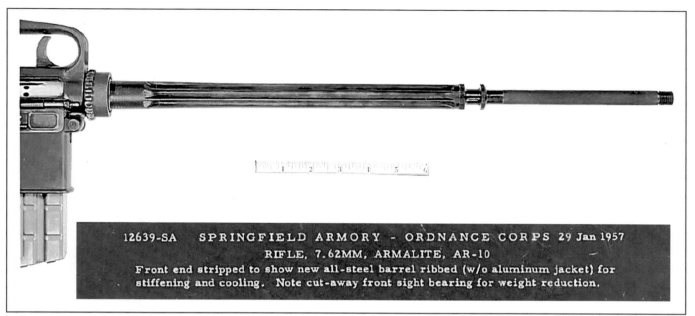

12639-SA    SPRINGFIELD ARMORY - ORDNANCE CORPS 29 Jan 1957
RIFLE, 7.62MM, ARMALITE, AR-10
Front end stripped to show new all-steel barrel ribbed (w/o aluminum jacket) for
stiffening and cooling.  Note cut-away front sight bearing for weight reduction.

85. Right side view of AR-10B no. 1002 fitted with its new ordnance steel barrel, machined and fluted to Stoner's specifications from an existing T44 rifle barrel blank by the

Mathewson Tool Co. of New Haven, Connecticut.
Springfield Armory photo no. 12639-SA
dated January 29, 1957, from Technical Report
No. TS2-2015, dated February 4, 1957

weight by reducing the amount of material present, but stiffen the barrel against flexion, which further improved accuracy. This innovation was already in use on the early composite barrels fitted to rifles nos. 1001 through 1005, and remains in wide use in civilian and military arms today; particularly those designed for light weight and accuracy.

## Concluding the Springfield Tests

In his internal memorandum concerning the Springfield Armory evaluation of the AR-10 dated March 30, 1957, excerpted in more detail below, George Sullivan stated that "rifle No. 1002 was returned to test one week from the time of the barrel failure with a barrel fabricated from 4140 steel of approximately the same weight as the barrel which ruptured."

On the subject of how the rebarreled rifle fared in the remaining tests, no report has come to hand, although it appears from Sullivan's confidential report to Richard Boutelle, dated March 30, 1957, further discussed and excerpted below, that such a report was produced. Sullivan states as follows:

86 (right). Left side closeup of AR-10B no. 1002 today.
Compare with fig. 64: note the lightening cut in the steel barrel, dividing the splined area over which the front sight tower attaches into two.
The front sights on these early prototypes were made of roll pins.                    Institute of Military Technology

*. . for all practical purposes, the test program was conducted in three separate, and apparently unrelated, phases. Phase I was terminated at our suggestion due to a minor problem encountered in the design of the muzzle brake . . Phase II was terminated when the stainless steel and aluminum barrel failed under extremely abusive cycling tests . . Phase III consisted of redoing certain of the tests encountered in Phase II, but using a steel barrel which we designed . . and manufactured in New Haven in minimum time.*

*Phases I and II were reported on by the Springfield Armory under the title "Project No. TS2-2105". . . Phase III of the test program carries no conclusions . . We know that the final phase, with certain very minor exceptions, was most successfully completed; evidently Ordnance chose not to draw a conclusion which might be in our favor . .*

## An Appreciation of the AR-10B Compensator

One of the many components of the AR-10B which performed well after the resumption of testing at Springfield in January, 1957 was the redesigned titanium compensator. However, as we shall see in Chapter Five, despite its ingenious construction and beneficial performance, particularly regarding recoil reduction, this was not to be enough to guarantee the longevity of such a necessarily large and obvious device in the face of military practicality, which placed a higher premium on versatility in such areas as bayonet mounting and grenade launching.

Nevertheless the design itself deserves a word of appreciation not only for its functionality, but particularly its ingenious use of space. The compensator as it first appeared on the AR-10A (fig. 52) was an extension of the barrel, with its entire length forward of its attachment point protruding ahead of the muzzle. In comparison, the AR-10B compensator was very largely configured as a sleeve or sheath which extended <u>backward</u> over the barrel, with nearly all of its length located behind the muzzle.

The key to this intelligent design is found in the muzzle cap, which was a cast component with an internal threaded ring, which mounted on the muzzle of the barrel. Ahead of the ring was an expansion chamber into which much of the gas, deflected by the muzzle cap, rushed and was forced rearward, where it was slowed and cooled before it exited from the perforated cylinder.

87. Interior view of the AR-10B compensator muzzle cap, showing the threaded portion which attaches to the barrel muzzle, and the expansion chamber ahead of it into which the gases flow upon firing, counteracting recoil.

Institute of Military Technology

# The Post-Mortem and Its Aftermath

It was determined by the testing staff that with their current design and materials, the AR-10 extractor springs were sufficient to last about 2,300 rounds, and the extractor itself would need replacing after every 4,500 rounds. This belied the fact that even though three extractor springs (a tiny number considering the number of rounds fired) and one ejector spring broke during testing, these breakages had not affected function, and rifles continued to fire without any stoppages or malfunctions until these issues were discovered during routine inspection. Furthermore, these findings were sharply rebutted in a report filed the next month by Charles Dorchester, who pointed out that parts changes on test versions of the

T48 and T44 were par for the course, and that the AR-10s had actually performed markedly better than its competition in this regard.

Backing up Dorchester's outrage was the incontrovertible proof in Springfield's own detailed report on the testing, which their leadership tried desperately to suppress. In the only extant copy of that report today, it is written plainly by none other than William Bonkemeyer himself, that of the 5,576 rounds in total fired by AR-10 no. 1002 (the one that suffered the catastrophic barrel failure), only 66 malfunctions of any kind took place, 53 of them during the cold test. Although the test report did not delineate when the malfunctions had occurred in the arctic conditions test, the observers on hand noted that all of them had happened in the stage where the rifle had been withdrawn from the cold chamber and allowed to accumulate considerable condensation, before being refrozen. This information made its way back to ArmaLite, where it was recorded. This meant that in all the other batteries of tests, which were much harsher than anything to which the babied T44 had been exposed, the AR-10 had experienced only 13 malfunctions; a number that was small even compared to the record of the T44 that had been refined for over a decade at the time. Even in the cold weather test, the malfunction rate of the AR-10 was considerably lower than the T48 in its cold-weather-modified version had shown back in December, 1953.

Exposing the duplicity of the test report even further was the later revelation that failures of the bolt to stay locked back after the last round had been fired were counted as "malfunctions", accounting for 43 of the 66 recorded malfunctions; an actual 65% of all claimed performance problems. Melvin Johnson had even discussed this possibility, and had concluded that such an issue could not reasonably be called a malfunction. Its occurrence did not affect rifle performance at all, and only added a small amount of time to the standard reloading process. It certainly did not constitute or induce a jam that had to be cleared.

As for the actual stoppages, all but three of the fifteen experienced were from the noses of bullets being stubbed, due to the extra-long and differently-contoured armor-piercing bullets used in the test ammunition. The remaining eleven malfunctions were of the trigger not resetting after a round had fired. As discussed in Chapter Nine, this recurring spring weight problem was the subject of a later design modification in serial manufacture, patented after Dutch AR-10 production had begun.

In a startling revelation, the observers who had been on hand for testing had been told that all Springfield Armory personnel, and anyone involved in the testing, was expressly forbidden to make any comparison, verbal or otherwise, between the AR-10 and the T44, and that to this end all comments regarding the AR-10's characteristics, performance, aesthetics, etc., were to be negative in nature. Bonkemeyer justified this with the rationale that positive comments would inevitably lead to comparison of weapons.

In direct contravention to the accusations made by Studler and Dr. Carten less than two months previously, the AR-10 did prove to be imminently controllable in testing. They had claimed that the weapon weight would have to approach nine pounds for acceptable rapid-fire controllability. However, in the report, it is stated that "recoil force of the rifle is low." Further increasing Dorchester's credibility is the result of high-speed filming taken of the rifleman firing the test rifles in first the semi-automatic and then the full-automatic fire modes. It was meant to show visually and empirically how much deviation of the rifle there was from its original alignment and orientation towards the target, and the degree of movement of the rifleman's body when subjected to the recoil forces. The video showed that there was almost no disruption of the rifle's orientation, and "little to no movement of the rifleman was disclosed." A stronger vindication for the combination of Stoner's direct gas impingement operation system, coupled with the straight-line recoil principle and the effective recoil compensator, could not be imagined.

## Springfield Concludes that the AR-10 is "Not Satisfactory as a Military Service Rifle"

The reasons why the AR-10 was ultimately declared to be "not satisfactory as a military service rifle in its present state" were delineated directly in the report. Both Charles Dorchester and George Sullivan critiqued each of the perceived deficiencies in internal

Fairchild correspondence, which is further excerpted below.

The first reason given was that the weapon was not safe, due to the likelihood of barrel failure, termed in the report as the potential "rupture of the composite steel and aluminum barrel." Simply in its own wording, this did not attempt to address the integrity of the newly-retrofitted steel barrel on no. 1002, which had not interfered at all with the AR-10's excellent reliability and accuracy, and actually improved its light weight characteristics slightly. This so-called "major deficiency" acknowledged by its own wording that it was no longer a valid complaint, since all future AR-10s would also utilize steel barrels. Dorchester also made a note in his rebuttal to this point that the very rifle that had suffered the barrel rupture and had been retrofitted, was currently being utilized for endurance tests for other interested military forces (in the process firing a substantially higher round count than at Springfield), and was showing no signs of wear or weakening.

Next in the list of alleged major deficiencies was that the rifle was too sensitive to "normal ammunition variation." Not only were the rounds fired in the test not standard issue ammunition, being of an experimental armor piercing type; but it had been uncovered during the first days of testing that the lot of ammunition assigned to the AR-10s had been rejected at Fort Benning as producing both excessive and inconsistent pressure and velocity readings. This ammunition lot was thus by its very status of having been rejected outside of acceptable standards, and not at all indicative of "normal ammunition variation." A better description was that the AR-10's performance could be hampered somewhat by "considerably abnormal ammunition variation," which could be said of any firearm ever made.

The third of these alleged deficiencies once again based its conclusion on facets of the AR-10 that had been changed prior to completion of the tests. It claimed that flash suppression was hampered by buildup of residue in the original muzzle device, which was difficult to clean under field conditions. This statement described the first model of recoil compensator, constructed of aluminum, which was present on the rifles when testing began in December, 1956. However, it was obviously false when applied to the modified titanium devices that Stoner designed and installed on the rifles before the January 8, 1957 resumption of testing. One of the express and specific purposes for the redesign of this unit was to allow its easy removal for cleaning in the field. This modification was successful, and the locking ring on the new model of recoil compensator could easily be removed by an infantryman, allowed simple and intuitive servicing and cleaning.

The following item on the list in the test report brought up the issue of poor functioning in the sub-zero temperature test. It was indeed true that performance had not been satisfactory during the portion of this test that included the refreezing of condensation inside and outside of the rifle. However, what Springfield failed to mention was that in the first phase of the cold weather testing, which did not include refreezing, the AR-10 had performed in a satisfactory manner. This was the same test that had stopped the T48 in 1953, and no other rifle had yet been subjected to the refreezing component. It was pointed out by the ArmaLite staff that the T44 and T48 would have fared similarly had they been exposed to this additional variable. Dorchester added to his internal correspondence that a modification that had been subsequently developed at ArmaLite's headquarters, and had been tested subsequently in some very cold environments with excellent results, even when the rifle was purposely subjected to condensation and refreezing.

The fifth complaint seemed to be stretching to the point of incredulity the term "major defect." Its entire substance was that the charging handle was difficult to operate while wearing thick winter mittens. According to Dorchester, not only was the AR-10's charging handle of reasonable size, but simply fitting a tiny crossbar on the face of the handle would allow it to be actuated from either side, making the weapon extremely easy to charge, even with one's elbow or forearm. It was emphasized that such an attenuated concern did not even approach any reasonable criteria for a "major defect," and could have been satisfactorily corrected as the minor complaint it was within a few days, and with the expenditure of almost no resources.

Once again, a defect that had been corrected by one of Stoner's December modifications, was claimed as being still extant in the Springfield Armory final report. It was charged that when removing and reinstalling the recoil compensator, the point of impact of the rounds fired changed relative to the sight alignment. The redesigned titanium muzzle device indexed automatically to its original position on the barrel and in relation to the front sight base, and when it was indexed correctly, which with this modification was done easily, the point of impact did not deviate as was claimed.

Finally, one defect was pointed out which was acknowledged by ArmaLite as having merit. When the weapon was left uncleaned overnight with a round in the chamber, the next day it would be very difficult to work the charging handle the first time and extract the chambered round. Dorchester concurred that this issue did exist, but clarified that the rifle had to be fired many hundreds of times with no cleaning, brushing, or oiling for it to take place, and then only by being left unattended and loaded for a very long period. For ArmaLite's internal consumption, the production manager also confirmed that the modification made to address the reliability issues suffered in the refreezing of condensation also had an excellent effect in alleviating this issue.

A litany of other "minor defects" was claimed in the report, all of which are, to an item, too petty and equivocal to merit discussion in detail. One must consider that when "difficulty of actuating the charging handle in thick winter mittens" was called a major defect, the minor defects bordered on the absurd. In a curt and forceful single sentence, Charles Dorchester dismissed categorically the defects delineated in the test report, and responded that all of the issues not only could be (and later actually were) eliminated in a matter of a few weeks, but concluded

that with the eminent elimination of the "so-called deficiencies," even Springfield Armory would have no choice but to categorize the AR-10 as "satisfactory as a military service rifle." ArmaLite even adopted the policy on February 15, 1957, that the conclusion of the Springfield Test report was actually that, with 13 minor design changes, the AR-10 was indeed "satisfactory."

The aluminum magazines, which had been the subject of much ridicule and sneering by the Armory staff, actually held up to the roughest testing, and were not found to induce any of the few malfunctions experienced during the final round of evaluations. Although later proving to be the one weak point of the rifle (when reloaded repeatedly and used far beyond their intended design life), the corrugated aluminum design, patented by Stoner, proved to be at least as rugged and sturdy as contemporary detachable magazines constructed of steel.

## ArmaLite Strikes Back

As noted, both Charles Dorchester and George Sullivan lambasted with fiery abandon each of the perceived deficiencies in internal correspondence classified as Fairchild Confidential, which was never intended for general distribution and, as such, bolsters the credibility of these rebuttals.

Charles Dorchester's internal report to the Fairchild board concluded defiantly as follows:

*. . The AR-10 has shaken hell out of Springfield and we are prepared to combat any negative report they may come up with with factual evidence of performance and sound engineering methods for correcting anything they may consider to be deficiencies . .*

*Despite all obstructions, negative attitudes, severe abuse given the rifles and slanting of Springfield reports, we are sure that the AR-10 is the outstanding weapon in the competition, not even considering the fact that we are providing the light weight mobility so greatly required by today's forward military planning.*

*Without any hesitation, we recommend that we offer to place the AR-10 into competitive testing with any other automatic rifle in the world in the hands of unbiased troops and we will come out head and shoulders above the best of them.*

*[signed] Charles H. Dorchester,*
*ArmaLite Division*

For his part, George Sullivan also critiqued the Springfield assertions in a seven-page rebuttal, excerpted as follows:

*. . This rifle [serial no. 1002] and barrel with over 10,000 rounds on the action including adverse condition tests is now being used as a demonstrator to interested foreign allied governments.*

*. . We have been informed, first by General Schromburg, Chief of R & D for the Dept. of Ordnance, that we should make the necessary fixes in accordance with the Springfield findings and return the weapon to Springfield to obtain final approval prior to testing at Aberdeen. This recommendation was also made by General Gavin. Col. Calland of the Marine Corps Equipment Board has informed our Alexandria office that the Marine Corps would like to test the AR-10, but can only do so after the weapon has passed the Springfield test.*

*Our personal feeling is that another Springfield test would be a waste of valuable time; actually the next tests should be made by the user, the Army Field Forces at Ft. Benning, Georgia; however, political and other factors may be such as to override this conclusion.*

### ArmaLite Conclusions

*1. The AR-10 is the only rifle which will meet the specification drafted more than ten years ago for the future standard light-weight basic infantry weapon. It represents the first new concept in firearm design in many decades.*

2. *The few valid deficiencies found as a result of the Springfield tests have been corrected— and thus eliminated as an argument against the AR-10 rifle.*

*Accordingly the slightly-modified weapons now being fabricated (even in the eyes of the Ordnance) <u>must be acceptable as a military weapon</u> . .*

## Time Wounds All Heels

The dishonesty and duplicity that would eventually come back around to play a large part in the ultimate demise of Springfield Armory as an institution, was present in spades in the AR-10 tests, so grudgingly performed only after being ordered to do so by the Continental Army Command. In addition to the usual bullying by the test staff, many of whom exhibited barely-contained contempt towards the reportedly-polite and professional ArmaLite observers, actual acts of sabotage were witnessed by those present. Charles Dorchester was interviewed just a few years later on this topic for an article in *Gun World* magazine that caustically lambasted Springfield Armory, its insular and wasteful development and evaluation programs, and in particular the M14, the eventual "winner" of the Lightweight Rifle Program. This article was representative of a movement in American military small arms thinking that would prove to be the sweet vindication of the ideas of ArmaLite and Eugene Stoner over the hidebound traditionalism of the Ordnance Corps. Among other transgressions, Dorchester reminisced how "[o]ne of the Springfield test staff fired the AR-10, and I was looking over his shoulder when he wrote on the test report that this was the best light weight automatic rifle ever tested in the Springfield Arsenal. But a few minutes later, his supervisor looked at the statement, and erased every word of it."

It should come as no surprise from this, that once testing was finished without further incident in February, Springfield passed along a recommendation to the Ordnance Corps that the AR-10 "would require approximately five years of development" before it could be considered sufficiently reliable and able as a standard-issue military rifle. This was directly in contradiction to the facts that the AR-10 had not only performed more reliably than the T44 and as reliably as the T48 in every round of their testing except the last, but actually did so with better accuracy, greatly superior rapid-fire controllability, and remarkably reduced weight.

Conveniently, as mentioned above, the actual official report containing the numbers and detailed data from the 1956-57 tests disappeared from Springfield's archives, and it is only thanks to the help of an anonymous benefactor in the Pentagon that the details of this dark period in the AR-10's history survive for the benefit of posterity.

## What Might Have Happened

With all the huffing and puffing, however, Fairchild's rejection of the unprecedented offer by the Ordnance Corps to develop the AR-10 under its own auspices deprived the fledgling rifle of the benefit of services and facilities far in advance of anything that ArmaLite could afford, or even envision. As William C. Davis stated, in a letter to the editor concerning the similar situation that later confronted the AR-15 (ironically due this time to Dr. Carten's perfidious denial that the AR-15 possessed enough merit to warrant such development) but equally applicable to the AR-10,

*. . Had that recommendation [for continued development under Army sponsorship] been ac-*

*cepted . . then the expertise and the resources of the Army technical establishment would have been made available . . for the engineering development and engineering testing which are invariably required to bring a weapon/ammunition system from initial design to a state of readiness for full-scale production and general deployment . .*

*. . While Stoner's design was excellent, there had been no significant engineering development of it . . Indeed none of the parties who had custody of the design . . had the facilities required for an adequate program of engineering development and engineering testing. In fact they had little appreciation of the <u>necessity</u> for adequate engineering and testing of military weapons . .*

# An Interesting Glimpse Inside a Working AR-10

Not all of Springfields' efforts were aimed at discrediting the AR-10 by nefarious means. A brief glimpse at an interesting study contained in the Springfield Technical Report of the AR-10 tests reads as follows:

### Kinematic Data

*Appendix G [not included] contains in detail plotted data pertaining to the cyclic operation of the breech mechanism [of the AR-10]. The data are based on high-speed filming of the mechanism while the rifle was shoulder fired from a bench rest.*

*The data presented were obtained with Rifle No. 1002 prior to modification. At that time the barrel port was measured as 0.125 inch diameter and the [composite] barrel had six grooves.*

*An analysis of high-speed films of the breech mechanism produced the following average measurements:*

*a. Start of recoil to end of dwell* . . . . . . . . . . . . . . . . . . *1.2 milliseconds*
*b. Start of recoil to end of unlock* . . . . . . . . . . . . . . . . *1.9 milliseconds*
*c. Overtravel time* . . . . . . . . . . . . . . . . . . . . . . . *17.2 milliseconds*
*d. Start of recoil to end of counter recoil* . . . . . . . . . . *66.4 milliseconds*
*e. Hammer fall and firing* . . . . . . . . . . . . . . . . . . . . *10.7 milliseconds*
*f. Total cyclic time* . . . . . . . . . . . . . . . . . . . . . . . *77.1 milliseconds*
*g. Cyclic rate of fire* . . . . . . . . . . . . . . . . . . . . . . . . . . . *778 spm*

# *Ordnance* Waits to Back a Winner: Mel Johnson's Belated Article

An article entitled "A New Automatic Rifle", by none other than Melvin M. Johnson, Jr., then a colonel in the U.S. Army Reserve (Inactive), appeared in the May - June, 1957 issue of *Ordnance* magazine, the official journal of the U.S. Army Ordnance Corps.

A sidebar to the published article introduces the author as follows:

*Colonel Johnson . . . is the distinguished inventor of Johnson arms. He is currently serving as consultant on research to the Operations Research Office of Johns Hopkins University and in a similar capacity to several industrial organizations including the Fairchild Engine and Airplane Corporation.*

The reason behind this anachronistic introduction, Johnson having already long since given up his advisory position at ORO by the time this appeared, was that his article had actually been written and submitted at least a year earlier, before the first demonstrations of Stoner's X-02 and AR-10A had taken place. Whether it would have made any difference in the consideration given the AR-10 had the article appeared earlier is debatable, but its long-delayed publication, coming on the heels of the AR-10's rejection in the Springfield trials, truly added insult to injury as yet another example of the forces of systematic bureaucratic betrayal that were aligned against the AR-10.

Not coincidentally, the appearance of Johnson's paean of praise for the AR-10 coincided almost to the day with the official anouncement on May 1, 1957 by Wilbur Marion Brucker, Secretary of the Army, that the Lightweight Rifle Program was over, and a new rifle, christened the "U.S. Rifle, 7.62mm, M14," had been selected for universal adoption by all U.S. Army forces. The M14 was slated to replace not only the M1 Rifle, but the M3 Submachine Gun, the M1918 BAR, and the M1/M2 Carbine. This had been an early goal of the Lightweight Rifle Program, and it had become dogma for the insiders of the tests, despite the fact that it was, by this time, pure fiction. As Eugene Stoner pointedly recounted decades later, "Anyone who knew anything in the firearms community knew that it would only replace the M1 Rifle."

In his article, Johnson gamely did his best to portray the AR-10 in the most favorable of lights. In three pages of text he sketched out some early history and then went on to extol not only the AR-10's performance and characteristics, but the ingenious inner workings of its mechanism, understandably putting great emphasis on the bolt-to-barrel locking system, first developed by Johnson himself. A brief excerpt from this article, written in Johnson's characteristically brusque style, is as follows:

*. . For over the past twenty years, with few exceptions, there has been little inventive activity in arms by American industry. Army Ordnance arsenals have had to carry the brunt of creation and conception. Industry has performed chiefly advanced engineering and production . .*

As discussed in Chapter Four, the Marine Corps had not gone along with the Army in rejecting the AR-10 out of hand, and would request samples for testing later that year, despite its not getting the nod from Springfield.

The Air Force never accepted the M14, instead pioneering the military acquisition of an even lighter AR-10 follow-on called the ArmaLite AR-15, another strange, futuristic rifle with an aluminum receiver and plastic stock, just two years later.

# Carrying On Regardless
## The 1957 ArmaLite AR-10 Brochure

88. The cover of the printed ArmaLite AR-10 brochure, copyrighted in 1957.

This spiral-bound, 12-page, two-color document put as brave a face as possible on the AR-10's future after the disastrous trials at Springfield Armory had concluded that the AR-10B was "not satisfactory as a military service rifle in its present state."

On the last page the printed address, 6567 Santa Monica Boulevard, was overpasted with a small sticker showing the new Costa Mesa address, to which ArmaLite relocated in August, 1957.

Charles Kramer collection

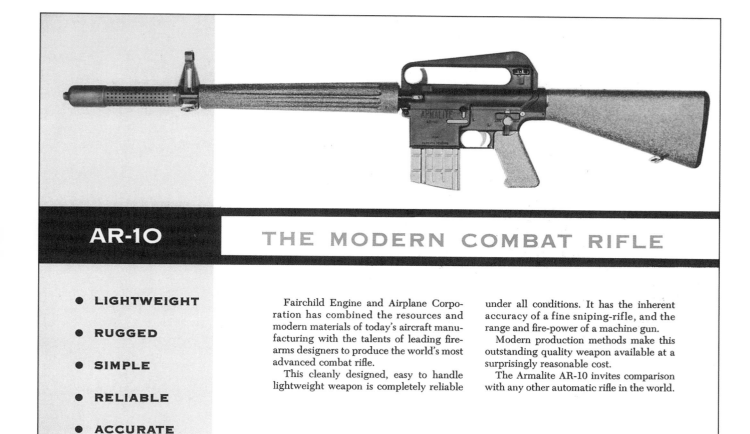

# AR-10

## THE MODERN COMBAT RIFLE

- ● **LIGHTWEIGHT**
- ● **RUGGED**
- ● **SIMPLE**
- ● **RELIABLE**
- ● **ACCURATE**
- ● **VERSATILE**

Fairchild Engine and Airplane Corporation has combined the resources and modern materials of today's aircraft manufacturing with the talents of leading firearms designers to produce the world's most advanced combat rifle.

This cleanly designed, easy to handle lightweight weapon is completely reliable under all conditions. It has the inherent accuracy of a fine sniping-rifle, and the range and fire-power of a machine gun.

Modern production methods make this outstanding quality weapon available at a surprisingly reasonable cost.

The Armalite AR-10 invites comparison with any other automatic rifle in the world.

COPYRIGHT
1957
ARMALITE

89. Page two of the 1957 ArmaLite AR-10 brochure, depicting one of the original (pre-Springfield) prototypes, and defiantly proclaiming that "The ArmaLite AR-10 invites comparison with any other automatic rifle in the world."                    Charles Kramer collection

90. Page nine of the 1957 ArmaLite AR-10 brochure, aimed squarely at the military market, and stressing the design simplicity of the rifle, both for manufacture and service.
Charles Kramer collection

## Looking Ahead to the Final Series of AR-10Bs - Serial Nos. 1006 - 1047

In retrospect it seems that while Fairchild had rejected the offer made to develop the AR-10 under the auspices of Army Ordnance, they actually managed to profit from the disastrous Springfield trials, audaciously treating them as a *de facto* program of production engineering development.

As further discussed in Chapter Five, by September, 1957 the final series of improved, hand-made ArmaLite AR-10B rifles had been produced, serially numbered 1006 through 1047, all fitted with barrels made of ordnance-grade steel.

## *Chapter Four*

# New Directions

## Two Divergent Paths for the AR-10

Although the crushing results of the Springfield Armory tests came as a heavy blow to the AR-10, especially in the eyes of the Fairchild board, all involved were still encouraged by its previous stellar record of performance, and were intent on developing it further and pursuing new opportunities.

With the door to U.S. adoption finally and firmly closed, the road to the AR-10's future reached a crossroads, and with the help of outside expertise, capital, influence, and experience, was able to branch out and continue in two new directions, achieving in one the most fantastic success a U.S. military rifle has ever attained, and in the other the quiet dignity and almost-forgotten legend as the most precocious and effective battle rifle of all time.

Both of these new and diverging efforts actually began before the AR-10 was ever brought to Springfield Armory; one from high in the ranks of the U.S. Army, and another at the behest of Fairchild, ArmaLite's parent company.

# Path I: the Small-Caliber, High-Velocity (SCHV) AR-10 Clone

In the rarified air at Fort Monroe, Virginia in March, 1956, when the AR-10A was so spectacularly demonstrated to the rapt attention of the headquarters staff of the Continental Army Command, a general who had just that month attained his fourth star was particularly impressed, but had it in mind to take the AR-10 in a different direction. The new chief of the Continental Army Command (for about 40 days as of the demonstrations) was a reformer, and did not approve of the fanatical pursuit of a new rifle in a full-power .30-caliber cartridge. It is for this reason that General Willard Gordon Wyman, while he had been behind the strongly-worded order that forced Springfield to halt the T44 adoption process and test the ArmaLite rifle, did not intervene more heavily in favor of the AR-10's adoption.

The progressive general actually paid a brief visit to ArmaLite on July 26, 1956, where he became better acquainted with the AR-10, and decided that it had the makings of the ideal SCHV platform. His staff had actually contacted George Sullivan just a few days before to inquire as to whether security clearances were required for the visit to ArmaLite's "factory," and whether the General's personal helicopter could be landed on ArmaLite's grounds. Sullivan's response was a polite "no" to both.

General Wyman had always been a strong believer in a lighter rifle, and had been a devotee of the SCHV concept since Project SALVO had begun. He saw in the AR-10 the perfect platform for a smaller and more easily-controllable cartridge. Although not involving himself in the conduct of the evaluations, the reports generated from them, or the final decision of weapon adoption, he had observed closely from the background the progress of this new wonder rifle.

ArmaLite had also been dabbling in some new ideas brought about by the SCHV concepts after being made privy to the results of the first round of classified SCHV testing in Project SALVO. Although no prototype had yet been planned or drawn up, the feasibility of adapting Stoner's design to a lighter cartridge was apparent to all involved in the crash development of the AR-10. To this end in February, 1957 General Devers, the previous holder of General

Wyman's job and currently employed as a Fairchild consultant, called on the Chief of Ordnance, Lieutenant General Emerson LeRoy Cummings, successor to General Ford, the more conservative proponent of the Lightweight Rifle Program. In this visit, in addition to attempting to smooth feathers following the acrimony engendered by the Springfield Armory tests of the AR-10, General Devers sought and received the classified data covering the second round of Project SALVO's SCHV tests.

Around the time of the announcement of the T44's adoption as the M14, General Wyman flew to California along with several members of his staff, and paid a surprise visit to the ArmaLite facility, where Eugene Stoner was hard at work. According to Stoner, the general simply walked in, surprising him once again, and asked, "How would you like to get in on a rifle program for me?" Obviously enthused by the new and unlooked-for opportunity that had literally just walked through his door, Stoner expressed his very keen interest, and sat down with Wyman. Stoner could not be told in detail about the current design goals, partly because the ArmaLite shop was considered an insecure location, and partly because the general himself was not authorized to

release such classified information. Instead, Wyman suggested that Stoner travel back to Fort Benning and meet with the Infantry Board once again to learn what he needed to know. Stoner agreed, and the two set up the meeting.

In order to prepare for Stoner's arrival, General Wyman provided the Infantry Board with a copy of a very interesting document, classified as CONFIDENTIAL, which he had found: one which Dr. Carten—Colonel Studler's successor and the mastermind behind Springfield's smear of the AR-10—had attempted to suppress. It was a proposal and funding request from Aberdeen Proving Ground's Development and Proof Services (D&PS) for the continued development of the .22-caliber military rifle projectile, in the words of William C. Davis, then chief of the Aberdeen Small Arms Branch, "by developing a boattail bullet of approximately 55 grains weight, at a muzzle velocity of approximately 3,300 fps, for use in a rifle substantially lighter than the T44/M14."

Dr. Carten had denied this request, as William Davis himself noted later, "for the very valid reason that the activity we proposed was not within our mission responsibilities."

## Flashback: the "Too Aerodynamic" .30 M1 Cartridge

The .30 M1 was an improved loading of the original M1906 Springfield cartridge, developed by Army Ordnance during the 1920s for long-range use in machine guns. The M1906 round had fired a flat-based 150-grain bullet at about 2,800 feet per second. The M1 cartridge used a considerably-heavier 174-grain bullet which was tapered at the base, forming a boattail. The increased weight and streamlined profile made this round significantly more aerodynamic than the M1906, and thus more effective at longer ranges. The maximum range of the M1 ball cartridge was officially listed at 5,200 yards.

Rather ironically, the only reason it was dropped less than a decade after its adoption was that it was considered actually to be too aerodynamic, and

its use created a safety hazard for persons who might be downrange of military training areas. Since the turn of the century military shooting ranges had intentionally been designed with a distinct amount of "buffer space" surrounding them, so that a bullet that missed its target at a very high angle could safely come to earth within the buffer area, and not in a populated locale. The greatly improved range of the M1 round extended well outside of these buffer areas, and so the military ordered a new cartridge, the M2 ball, with the specific goal of shortening the maximum range of the M1 loading. The .30 M2 served as the standard round for the U.S. military through WWII and Korea.

## Back to the Story

Upon arriving in Georgia, Stoner was shown the Aberdeen proposal, and informed of the details of the Army's current SCHV initiative. Their plan was to acquire a rifle that would weigh six pounds or less when loaded with a full 20-round magazine of ammunition, and which would fire a to-be-determined

.22-caliber high-velocity round, which would be effective out to 300 yards.

The Ordnance staff assigned to Fort Benning had fabricated a few makeshift test rifles, and had been experimenting with cartridges of different potential to achieve the desired balance of power, recoil, and weight. For their part, Gerald A. Gustafson and

William C. Davis of the Development and Proof Services (D&PS) at Aberdeen, had themselves created a .22 caliber SCHV cartridge by reducing the length of the .222 Remington cartridge case, loading it with a bullet of extremely light weight, fitting an M2 carbine with a barrel that would chamber the stopgap sub-caliber round, and opening up the cartridge seating area on the bolt face to accommodate the larger base diameter of the shortened .222 Remington round. On the other end of the spectrum, Project SALVO engineers had been experimenting with a .22-caliber High Velocity cartridge produced by necking down the T65E3, which had by then been officially adopted as the 7.62x51mm NATO round, loaded with a 68-grain "homologue"—a bullet of exactly the same proportions as its inspiration, the .30 M1 bullet—but decreased in size. This had been designed at Aberdeen in 1954.

The superior .30 caliber M1 bullet configuration was well remembered within the military arms design community, and in creating its .22 caliber homologue the exact angles of curve, amount of taper, weight and length-to-diameter ratios were kept from the .30 M1, and the result was a 68-grain .22-caliber bullet. This bullet possessed the remarkable ballistic efficiency of its parent projectile, and when propelled to the blazingly-fast velocities made possible by the large powder capacity of the necked-down NATO cartridge case, gave blisteringly-efficient performance, retaining a flat trajectory out to very long ranges, all while producing relatively little recoil, at least when compared to the full-power NATO round. To fire it, one of the T48 FAL rifles left over from the Lightweight Rifle Program was rebarreled, and its gas system reworked to fire the new round.

## "Too Little" and "Too Much" - the Search for Baby Bear

By the time of Stoner's arrival, it had been conclusively determined that the .30 Carbine SCHV variant was too weak for serious military consideration, and the 7.62mm NATO round necked down to .22 caliber, while possessing excellent performance, required too heavy a weapon to handle its large cartridge size, high chamber pressure, and comparatively violent recoil.

The Aberdeen proposal which Dr. Carten had denied stemmed directly from this determination, and had settled on a bullet weight of approximately 55 grains. Stoner was instructed to formulate not only one or more rifle prototypes following these specifications, but to design the actual cartridge.

In a typical example of military buck-passing, the Infantry Board, being most in tune with the needs of the individual infantryman, had originally settled on a 300-yard range requirement, which had been found by numerous studies conducted by several forces in the two World Wars and Korea to be the maximum effective engagement range for infantry using standard, non-scoped rifles, However, after Stoner had left to return to California, they had

second thoughts about how this range requirement would look to the more conservative types at Fort Monroe, and increased their recommendation to 400 yards. Continental Army Command at Fort Monroe, in turn, felt that the still-more-conservative Ordnance Corps would find 400 yards cause for objection in what they were already compromising to accept, and thus settled on 500 yards as the final effectiveness requirement for the new, prospective .22-caliber round.

The exact requirement characteristics were then formulated and communicated to ArmaLite. They were that at 500 yards, the new round should be able to penetrate a standard steel infantry helmet, body armor, or a sheet of 10-gauge steel. It also was required, at the given range, to have a ballistic trajectory as flat as or flatter than, and accuracy as good as or better than, the .30 M2 round fired from the M1 rifle. It was decided that the new round also must have terminal ballistics (effect on a living target, calculated by performance in clay or ballistic gelatin) equal to or better than the .30 Carbine cartridge at the 500-yard mark.

## The ArmaLite .222 Caliber "Stopette" (the AR-11)

Before these revised requirements could reach ArmaLite, however, the first working prototype had been completed based on the original set of specifications, which had specified a maximum range of 300 yards. This rifle had a far more traditional appearance than the AR-10, and showed influences

both from Stoner's AR-3, and the AR-5. In fact, Colonel Enewold of the Air Force even consulted once again with his old friends at ArmaLite on the stock of what became designated the AR-11. Given the original, relatively short 300-yard range requirement, it was felt that a new cartridge was not neces-

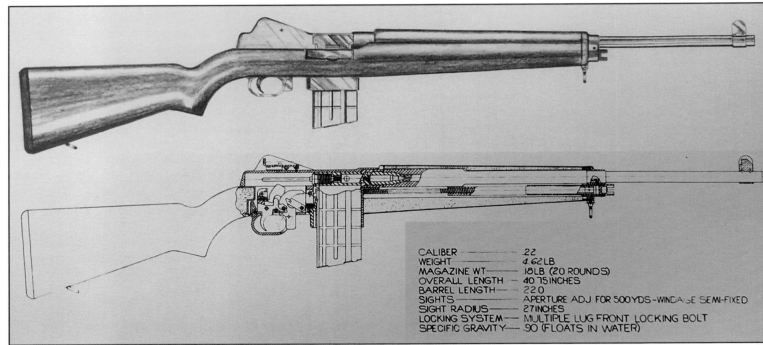

CALIBER —————— .22
WEIGHT —————— 4.62LB
MAGAZINE WT —— .18LB (20 ROUNDS)
OVERALL LENGTH — 40.75INCHES
BARREL LENGTH —— 22.0
SIGHTS —————— APERTURE ADJ FOR 500YDS-WINDAGE SEMI-FIXED
SIGHT RADIUS —— 27INCHES
LOCKING SYSTEM — MULTIPLE LUG FRONT LOCKING BOLT
SPECIFIC GRAVITY— .90 (FLOATS IN WATER)

91. A drawing, presumably hand-done by the designer, Doc Wilson, showing two right side views of the conventionally-stocked ArmaLite AR-11, nicknamed the "Stopette". The gas piston, located under the barrel, is indicated in the lower sectioned view.

This is the only surviving record of the AR-11. It was included in ArmaLite's Activities Proposal for 1957, and appears to be nothing more than a scaled-down AR-3, chambered in .222 Remington. It had a proposed weight of just 4 lbs. 7 oz., and as indicated in the legend at bottom right, it had a specific gravity of .90, meaning that it would float in water.

Only one example of the AR-11 was constructed, for Stoner's demonstration at Fort Benning.

Smithsonian Institution collection

sary, and the AR-11 prototype was chambered for the existing commercial .222 Remington varmint-hunting round.

Nicknamed the "Stopette," after a popular, contemporary type of spray-on deodorant which came in a light plastic bottle, no photographs of the single example made of the AR-11 rifle have survived.

When presented at Fort Benning, not only did the existing commercial round prove inadequate, especially with its aerodynamically inefficient hunting bullets, but the rifle on the whole was considered not ideal for the same reasons that the AR-3 had been rejected. It boasted a very high cyclic rate of automatic fire which, when combined with its light weight and traditional stock design, was utterly uncontrollable in rapid and automatic fire. It was suggested by the Infantry Board, and simultaneously decided by ArmaLite, that a scaled-down version of the AR-10 could achieve controllability at the same light weight, but with an even more powerful cartridge; one that could meet the increased range requirements.

According to L. James Sullivan in Part I of his interview with Dan Shea, published in the March, 2008 issue of *Small Arms Review*, the barrel extension in the AR-11 was not strong enough, and after testing at Fort Benning it met a sad fate when during further firing tests at ArmaLite's headquarters, it suffered a catastrophic failure and exploded. As Sullivan recalled, "The [barrel] extension broke and the head of the machine shop [who] was firing it was creased on the top of his head, and required getting patched up. The Stopette was originally designed by somebody who wasn't there any more [named] Doc Wilson."

All that remains of the sole copy of this evolutionary link is the hand-drawn diagram shown above.

# The First Step to the AR-15: the "XAR-1501"

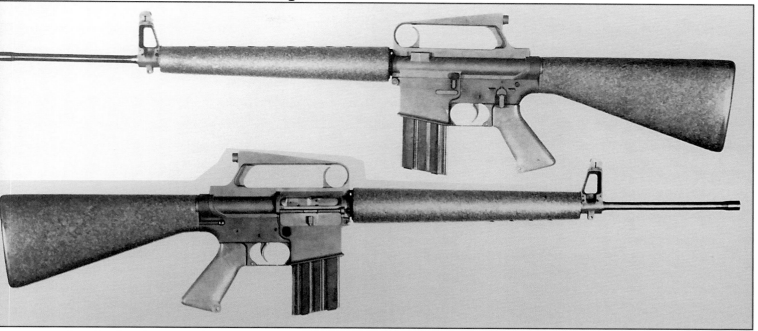

92. Left and right side views of perhaps the earliest proto-type of the scaled-down AR-10, called the XAR-1501. These views were also included in ArmaLite's Activities Proposal for 1957.

The XAR-1501 was designed at Stoner's behest by Robert Fremont and L. James Sullivan.

Note the unusual configuration of the carrying handle and fixed rear sight, the 25-round magazine, and the plain muzzle. Institute of Military Technology

Splitting his attention between his new AR-10 development program and the SCHV initiative, Stoner set about envisaging a version of the AR-10 that could fire a .22-caliber round. The .222 Remington sporting round had proven inadequate, but Stoner decided that a better-designed bullet traveling at a slightly higher muzzle velocity would be just the ticket, so this cartridge would be an excellent starting point. Stoner, who was the head of the ArmaLite design department by this time, directed his two main subordinates, Robert Fremont and L. James Sullivan, to create a scaled-down AR-10 to fire the .222 Remington round, just to make sure that it would indeed tame the recoil to an acceptable level.

The prototype of this new weapon, named the XAR-1501, was demonstrated for General Wyman a few months later at Fort Monroe, and he was so impressed that he immediately sought and received funding for ArmaLite to build ten more of these new rifles, with the stated objective of replacing the four-month-old M14. What remained to be done before military development and trials could begin was now the production of a better-performing cartridge.

Stoner had already concluded that the .222 Remington cartridge itself was not necessarily the problem, but that a more aerodynamic bullet was needed. All previous loadings in this cartridge used relatively blunt soft-nosed, flat-based bullets, which were ballistically inefficient. He thus drew up plans for a 55-grain bullet, essentially a homologue of the projectile used in the standard 7.62x51mm NATO M80 ball round. It had a very sharp pointed tip, and a heavily tapered, boattailed base. He presented his plans to the Sierra Bullet Company, located in nearby Whittier, California, and they made up a substantial quantity of bullets from his blueprints. He then approached Winchester and Remington—both prominent firearms and ammunition manufacturers—about the feasibility of loading existing .222 Remington cases with his new bullet and slightly more powder, to produce higher velocities. They both balked initially, as the bullet's shape and dimensions would result in chamber pressures that were too high, and thus unsafe for use in existing .222 Remington rifles.

Remington, however, agreed to design a new case that would achieve Stoner's desired velocity with his new bullet, and did so by simply lengthening the .222 Remington case from 1.700" (43.18mm) to 1.7650" (44.7mm), to allow for a bit more powder

93. Right side view of a slightly later XAR-15 from the Institute of Military Technology collection. The fiberglass stock has been fitted with a lace-on aftermarket leather

"boot".

The front sight blade is missing, and an open three-prong flash hider has been added to the muzzle.

photo from *Small Arms Review* Vol. 11 No. 6, courtesy Dan Shea and Robert Segel

capacity. The new round was originally designated the ".222 Special", soon marketed commercially by Remington as the ".223 Remington", both to allow civilian shooters to take advantage of its increased performance, and to make sure that no one would attempt to load the new round into a .222 Remington-chambered rifle. The new bullet/cartridge combination achieved in a 20" barrel the desired performance at 500 yards toward which ArmaLite was striving.

With the enthusiastic approval of General Wyman and the Infantry Board, ArmaLite set about building ten (eventually seventeen) of these new rifles, and procured 100,000 rounds of .223 caliber ammunition from Remington. Seeing the fantastic potential of this new version of their space-age battle rifle, this miniature version of the AR-10 was designated as the AR-15.

# The AR-15 Circa 1958

94. A range test of the AR-15 as it had evolved by 1958. The original caption reads as follows:

"On the range, Maj. Eugene M. Lynch instructs Lt. Col. Robert Vallendorf on how to fire the new ArmaLite AR-10 [sic] rifle now being developed by the Combat Develop-

ment Experimentation Center, 31 March, 1958."

Note the original 25-round magazine, the cocking handle located within the carrying handle, the round, one-piece fluted fiberglass handguard, and the plain muzzle.

photo from the Dr. Edward C. Ezell Archive

Although not the topic of this book, the story of the AR-15 is inexorably linked to that of the AR-10. Suffice it to say, the AR-15, later given the military designation M16, has become the longest-serving standard issue infantry rifle in American history, burying the rifle that had unfairly beaten the AR-10 at Springfield, and making the M14 the shortest-serving standard U.S. military rifle in history.

The unprecedented success of the M16/AR-15, both in the militaries of numerous nations and in civilian-legal semi-automatic-only format as "America's Rifle," is a testament not only to the genius of Eugene Stoner and the team at ArmaLite, but to the true masterpiece that was the original AR-10.

# Path II: The Dutch Partnership with ArmaLite

## Fairchild, Fokker, and the "Friendship Liner"

Before the AR-10 had been delivered into the prejudicial hands of Springfield Armory, a very interesting agreement had been reached between two unlikely partners. While maintaining great personal interest in the firearms branch of his company, Richard Boutelle was still mostly occupied with the aircraft side of Fairchild's business. In December, 1955, however, discussions between Fairchild and another preeminent aircraft company would see these two facets of Boutelle's diversified empire collide. Fokker Aircraft of Holland, founded by Anthony Fokker, who had designed the famous WWI German fighters which bore his name, had licensed plans for a new high-capacity (for the time) airliner dubbed the F-27

to Fairchild in 1952; a relationship that called for ongoing discussions and research.

The F-27 "Friendship Liner" was still very much a work in progress when Fairchild signed their development and manufacturing contract with the famous Flemish fighter designer on August 29, 1952. Airframe and materials research still needed to be conducted, production drawings required overhaul, contractors had to be found, the assembly line had to be set up, and customers had to be sought. Over the succeeding years, it became settled between the firms that 1957 would see the first standard-production F-27s, and buyers were solicited from all over the world.

## First Contact with Artillerie-Inrichtingen

In the fall of 1955 final talks were taking place at Schiphol (Holland's major airport, serving the Amsterdam area), and on November 24, 1955, Boutelle flew over to join his colleagues. There, from some like-minded Dutch counterparts, he learned that the Dutch military had been testing the FN FAL since 1954, but were not overly pleased with it. Furthermore, Frits J. L. Diepen, one of Fokker's chief negotiators, was a former high-level employee of Artillerie-Inrichtingen, the Dutch State Arsenal, and was still a close friend of its director, Friedhelm Jüngeling. The personal contact formed between Diepen and Boutelle at Schiphol would prove to be one of the most critical turning points for the AR-10's later development.

Just over a year later, Jüngeling, a connoisseur of current events from around the world, was reading the December 3, 1956 issue of TIME magazine. To his surprise and delight, he witnessed in it a large photograph showing a man firing a radical and futuristic new rifle. Reading English well, the director of the Dutch national arsenal gleaned the startlingly-attrac-

tive statistics of this new military rifle. He also recognized the name, if not the face, of Richard Boutelle, the American aircraft executive who was shown firing the weapon.

Jüngeling had been well informed of the progress in his nation's search for a new rifle to replace the American surplus M1 Garands that had been standard equipment since the end of WWII, and had also been kept well apprised of the developments in rifle adoption in the United States, through his friends and competitors at Belgium's Fabrique Nationale. Just like the forces he worked to equip, he was not overly impressed with either the Belgian FAL or the Spanish CETME, both of which had been trotted out for demonstrations in Holland. In the AR-10, however, he saw something truly novel and promising, and he asked Diepen to put him in touch with Boutelle.

In their correspondence, Jüngeling learned of ArmaLite's special status as a design house, without the funding or the charter to actually undertake quantity production of its inventions. The entity of which he was a director, Artillerie-Inrichtingen of Hembrug-Zaandam, had been made a *Staatsbedrijf*, or state-owned enterprise, in the period leading up to WWI. This status in Dutch industry was unique, in that while the government maintained a controlling interest in such organizations, they received most of their funding from sales of their products made on the open market or to the government, and did not rely on operating subsidies like most state-owned enterprises.

Having been established just thirty years after Holland gained independence from Spain in 1648, Artillerie-Inrichtingen, located by the start of WWI at Hembrug-Zaandam, had been producing arms for the Dutch forces since the mid-Renaissance. Its long and storied history gave the Fairchild board and ArmaLite staff confidence in its ability to economi-

cally produce high-quality copies of the AR-10, and the deal was sweetened by assurances on Jüngeling's part that if it were produced domestically, the Dutch military would almost certainly adopt the AR-10.

Originally established to produce and sell infantry rifles as well as ammunition and artillery pieces to the Dutch government, the firearms manufacturing arm of the concern had fallen on hard times since the end of WWII, with only the artillery and ammunition branches receiving attention. There was, consequently, a vast store of personnel skilled in small arms production who had either been idle, or involved in other aspects of the company outside the area of their original expertise. In fact, even the ammunition and artillery divisions had had to supplement their work with agricultural implement production.

# Licensing Production of the AR-10 in Holland

Sensing a golden opportunity, Jüngeling sought to bring the firearms division back into the fold and unite all three A-I branches more closely, by acquiring the production rights to the AR-10. The discussions between himself and Boutelle progressed smoothly through the winter and spring of 1957, and by May things were looking quite promising for the near future. At this time, Boutelle received approval from Fairchild's board of directors for a deal between himself and Jüngeling, under the terms of which the Dutch State Arsenal would receive the sole worldwide production license (excluding rifles built by ArmaLite itself) for the ArmaLite AR-10. The new agreement reached was announced through a press release on June 4, 1957.

The timetable for the beginning of regular production was set as January 1, 1958. The Fairchild press release that summer actually stated that in addition to sole worldwide production rights (aside from ArmaLite-Fairchild itself), Artillerie-Inrichtingen would have sales rights in the Netherlands, Germany, France, Italy, Austria, Belgium, Luxembourg, Spain, and Portugal.

This would change, however, as the Dutch manufacturing concern, living on a tight budget under the watchful eye of its majority shareholder, the Dutch government, would realize that it did not possess the staff, facilities, or know-how to mount a sales and demonstration campaign sufficient to satisfy the lofty goals of the agreement. This was all taken into account in the following month of polishing of the deal, which was finally and formally signed in Paris on July 4, 1957, under which a new sales

95. The trademark logo of Artillerie-Inrichtingen, emblazoned on the main building of the works when they were rebuilt just the year before the AR-10 production deal was worked out and the Dutch began gearing up for series manufacture.                    Marc Miller collection

arrangement was to be worked out subsequent to the signing of the deal.

The actual signing ceremony, on American Independence Day, was conducted with suitable pomp and circumstance at an historic site. In the former apartment of the legendary French actress Sarah Bernhardt of the late 19th century's *Belle Epoque* artistic period, both Boutelle and Jüngeling signed the document with a custom-crafted golden pen, on which were engraved the twin emblems of Fairchild (Pegasus in flight) and Artillerie-Inrichtingen (a stylized 'A' and 'T' joined together in a delta-shape with rounded corners). This pen was later presented as a gift to George Sullivan, to commemorate the successful conclusion of this important step forward for ArmaLite.

## A Gift for Crown Prince Bernhard

PRESENTED
TO
H.R.H. PRINCE BERNHARD
BY
ARMALITE
HOLLYWOOD. CALIFORNIA.

96. Right side view of an ArmaLite AR-1, sporting a tan stock and matte-finished grey aluminum receiver. Compare with fig. 13: the promise of this alloy receiver development, stated in the brochure (fig. 2), had by this time been fulfilled.

Inset, right: inscription engraved on the top of the barrel denoting presentation to Prince Bernhard of The Netherlands in 1957.                Institute of Military Technology

97 (below). Grinning from ear to ear, Crown Prince Bernhard inspects the new, custom AR-1 presented to him by Fairchild president Richard Boutelle at the re-opening ceremony of the Small Arms Division of the Artillerie-Inrichtingen works on August 12, 1957.

Smithsonian Institution collection

## Unlimited Possibilities for the AR-10 Seen in the New NATO Alliance

The possibilities stemming from large-scale AR-10 production in Europe seemed practically unlimited. Most of the nations in the new NATO alliance had not yet selected a service weapon in the new universal caliber, and were still issuing obsolescent bolt-action rifles dating from the turn of the century. The last standard equipment the Dutch themselves had produced was a series of Model 1895 Mannlicher 6.5x54mm bolt-action infantry rifles and carbines, which had been manufactured by Artillerie-Inrichtingen.

In fact, only Britain, Belgium, Canada, and the United States had adopted new standard infantry arms by the time the AR-10 deal was finalized between Fairchild and Artillerie-Inrichtingen in 1957. Britain and Canada had chosen inch-measurement versions of the FN FAL, with Belgium, of course,

sticking with the FAL in its original Metric form. The U.S., as discussed above, had adopted the M14.

With the other members of the new NATO alliance unsure about which weapon to procure, Boutelle and the ArmaLite staff reasoned that production at the Hembrug-Zaandam plant would lend an air of credibility and respectability to the AR-10 which would eclipse the prestige of all other European designs. Historically speaking, the famous and vaunted Fabrique Nationale, one of the best known arms manufacturers and designers on the continent, was but a mere infantile upstart compared to Artillerie-Inrichtingen. Indeed, the Dutch state arsenal had been turning out military equipment three hundred years before FN had even been a gleam in the eye of King Leopold II.

## A Brief Modern History of A-I and the Dutch Military

Excerpts from an anonymous monograph held in the archives of the Dutch National Defense Museum, probably written by a former employee of the arsenal, briefly describes the history of Artillerie-Inrichtingen during and shortly after WWII as follows:

*When the Netherlands were overrun by the Germans in May, 1940, it was obvious that A-I was potentially an attractive spoil for the enemy, and the employees were successful in turning the company into a manufacturer of agricultural tools and machinery, in order to distance itself from becoming a part of the enemy's war production effort. Later, in 1944, under pressure from the Germans to produce weaponry, the entire roster of personnel went into hiding, so that at the end of the war in 1945, the company had to be rebuilt. However by that time the Dutch armed forces, although willing to resume their old relationship with A-I as the State arsenal, could get all their needs for little money from the huge Allied army dumps, and so the willingness to restore relations could not immediately be turned into orders, and A-I remained a producer of tools and agricultural machinery. The general management even claimed that it never again wanted A-I to produce military weaponry, which was understandable so shortly after the devastation of the war, but this attitude caused some anger among the armed forces, and went on to influence the relationship between A-I and the military.*

*Nevertheless the armed forces needed capacity in order to help them to recover and rearm, and since A-I was the only Dutch company which employed such specialists, they reluctantly began to focus on arms production, and also began to produce ammunition as well, so that by 1957, when the AR-10 became a candidate for adoption by the Dutch forces, it seemed that the former anti-militarist sentiments had been forgotten.*

Meanwhile, new NATO members and other staunchly anti-communist states such as Austria, France, Germany, Norway, Portugal, Italy, Denmark, and of course Holland, were all either testing different rifles or entrenching themselves behind their own new domestic designs. Each of these countries, as well as the newly-independent nations of Southeast Asia and Africa who were throwing off the chains of colonialism in the aftermath of the war, would need new, advanced weaponry with which to equip their armed forces.

In this heady atmosphere, ArmaLite and Artillerie-Inrichtingen moved forward enthusiastically to perfect the AR-10 and offer it to all potential buyers. Although disappointment would dog the AR-10 in this endeavor as well, the successes it did enjoy were resounding, and in fact it could be said that the small number produced, and consequently the extremely élite and secretive units to whom they were issued, lend a great aura of mystery and legend to the later history of this fantastic rifle.

# The Quest for a "Family" of AR-10s

Although possibly no one but Eugene Stoner himself knew for sure when the idea of the AR-10 becoming a "family" of weapons first dawned, this effort was well under way by the middle of 1957 with working (though not necessarily reliably in all cases) first prototypes of some models having been produced.

## The AR-10 as a Light Support Weapon

At the time, the U.S. Army was not alone in seeking a more versatile, multi-use weapon to fulfill all or most of the basic infantry functions. The Belgians were offering their FALO (*Fusil Automatique LOurd*, or Heavy Automatic Rifle) to fill the squad automatic rifle role. The FALO was a standard FAL (*Fusil Automatique Leger*, or Light Automatic Rifle) with a much heavier barrel and more robust muzzle brake, fitted with a folding bipod. Every other part was exactly the same, except for the magazines which, while interchangeable with regular 20-round FAL magazines, were longer, holding 30 rounds each.

The British and Canadians followed suit with heavy-barrel versions of their own adopted FAL variants, but despite their extra weight and greater magazine capacity, their performance was disappointing, with full-automatic fire being largely inaccurate and thus wasteful.

In comparison, the controllability and low recoil of the AR-10, thanks to the admirable concentricity of its action, basic straight-stocked design, and efficient muzzle brake, meant that it actually *could* fulfill the requirements for both a standard infantry rifle and a light support weapon, and Stoner set about devising the most reliable way this could be done, with the secondary goal of making the rifle capable of firing either from a belt or magazine, without modification.

Several different means of allowing sustained belt-fed fire would be explored over the next four years, in both California and Zaandam.

## The Mechanical Ammunition Supply System (MASS)

Formulating this concept of AR-10 modularity even before the standard rifle version had been presented to Springfield Armory, ArmaLite set about reworking the AR-10 in mid-1956 to adapt it to an entirely new and ingenious system of feeding.

Seeing the potential disadvantages of modifying the bolt and carrier to perform the indexing actions necessary in a belt-fed firearm, Stoner first opted to try to avoid this engineering complication. In order to eliminate the drawbacks of both standard magazine and belt feeding, ArmaLite opted for a wholly new loading apparatus, for which no precedent then existed in small arms design.

Desiring to use the standard magazine well, thus retaining the ability to load the rifle with standard magazines if the primary feed system were damaged, ArmaLite subcontracted this experimental development to Armament Components Incorporated of nearby Santa Ana, California.

98 (right). An artist's concept of the MASS system as applied to the AR-10.

From Report No. 2-57, "Mechanical Ammunition Supply System for ArmaLite AR-10 Weapon". Armament Components Inc., January 4, 1957.        editor's collection

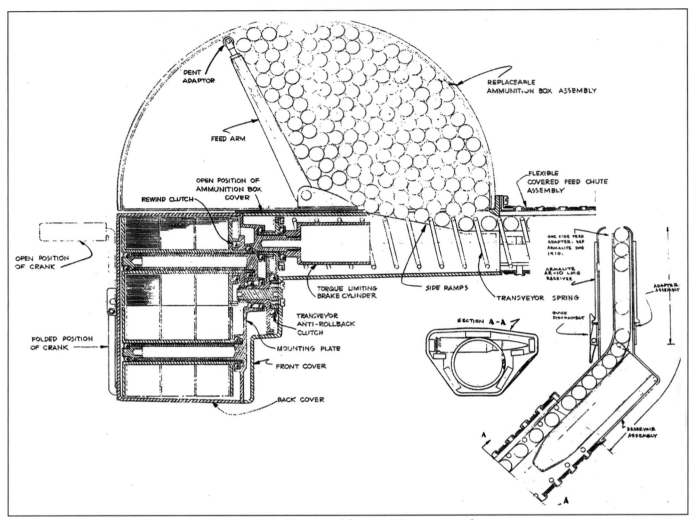

99. A diagrammatic view of the MASS system, with components labeled.

According to the report submitted by Armament Components, Inc., dated January 4, 1957, the MASS system "incorporates in one man the capabilities of a light machine gun squad without the need for using belted or linked ammunition."                    editor's collection

The proposal developed under this arrangement, presented in a report dated January 4, 1957, was known as the Mechanical Ammunition Supply System, or MASS. This complex assembly consisted of a flat, half-circular drum, intended either to be carried on the soldier's back in a harness, rather like a backpack, or placed on a specially-designed tripod, in which "250 rounds or more" of non-linked 7.62mm ammunition were stacked against the flat bottom and curved top, with all of them pointing in the same direction. At one corner of the semi-circle was a funnel-shaped opening through which rounds would enter the "flexible feed chute", which connected the ammunition box to the magazine well.

The feed lever, also known as the "feed arm," controlled by a powerful driving spring inside the mechanical housing which sat against the flat side of the ammunition chamber, had its hinge and axis on the flat bottom of the ammunition compartment, and under spring pressure pushed from the side opposite the feed chute, forcing the rounds that way and moving closer to that end as the number of cartridges inside the chamber decreased.

The feed chute was covered with a flexible sleeve, intended to keep dirt or other contaminants from entering and clogging it. The end of the feed chute opposite the ammunition box was fitted with a cutout and feed lips which were inserted into the

AR-10's magazine well, and held in by the rifle's standard magazine release. Thus the weapon could feed just as it did from a magazine, while the spring-loaded lever in the box would force rounds into the funnel and through the feed chute. Once inside the feed chute, a coiled spring that ran the entire length of the chute would catch individual rounds between its coils. The main driving spring inside the feed box was wound up to cause this long rotating spring to

work just like an Archimedes' Screw, pushing the rounds towards the weapon, regardless of its position or alignment. This rotation provided the pressure that the spring and follower would in a conventional magazine, and conveyed fresh rounds from the vastly increased supply carried in the ammunition box to the top of the chute, to be stripped into the chamber by the bolt.

## An Improved Conveyor for Belted Cartridges

The proposal for the ingenious but extremely complex and unwieldy MASS system was superseded by another much simpler development which consisted of a flexible conveyor constructed of a series of articulated metal links, designed and patented by a Ralph A. Van Fossen of Los Angeles, and assigned to Armament Components Inc. of Santa Ana. A patent, applied for on July 26, 1956 and granted as U.S. patent no. 2,838,154 on June 10, 1958, titled "Flexible Conveyor Chute", described this as an articulated pathway for cartridges held in standard non-disintegrating link belts, guiding them from a boxlike cartridge container mounted on the soldier's back into the left side of a specially-modified AR-10 rifle with a belt feed apparatus installed.

In the words of the patent, "In certain applications and uses of articulated conveyor chutes, it is desirable or essential that the chutes be capable of substantial flexing without impeding the movement of the articles conveyed."

The chute was made up of a series of rigid, interconnected metal links, hinged to one another, which allowed "substantial lateral and longitudinal flexibility", while still providing a rigid channel through which a belt of linked ammunition could flow unimpeded from the feed box into the weapon.

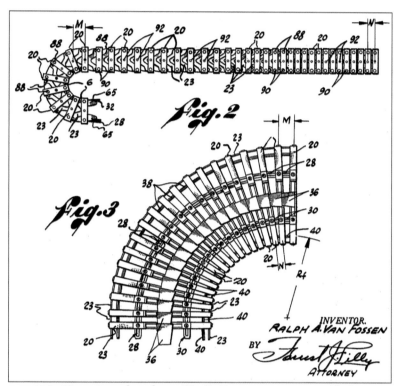

100. Figs. 2 and 3 from U. S. Patent 2,838,154 titled "Flexible Conveyor Chute", granted on June 10, 1958, showing how the chute components could flex along both of their axes without compromising the feeding of the belted cartridges. U.S. Patent Office

101. A still from the promotional film produced by the ArmaLite Division of Fairchild Engine and Airplane Corporation titled "The Modern Combat Rifle", showing Gene Stoner firing an AR-10B rifle modified to feed ammunition held in a non-disintegrating belt through the patented conveyor system shown above.

Note the emptied links, falling to the ground through the aperture cut in the right side of the lower receiver.

editor's collection

## Further Versatility: Scoping the AR-10 "Sniper Rifle"

By virtue of its remarkably light weight, even the standard-length AR-10 could be said to fill much of the role formerly held by the M1 Carbine, and with its extremely high rate of fire and excellent controllability (for a weapon firing a full-power rifle cartridge), it could foreseeably also supplant the submachine guns that were at the time ubiquitous in the world's armed forces.

Not content with merely creating a variation that could do all the jobs that its boosters claimed (though probably never actually believed) the M14 could do, however, Stoner, with input from George Sullivan and Charles Dorchester, both proponents of the utility of scoped weapons for greater accuracy,

quickly realized that with its rugged carrying handle, and ejection port located on the side instead of the top, the AR-10 was singularly well-suited to being adapted for use as a sniper rifle.

While some nations like Germany, the Soviet Union, and Britain had long made organized and effective use of snipers on a strategic scale, the United States had never done so. While the Soviets in particular had fielded entire units of snipers on the Eastern Front against the Germans during WWII, such specialists were only employed in a tactical capacity in the U.S. Army, with a single infantryman known as a "designated marksman" being attached to certain infantry units to provide long-range, accu-

102. Left side view of an early ArmaLite AR-10B, serial no. S1004, with carrying handle modified for the attachment of a telescopic sight. The "S" stood for "sniper," indicating that this model had received its own special designation.

Note the rear iron sight and its elevating wheel have been removed.                    Institute of Military Technology

rate fire in support of his team. It was for this reason that a dedicated sniping rifle had not been seriously sought or designed for U.S. armed forces.

The term "sniper rifle" is used far more frequently than is warranted today, and even within the community of firearms enthusiasts the exact definition of this term is not well understood. It is common for anyone from laymen to experienced hunters and policemen to pronounce any rifle fitted with a telescopic sight a sniper rifle. In actuality, a sniper rifle must, in addition to being capable of mounting a telescopic sight, fire a military rifle round, and be capable of uncommonly good accuracy. Although there is no hard-and-fast rule of what constitutes "sniper rifle accuracy," it is well established that one-and-a-half minutes-of-angle or arc-minutes is the absolute minimum accuracy required. This means that the rifle is capable of firing strings of consecutive shots, usually of between five and ten rounds each, and have all the shots hit the target and form a group that constitutes 1.5 minutes-of-angle for the distance fired. In other words, a rifle capable of 1.5-MOA accuracy will be able to fire at least five consecutive shots at 100 yards into an area with a diameter of one and a half inches or less. This is the absolute recognized minimum, and most every sniper rifle today is generally capable of considerably better accuracy, sometimes approaching ¼-MOA.

Although the idea of the sniper as a strategic force multiplier on the battlefield was gaining traction, the concept of equipping him with a versatile automatic rifle instead of a bolt-action weapon had not really taken root in America. However the AR-10, with its novel direct impingement gas system, which did not employ a piston or operating rod, was in many ways capable of attaining an equal or better pattern of predictable barrel harmonics than a bolt-action weapon, and thus was a potential candidate for being adapted to the sniper role. Although it is a story for a later chapter, it can be said here that the first dedicated semi-automatic sniper weapon ever fielded by the U.S. Armed Forces, which was adopted in 2008, bears a striking resemblance to the AR-10 rifle of the 1950s.

Fully aware of the American military establishment's intransigence, but also hopeful at the potential for progressive attitudes in the burgeoning military forces around the globe, Stoner and the ArmaLite team decided to add one more entry to the list of specialist small arms roles that could be effectively filled by the AR-10.

The process of converting one of their AR-10Bs to mount a precision telescopic sight was sublimely simple. All that had to be done was the cutting down of the high sides of the carrying handle, and the machining of two attachment points where scope rings could be firmly secured to the resulting flattened portion on top of the handle. It was never intended that this would be the final and permanent means of adapting an AR-10 to the sniper role, but it was done simply to show that it *could* be accomplished.

The modified AR-10 "sniper rifle" was provisionally fitted with a Bushnell sporting rifle scope, much like those that had topped Sullivan's and Dorchester's AR-1s during the early pre-ArmaLite days of S-F Projects. Just as a newer, sturdier, and more production-friendly means of attaching a scope was intended for the future, a more rugged, compact, lower-magnification, higher-end military telescopic sight was slated to take the place of the sporting scope. The Bushnell optic sat in its place of honor atop the first scoped AR-10 as an indicator of how easily and economically the standard AR-10 could be turned into a designated sniping weapon with no further modifications being necessary. Although no precise results of contemporary accuracy tests of this rifle can be located today, it was soon claimed by Fairchild that "firearms experts are much impressed by the fact that this weapon, with the firepower of a machine gun, has all of the inherent accuracy of a fine sniping rifle."

## The "Family" Concept is Announced

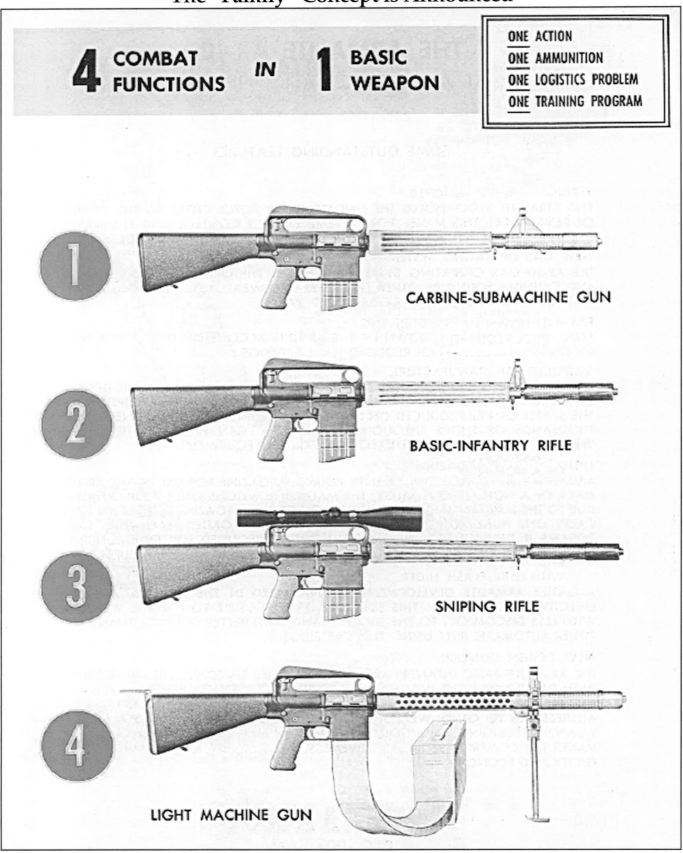

**4 COMBAT FUNCTIONS** IN **1 BASIC WEAPON**

- ONE ACTION
- ONE AMMUNITION
- ONE LOGISTICS PROBLEM
- ONE TRAINING PROGRAM

1. CARBINE-SUBMACHINE GUN

2. BASIC-INFANTRY RIFLE

3. SNIPING RIFLE

4. LIGHT MACHINE GUN

Within just a few short months of the U.S. Army's adoption of the M14 in May, 1957, the plucky and undeterred design team at ArmaLite had expanded their singularly-advanced battle rifle into a full complement of infantry weapons. It was announced in the sales literature which began to find its way to military institutions all around the world that the AR-10 really could do it all. Its characteristics of low weight, recoil control, and modularity lent them-selves to the extreme adaptability of what was at the time a new concept in small arms; the "weapons platform."

As shown in fig. 103, an advertising brochure published in the fall of 1957 displayed ArmaLite's optimistic attitude concerning the precocious set of designs they had developed, showcasing the subset of the young company's weapons that were designed specifically for military applications.

## Some "Family" Developments Still Only on Paper

It is unclear whether any but the standard infantry and scope-mounted sniper versions of the multi-role AR-10s trumpeted by Fairchild in 1957 were ever actually functional. Certainly the whereabouts of any actual examples of these variants remains unknown. They have never been confirmed as fired, and no other photographic or documentary evidence of their existence has ever come to light.

It is clear, for example, that the light machine gun version, at least as it was depicted in the litera-ture, would not have worked. The carbine/subma-chine gun model would appear to have had the same-length barrel as its standard counterpart, but with a shortened handguard and a modified AR-10A recoil compensator set back along the barrel.

The actual sniper weapon does exist, and is still in excellent condition today. It appears different from the artist's rendering, in that it did not have the shaved front sight base as shown, and was intended to be able to accept a variety of standard sporting telescopic sights.

The ArmaLite staff was subsequently able to modify the bolt carrier of an AR-10 with a cam track and add a belt-feed mechanism which would allow the rifle to be converted to belt feed. However, at the time this sales brochure was put together, internal company memoranda confirm that this effort had not yet been seriously undertaken.

# Formulating the AR-10 Sales and Marketing Scheme

With the licensing agreement finalized with Ar-tillerie-Inrichtingen, and large-scale production of the AR-10 well on track, Fairchild began to contem-plate how it would go about demonstrating and selling the AR-10 to the militaries of friendly nations around the world. With the shock of the U.S. military rejection of the AR-10, along with the announcement of M14 adoption in May, 1957, Fairchild was left in the lurch, with Boutelle guardedly admitting to

03 (previous page). A page from the ArmaLite brochure published in the fall of 1957, displaying all four members of the AR-10 "Family."

This marks the first time that all of the incarnations of the AR-10 weapons platform were publicized. The carbine 1) and sniper (3) versions were straightforward modifica-ions of the basic rifle (2). The actual sniper version of the AR-10B (fig. 102) appears rather different from that shown here, as it has a normal front sight base.

The LMG version (4), shown feeding from a standard ammunition box through a chute that connects to the magazine well, would have been a nonfunctional artist's conception only.          Institute of Military Technology

friends that there was absolutely no concrete plan in place for how the company would go about market-ing the AR-10 abroad. They had been so encouraged by the *tours de force* of 1955 and 1956 that they had essentially put all their eggs into the basket of Ameri-can military adoption, instead of thinking ahead to unfamiliar and rapidly-evolving foreign markets. Af-ter all, ArmaLite was not a large part of Fairchild's business, and being so divergent it was often seen as an afterthought by much of the management and board, who were aviation people at heart.

## Courting the Merchants of Death

Once again bridging the gap between the worlds of aircraft and firearms with his personal interests and connections, was Fairchild president Richard Boutelle. Being a dyed-in-the-wool firearms enthusi-ast in his free time, Boutelle would spend his week-ends shooting at his farm in the hills of Maryland near Hagerstown. Guests were often invited, and the fact that many of these were well connected in the business and defense worlds was probably more of an afterthought to Boutelle who, like many readers of this book, just liked going shooting with some friends on nice days.

It was just one such afternoon outing in the spring of 1957 that drew together the paths of Boutelle and another figure who was destined to become relatively famous, or infamous, depending on whom one asks. This man was Samuel Cummings, who found his way to the Boutelle estate by a roundabout route that had begun a year earlier.

Another early devotee was William B. Edwards, the editor of the influential *Guns* magazine, who had become familiar with the AR-10 from its earliest days. Eugene Jaderquist, described as *Guns* magazine's publicity writer, had recently visited ArmaLite's machine shop in Hollywood, and had drafted what would become the most famous introduction of the AR-10 to the American shooting public. In his article titled "Is This the Next G.I. Rifle?", which appeared in the March, 1957 issue of the magazine, Jaderquist made it perfectly plain that he felt that the AR-10 was the most prochronistic military rifle in existence, which opinion still stands today, over half a century later.

Interestingly, it has been theorized that "Eugene Jaderquist" was actually Bill Edwards himself, using one of his many pen names. Although *Guns* magazine became a very well-regarded and authoritative journal, it was then still in its infancy and did not have the budget to hire a large staff. It became common knowledge in later years that most, if not all, of the articles published in the magazine, certainly during its first few years of existence, were actually penned by Edwards, using various pseudonyms.

## Introducing Samuel Cummings, and the Interarmco Phenomenon

Edwards was entranced with the AR-10, and when corresponding with his contact from an article he had written the previous year on the successful upstart arms trading firm International Armament Corporation (Interarmco), the ArmaLite rifle was still fresh in his mind. He freely admitted years later that the 1955 Interarmco article was "a puff piece," meant to bring Interarmco to the attention of potential foreign clients. His contact at the new firm was none other than its "vice president," Samuel Cummings.

In his trademark sly and self-deprecatory style, Cummings later explained to interviewers that he had taken the title of vice president in order to give the false impression that the company he had founded in 1953, while he was still working for the CIA, had more employees than just himself. Cummings, who would later go on to be recognized as the top international arms merchant of the second half of the 20th century (partly because of his habit of working within the law and not doing under-the-table deals), was steadily building his fortune and reputation not only by buying and selling military surplus but by his cunning eye for new and promising weapons.

Cummings had become an enthusiast of small arms at the age of five, when he discovered and acquired with the help of his parents an original Maxim MG08; the German version of the world's first self-loading truly automatic weapon. Over the next few years, being a precocious early reader, he became well versed in the weapon's operation, restored it, and by the age of 10 was operating it at local firing ranges. After attending Quaker high school he served in the Army during WWII as a small arms analyst at Fort Lee, Virginia, and then attended George Washington University, graduating in 1949, after which he needed a job to see him through law school.

By this time the Korean War had broken out, and the fledgling CIA realized that they needed weapons analysts to examine and identify the Soviet ordnance that was being captured from the communists in Korea. Cummings, who had amassed an impressive fund of knowledge of small arms through his avid interest in military history and his growing gun collection, was interviewed by a panel of three officers and hired on the spot. He was given a full-time job in the small arms division of the Office of Scientific Intelligence, where he was needed to identify arms captured in Korea.

The following brief excerpts are taken from the captivating 1983 biography of Cummings and his company titled *Deadly Business* by Patrick Brogan and Albert Zarca:

> . . [Cummings] would study photographs, or the weapons themselves and would be able to tell at once what they were and how old they were, and then to work out where they came from . .
>
> At one stage, a note landed on his desk containing a report on the subject [of surplus weapons left over from World War II] prepared with the aid of

*American military attachés in all the embassies in western Europe.*

*The attachés had been asked whether there were considerable quantities of German weapons left in Europe [and] had unanimously replied that there was no German surplus left in Europe.*

Cummings knew better. He had visited Europe on his summer vacation with some friends while still a college student in 1948, and had personally noted that

*. . the wreckage of war was everywhere. In the Falaise Gap, six German divisions [had] abandoned all their equipment as they fled, and it was still there, four years later. In the bunkers of the Atlantic Wall, where Cummings and his friends stayed when they visited the beaches, the guns were still in place, and there were cases of shells, cartridges, grenades and rockets stacked from floor to ceiling . .*

*. . The fields were littered with small arms, rifles, pistols, and machine guns that the French Army had not yet collected . . This prodigious harvest of the war lay undisturbed . .*

Cummings was therefore surprised by the claims of the U.S. military attachés that there were no existing stocks of surplus weapons in Western Europe:

*. . Cummings knew that [the attachés] were wrong. He had seen the stuff himself, three years earlier, stacked up in the fields. His studies in the CIA's archives had furthermore shown that great quantities of German weapons were still in the hands of the various western European armies or in the depots where the Germans had left them in 1945. He therefore composed a short memorandum stating all this, politely but firmly pointing out that the attachés were wrong . .*

*A few weeks later Cummings was abruptly summoned to explain himself [to] no less a personage than Allan Dulles himself, the deputy director of the CIA.*

*. . Cummings was treated courteously and listened to attentively. Dulles was a shrewd judge of men. For the first time, Cummings had to sell himself to a man of real seniority and importance, and succeeded triumphantly . .*

His 1950 report on the actual state of military stocks left behind from the war was paradigm-shifting, and it was not only the military and intelligence authorities who took note. Cummings immediately saw the commercial value of these finds, and although he was not yet financially in a position to purchase them himself, he was sent by the CIA on an undercover mission to Europe, posing as the assistant to a Hollywood director of photography. Although he knew nothing of firearms, this director, a part of his cover, actually traveled with him, and they went about buying up stocks of surplus weapons with the stated objective of using them as props for filmmaking. This CIA operation was actually intended to acquire large amounts of weaponry with which to re-equip the Chinese National Revolutionary Army, the armed forces of the Republic of China, which had just the previous year been routed by the victorious communists, and forced to seek refuge on the island of Taiwan. The stated aim of the Republic's ruling Nationalist Party, which never came to pass, was to rearm and return to the mainland as quickly as possible, to drive Mao Tse-tung from power.

After the success of Cummings's "filmmaking" adventure, together with another coup in Costa Rica where he acquired more armament for his parent agency, Cummings made up his mind that he would enter the international arms business for himself.

In 1953 Cummings declined a career position with the CIA and started his own company, International Armament Corporation, through which he contacted military, political, and police officials all over the world, offering to purchase their stocks of surplus weapons. Within a year he had made his first deal for a cache of confiscated weapons in Panama, on which he made $20,000 for a week's work. From there, his business began to boom, especially from his contacts within the military juntas ruling much of Latin America, and he quickly became a celebrity in the U.S. arms business.

Remaining friends after their several meetings in the course of editing and publishing the "Jaderquist" article, Bill Edwards continued to keep in touch with Cummings. In some of their discussions in December, 1956, certain characteristics of the AR-10 were mentioned, which immediately piqued Cummings's interest. He was himself a great devotee of the efficacy of the German FG42, with its straight-line stock and modular design, and from even a cursory oral description, recognized the similarities between the two weapons. Their conversations continued over the next few months, and Cummings took it upon himself to learn more about this new rifle that fascinated him so.

It is not clear whether the fortuitous shooting trip that brought Cummings irretrievably into the orbit of the AR-10 took place just before or just after the U.S. Army's announcement that it had adopted the T44 as the M14, but it is known that it occurred in the spring of 1957, after the Springfield Armory tests of the AR-10 were concluded, but before the licensing deal between Fairchild and Artillerie-Inrichtingen was finalized. Cummings, along with several other notable figures in the military and civilian firearms fields, received invitations at that time from Richard Boutelle to participate in an afternoon of shooting at his property outside Hagerstown, Maryland.

Edwards recounted for the *AR-10'er* newsletter in April, 1984 how he and Val Forgett III, a veteran and recognized world expert on automatic weapons, picked up Cummings and Marine Colonel George M. Chinn, and drove to Maryland to join the party at Boutelle's farm. Cummings had brought with him his own personal FG42 to test against the mysterious AR-10, while Forgett had stowed in his trunk an 1889 "world standard" Maxim gun, serial no. 525, which he had bid for and won at the estate auction of the inventor's brother Hudson Maxim in June, 1953.

Cummings also packed a German MG42 medium machine gun and some experimental German explosive ammunition, just for the sake of an exciting day at the range.

At the Boutelle manor, the awestruck guests looked over a full complement of ArmaLite weaponry, including several AR-1 rifles with blue-colored receivers, an AR-5, and an AR-10. These pieces were presented by none other than Charles Dorchester and Eugene Stoner, who were on hand for the firing demonstrations. Colonel George M. Chinn, who had written the classic, multi-volume treatise *The Machine Gun*, the definitive compendium of automatic weapons, published by the U.S. Navy Department's Bureau of Ordnance, was then perhaps the world's greatest living expert on machine guns and automatic rifles. Chinn was so humbled by his examination of the AR-10 that he remarked to Edwards that "[he] would have walked on [his] hands and knees up here to see this."

All of the men present were intimately familiar with at least a subset of the burgeoning new genera-

tion of automatic rifles, and each compared his knowledge of these types to the AR-10. Cummings felt that the ArmaLite weapon outclassed his darling FG42, not only in weight and ergonomics, but actually in rapid-fire controllability. The bottom-mounted magazine doubtlessly helped make the AR-10 balance better than the German weapon, and the more efficient recoil compensator and more pronounced straight-line stock helped the AR-10 stay on target better, even considering its significantly lighter weight. Later in his life, when asked about what he thought the best weapon he sold was, Cummings would staunchly aver that the AR-10 was the finest infantry rifle ever devised.

Edwards received an impression every bit as favorable as had Chinn and Cummings about the AR-10, particularly when he compared it with his prior experience with the U.S. version of the FN FAL, known as the T48 (fig. 44), which had been undergoing small-scale troop trials at Fort Benning. Edwards had been one of the fortunate few journalists who had been granted the privilege of examining, testing, and writing about the equipment under consideration in the Lightweight Rifle Program. The T48 was the heaviest entrant in the trials, and given its somewhat more ergonomic stock and separate pistol grip, though not a true in-line design like that of the AR-10, it was universally hailed, even chambered for the powerful T65 cartridge, as being the easiest weapon to control—prior to the appearance of the AR-10. The usefulness of the T48 in keeping strings of rapid fire on target when fired as an automatic weapon was accepted by all who tested it as being far superior to that of the T44. However, when comparing his impression of the T48's controllability in the automatic mode to that of the AR-10, Edwards felt that there was no contest. He recounted for AR-10 enthusiasts in an interview three decades later how automatic fire control even with the FAL took a lot of practice, as its tendency was to rear up and start spraying the sky. The AR-10, however, stayed on target for him with almost no effort.

# Fairchild's Plan to Divide Up the Free World

In the late summer of 1957, with the manufacturing licensing contract signed and Artillerie-Inrichtingen gearing up to begin series production of the AR-10, Fairchild aggressively began to pursue strategies for world marketing, with Boutelle taking the lead. Al-

though the original press release had stated that Artillerie-Inrichtingen, as the licensed manufacturer, would have sales rights in a large portion of Europe, this arrangement had as mentioned been altered so that the Dutch factory would now concentrate on the production of the AR-10, and leave the sales to others.

Boutelle was already aware that Sam Cummings of Interarmco was intensely interested in the AR-10. In fact, typical of his aggressive "all or nothing" approach to business, Cummings had already requested that his company be granted the sole and exclusive worldwide sales rights to the new rifle back in January, 1956, before he had even had the chance to shoot one himself. Instead, Boutelle formulated an innovative technique that would ensure more exposure for the AR-10 in a number of different markets by dividing the world into sales areas, and allowing the firm with the most experience in that geographical region a monopoly in that area, which he felt would concentrate the efforts of each agency in the areas they knew best.

Meanwhile, in addition to Sam Cummings, subsequent demonstrations, both at Boutelle's farm and elsewhere had attracted other sales talent, such as Robert "Bobby" MacDonald of Cooper-MacDonald, Inc., an arms sales firm based in Baltimore, Maryland, with offices in Singapore, whose specialty was Southeast Asia; and Jacques Michault of SIDEM International, who has already been mentioned in Chapter One in connection with the AR-10 story.

## More on Jacques Michault

Born and educated in Belgium, Jacques Michault emigrated to America in 1938 and became a U.S. citizen. He served in the U.S. Army during and after WWII, his last posting being a tour of duty in occupied Japan.

After leaving the military he returned to Belgium and began a career as an arms merchant, concentrating on selling commercial arms to U.S. servicemen stationed in Germany. A short notice carried in *The Stars and Stripes* of January 9, 1950 reads as follows:

### Rifles Available in Brussels

*Rod and gun clubs having difficulty in securing rifles and accessories will be interested in knowing that the Sidem SAB plant in Brussels is anxious to contact all American sportsmen groups in Germany who wish to obtain European sporting weapons. Write to J. S. Michault, 18 Avenue des Nenuphars, in Andeghem, Belgium for prices, catalogues, etc.*

Five years later he moved to Bonn, the better to concentrate on the market provided by U.S. servicemen stationed in Germany. A further clipping from *The Stars and Stripes* of March 25, 1955 is excerpted as follows:

*J. S. Michault, export manager of a famous sporting arms outlet, writes the following:*

*"I want you to know that the sales of Firearms International products (FN Mauser, Sako Mauser actions and barreled actions; Star and Unique handguns) to U.S. personnel in Europe, for whom we are exclusive distributors, have been transferred from Brussels to our Bonn office . ."*

**J. S. MICHAULT**

Mr. Michault is Chairman of the Board and our Executive Officer. He was educated in Europe and served in the U.S. Army during the war as an Infantry Officer. He co-ordinates the world wide activities of the group and specialises in small arms and ammunition.

104. A photo of arms dealer Jacques Michault, taken from a 1962 SIDEM brochure.

As we shall see, Michault and Cummings did not always see eye to eye regarding the sanctity of the contract sales territories for the AR-10 assigned to them under the agreement with Fairchild.                                    courtesy Vic Tuff

# Defining "Contract" and "Standby" Territories

The focus of the sales representation contracts was, with some exceptions, worldwide. Two categories of sales rights were assigned: "contract" and "standby." The "contract" areas were those for which the individual firm had negotiated and received the exclusive sales rights.

These did not cover all the available potential customers, however, and Fairchild's Paul Cleaveland had realized that different salesmen attempting to sell the AR-10 in the same country and making conflicting claims of exclusive sales rights would create chaos, which would be detrimental to the successful marketing of the rifle. He therefore formulated the concept of "standby" areas, which consisted of all the remaining areas of the world where Fairchild would allow sales. In addition to the "no-go" areas covered by governmental restrictions on trade, it was recognized that the Soviets, Red Chinese, and North Koreans would almost certainly not be interested in equipping their forces with a rifle chambered for the NATO cartridge.

The "standby" countries were divided evenly among the three sales agencies, based on proximity to their assigned "contract" sales territories, and their previous experience or connections in these areas. The way this system worked was that the salesman who was assigned a certain country as a standby territory had the right to market and sell the AR-10 there unilaterally as if it were contract territory. Unlike a contract territory, however, a different salesman could operate in another's standby territory, as long as Fairchild, ArmaLite, Artillerie-Inrichtingen, and the other sales firms were notified, and gave unanimous consent. This consent was anticipated as being reciprocal, with the party seeking to enter another's standby territory either offering access to another standby area in return, or some other type of consideration, to be decided in the event these rights were exercised.

The "standby" territory system was also meant to deal with the prospective, and indeed inevitable, shifting of national borders and the establishment of new states. This agreement was drawn up during a period of extremely rapid decolonization, where former territories of the European powers were asserting their independence, some with the consent of their erstwhile mother countries, and some through bloody revolution. It was recognized, correctly, that these areas of the globe would provide exceptionally fertile ground for sales of the AR-10, as indeed they did, going on to host almost all of the subsequent combat use of the AR-10.

Cummings, because of his extensive record of successful dealings in Latin America, received as his contract territory exclusive rights to all of Central and South America, along with the island nations of the Caribbean. In Europe, he received exclusive rights in Sweden, Finland, and Norway, which was a member of the NATO alliance. His contract territory also extended somewhat into Africa, with Ethiopia and South Africa rounding out the sales areas for which he had negotiated. For his standby territory, he received many of the remaining African nations, and prospective nations. The already independent countries of Ghana, Liberia, and Nigeria went to Cummings, as well as the Belgian Congo, Portuguese Angola and Mozambique, Rhodesia, Madagascar, and most of West Africa.

Jacques Michault contracted on behalf of his firm SIDEM International, now headquartered in Bonn, West Germany. As he had in the preceding decade become the successful, exclusive marketer for prominent arms manufacturers in such places as Spain, Denmark, Sweden, and France, his reputation was strong as an inside man in the especially promising and lucrative NATO market. Initially, the Dutch were slightly apprehensive about granting the license to SIDEM, but a social visit and shooting trip to allow Michault and Jüngeling to become acquainted solved this. It turned out that Michault was never the problem, but the Dutch, still actively rebuilding their factory from what the Germans had done to it scarcely more than a decade earlier, had had some justifiable trepidation about granting an exclusive license to a German firm.

Once this was smoothed over, Michault's company was awarded rights as the exclusive sales agents for all of Free Europe, except for the portion of Scandinavia already staked out by Cummings's Interarmco. Michault also received the juicy prospect of the Kingdom of Saudi Arabia as his last contract territory. SIDEM's standby territory was first composed of the unassigned nations of Europe and Eurasia, consisting of Greece and Turkey, and went on to encompass the whole of North Africa, starting in the Spanish-controlled Western Sahara, and extending across the Maghreb (with the exception of Libya, which was part of Cooper-MacDonald's contract territory) east to Egypt, the Sudan, and Eritrea, reaching the border of Cummings's East Africa, which started at Ethiopia, just south of Eritrea. Finally, SIDEM's standby territory included the former French, Portuguese, and German colonies of West Africa. Michault's longstanding relationship with the management of Fairchild, and his intellectual and fiscal involvement in the foundation of S-F Projects before it had even became part of Fairchild (or known as ArmaLite) was another reason, in addition to his connections, that his firm of SIDEM International received what was considered the best of all the world markets.

The militarily and politically volatile world of the Far East could not be disregarded either, and opportunities seemed to abound there, as more nations were emerging from colonialism, and with considerably stronger economic, intellectual, and military traditions and foundations than their African counterparts. To represent their interests in this cryptic corner of the globe, ArmaLite signed the firm of Cooper-Macdonald of Baltimore. Boutelle knew of them, and particularly how their chairman, "Bobby" MacDonald, had for the last nine years been representing Colt's Patent Firearms and Remington Arms out of his Asian headquarters in Singapore, selling pistols for the former and rifles, shotguns, and ammunition for the latter, both into civilian markets and to the friendly governments of the region. His expertise and connections in Asia made MacDonald's firm the logical choice for getting the AR-10 into the hands of the military decision-makers of the Orient. MacDonald received contract rights to all the nations of the Far East, with the obvious exceptions of North Korea and the People's Republics of China and Mongolia. He also took exclusive rights to the formidable and advanced militaries of Australia and New Zealand. His standby territory, by virtue of geographic proximity to his contractual territory and his standing in the area, was made up of the independent states of western Asia and the Near East, containing Afghanistan, Ceylon (present-day Sri Lanka), India, Iran, Iraq, Jordan, Oman, Pakistan, Syria, and Yemen.

Finally, to round out the world, Fairchild kept the U.S. market for itself, although there is evidence that in their subsequent attempts to urge selection of the AR-10, first by the Marine Corps and then once again by the Army, they employed Cooper-MacDonald for the actual marketing and demonstrations. This is particularly likely considering what an integral role this firm would play in marketing the AR-15 in the next two years for Fairchild, and in arranging the sale of certain ArmaLite intellectual property to Colt's.

Although the distributorship agreements were not finalized until November, 1957, the initial contract territories began to be put in place as early as that January, when Cummings was pressing hard for an exclusive sales contract. By February, Europe was conclusively set as the personal fiefdom of Jacques Michault, just as Latin America, the Caribbean, and Scandinavia belonged to Cummings.

# Each Salesman Buys a Rifle

Meanwhile, although the sales contracts were not signed until the late fall of 1957, earlier in the year an interesting incentive program had been devised by Boutelle. Fairchild was relatively cash-strapped, and in order to make the ArmaLite division's continued viability look more attractive to the board of directors, the president hit on a way to have it generate some quick capital. As part of the contractual agreement for the rights to sell the AR-10s and take commissions, each salesman agreed to buy one hand-made AR-10 from ArmaLite's Hollywood machine shop for the princely sum of $1,000 (the equivalent of over $8,000 today).

## Sam Cummings Opts for AR-10B No. 1001, with Composite Barrel

Sam Cummings, showing his eye for a good investment, insisted that he wanted serial no. 1001 (fig. 105), which as the first numbered example he saw as becoming the most valuable AR-10. None of the other salesmen or ArmaLite personnel contested this, and so no. 1001 went to Sam Cummings.

Interestingly, in spite of the dire omen provided at Springfield Armory when serial no. 1002 had experienced the fateful catastrophic barrel failure earlier in the year, Sam Cummings elected to keep the original composite stainless steel and aluminum barrel on serial no. 1001, and went on to fire an estimated 70,000 rounds through it without failure in various demonstrations in many corners of the globe.

This rifle, today in the Institute of Military Technology collection, is still equipped with its original composite barrel.

105. Left side view of ArmaLite AR-10B no. 1001, which was purchased by Sam Cummings.

Note this rifle is still equipped with the very first style of cylindrical handguard. The barrel was and remains to this day the composite aluminum type designed by George Sullivan and so vehemently opposed by Eugene Stoner. Its second-pattern recoil compensator (fig. 77) was the only updated part of this weapon.

Institute of Military Technology

| FORM 3 (FIREARMS) REV. MARCH 1955 | U. S. TREASURY DEPARTMENT - INTERNAL REVENUE SERVICE<br>RETURN OF FIREARMS TRANSFERRED OR OTHERWISE DISPOSED OF<br>(Chapter 53, Internal Revenue Code - See Instructions at bottom of form) | | | | | |
|---|---|---|---|---|---|---|
| TO BE FILED WITH THE DIRECTOR, ALCOHOL AND TOBACCO TAX DIVISION, IMMEDIATELY UPON THE TRANSFER OF FIREARMS | | | | 1. FIREARMS TRANSFERRED OR OTHERWISE DISPOSED OF *(Date)* 4/19/57 | | |
| 2. DESCRIPTION OF TRANSFER | | | | | | |
| AUTHORITY FOR TRANSFER 1/ (a) | PURCHASER OR TO WHOM TRANSFERRED | | KIND OF FIREARM (d) | MODEL (e) | CALIBER (f) | SERIAL NO. (g) |
| | NAME (b) | ADDRESS (c) | | | | |
| 491778 | International Armament Corporation<br>Post Office Box 3722<br>Washington 7, DC<br>Attn:  Samuel Cummings,  Vice President | | Automatic Rifle | AR-10 Armalite | .30 | 1001 |
| | THIS RIFLE TO BE USED BY INTERARMCO FOR DEMONSTRATION PURPOSE THROUGHOUT THE WORLD AND WILL BE CONTINUALLY IMPORTED AND EXPORTED.  STATE DEPARTMENT AUTHORIZATION FOR THIS IMPORTATION AND EXPORTATION HAS ALREADY BEEN OBTAINED PER COPY OF LETTER ENCLOSED. | | | | | |

1/ Indicate date of approval by Director, Alcohol and Tobacco Tax Division, of Forms 4 or 5 (Firearms) or date of application for exemption, Form 5 (Firearms), or in the case of a transferee registered as a special-tax payer (Firearms), the serial number of his special-tax stamp.

I declare under the penalties of perjury that the above statements are true and correct to the best of my knowledge and belief.

| 3. NAME AND ADDRESS OF SPECIAL-TAX PAYER *(Transferor)* *(Number and street, city or town, zone, and State or Territory)*<br><br>INTERNATIONAL ARMAMENT CORPORATION<br>POST OFFICE BOX 3722<br>WASHINGTON 7, DC | 4. SIGNATURE | 6. DATE 4/19/57 |
|---|---|---|
| | 5. TITLE OR STATUS *(State whether individual owner, member of firm, or if officer of corporation, give title)*<br>Vice President | 7. SPECIAL-TAX STAMP NO. 491778 |

INSTRUCTIONS

1. PERSONS REQUIRED TO FILE RETURNS.—Immediately upon the transfer or other disposition of any firearm, every manufacturer, importer, and dealer (including pawnbroker) shall execute an accurate return on this form, in duplicate, setting forth the information called for in paragraph 2 below.  All transactions occurring during a single day may be included in one return filed at the close of that business day.  These returns shall be filed with the Director, Alcohol and Tobacco Tax Division, Washington 25, D. C.  The duplicate will be retained by the person making the return for a period of 4 years, and be at all times readily accessible for inspection. (Sec. 5842, Internal Revenue Code of 1954.)

2. PERSONS REQUIRED TO KEEP RECORDS OF FIREARMS.—Every manufacturer, importer, and dealer (including pawnbroker) shall make and keep at his place of business a record showing (1) the sale or other disposition of all firearms

taxable under chapter 53, Internal Revenue Code of 1954, (2) the date of such sale or other disposition, (3) the serial number, model, caliber, and trade name as well as other marks identifying each firearm, and (4) the name and address of the person to whom any firearm is sold, transferred, or otherwise conveyed.  This record must be preserved for a period of at least 4 years from the date of disposition of the firearm, and be at all times readily accessible for inspection.  (Sec. 5842, Internal Revenue Code of 1954.)

3. SCOPE OF RECORD AND RETURN PROVISIONS.—The provisions of section 5842 of the Internal Revenue Code of 1954 relating to records and returns are applicable to persons engaged in business within the States of the United States, the Territories of Alaska and Hawaii, and the District of Columbia.

NOTE:- For failure to file this return, a penalty of not more than $2,000, or imprisonment for not more than 5 years, or both, is provided by law.

(If additional space is needed, use reverse or additional sheets)

106. The Treasury Department Form 3, used to record tax-free firearm transfers between NFA licensees, dated April 19, 1957, covering the transfer of ArmaLite AR-10B serial no. 1001 to Sam Cummings's International Armament Corporation. As noted, this rifle was to be "used for demonstration purposes throughout the world and will be continually imported and exported."

Note Cummings still retained the title "Vice President," which he had taken in 1953 in order to give the false impression that the company had more employees than just himself.

Michael Parker collection

## Charles Dorchester Buys AR-10B No. 1002

Charles Dorchester, who was at the time the production manager of ArmaLite, and later its president, received serial no. 1002. As described and depicted in Chapter Three, this rifle had experienced a catastrophic barrel failure at Springfield Armory earlier in the year and had had a new, sturdy steel barrel retrofitted (fig. 85).

## Jacques Michault Buys AR-10B No. 1003

For use in demonstrations in his assigned territories, as discussed below, Jacques Michault of SIDEM International acquired AR-10B serial no. 1003.

## Bobby MacDonald Buys AR-10B No. 1004

107. Left and right side views of ArmaLite AR-10B serial no. 1004, after its conversion to a heavy-barreled squad support weapon. Sold to salesman Bobby MacDonald, this is the only known example of its kind ever made.

Note the ventilated steel handguard, bipod, heavier front sight base, and single-piece recoil compensator.

Institute of Military Technology

Perhaps most remarkable of the weapons sold to salesmen and ArmaLite notables was serial no. 1004, also described and depicted in its original form in Chapter Three, which was sold to Robert MacDonald. It shows the constant tinkering and experiments made necessary by the very small number of AR-10Bs then in existence.

This was the very same rifle that had been tested at Springfield Armory along with no. 1002. After the trials had concluded, instead of simply having its barrel changed out, the entire forward portion of its upper was replaced, encapsulating the entire barrel in a perforated steel handguard that protected the side-mounted gas tube. This sleeve also served as an attachment point for a bipod, making this ArmaLite's attempt at a fifth function for the AR-10; that of the squad support weapon. Something of a cross between a rifle and a belt-fed machine gun, the squad automatic is lighter than a machine gun (although many modern LMGs can be considered squad support weapons as well), which provides fire support but still generally uses magazines.

# Richard Boutelle's Rare Pre-Production "Ghost" AR-10B

108. Left and right side views of Boutelle's unregistered "ghost" AR-10B, which was seized following Boutelle's death and currently resides in the BATF firearms collection.

It is the only one of its kind, and as shown in the following illustrations it differs from the early production AR-10Bs in several ways.          BATF collection

110. Left side closeup of the muzzle end of Boutelle's one-off, pre-production "ghost" AR-10B, showing the blued muzzle flange to which the recoil compensator would be attached, the fluted aluminum barrel sleeve, and the early design of front sight tower, clamped to the barrel

by means of a hefty vertical hex nut located on the flat ahead of the integral front sight hood.

This was the only U.S.-produced AR-10 to feature a bayonet lug, to which an M1 rifle bayonet could be mounted.                                    BATF collection

Richard Boutelle would end up getting the drop on everyone in the game of rare AR-10 variations, as was only discovered when he unexpectedly passed away just a few years later, on January 15, 1962. At that point, as his family set about selling some of his gun collection, an early pre-production prototype was discovered, which would turn out to be the first of the AR-10B series ever made.

The attempted sale of this rifle came to the attention of the Treasury Department, which handled automatic weapon registration before the Bureau of Alcohol, Tobacco, and Firearms was set up in 1972. On inspection, it was discovered that this AR-10B did not have a serial number, and had never been registered with the federal government, which was a violation of the National Firearms Act.

The weapon was thus seized—according to a note in Michael Parker's Interarms files from 1990, from one Lawrence R. Smith, a machine gun dealer in Cabin John, Maryland—and marked with an electropenciled serial number "IRS 3039". It has ever since been kept in the ATF vault, where its current whereabouts were discovered by the intrepid AR-10 researcher Marc Miller, in 1984. Without his diligence, this fascinating lone branch of the AR-10

111. Left front three-quarter closeup of the front sight tower on Boutelle's rifle, showing the gas adjustment screw configured to take a cartridge rim.

The front sights on the early AR-10Bs were actually roll pins.                                    BATF collection

09 (preceding page). Left side closeup of the action of outelle's unserialled AR-10B, now with the number "IRS 039" added.

It appears this pre-production rifle, the only one to be ctory-marked "AR-10B," was built on a set of receivers achined from bar stock.

Note the distinctive square-backed carrying handle, hich gave this model the nickname "humpback." The ar sight elevating wheel is as yet unmarked with sight adations.                                    BATF collection

112. Top view of the bolt assembly from Boutelle's pre-production AR-10B.

This may be the first example of the bolt carrier ever constructed without the buffer mechanism attached, as shown in fig. 51.                                   BATF collection

113. Side and front views of the bolt assembly from Boutelle's rifle. Aside from the later additions of the gas rings, it seems almost identical to later iterations of this component.                                   BATF collection

family tree would likely have been lost forever, as no record of this rifle's existence ever appeared in the ArmaLite records. It appears that Boutelle had arranged for ArmaLite to hand-build (as they did all of the U.S.-produced pieces) a "custom" weapon for his own personal use, the existence of which only became known after his death.

Boutelle's unnumbered AR-10 also provides an interesting look at the transition between the AR-10A and the AR-10B, which as noted earlier was the model tested at Springfield Armory. As the "B" model was the first to use the front sight base to house the front end of the gas tube, it makes sense that the first attempt at a multifunctional sight tower would appear cruder and thicker, as does Boutelle's, compared to the models that were tested by the U.S. Army. This type of sight, albeit finished, has only been seen once elsewhere; on the modified serial number 1004 squad support weapon.

Also, the fluted aluminum sleeve covering the front end of the composite barrel of the AR-10A is carried over on Boutelle's rifle, whereas on later examples of this model the exposed front portion of

114. Right side closeup of the action of Boutelle's pre-production AR-10B, with bolt retracted.

Note the gas tube extension, which nested inside the bolt carrier, and the early version of the ejection port cover catch, configured as a simple rolled piece of spring steel riveted to the cover.                                   BATF collection

the barrel is thinner, unsleeved, and completely obscured by the recoil compensator.

The bayonet lug did not appear on the later ArmaLite AR-10s, but one is apparent on the Boutelle rifle, to which an M1 rifle bayonet could be mounted.

As discussed below, Boutelle's personal "ghost" AR-10 actually featured in a less-than-successful evaluation of the ArmaLite design by the U.S. Marine Corps.

## The AR-10B Becomes Simply the "AR-10"

Finally, although the series tested by Springfield Armory was collectively known as the AR-10B, ArmaLite made the decision in late 1956 to redesignate it simply as the "AR-10," using this for the second time (the X-02 prototype had previously been referenced simply as "AR-10"). Therefore, this unnumbered example of the very first AR-10B ever produced is the only one to actually be stamped "AR-10B," with serial number 1001 going to Cummings with only "AR-10" stamped on the receiver.

Somewhat confusingly, ArmaLite would continue this tradition of reusing designations, with the last U.S.-made AR-10 of 1959 also named the "AR-10A" (fig. 304). In addition, the modern AR-10s manufactured by the current owners of the ArmaLite trademark are marketed as the "AR-10B" and "AR-10A"; the third use of this nomenclature.

## Preparing for Series Production in Holland

With the sales and marketing scheme in place and the sales contracts signed, the full focus of ArmaLite (that which was not hard at work on the AR-15) could now turn to helping their new Dutch partners gear up for the efficient series production of the AR-10. The fact that the AR-10 could be produced by hand in a tool room so economically was a peerless engineering achievement, but an assembly line working from newly-drawn and -toleranced Metric blueprints was an entirely new arena, one with which the ArmaLite personnel were not familiar. Their technical expertise, however, when combined with the manufacturing experience of their new colleagues, would overcome innumerable challenges.

# The Marine Corps AR-10 Tests Go Awry

Meanwhile, beginning on August 12, 1957, a dark and little-known episode in the history of the AR-10 had unfolded in the steamy pine forests of northeastern Virginia. Not welcome within the sanctum sanctorum of the Army's Lightweight Rifle Program after a cloistered decade of history had been the United States Marines. Viewing the Springfield Armory rifle evaluations every bit as much in the capacity of spectators as had ArmaLite or any other interested party, this élite assault infantry branch of the U.S. Navy was not quick to jump on the bandwagon when the M14, which appeared to the Marines to be a barely-modified M1 Garand, was adopted.

ArmaLite learned of this attitude, and realized that they had just the tool to exploit Marine recalcitrance to the Army's "new" infantry weapon. Melvin Johnson, their vitally important and hardworking consultant, had been through the same frustrating experience with Army Ordnance fifteen years earlier while attempting to displace the Garand with his own M1941 rifle, as well as in trying to get his light machine gun adopted over the aging M1918 BAR. When he was stymied by the insularity of the Army, which was not nearly as entrenched in 1941 as it had become by 1956, Johnson had turned successfully to the Marines. In communication with Boutelle, Johnson explained how this branch of the armed services was both a maverick in trying new equipment and tactics, and not as hidebound as the Army in its procurement methods. While failing to get his weapons adopted, Johnson had succeeded in selling large quantities of them to the Marine Corps, for which he was still very grateful, and hopeful that it could happen again with the AR-10.

In the spring and summer of 1957, through his personal friendship with Marine Commandant General Randolph Pate, Melvin Johnson managed to schedule a Marine Corps evaluation of the AR-10.

Despite being more open-minded to new weapons than the Army, the Marines were still a conservative organization, and as such would need convincing by more than the publicity and rumors concerning the AR-10, which had of course reached them.

The Marine firing trials organized for the AR-10 were similar to those employed at Springfield earlier that year, but were more realistic and seemed less purposefully designed to engender the maximum number of malfunctions attainable. One rifle was assigned to accuracy, performance, and adverse-condition function testing, while another was set aside to be put through a series of endurance trials.

On August 10, Johnson arrived in Quantico, Virginia with the two rifles which were to be evaluated, and that Monday, firing commenced. As the tests began, Charles Dorchester was hastily making his way from Hollywood to the east coast in order to personally observe the testing. When he arrived on Tuesday morning, however, he was surprised to learn that testing had already been stopped, although only 2,240 rounds had been fired in the rifle designated for endurance testing.

The immediate reason for the cessation of testing turned out to be that the barrel on the weapon assigned to endurance testing was starting to bulge noticeably, a very strong warning sign that a catastrophic failure of that part was imminent, although in a different manner from the failure at Springfield. Dorchester immediately sent a message back to ArmaLite to update them of the situation. He informed Captain Edwards, the Marine officer responsible for the conduct of the tests, that he had brought another example of the rifle, and requested a meeting to discuss the numerous deficiencies of the AR-10 to which Edwards had alluded in their interactions which, aside from the obvious and serious problem of the bulged barrel, concerned gas leakage from the front sight base, and some parts breakages.

## The Marines Critique the AR-10

Captain Edwards, the judge of the contest, laid out for Dorchester what he personally considered to be serious deficiencies within the basic design of the rifle. As one will see in the following analysis, many of these objections were simply ridiculous or plainly in error, while a few were sound and stemmed from the peculiar rifle marksmanship doctrine employed by the Marines.

The Marine Corps had in place at this time a mature and well-established tradition of focus on individual rifle marksmanship that bordered on the fanatical. In developing their standards of world-class rifle accuracy, the Marines had adopted very particular methods of firing a rifle for accuracy. These led to several of the perceived deficiencies claimed by Edwards in his meeting with Dorchester.

The first of these, fourth overall on Captain Edwards's list, was that the trigger should be located further from the grip, so that only the very tip of the finger would naturally fall on its face. This characteristic of having what is essentially an oversized gripping surface was well known in the sport-shooting world, and essentially involved having the grip area take up more of the shooter's hand so that the tip, the most sensitive and precisely-controllable part of the index finger, is the only point to contact the trigger. This configuration was, however, not accepted on standard military equipment, as it prevented shooters with smaller hands from properly gripping the weapon. The problem of naturally placing a less ideal part of the finger on the trigger could be corrected through proper training, while still providing a weapon that allowed soldiers with hands of more widely-varying dimensions to fire it comfortably. In response to this criticism, Dorchester pointed out that the distance from the grip to the trigger on the AR-10 was the accepted military standard, being the same as that on the M1, the M14, and the FAL.

Also based on this tradition of target marksmanship was the critique (number five of thirteen overall) of the trigger itself having a flat face with squared edges, instead of being rounded. This seemed rather arbitrary and potentially a means of maligning the rifle, as flat triggers were known to provide more consistently repeatable finger placement, helping accuracy. This was well known in the shooting community, with the standard for dedicated target rifles being flat-faced triggers, as well as the specification desired by the Army in the Lightweight Rifle Program.

Deficiency number seven spelled out the Marine requirement that any rifle issued by them would have a rear sight adjustable for windage, as well as elevation. Although he did not agree with their inclusion, items five and seven were acknowledged by Dorchester as being easily modified if the Marines chose the weapon.

Deficiency nine commented on the sight radius, which was considered too short by Captain Edwards. The current-issue M1 rifle did indeed have its front sight mounted farther forward on the barrel than did the AR-10, but Dorchester explained that there would be a significant handling disadvantage to moving the front sight assembly further forward on the AR-10, because of its relative weight necessitated by the role it played in the gas system. The M1 rifle's front sight was just that: a sight. It did not perform any other function, and was also very short, so little material was needed to construct it. The AR-10s of this time also boasted a sight constructed out of steel (compared to aluminum on current production models of the rifle), which had to be taller, thus heavier, in addition to their secondary function as gas ports and housings for the gas tube, recoil compensator, and handguard. Moving something so heavy forward on the barrel would increase its mechanical advantage from the position of the shooter, making the rifle more ungainly, thus requiring more effort to swing between targets, and to keep levelled at a target.

Finally, the unique way Marines were trained to position their head on the stock in relation to the rear sight of a rifle caused additional misunderstandings. In critique twelve, Edwards commented that the upper receiver had the tendency to strike the shooter's face during recoil. The preferred Marine head position for rifle sighting, particularly in the prone position, is with the cheek very far forward on the stock, so as to give the eye the closest possible

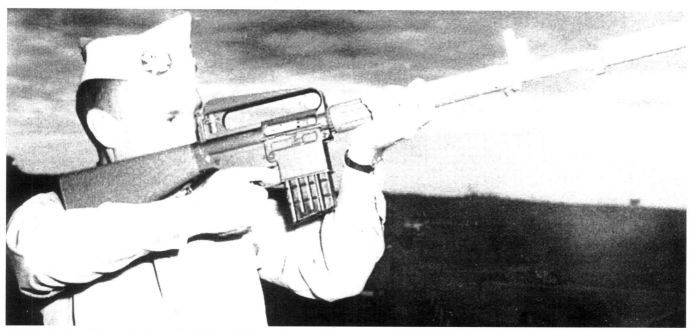

115. A Marine firing the Boutelle "ghost" AR-10B, part of the vanishingly scarce original documentation proving that this round of USMC testing actually did take place.

This rifle is none other than Richard Boutelle's personal AR-10B (perhaps it became so after its use here). Note the distinctive front sight post, the bayonet lug, and the recoil compensator, which is mounted significantly further forward than on the standard models tested at Springfield Armory.

Note also the distinctive Marine Corps firing stance, with the face very far forward on the stock, so as to give the eye the closest possible view of the rear sight. The Marines complained that when using this head position the upper receiver had a tendency to strike the shooter's face during recoil. They also complained that the sights were too high, and that they should be brought down closer to the axis of the barrel.      Marc Miller collection

view of the rear sight, thus increasing the precision with which the rifle could be aimed. In points two and thirteen Edwards also complained that the sights

were too high, and that they should be brought down closer to the axis of the barrel.

## Charles Dorchester Responds

Dorchester opined, and logic tends to agree with him, that in making these suggestions Edwards was confusing the issue of the exact head position with what it was meant to achieve. On the rifle around which the Marine Corps' marksmanship tradition was built, the bolt-action M1903 Springfield, the rear sight was initially located on the barrel forward of the action, and so moving one's head as close to the receiver as possible would indeed make the rear sight's alignment with the front sight more precise. On the AR-10, however, the rear sight is located at the very back of the receiver, above the action, and even further back in relation to the stock than the receiver-mounted aperture sight on the M1, to which the keepers of the Marine marksmanship flame were only starting to warm. When firing the AR-10, holding one's head in a position on the stock designed for making use of the sights on an M1903 or M1, as the rifleman is doing

in the above illustration, would indeed cause the shooter to stub his nose on the rear of the carrying handle base, and to view the sights from too low an angle to be comfortable. What Dorchester pointed out, however, was that the AR-10 eliminated the need for such strained head positioning, and that the shooters should simply place their cheeks further back on the stock. Suggesting that infantrymen break a traditional protocol or method that had been deeply ingrained in them during their training and service in one of the finest and proudest fighting forces on earth, however, did not go over at all well, and Dorchester's advice was rejected.

On a personal note, this author experienced the same phenomenon that was described in the Marine test report when he first began using his AR-10 in precision marksmanship competitions. The first rifle that he ever owned, and the one with which he learned the art of accurate and effective rifle shooting, was a German *Gewehr* 98, of which the M1903 Springfield was largely a carbon-copy, with only the caliber changed. Receiving instruction in rifle marksmanship, including proper head placement on the stock, by a Marine veteran, and using this rifle in competition for five years, it came as a shock to the author that when this method was employed in firing his new AR-10 from a prone position, the portion of the upper receiver that houses the rear sight did indeed smack him on the nose with the recoil of every shot. It took some time and practice, as well as quite a few competitions, to learn the acquired skill of proper head placement on a rifle like the AR-10, with its raised sights and straight stock, instead of the traditional bore-level sights with a drop-heel stock. He can indeed attest that this means of holding the rifle and sighting are quite different, and take consid-erable practice before they feel natural for one thoroughly trained on a traditionally-stocked rifle.

In his refutation of these deficiencies claimed by the Marines, Dorchester also pointed out that lowering the sights would necessitate a redesign of the stock to a traditional drop-heel design, which would completely negate the AR-10's peerless rapid-fire controllability, at the heart of which was its straight-line stock principle. Doing just what the Marines suggested had been tried before, in the form of the AR-3, and the results had been exactly as Stoner had predicted: the drop-heel of the stock provided a pivot point around which the rifle rotated upward under recoil, and the weapon, particularly with its light weight, which the Marines loved, would be impossible to keep on target in automatic or fast semi-automatic fire.

## Further Marine Complaints

The remaining deficiencies, against which Dorchester railed in internal communication with Fairchild's head office and George Sullivan back at ArmaLite, consisted of completely arbitrary personal proclivities and outright errors. The first claimed that the stock was too short, even though, as Dorchester correctly pointed out, it was identical in length to that of the M1 and M1903 rifles, as well as that of the FN FAL and the newly-adopted M14.

The third claimed that the weight of the AR-10 was 7.7 lbs, when it was in fact exactly seven pounds even. This was a clear mistake, just as was the sixth criticism, in which Captain Edwards claimed that the weight needed to pull the trigger varied from shot to shot. Subsequent actual measurement with a trigger scale showed that this was demonstrably untrue, but the assertion remained just the same.

The eighth alleged deficiency was merely the existence of the carrying handle. Not swayed by Dorchester's explanation of how this was important for protecting the charging handle and rear sight, in addition to providing a convenient means of carrying the rifle and mounting a scope, Edwards flatly stated that "Marines just don't need a carrying handle."

Item eleven claimed that after firing a large number of rounds and letting the rifle sit for a few minutes, heat transferred from the barrel, causing the receiver to become warm. Dorchester frustratedly explained to Captain Edwards that this was by design and was a good thing, as the receiver did not become uncomfortably hot, and was actually serving the purpose of wicking heat from the barrel, dispersing it into and through the aluminum, and because of the greatly lower specific heat (potential for holding heat or resisting heating) of the aluminum compared to the barrel steel, releasing it harmlessly into the air. Testing actually showed that the AR-10 was able to resist heating better than any other military rifle of its time, taking the largest number of rounds fired to reach a heat at which a round would "cook off." This cook-off test, performed by Springfield, ArmaLite themselves, and several other nations soon to evaluate the AR-10, was the best way of testing how well a rifle controlled temperature, by measuring how many rounds had to be fired in a short time before the powder in a chambered round reached the igniting point from the heat it absorbed from the chamber, and detonated on its own.

The last of the frivolous and erroneous deficiencies claimed at Quantico was that the AR-10's accuracy was simply not as good as that of the T44. When Dorchester provided the actual 100-yard test results from Springfield Armory, which proved conclusively that the AR-10 had handily outmatched both the T44 and T48 in accuracy, Captain Edwards responded that "[a]ccuracy at 100 yards has no bearing on what it is like at 300 yards." This understandably enraged Dorchester, according to all accounts, although he did not show it at the time.

## Mel Johnson Steps In

Following Dorchester's fiery and defiant rebuttal of the criticisms resulting from the Marine tests, Melvin Johnson also submitted his own commentary, which agreed with Dorchester's on all of the objective items but was far more subdued in tone and on what course of action was best at that point. Johnson's analysis focused far more on the shortcomings of the particular examples of the AR-10 which were provided to the Marine Corps, and provided a critical look at why the ArmaLite had failed so miserably at Quantico, adding the faults of ArmaLite to Dorchester's claims of a conspiracy by jaded Marine officers who had it out for the new rifle from the start.

The actual weapons that were furnished for the tests, as Johnson discovered in the days following the halt of firing, were actually two of the earliest prototypes, including Richard Boutelle's personal "ghost" AR-10B, the one with no serial number and the strange front sight assembly. More to the point, however, neither of these rifles had been retrofitted with any of the updates that had been made to correct the problems encountered during testing at Springfield. Boutelle's rifle still had its original composite barrel, although the other example used in the Marine tests had been fitted with a steel barrel which ironically, it was found, had contributed to the most serious malfunction—that of the bulged barrel. It was determined that this had occurred because the incorrect chrome-vanadium alloy had been used in its production, and that the proper heat treatment had not been applied. This mixture of iron, chromium, and vanadium was quite new, and its precise proportions were not well known at that time, leading to mistakes like this.

There was also the issue of gas leaking at the juncture of the front sight base and gas tube. This was traced by Johnson to improper manufacture and heat treatment of the sight housing and gas tube, and to their faulty installation. None of these issues had ever surfaced before, and so they came as a surprise at a very inopportune time.

The attendant accuracy problems were also attributed to the substandard material and heat-treating process of the barrel, and the inexact tolerances in the fitting of the front sight base to the barrel. The same manufacturing errors that allowed gas to escape on firing also caused the front sight base to vibrate in relation to the barrel, instead of staying tightly attached to it. This phenomenon led to a serious degradation in barrel harmonics, which were responsible for the AR-10's poor accuracy perform-

ance in the Marine tests, compared to its usually extremely high standard.

In his conclusion, largely attributable to his longstanding and cordial relationship with the Marine Commandant and the organization in general, Johnson urged ArmaLite to withdraw the rifles and placate the Marines. He wrote that he felt the AR-10's dismal performance had "alienated a friendly service branch," and that this needed to be remedied. Johnson also made the point very strongly that the test report ought to be buried, and its distribution prevented. He asserted that with his connections within the Marines, he would be able to accomplish this.

To others within the Fairchild organization, Johnson's advice came across as apologistic, motivated more from a desire to maintain his professional connections and friendships than from a real desire to help the AR-10 overcome these obstacles. On the other side of the coin, Dorchester fumed in his report about the open hostility he and the AR-10 had encountered at Quantico, and hammered home his position that there was a strong institutional bias against the AR-10, not within the entire Marine Corps, as it had been within the Studler-Carten Ordnance Research and Development Department, but on the personal part of Captain Edwards. He reported in his conclusions not only how he heard Edwards remarking derogatorily about the AR-10 to the riflemen whom he outranked, but how Edwards, even in the presence of superior officers in charge of the tests, flatly refused to hear accurate and sound arguments concerning the merits of the rifle.

In Dorchester's meeting with Captain Edwards and Lieutenant Colonel Bell on the day of his arrival, he had pointed out that the rifle he had brought with him had all of the updated parts, and one in the same configuration had just fired in excess of a recorded 20,000 rounds without any breakages or serious malfunctions. He apologized for the poor performance of the rifle assigned to endurance testing, explaining that incorrect heat treatment had caused the issues, and that ArmaLite's 600-round test firing that had been done on the rifle immediately before its shipment to Quantico had not turned up a malfunction of any kind. Over Captain Edwards's objections, Lieutenant Colonel Bell told him that he could make his case on the continuation of testing before the President of the Weapons Evaluation Board of the Quantico Marine School, Colonel Crocket.

That conference, which took place on the same day, went well for Dorchester, as Colonel Crocket agreed to resume the testing with the new rifle he had brought if this were confirmed by Colonel Durand at Marine Headquarters. The next day, when no word from Durand had been received, he was granted permission to call on the colonel, and tried to do so, finding him busy. The next day, August 15, he received the surprise word from Fairchild's Warren Smith that Durand had denied his requests, either to allow the tests to continue further, or to permit the substitution of the properly-outfitted rifle.

Adding insult to injury, when Dorchester took this matter back to Lieutenant Colonel Bell, Edwards, who was also present, told him in front of his superior officer "I don't care if the rifle fires 10,000 rounds without malfunction or breakage, we're just not buying this design—you can't prove a thing by further testing." In that moment, Dorchester saw his suspicions proven, which he made very clear in his report. Citing Edwards's calling the AR-10's sights "mickey mouse sights" to the riflemen, among other things, and his asinine comment about 100-yard accuracy having no bearing on 300-yard accuracy, Dorchester concluded that the spirit and intent in which the testing was devised by Marine Headquarters had been subverted and violated by the prejudiced Captain Edwards. He concluded that "these tests were conducted by a highly opinionated, self-styled 'guns authority' who stated that he had his own ideas of what a service rifle should be like and that the AR-10 was not it." His recommendation for policy at this point was to double down and lobby hard for the Marines to give the AR-10 another chance.

In the end, Johnson prevailed, repairing his personal relationship with the Marines, even with the alleged saboteur Captain Edwards. Talks went on through the next month, and Fairchild ended up not pursuing a second test by the Marines. One must not forget the number of formal evaluations that the T44 and T48 were permitted before an adoption decision was made.

The salient point in all of this, to which it seemed that no one wanted to own up, was that the early rifle initally submitted for the Marine test had been defective in both materials and manufacture, and so the blame for its dismal performance rested squarely and legitimately with ArmaLite. Disregarding this, some complained, rather fairly, that if Johnson had really been dedicated to the AR-10, he could have easily used his significant influence within the Marine Corps to convince them to give the ArmaLite weapon another chance. All of the available evidence seems to suggest that Johnson would likely have been able to accomplish this, and so one must ask what might have changed had he done so.

## Johnson Has the Last Word

Johnson did accomplish one thing, however, which proved both his allegiance to the AR-10 and his strong influence within the Marine Corps: he completely quashed the distribution of the test report on the AR-10, making it as if the Marines had never evaluated such a weapon. In fact, so successful was he in removing this black mark from the AR-10's record that in all of the contemporary sources regarding the history of the AR-10, no mention was ever made of the Marines considering the adoption of the AR-10. One can imagine this author's delight upon rediscovering this totally blacked-out portion of the AR-10's history from an unlikely and esoteric source.

# Part II: The "Prime Time" AR-10

*Chapter Five*

# The Switch to Dutch Production

While brass clattered to earth near Hagerstown, and the wheeling and dealing of the salesmen began, ArmaLite and Artillerie-Inrichtingen engineers worked tirelessly to prepare the AR-10's revolutionary design for series production on the assembly lines in Zaandam, both before and after the catastrophic Marine trials. While the deal for production was still in the works, and before the final licensing and sales agreements had been signed, the ArmaLite engineers had once again begun making some substantial changes to the AR-10B design, just as they had to the AR-10A in 1956 before the Springfield tests. These initiatives continued after June 4, 1957, with the Dutch quickly brought into the fold.

# More ArmaLite Improvements

The final series of serially-numbered ArmaLite AR-10s, manufactured during the fall of 1957, were numbered from 1006 through 1047. These illustrate the radical design evolution undertaken by ArmaLite in the aftermath of the Springfield trials, and preceded the start of production in Holland.

### A New Front Sight Alignment System

In addition to the redesigned titanium muzzle compensator and improved furniture, chrome-vanadium steel barrels were fitted to all the rifles in this latter series, and as illustrated a new method of front sight/gas block alignment was introduced at serial no. 1006.

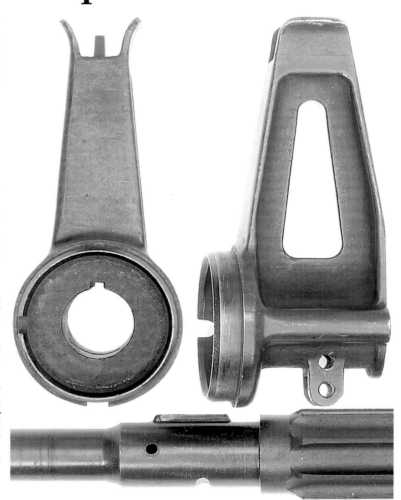

116 (right). Front and left side views of the redesigned front sight tower from AR-10 no. 1006, with a portion of its fluted chrome-vanadium steel barrel shown below.

Compare with fig. 86: note the front sight is now an actual blade, as opposed to the roll pin previously used, and the sight is open rather than hooded ('B', fig. 63).

The method of attachment is also different, with a key fitted into the top of the barrel aligning with a slot cut into the interior of the tower. The tower is affixed by means of a transverse pin driven into the hole above the sling swivel mounting hole, locking into the groove in the barrel.

Institute of Military Technology

117. Left side view of AR-10 no. 1008, the third example produced with the new open front sight and method of retention of the front sight tower shown in fig. 116.

Unlike the rest of this later series, no. 1008 is fitted with an open three-prong flash hider, a harbinger of the design to be used on the first Dutch production version, the Cuban model AR-10.　　Institute of Military Technology

118. Left side closeup of AR-10 no. 1008, showing the open three-prong flash hider.

Note the transverse pin above the sling swivel, which locks the front sight tower to the barrel.
Institute of Military Technology

# Further ArmaLite Belt-Feed Experiments

Meanwhile, ArmaLite efforts between February and June, 1957 had concentrated on increasing the AR-10's modularity, particularly in the direction of the light machine gun role, and the concept of feeding interchangeably from magazines or belts of ammunition. It was quickly realized that guiding a flexible belt of cartridges up through the magazine well, as had been shown in the artist's rendering in the early brochure (4, fig. 103), would be impracticable.

The solution to this was to introduce new feed and belt ejection openings, the former cut into the left side of the rifle's upper and lower receivers to receive components designed to hold and index the belt, and the latter cut into the right side of the lower receiver below the cartridge ejection port. This was not only the configuration normally employed on belt-fed weapons, but it solved the problem of ejecting the empty belt once the cartridge each link contained was thrust forward into the chamber. The bolt carrier was modified on the left side to interact with the belt feed device, camming the belt into the receiver with each reciprocation.

# Introducing the Top-Mounted Gas Tube

There was one big obstacle to the efficacy of left-side belt-feeding: the side-mounted gas tube. Critical in Stoner's gas system was the extended gas tube that protruded well into the receiver, and which entered the left side of the bolt carrier. The existing location of the gas tube made opening up the left side of the upper receiver and feeding ammunition through it a physical impossibility. Instead of scrapping this idea of belt-feed modularity, however, ArmaLite searched for a way to preserve all of the advantages of Stoner's ingenious gas system, while also freeing up the left side of the rifle for ammunition feeding. The solution became clear; that the gas tube had to be moved to the top of the bolt carrier.

However, this change was easier envisaged than accomplished. Critical to the efficiency and cleanliness of Stoner's direct gas impingement mechanism was that the gas was fed directly into a chamber in the bolt carrier, creating a closed system, and thus allowing a miniscule amount of propellant to be used to power the action. Moving the gas tube hole to the top seemed at first to be impossible, because the cam pin that guided the bolt through locking and unlocking had to be on top, and could not be moved to the side or bottom, where it could not be controlled during bolt reciprocation. The cam pin also had to be at the front, as the gas expansion chamber needed to be at the very back of the bolt tail. Therefore, a way had to be found to locate both the gas tube and the cam pin on top of the bolt carrier.

## The Evolution of the Top-Mounted Gas Key and Tube

The story of how the gas tube was relocated from the side to the top of the receiver of the AR-10 is kindly related in his own words by the designer, L. James Sullivan, as follows:

### The Genesis of the Top-Mounted Gas Tube
#### by L. James Sullivan

*When I first went to work at ArmaLite in June of 1957 I was hired as a draftsman assigned to help an engineer, John Peck, who was Gene Stoner's principal design assistant on the AR-10 starting with the first aluminum receiver designs. Peck had started to redesign and reposition the Dutch-made AR-10 gas tube from the side to the top. I redrew the bolt carrier, replacing the gas tube hole on the left side with a slot on the top to fit an as-yet undesigned carrier key with a snout that would slide over a top-mounted gas tube. The problem was that the snout would be over the cam pin, blocking its removal, which I believe caused an argument between Stoner and Peck. In any case Peck resigned in September or October, 1957, shortly before ArmaLite moved from Hollywood to Costa Mesa. I inherited the top gas tube project and, as directed by Stoner, designed the key with the now-familiar shape, but with a relief cut on the left side that allowed the round-headed cam pin to be removed when the bolt was rotated rearward.*

*I then started designing the bolt carrier group and upper receiver of the AR-15, which was primarily a scale-down of the AR-10. Stoner didn't believe in the small-caliber AR, and was starting to design another 7.62mm rifle (the AR-16, prede-cessor of the 5.56mm AR-18), so I was pretty much left alone until Bob Fremont joined the company about two months later. Bob and Gene Stoner had met while both were working as tool-and-die men at the Whittaker Aircraft Controls Company.*

*In the meantime, word from Artillerie-Inrichtingen about the top-mounted gas tube was that it got the cocking handle so hot people were complaining, so I came up with the rear cocking handle for both ARs, and closed the cocking handle slot on both upper receiver drawings. The problem then was that Harvey Aluminum, the supplier of the receiver forgings, had already made the first AR-15 upper receiver forgings and couldn't make more than 15 or 20 without reworking the initial soft forging dies, and our machine shop had already cut the slot in the top, so I designed an interim rear cocking handle with a little "T" bump to reach up through the receiver slot to hold the front of the handle up. That's why you see two models of AR-15 prototypes, some with the front cocking handle and some with the rear, but all with top slots. The interim "T" bump can barely be seen on serial no. 000002, illustrated on pages 67 and 68 of* The Black Rifle.

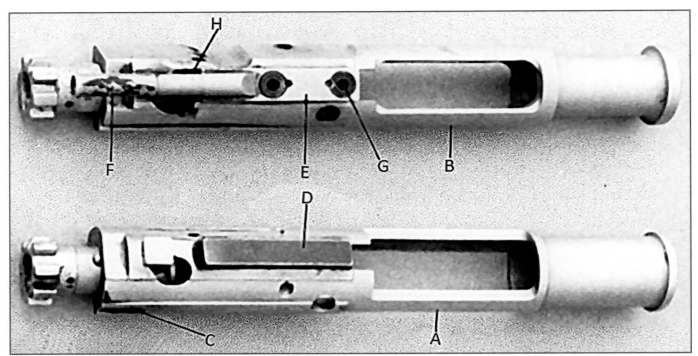

119. Top comparison view of two AR-10 bolt assemblies.

Above: the redesigned bolt carrier, which is essentially identical to the design currently in use in modern-production AR-10s and AR-15s. The gas key (E) and its snout (F) protrude from the top of the carrier, where it is attached by two heavily-staked set screws (G). The snout of the key (F) fits over the rear end of the main gas tube, forming a seal that only breaks when the carrier has moved backwards and the attendant pressure in the barrel and gas tube has dropped to safe levels. One can see the cam pin (H) riding under the gas key tube (F), where its wider body is held in place during the cycling of the action by the lower portion of the charging handle guide rails in the upper receiver.

Below: the original AR-10B carrier, with side-mounted gas tube orifice in bolt carrier (C). Its gas tube interaction necessitated it being built with the flat-top stabilization ridge (D) which rode inside the charging handle.

photo from Marc Miller collection

*The next word from A-I on the top gas tube was that the cam pin relief cut on the snout of one of their carrier keys had blown through in firing tests. So I had the idea to make a rectangular-topped cam pin that quarter-turned to remove, which eliminated the need for the relief cut. Without the rear cocking handle, and the flawed carrier key corrected with the quarter-turn cam pin, the top gas system wouldn't work. This combination got me promoted to engineer, and represented the only development work I did on the AR-10.*

Transmitting the gas itself directly into the expansion chamber eliminated the need for the receiver-mounted gas transfer bracket, which itself had been found prone to breakage during drop tests or extended firing. The top-mounted gas key subsequently proved to be extremely robust compared to the transfer bracket's relative fragility, and is still in use in the bolt carriers of all modern AR-10s and AR-15s.

The gas key was bolted securely to the top of the bolt carrier, behind the cam pin. Its forward end was configured as a raised snout which fit perfectly over the end of the gas tube, now mounted directly above the barrel, where the two together formed an essentially air-tight seal. This snout channeled the incoming gas back and at a downward angle through a hole in the gas key's underside that led directly into the internal expansion chamber between the bolt and bolt carrier, so that with the forward hole into which the earlier side-mounted gas tube had fit closed up, this was the only way (when the bolt was closed) for gas to enter or leave the chamber.

The "gas cutoff" function was continued by the rearward movement of the bolt carrier after the bolt began to unlock from the barrel extension, which limited the time during which there was a connection between the gas tube and the expansion chamber. This accentuated the Stoner operating principle, whereby only a very small amount of the expanding propellant gases was needed to power the action, by preventing any more of this already very fine and low-volume stream of gas than was actually needed from reaching the interior expansion chamber. This innovation further improved the AR-10's ability to function longer without cleaning or lubrication than any other automatic rifle of its day.

As the bolt assembly reciprocated, the raised front snout and body of the gas key rode in a longi-tudinal channel in the cocking handle shaft, while the bolt was retained in its forward, or unlocked, position by the cam pin riding between the guide rails on both sides of the upper receiver. These kept the cam pin vertical until the bolt had reached the end of its forward movement in counter-recoil, and was safely through the slots in the barrel extension.

## More Positive Ejection Port Cover Operation

This redesigned bolt carrier also incorporated an-other design change introduced by ArmaLite during these four months, which resulted in a better system of unlocking the ejection port door. In the original design, shown in figs. 60 and 114, the ejection port door was controlled by a simple rolled sheet-steel spring, riveted to the inside of the cover, which fit into a shallow slot in the bolt carrier. When the carrier moved and the slot and spring were no longer in alignment, the thicker portion of the bolt carrier body would cause the door, already under spring pressure, to snap open and stay that way. With this original design it was possible, if done so carefully by hand, to retract the carrier without opening the port door. The designers counted on the residual gas vented through the vent holes in the right side of the bolt carrier to assist the spring-slot interaction in forcing the door open. This by all accounts functioned ex-actly as intended. However, the engineers came to the conclusion that in cases where the weapon was very dirty and either carbon fouling or mud clogged the vent holes and adhered to the door, making it stick in the closed position, the carrier's movement might not be sufficient to knock the door open, and thus the spent shell casing would be trapped inside the receiver, causing a jam. The solution to this was extremely simple: a new assembly consisting of a block with a spring-loaded pin replaced the simple sheet-steel spring on the port door (fig. 126), and the slot on the bolt carrier was enlarged so that the bolt carrier could not move backward at all without con-tacting the port door and snapping it open.

This change in the ejection port door/bolt carrier interface actually took place slightly prior to the relocation of the gas tube and the addition of the gas key on the bolt carrier, and as such both had already been debugged (the former less so than the latter) when the Dutch added their expertise to the project.

One final and likely-unintended latent benefit of the migration of the gas tube to the top of the weapon was that the rifleman now would experience far more comfort when reloading a weapon that was in the midst of rapid strings of fire. The original side-mounted gas tube had reached its terminus in the exposed gas transfer bracket, located on the left side of the receiver, that directed the propellant through the secondary gas tube inside the receiver, which when the bolt was closed was housed in the hole in the left front of the bolt carrier. With extended firing the gas transfer bracket area, coincidentally the exact spot where the thumb of a right-handed shooter's support hand would come to rest while inserting a fresh magazine and pressing the bolt release lever, would become extremely hot. With the gas tube now mounted on top, the heat was dissi-pated into a thicker part of the upper receiver, which the rifleman would not normally be inclined to touch during the normal operation of the weapon.

## The First Working Belt-Fed AR-10, No. 1025

### The First Experimental Rear-Mounted Charging Handle

The top-mounted gas system was first incorporated in two examples of the small but steady stream of AR-10Bs which ArmaLite was handcrafting in their Costa Mesa workshop. These rifles bore distinctive cuts on the left sides of their upper and lower receiv-ers, which allowed the mounting of an assembly that connected to cams cut in the bolt carrier to stabilize and draw in ammunition in linked-belt form, as well as belt ejection slots in the right sides of their lower receivers.

Both of these weapons, differing considerably from one another, showed several interesting and unique characteristics. The first of the two (serial no. 1025) resembled quite closely a rifle, not being equipped with a bipod. It utilized an AR-1 flash hider, and sported a stamped-steel handguard that enclosed the very thick and rigid gas tube. The shooter's hand

120. Left side view of the first working belt-fed AR-10, serial no. 1025, with an AR-1 flash hider on the barrel muzzle. It is fitted with a detachable belt-feed device mounted on its left side. Note the experimental rear-mounted charging handle. There is no slot in the top of the receiver where the traditional charging handle would have been located.

This is the first appearance of the mysterious transverse hole in the upper receiver under the carrying handle, which is discussed further below.

Institute of Military Technology

was protected from the steel by a ridged wooden handle, mounted on the underside of the handguard.

The most distinctive feature found on rifle no. 1025, aside from its belt-feed device, was its experi-mental, rear-mounted charging handle. This means of charging the weapon would not be seen again until much later on the AR-10, but would become standard on both modern AR-10s and AR-15s.

## A Heavy-Barrel AR-10 Light Machine Gun, No. 1026

121. Left side view of the initial heavy-barrel LMG version of the AR-10, serial no. 1026, with bipod folded. It is shown configured for magazine operation, in that the belt feed device is not mounted. It is fitted with a special port door that covered up the opening for the belt-feed unit when it was not in use.   Institute of Military Technology

The second variant (serial no. 1026) looked more like a machine gun, but paradoxically went back to the original charging handle configuration. It was fitted with a bipod, which is almost an essential feature on an LMG. The muzzle of the heavy, quick-change barrel was fitted with a simple open three-prong flash hider.

Both of these prototypes featured a foldable shoulder rest built into the buttplate. When firing from the prone position, this rest was opened and placed on top of the gunner's shoulder, preventing the symptom of recoil in which the muzzle rises and the buttstock correspondingly migrates down the shoulder into the armpit.

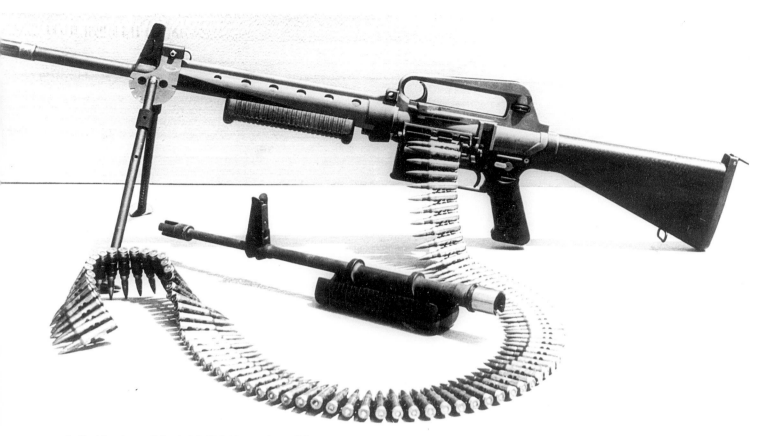

122. Left side view of the initial LMG version of the AR-10, serial no. 1026 configured for belt feed, with a non-disintegrating belt of 7.62mm NATO cartridges loaded into the opening of the belt feed device.

This model, which featured a quick-change barrel system, is shown with its heavy barrel in place and a lighter barrel in the foreground. The bipod allowed the weapon to be fired from the prone position without the need for a sandbag or other type of support.

Note here and in fig. 121 that the small circular opening in the upper receiver has been moved forward from the position shown in fig. 120.

photo from editor's collection

# Two Late U.S. Modifications

Two modifications of note were incorporated during this period of Old World-style craftsman production, ironically the side of AR-10 development actually taking place in the New World. The first was one which was actually an important step on the path not only to more sturdy furniture for the AR-10, but towards the light, precision-contoured, and ex-tremely robust and tough polymer stocks of today. The other—the small transverse hole in the upper receiver—was more of a curiosity than an actual important development in the AR-10's history, the purpose of which has served to perplex enthusiasts of the AR-10 for years.

## Strengthening the Fiberglass Furniture

Meanwhile, Tom Tellefson had improved the dura-bility of the furniture of the AR-10B by helping to develop a proprietary type of epoxy which, when added to the fiberglass shell and interior foam plastic of the AR-10's stock and handguard, increased their tensile strength. The furniture of the AR-10, while not necessarily the most-touted feature of the rifle, not only contributed to its light weight, but gave it its signature "touch."

In a final change that was communicated to the Dutch on September 23, 1957, Tellefson, working behind the scenes as he so often did, announced that he had formulated yet another new means of strengthening polymer rifle furniture. Whereas be-

123. Left side view of ArmaLite AR-10 serial no. 1036, fitted with the improved woven fiberglass handguard.
    The experimental belt-feed attachment hole appears on all of these late U.S.-made AR-10s.

photo from Charles Kramer collection

124. Left and right side views of ArmaLite AR-10 serial no. 1038.
    Note that even though it bears a later number than no, 1037, all three components of the spatter-painted furniture are still the original foam-filled shells.

courtesy James D. Julia auctioneers

fore he had introduced an epoxy to make the fiberglass stronger for the AR-10B's debut, he now combined this epoxy with a novel structure for the fiberglass. Instead of simply forming a shell out of the fiber material, as on the original AR-1 "shellfoam" stocks, the furniture on later examples of the U.S.-produced AR-10 featured stocks and handguards that were produced by combining the special graphite epoxy with the fiberglass shell, in which the fibers were actually woven together like a basket, instead of simply formed into a smooth skin. This resulted in not only a stronger product, but in an extremely unique appearance, distinctive solely to the later U.S.-made AR-10Bs.

Tellefson further improved this procedure by developing the best and most durable means of constructing the handguard and buttstock on the soon-to-be-produced Dutch rifles. Adding to his original improved shell made of woven fiberglass cloth of a thickness of between .03 to .04" with the added graphite epoxy resin, he replaced the original foam plastic with glass-filled phenolic compression moldings. This consisted of a type of resin, very much like that used to reinforce the fiberglass cloth of the shell, with loose strands of hard fiberglass mixed into it before molding. These strands acted much the same way as the grass or straw that ancient brick makers would sometimes incorporate into mortar or brick material, to increase its cross-sectional strength and integrity. This resin, which went into the fiberglass shell as a liquid, cured into a dense foam that formed a much harder and more durable solid than the foam plastic, while only weighing a few percent more by volume than George Sullivan's original formula, which it replaced.

The pistol grip was not neglected in this radical improvement over the original set of furniture. While the original had been a simple molding of plastic foam, Tellefson's new procedure called for its construction of phenolic resin inlaid with fiberglass rags, and compression-molded. Just as with the handguards and stocks, the phenolic resin cured into a very hard, durable, and strong solid, the increased integrity of which was even more important on this part, which did not have the protection of a hard fiberglass shell. The addition of fiberglass rags acted just like the fiberglass strands in the handguards and stocks, but instead used fully-woven bits of fiberglass cloth that had been cut up prior to being mixed into the epoxy. Both by virtue of their larger size and more integral woven structure, these added an even greater amount of strength to the pistol grips into which they were incorporated than the strands did in the other furniture components, making up for the lack of an outer shell.

Finally, the compression molding process by which the fiberglass rag-impregnated phenolic resin was formed into the grips created a stronger product than simple molding. Molding, in its basic form, simply involves pouring a liquid into a mold, and allowing it to cool or cure into a solid. It is the plastic equivalent of casting, while pressure-molding is more like the metal-processing technique of forging, which is how the AR-10's receiver halves were pro-

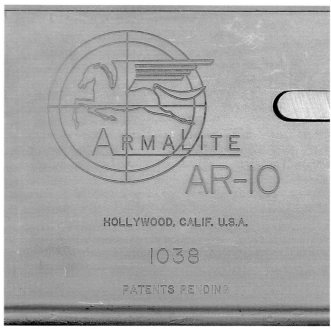

125. Left side closeup of the receiver of ArmaLite AR-10 serial no. 1038, showing markings.

The Pegasus logo appeared on all the later series of numbered AR-10s.    courtesy James D. Julia auctioneers

duced. The new process added pressurizing to the molding, which forced the soft material before its hardening into a thicker, denser structure free from inclusions and defects, just as the hammering process in forging removes flaws in steel or aluminum by compacting the metal.

Although Tom Tellefson's contributions to the AR-10 are not the most widely remembered, they were absolutely integral in creating the perfect combination of light weight, ease of handling, modularity, and firepower that made the AR-10 what it was. The evolution of polymer stocks would be critical to every form the AR-10 and its more compact descendant the AR-15 would take over the next half-century, and would pave the way for the great leaps forward that have been achieved in weapon furniture design and materials, which grace even the humblest hunting rifles and shotguns today.

## Explaining the Experimental Belt Feed Attachment Hole

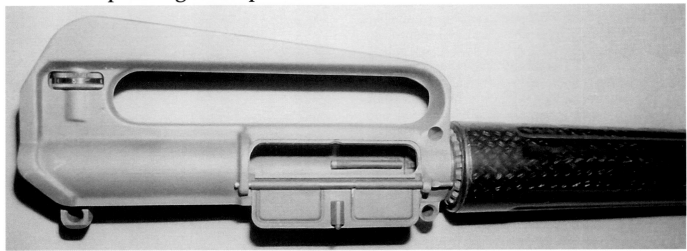

126. Right side view of the detached upper receiver group from ArmaLite AR-10 no. 1039.

Note the woven chocolate-brown-colored fiberglass of the strengthened handguard, and the mounting hole for the feeding device drilled through the area under the very front of the carrying handle.

Also shown is the gas transfer tube inside the receiver, which nests inside the bolt carrier, and the final form of the spring-loaded plunger assembly on the ejection port cover.                              photo from Marc Miller collection

The mysterious "tweak" ArmaLite introduced in the final series of U.S.-built rifles was to drill a transverse hole through the top of the upper receiver, just above the barrel. The surfaces of this hole were finished just like the rest of the receiver, but it seemed to serve no purpose. It was revealed later by those who knew Stoner well, that this hole was intended as a mounting position for a detachable feed device that would allow a belt of ammunition to be loaded into the left side of the receiver. Its presence was purely experimental at this stage, however, since not only were the receivers of some of these later ArmaLite AR-10s produced with a closed left side, making belt feeding impossible, but the traditional gas transfer bracket and gas tube were still located on the left side, making feeding anything from that side, even if the receiver were cut open, a patent impossibility. It is likely that this hole was drilled in these rifles simply to confirm that it would not have a deleterious effect on the structural integrity of the receiver, saving future tests with more expensive, heavily modified belt-fed prototypes.

# The Dutch and Americans Join Forces

## The Formation of Fairchild Arms International, Limited

A new division of Fairchild Engine and Airplane Corporation had been approved by the board of directors on June 19, 1957, shortly before finalization of the production agreement between Fairchild and Artillerie-Inrichtingen. Fairchild Arms International would be in charge of all facets of the Dutch AR-10's production and sale: refereeing possible territorial disputes between salesmen; providing resources, support, and expertise to the Dutch; arranging commissions for the sales staff and producers; supervising all sales calls and evaluation testing to ensure these were done properly and fairly; liaising with foreign governments and militaries, and providing product information to interested parties. This new branch of the Fairchild organization was incorporated in Toronto, Ontario, under Canadian law.

This new division was spearheaded and championed by Richard Boutelle, who had managed to convince the Fairchild board that the Springfield Armory tests had been an unfortunate case of bad luck, and that with large-scale production soon to become a reality in Holland, coupled with the nearly limitless demand for new military rifles around the world, the AR-10 was sure to become a major cash cow for the firm. He would himself pay several visits in the following year to Artillerie-Inrichtingen, both to check up on how things were going, and to provide encouragement and boost morale among the Dutch factory workers, who after all were building the finest infantry rifle ever devised.

## Head to Head: the Americans Bring Samples

The following month, July, 1957, before the sales arrangement had been finalized and the territories assigned for marketing the AR-10, designer Eugene Stoner, production experts Charles Dorchester, L. James Sullivan, Leo Killen, and Fairchild representative Horace Kennedy flew over to Holland with design drawings and three examples of the ArmaLite-produced AR-10, to begin the process of working up production parameters, as well as modifying the AR-10 to more closely mirror stated military requirements. There they met with their Dutch counterparts J. E. Manus Van Der Jagt, F. W. Spanjersberg, and M. A. Bakker, who were to take the lead in perfecting the AR-10 for different world militaries while the Americans concentrated on the AR-15 project. Unfortunately, no record has been found of the serial numbers of these rifles, and over the years all three of these U.S.-produced AR-10s, as well as the set of design drawings taken to Holland, have disappeared.

Horace Kennedy, an unsung member of the AR-10's entourage, hailed from the newly-established Fairchild Arms International, Limited. Leo Killen, on the other hand, not being part of either ArmaLite or Fairchild Arms, was still under the direction of the Fairchild Engine and Airplane Corporation's main office. During these summer months he was actually ordered to return to Maryland to resume his role at the aircraft division, but the encouragement of Paul Cleaveland, as well as Killen's dogged devotion to the mission, prevailed, and he was allowed to stay in Holland and continue the process of streamlining the AR-10 for production.

## Converting the Drawings to the Metric System

The workload facing the two sets of engineers in the late summer of 1957 was daunting, and it started with a task that seems simple on the outside, but which actually perplexes engineers and production experts whenever they encounter it. This Labor of Hercules was the necessary conversion of the measurements and tolerancing on the AR-10's design drawings from inch to metric units. Stoner and indeed ArmaLite had always produced all of their design drawings using inches as the standard measurement on their first-angle projection drawings, as was the accepted practice in American industry. Europe, on the other hand, as well as most of the rest of the world, uses metric units for all dimensions, and blueprints are drawn showing third angle projections. This means that on an American drawing showing a typical side view of a component, the first angle projection shows the top view of that component; while on a side view drawing of the same component done in the metric system, the third angle projection shows the bottom.

This process did not simply involve using a conversion formula to convert inches to centimeters and millimeters and writing the new numbers on the blueprints; the compound nature of different measurements interacting in an assembly called for an entire rework of the design drawings.

ArmaLite's Arthur Miller would also make a serious name for himself spearheading the project of conversion of the AR-10 design drawings. Extremely critical in the task of conversion was to maintain complete and perfect component interchangeability within individual models of the AR-10. This was not only important as a mechanical achievement that could be touted to potential buyers, but was critical in the testing of new, European-produced parts and components on existing U.S.-built rifles to make sure they were correct in every parameter. By all accounts, Miller did a superlative job of this conversion, and did it in record time, having the final draft of the metric plans prepared within a month from start to finish.

# Analyzing the Components for Serial Manufacture

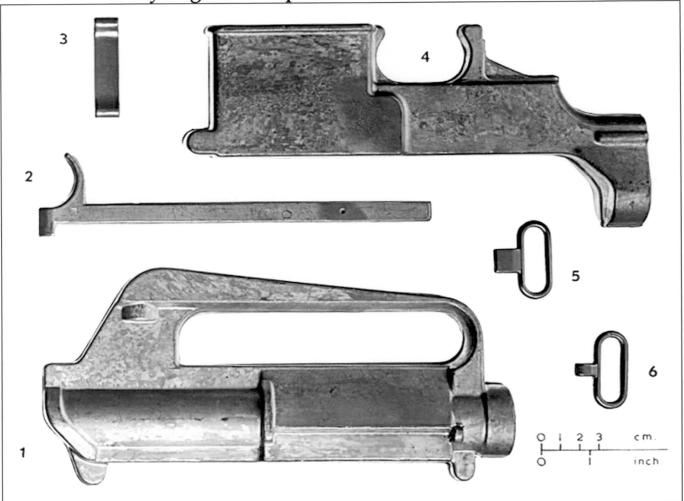

127. A view of the components of the AR-10 manufactured from forgings, showing the state of these as raw forgings before machining. Note the upper receiver (1) has been updated for use with the gas key and top-positioned gas tube.

Compare with fig. 129.

photo from Charles Kramer collection

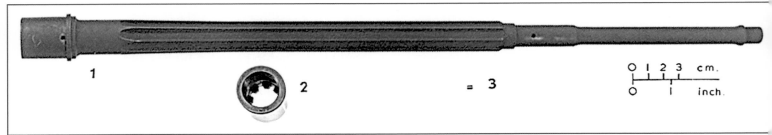

128. The hammer-forged and fluted steel barrel (1), shown fitted with the barrel extension and pin.

Below: the barrel extension (2) and locking pin (3).

photo from Charles Kramer collection

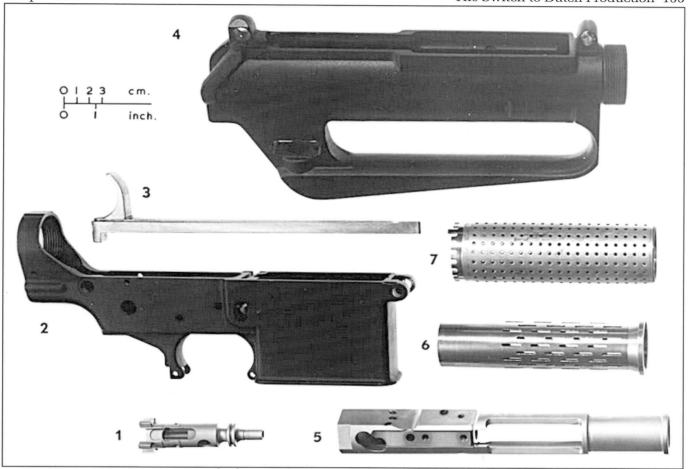

129. The forgings shown in fig. 127 after machining.
In addition, the inner and outer perforated sleeves of the recoil compensator/flash hider are shown, together with the bolt and carrier.
photo from Charles Kramer collection

One of the first orders of business at Zaandam was to analyze all the parts of the American AR-10B in order to decide how they best should be manufactured, whether by machining from forgings or bar stock, casting, or stamping, and if necessary find appropriate subcontractors to produce them.

As shown, those parts selected for production in-house included the major components of the receiver groups, flash hider, barrel assembly, and small fire control parts.

130 (right). The AR-10 components manufactured as stampings.
Clockwise from left: the buttplate (1), butt trap door (2), ejection port cover (3), and auto sear (4).
photo from Charles Kramer collection

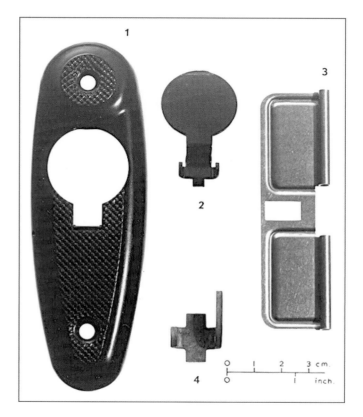

131. The AR-10 parts produced by lost-wax casting, shown as they came out of the molds.

From top left: front sight tower (1), recoil compensator seat (2), muzzle cap (3), hammer (4), trigger (5), sear (6), change lever (7), magazine release (8), bolt holdopen (9).

photo from Charles Kramer collection

132. The same parts as shown in fig. 131, after final machining operations had been performed.

photo from Charles Kramer collection

## Adding the Top-Mounted Gas System

At the same time, further changes had to be implemented to prepare the rifle for series production. Although the top-mounted gas system had been perfected and put on paper by L. James Sullivan in the spring of 1957, aside from the two experimental belt-feed models shown above in figs. 120 - 122, the ArmaLite shop in Hollywood had not actually produced any examples of a standard AR-10 incorporating this design change. Therefore, this mechanism had to be incorporated without even a working example of the finished design from which to make measurements, which was the main means of converting the dimensions of the other parts of the rifle.

## Further U.S. Experiments

Concurrently with the Dutch preparations for series production, ArmaLite's humble three-room machine shop continued to act a test bed for many small modifications, cranking out hand-made copies of the AR-10 and subtly tweaking the design with each successive example.

Although this was intended to improve performance and to test out different methods for increasing modularity, its main purpose was to equip the sales agents in their sojourns around the world with sample rifles. Time was of the essence, and while Artillerie-Inrichtingen geared up to begin full-scale production, it was important that the salesmen should have something more than just illustrated brochures to show to interested foreign parties. Fairchild also wanted ArmaLite to have a quantity of rifles on hand to send a potential buyer a sufficient quantity for small-scale troop trials. Hence, as mentioned above, production continued during the fall of 1957 until a total of 47 numbered examples of the ArmaLite AR-10 (plus Richard Boutelle's unnumbered "ghost" rifle) had been produced.

## Versatility Comes with a Price

In readying the AR-10 for its European debut, the engineers realized that they would no longer be dancing in the chains of the now-completed U.S. Army Ordnance Lightweight Rifle Program, and that they thus had a bit more freedom to adjust weight and other parameters. Fairchild Arms therefore sent their representatives Horace Kennedy, Leo Killen, and Edward Bishop to observe the military demonstrations which the sales organizations were already scheduling, and to interview the foreign military observers as to what functions and characteristics they desired on a rifle.

These sales trips not only proved fruitful and full of interesting stories which will be told in the next chapter, but they also revealed a treasure trove of information concerning what characteristics this "prime-time" AR-10 should feature.

At this time the NATO and other European countries on the free side of the Iron Curtain were quite keen on the ability of rifles to launch grenades, as could be done from M1 rifles and carbines fitted with accessory grenade launchers, which capability had impressed America's allies during WWII.

Many were also, like their U.S. counterparts, still intent on a rifle's ability to mount a bayonet, although this was deemed a secondary consideration, and the Americans and Dutch agreed that a bayonet-and-mount arrangement would be worked out on an order-by-order basis, and that this was therefore not necessary on the first generation of production weapons.

Although light weight was universally in demand, the ludicrous requirement of a weapon weighing seven pounds was not on the minds of European ordnance officers at the time, who understood, just as the Americans had learned, that it was simply not feasible in a conventionally-stocked shoulder rifle firing such a powerful cartridge. Along these lines, the engineers resolved to keep and accentuate the already-inherent feature of remarkably light weight in the AR-10 design, but not to do so to the extent that the long-term integrity of parts was threatened. The idea was to aim for reliability and durability first and foremost, and work to keep the weight as low as possible within those more-important parameters.

# Ominous Opposition

Once the production drawings were satisfactorily converted to the metric measurement system, the challenge of designing and procuring specialized machine tools for mass production had to be undertaken, and subcontractors who would manufacture many of the rifle's more specialized parts had to be located.

Unfortunately, the political factionalism within the Dutch government, while rampant, was not well understood or appreciated outside the Netherlands at this time. Unbeknownst to Boutelle or indeed to anyone in his organization, it transpired that Friedhelm Jüngeling, the Director of Artillerie-Inrichtingen and the Dutch Military Defense Materiel Organization, and Cornelis Kees Staf, the Dutch Minister of Defense, loathed one another. Thus Artillerie-

Inrichtingen encountered some particularly difficult opposition within the Dutch Defense Department, which was just the first sign of a caustic feud that would continue to dog the program and culminate just a few short years later in the termination of Dutch AR-10 production.

In the acquisition of proper tooling for production, this rivalry between political figures caused particular difficulties. Time and time again, requests for industrial machinery and orders for specialized tooling were either denied for suspect reasons, or delayed for unreasonably long periods of time. Defense Minister Staf did not try to hide his attempts to stall the work of Artillerie-Inrichtingen, and openly laid a plethora of bureaucratic obstacles and red tape in the way of Jüngeling and his organization.

## The Search for Competent Subcontractors

Comparably difficult, but not stemming from the bad faith and petty spite of the Dutch Minister of Defense, was the process of locating, evaluating, signing, educating, and equipping the needed subcontractors.

An item of particular importance was the requirement for hammer-forged steel barrels, with the exterior fluting cuts Stoner had introduced added to keep the barrel light and increase its strength, which would both help make the rifle extremely accurate and keep its overall weight down. As described on page 370 of *The Black Rifle*, the hammer-forging process consists of taking a carefully honed, smooth-bore "blank" and passing it through the powerful hammer forging machine, which exerts tons of pressure on the outside surface in a continuous and extremely rapid multi-hammering process, drawing the blank out to roughly one-and-a-half times its original length and cold-forging it down around a hardened mandrel. The mandrel, about six inches long, is perfectly formed in the shape of the chamber, leed and the first few inches of rifling. Concentrating initially on the rifling section, the blank is slowly rotated as it is hammered, the mandrel being simultaneously withdrawn until the end of the blank approaches, whereupon the mandrel stops while the hammers continue, thus cold-forming the complete bore and chamber configuration, all in one pass.

This was a very new concept in 1957, and industrial processes had simply not been developed to accomplish it in Holland. An agreement was reached with the *Gesellschaft für Fertigungstechnik und Maschinenbau* (GFM) company of Austria, which had pioneered the hammer-forging process, whereby the first-ever vertical hammer-forging apparatus was built and delivered to the works at Zaandam.

For a rifle barrel to be accurate, its internal dimensions must have the utmost uniformity and exactness and, as such, this process was predicted to produce barrels that boasted superior accuracy. The hardest, sharpest, and most exact tooling was necessary for this process, and so the initial startup costs were high.

Subcontractors were also necessary to produce other components for the rifle that the factory had determined were better outsourced than produced in-house. Finding a firm to manufacture rifle stocks made of plastic material, for example, turned out to be one of the more daunting challenges. In a communiqué thanking Fairchild personnel for their invaluable help in getting production moving, the Dutch referred to the search for a subcontractor to fabricate the furniture as "the plastic stock flap". This difficulty stemmed from the fact that, compared to the United States, the European plastics industry was still in its infancy. Many subcontractors misinterpreted instructions, mixed up the required chemical formulae, and in general caused headaches for the ArmaLite and Artillerie-Inrichtingen staff.

Finally, by the end of 1957, a suitable subcontractor for the furniture had been found in Dynamit Nobel A.G. of West Germany, which had been founded in 1865 by Alfred Nobel. Originally in the chemical business, particularly explosives, Dynamit Nobel had been involved with plastics since 1905, and had expanded greatly into this industry after the end of WWII, making them the natural best choice for working with new polymers. They correctly applied the chemical formula developed by Tom Tellefson in his ever-evolving quest for the perfect synthetic gunstock, and thus received the contract for manufacturing the AR-10 furniture for Artillerie-Inrichtingen.

The other major components which Artillerie-Inrichtingen concluded could not be as economically produced at their works as elsewhere were the magazines and telescopic sights. They also decided that if bayonets of a particular design were required by a purchaser, these would be subcontracted out as well.

## Initial Notes on Dutch-Made Magazines

Although ArmaLite had originally envisioned its aluminum magazines as throw-aways, discardable after use in the field, the standard European and American tactical doctrine for use of weapons that fired from detachable magazines (an area that was even newer to most continental forces than it was for the Americans) was to retain the empty feeding devices when at all possible so they could be refilled and reused. To this end, the Dutch instruction manuals issued to purchasers of the new AR-10s described how to change a magazine quickly without leaving it behind. This was done by using the support hand to catch the ejected magazine as it fell from the lower receiver when the trigger finger depressed the release button.

This was particularly stressed in what is known as a "tactical reload," in which a rifleman removes a partially-expended magazine from his rifle during a lull in combat or while in a position of relative safety, and replaces it with a full one, reserving the replaced magazine and its remaining ammunition for later use.

An "emergency reload," in contrast, is when a soldier is forced to change magazines because the current one is empty. In these cases, when the rifle had to be brought back to bear on the target as quickly as possible, magazines were usually let fall while the support hand withdrew a fresh one from the bandolier, saving the precious few seconds that would have been necessary to catch and stow the ejected magazine.

As further discussed in Chapter Ten, in order to improve durability, the magazines had been slightly redesigned by ArmaLite in coordination with Artillerie-Inrichtingen during the run-up to production. In each of the previously-smooth feed lips, a straight cut was made with a semi-circular "smile" or flange sticking up (fig. 294), which would contact the inner edge of the upper receiver and prevent the magazine from being inserted too far into the magazine well, which would cause jams if it were done on an open bolt (the only time magazine over-insertion was possible).

## Sourcing the Magazines

Since this semi-circular "smile" modification had already been implemented, the magazines were the perfect candidate for production by a large industrial metalworking facility. Bids were sought, and the Dutch subsidiary of the well-known German Rheinmetall firm, Nederlandsche Metaalfabrik, won the bidding process. They proved to be a model of smooth and transparent dealing, high manufacturing standards, efficiency of production, and excellent quality control throughout the production history of the Dutch AR-10, which was all-too-often a quagmire of inefficiency, government meddling, shortages of materials, missed quotas, and poor quality control.

In their first trial production runs, however, this company also produced magazines according to the original design, with smooth feed lips that lacked the "smile" cut. They also made a stamping error in the very first run of production, engraving "ARMALITE PATENT PENDING" on the bottom of the floorplate (fig. 295), and when discovering their mistake, added an "S" between the words "patent" and "pending," so on the earliest of all Dutch-produced AR-10 magazines, there is no space between the last two stamped words.

## The Rare Dutch-Produced AR-10 Magazine Loader

To accompany the magazines, a loader was also devised, which would fit over the top of the magazine and allow easy loading in the field from ammunition stored on standard five-round charging strips. While cosmetically almost identical to the relatively common FAL magazine loaders seen today, the loaders designed and produced by Nederlandsche Metaalfabrik for the AR-10 were uniquely dimensioned to fit perfectly atop the ArmaLite-designed magazines, and can easily be identified by the small transverse hole in both sides of their bodies.

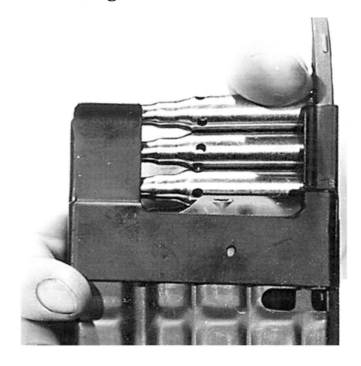

133 (right). An illustration from the first Artillerie-Inrichtingen handbook on the AR-10 demonstrates to prospective customers how to make use of the AR-10 magazine loader by pushing rounds from a stripper clip down into the magazine.

Note the transverse hole, a feature in both side walls of the purpose-built AR-10 loader.          author's collection

# The Home Stretch

With the proper subcontractors sourced, signed up and trained, it was now time for Artillerie-Inrichtingen to finish testing all of the mass-produced parts, and prepare for the onset of actual series production, which was scheduled to begin in January, 1958. As one might expect, there were delays and unexpected obstacles, but Jüngeling and his staff were fanatical about meeting their required deadline and being able to supply a world market hungrily awaiting the arrival of this fantastic new rifle. The publicity campaigns of both Fairchild and Artillerie-Inrichtingen were in full swing, and the sales and demonstration trips by the licensed salesmen were proceeding extremely well, as will be explored in the next chapter. All of the excitement these endeavors were generating added fuel to the fire of Dutch industriousness at the state weapons factory.

The problem of delays in the arrival and supply of tooling, materials, and machine equipment continued, but as the months progressed, these necessities began to trickle through in large enough quantities to allow production of some units to be realizable in the near future. The production goals set forth in the agreement finalized in July, 1957, called for the production of 10,000 rifles per month, beginning in January, 1958. By December, 1957, however, when the ArmaLite division was preparing its end-of-year progress report to the Fairchild board, the management admitted that there was "considerable doubt that [the] schedule will be met." They forecasted that production would more likely be closer to 2,500 units per month for a transitional period, until the full quota could be met. They also pointed out that production could also be quickly increased to a great degree to meet orders that were forthcoming.

As the deadline for the commencement of production was closing in, one final teething problem beset the Artillerie-Inrichtingen staff. The first barrels being cranked out on their Austrian hammer-forging machine were tested for quality control in December, 1957, by being installed on examples of the hand-made, U.S.-produced AR-10Bs, which continued to evolve as their production number increased. To the dismay of the engineers, test firings of these barrels produced extremely poor accuracy, which was not a possibility anyone had foreseen. The AR-10 had always exhibited peerless accuracy since its very first prototype phase, and at first it was not clear what was causing this problem.

Jacques Michault recounted later that all the newly-forged barrels, ready and waiting to be installed in the first rifles, had to be recalled. It was discovered in due course of cooperation between Artillerie-Inrichtingen and ArmaLite that the hammer-forging process, as well as the subsequent machining of the external barrel flutes, had highly stressed the barrels. This made the metal take on unpredictable and chaotic resonant and harmonic qualities, to the detriment of accuracy, which is so dependent on uniformity and consistency in the metal throughout the barrel. The stress areas, which would pop up in unexpected and unpredictable places, were essentially parts of the steel that were more unnaturally compressed than others. This stressing was a disadvantage of cold-forging that had been overlooked, both because this technology was rather new in its application to firearms components, and because the exterior fluting process had exponentially increased the stress load, making the ill effects this phenomenon had on accuracy far more pronounced.

The solution at last became obvious: the barrels had to be stress-relieved after hammer-forging. This process involved heating the finished barrels and then slowly returning them to normal temperature. This practice allowed the more "tightly-wound" stress areas to relax somewhat, and made the metal grain structure within the barrel more uniform.

It was known, however, as a matter of both common industrial knowledge and common sense, that this heating and stress relief could slightly alter the dimensions of the finished product, the exactness of which was extremely important within the bore. To correct these issues, a process known as "lapping" was developed. This method of perfecting interior barrel uniformity, still used on ArmaLite's production AR-10s today, consists of forming a plug of hot lead inside the barrel and allowing it to completely fill the interior of the bore for a length of around a centimeter, and harden there. Once it is perfectly formed to the inside of the barrel, creating an airtight seal, the lead plug is forced through the entire length of the barrel, and out the other end. It is then examined with machine tools to determine any areas where it has been dented or scored.

The lead plug was much softer than the barrel steel (far too soft to damage or wear it during this process), and thus an imperfection or swell on the interior surface of the barrel would scratch or deform it. Once the plug was analyzed and any dents discovered, an abrasive compound and lubricant was poured over the plug and allowed to harden. Then the plug was pushed back and forth through the barrel, where the lubricant allowed it to move through the smooth areas, while the abrasive compound wore down the swells and imperfections. Finally, and usually after several plugs had been formed and used, the resistance throughout the bore became the same for the entire length, indicating that the barrel was perfectly uniform inside.

These two procedures—stress-relieving and lapping—were performed on all of the barrels that had been recalled from the line, and when samples of them were installed in the test rifles and fired

again, they displayed the AR-10's signature superlative accuracy. In pioneering these processes in their application to military rifle barrel production, Artillerie-Inrichtingen not only produced barrels that achieved a new standard in infantry rifle accuracy, but blazed the trail for what was to become the accepted practice in the manufacture of high-quality barrels. This industrial regimen would even be adopted by the Soviets, who were not only notorious for their poor quality control and lack of attention to material consistency, but whose infantry doctrine called for rapid advances with massed automatic fire, de-emphasizing the importance of the accurate, deliberate fire of individual riflemen.

## The Final Touches Before Dutch AR-10 Production Begins

As the bugs in the newly-manufactured parts were progressively worked out in Zaandam, representatives of both ArmaLite and Artillerie-Inrichtingen aggressively solicited input from military ordnance officials in prospective countries, to ascertain the characteristics their forces would desire in a new rifle. They intended to tailor the first production series of rifles, which would not only be used for demonstration, but hopefully sold to different nations in small quantities for troop trials, to the standards that the militaries expected, hoping thereby to not only make the AR-10 more attractive, but to streamline its production and minimize the number

and complexity of the changes they would have to make to the design later.

Naturally, the main source of military input during the ramp-up to production was from the Dutch forces. This was an obvious first choice, because the Netherlands, as was promised by Jüngeling, would likely adopt the rifle straightaway if it were produced domestically, and second, because Holland, being a member of NATO, would closely mirror the acceptance requirements of other countries in the alliance. The NATO nations were seen as the potential bread-and-butter for the AR-10 at this point, and it was thought that adoption by one prominent member would prompt others to follow suit.

## ArmaLite Muzzle Variations

Along these lines, several changes were made, marking the next step in the evolution that had taken Eugene Stoner's M-8/X-01 to its current incarnation as the AR-10.

The first alteration focused on the muzzle of the weapon. NATO had settled on the 22mm rifle grenade as standard equipment for the entire alliance. Artillerie-Inrichtingen initially followed Stoner's lead in attempting to solve this problem, as the AR-10's trademark recoil compensator would never be able to accept the 22mm mounting chamber of the rifle grenades, even if modified greatly.

The solution developed by ArmaLite was to modify several different types of devices that could all be interchanged with about the same ease as removing the improved pattern compensator, with its locking ring and pin. These consisted first of an integral bayonet lug that was welded to the bottom

of the barrel, forward of the front sight base, and about one third of the distance to the plain muzzle (fig. 134, center). This lug was short enough that the compensator could be screwed on over it without the two accessories making contact and having a deleterious effect on barrel harmonics.

The engineers scoured the massive international stockpile of surplus WWII bayonets in search of one whose muzzle ring would fit perfectly and snugly onto the naked muzzle of the AR-10's barrel when the recoil compensator was removed. They found it in the unique Italian folding-blade bayonet designed and manufactured specifically and solely for the Model 1938 Mannlicher-Carcano short rifle version of the Italian standard infantry rifle, which fit tightly on the crown of the AR-10B's muzzle, and the bayonet lug was thus positioned and dimensioned to allow a perfect lockup of this piece.

Addressing the second function desired by world militaries, the ability to fire rifle grenades, some copies of the ArmaLite AR-1 flash hider were reworked to screw onto the barrel threads of the AR-10's muzzle. Unlike the integral bayonet attachment, however, the AR-1 muzzle device could not easily be replaced by the standard AR-10 recoil compensator in the field. Also, it was noted that it would be awkward for infantry to have to remove their (possibly extremely hot) recoil compensators in the middle of combat in order to fix bayonets or launch grenades.

## The New Steel Buttplate and Trap

In a final change to the AR-10, the necessity for which ArmaLite personnel noted that they had never once predicted, the buttplate was strengthened. The early rifles had been assembled with a thin aluminum buttplate. In Europe, however, Stoner, Sullivan, Kennedy, and Dorchester noticed that European troops, from nations as diverse as Norway, Austria, Italy, and Spain, would keep time during long-distance cadenced marching by tapping the butts of their rifles on the road. Wanting to avoid large-scale complaints from soldiers who bent or broke their stocks or buttplates while marching, Artillerie-Inrichtingen changed the material from a thin aluminum sheet to a more robust steel stamping. This certainly did not make felt recoil any easier on the shoulder, but it did protect the buttstock against abuse.

Further, the Dutch forces requested that the rifle should be able to pack its own cleaning kit. The stock was obviously the only place where such an apparatus could be stored, and so the factory staff set about designing a compartment within it. The improved resin design of the furniture, which gave increased

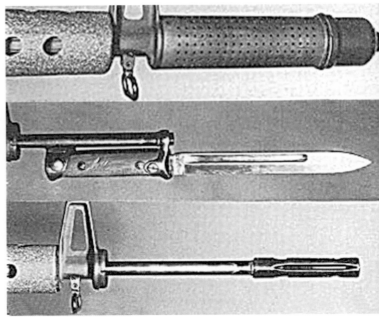

134. A photo montage, part of the ArmaLite promotional pamphlet put together in late 1957 (figs. 88 - 90), meant to demonstrate the modularity of the AR-10 in its ability to take several muzzle devices.

Above: with the standard recoil compensator installed.

Center: demonstrating the ability to mount the Italian Model 1938 bayonet on an attached bayonet lug.

Below: with a modified AR-1 flash hider attached to the barrel, making it possible to launch rifle grenades.

Institute of Military Technology

load-bearing strength and rigidity to the polymer parts of the weapon, allowed for the stock to retain its durability with a hollow compartment within it, located underneath the buffer tube assembly. To allow convenient access to the kit, a hinged, spring-loaded trap door was added to the new stamped steel buttplate. These components are both illustrated in fig. 130.

## ArmaLite Develops a Dedicated Grenade Launching Capability

135. Left side view of ArmaLite AR-10 no. 1017, fitted with a specially designed flash hider capable of launching rifle grenades.

It also utilized a bipod, in an attempt to allow the shooter to fire more effectively from the prone position.

Institute of Military Technology

136. Side view of an example of the first model of Tom Tellefson's improved handguard, as shown installed on ArmaLite AR-10 no. 1017 in fig. 135.

This updated ribbed design provided additional grenade-launching strength to the fiberglass shell filled with isocyanate foam.                     Charles Kramer collection

137 (right). Closeup view of the receiver end of the above handguard, intricately-machined by Tom Tellefson, dimensioned to fit around the barrel nut and accept the side-mounted gas tube.          Charles Kramer collection

138. Eugene Stoner test-firing training grenades from the experimental flash hider/grenade launcher fitted to AR-10 no. 1017 (fig. 135), shown with its bipod folded. These evaluations confirmed that not only could the AR-10 serve as a versatile grenade-launching platform, but the unique gas system would actually cycle the action after each grenade, allowing the rifleman to simply attach another grenade and fire, without having to work the charging handle as on all other grenade-launching automatic rifles to this day.          Smithsonian Institution collection

During the same time period, ArmaLite separately undertook another project to make the AR-10's muzzle functionality more "traditional," for those customers who might not consider rifle grenade launching a necessity. Back in California, the ArmaLite staff set up Richard Boutelle's personal, un-numbered "ghost" AR-10B so it could be fitted with a bayonet. On the already-exposed area of the barrel, they installed a large barrel ring which had at its bottom point a bayonet lug (fig. 110), located the correct distance from the recoil compensator to allow for the standard M1 rifle bayonet to be mounted directly onto the AR-10 with its compensator intact, so that the rear locking area locked onto the lug, and the muzzle ring fit around the cap at the front of the

muzzle device. There unfortunately exist no pictures of the bayonet actually being mounted on this particular rifle, the only one of the U.S.-made AR-10s to be so modified, but the rifle is pictured in fig. 115 with the compensator and its lug installed.

Finally, in a parallel project back at Artillerie-Inrichtingen, a way was sought to mount a bayonet without getting rid of the existing recoil compensator. As shown in fig. 139, the result was both simpler and more efficacious than ArmaLite's design, and consisted simply of mounting a bayonet lug directly to the bottom of a standard AR-10 second-model recoil compensator. The standard U.S. M1 bayonet would lock onto this lug, and its muzzle ring would fit over the end of the compensator cap.

# Proof of the Pudding - the First Completely Dutch-Made AR-10

139. Right side view of the very first assembly line-produced AR-10, this un-numbered, Dutch-fabricated AR-10 was the first of a great many evolutionary links between the distinctive and (relatively) well-known models of ArmaLite's superlative battle rifle. This is the first standard

AR-10 rifle to be fitted with the top-mounted gas tube.

Note the bayonet lug on the bottom of the recoil compensator, dimensioned for the standard U.S. M1 bayonet, the muzzle ring of which would fit over the end of the compensator cap.     Dutch Military Museum collection

This Dutch initiative, although small and not particularly significant in the path towards the AR-10's final, formidable incarnation, marked a huge historic milestone for the Artillerie-Inrichtingen plant, because to complete their bayonet-mounting experiment, they actually manufactured their very first AR-10. The existence of this rifle has been overlooked in the sparse historical records of the AR-10's history, but in December, 1957, before actual serially-numbered production began, a working AR-10 was assembled at Artillerie-Inrichtingen, completely from mass-produced parts. Making this even more of an achievement, one change was made to ArmaLite's AR-10B to allow Artillerie-Inrichtingen to use all of the important subcontractor-sourced parts; the gas tube was

moved to the top and the new-model bolt carrier with gas key was utilized.

Though appearing almost identical to its American predecessors, the higher ridge on the upper receiver inside the carrying handle indicates the use of the new top-mounted gas tube and charging handle, as well as the bolt carrier with gas key.

This un-numbered rifle marked the point of transition when the Dutch switched from simply installing and testing their own and subcontractor-produced parts in existing, hand-made AR-10s from the U.S., and actually started producing the entire rifle themselves. This was the very first industrially-manufactured AR-10 in history, and can rightfully be considered the seminal moment in military fabrication of this most wondrous rifle design.

## Sacrificing the Recoil Compensator

As these several programs reached fruition, the Royal Dutch Army and Marines informed the representatives of Artillerie-Inrichtingen, with whom they had been liaising, that the rifle they adopted had to be able to fire the NATO-standard grenade in its original configuration, without any disassembly or exchange of accessories. The 22mm interior diameter of the grenade tube was, as noted, much smaller than the diameter of the AR-10's recoil compensator, and so, after much deliberation, the decision was made that this feature would have to be eliminated from the production version of the rifle.

There was a silver lining to the move away from the original muzzle device. The reason it existed in the first place was to tame recoil in an almost impossibly-light weapon. Unlike the U.S. Ordnance Department, however, neither the Dutch nor the other potential European buyers included a ridiculous weight requirement in their selection criteria. Thus, while they were initially amazed at the AR-10's naturally light weight, compromises in parts integrity and longevity did not need to be made in a slightly heavier weapon. Accordingly, portions of the weapon such as the buffer tube were beefed up, making the weapon generally sturdier and better able to handle the increased recoil produced when launching rifle grenades.

Nevertheless, the decision to eliminate the very effective recoil compensator was a hotly-debated one, as felt recoil did considerably increase without it. In direct comparison with the sample M1 Garands and FALs that Artillerie-Inrichtingen had on hand, the felt recoil on the shoulder was considered by many of the staff to be higher with the AR-10, but control was still leagues ahead of its competition, for even without the muzzle device the straight-line stock, the concentric direct gas impingement system, and the strong action spring kept the rifle remarkably steady, even during rapid fire.

## Modifying the Rear Sight for Metric Measurements

The rear sight elevation wheel was also slightly modified from its original American form to provide accurate range settings in the European-standard metric system. The new dial for setting range was numbered sequentially from "2" to "5," with a few clicks made possible beyond "5" (the sight setting for 500 meters), bringing the maximum range with iron sights of the production AR-10 models to 540 meters when the trajectory was measured mathematically.

## Adding the Gas Adjustment Screw

Also incorporated into this first A-I model was an adjustment screw located in the front of the front sight base, which allowed for the amount of gas taken off to be adjusted. This ensured that the AR-10 would function well in a variety of climatic conditions and states of cleanliness, and also accommodate any low-quality or inconsistent ammunition that might be issued, particularly in the poorer areas of the world where the rifle might be used.

The early instruction handbooks for the AR-10s with this function (all the Dutch production models possessed some form of it) stated that gas adjustment could be made to allow less gas into the action for very hot environments, or when using ammunition that generated pressures significantly higher than the specifications for the NATO round. The intermediate setting, on which the rifle was kept as a default, would allow the best functioning in normal weather with ammunition that generated the correct amount

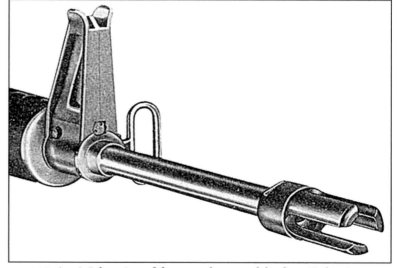

140. An A-I drawing of the muzzle area of the first (Cuban) production version of the AR-10. Note the gas adjustment screw in the front sight base.          author's collection

of pressure. The widest setting let an extra-large quantity of gas into the action, to allow proper functioning in very cold environments and with ammunition that was underpowered, or when the weapon's interior parts were very dirty and there was thus more friction in the action. What most soldiers and subsequent collectors of AR-10s found, however, was that the rifles, with their ingenious Stoner-designed gas systems, functioned perfectly, even in cold weather and with relatively low-pressure ammunition, with the gas adjustment set rather low. This was preferred when it still provided enough gas power to actuate the operating system, because the lower amount of propellant used to operate the action meant a noticeable reduction in recoil, lessened wear on the parts, and a smaller amount of carbon fouling in the receiver and on the bolt carrier that necessitated cleaning.

A means for cooling the barrel and protecting the rifleman's support hand was another consideration that was aided by the dropping of the stringent weight requirements, and a new type of handguard was designed which contained a sheet-aluminum liner attached to its inner surface, shaped to the exact contours of the inside of the handguard. This acted as a sink which absorbed radiated warmth from the air surrounding the barrel and gas tube, and allowed it to dissipate at a much greater rate than in a handguard made simply of fiberglass and foam plastic. Adding the rigidity and massively-improved heat dissipation properties of incorporating this aluminum liner to Tom Tellefson's already-improved construction formula for the handguard, which he had updated in August, 1957, a truly modern and rugged piece of furniture was realized.

## The First Numbered Dutch AR-10, Serial No. 000001

142. An illustration from the undated first A-I handbook titled "Description of the ArmaLite AR-10 Basic Infantry Weapon Caliber 7.62 NATO," a left side closeup of serial no. 000001 shows the standard receiver markings as used on the first several series of Dutch production weapons.
Michael Parker collection

143 (right). The A-I logo, as embroidered on the pocket of a factory workman's smock.
Institute of Military Technology

When the clock in the Artillerie-Inrichtingen plant struck midnight on New Year's Eve, 1957, all of the necessary parts, tested for quality, lay ready to pro-duce a working rifle of the first series. Within a few days, Mr. Hilary, a technician working on the end of the assembly line, proudly grasped in his hands ArmaLite AR-10 serial number 000001. Ceremoni-ously, he departed the line as the other workers watched and applauded, and proudly made his way to the management office, where Director Friedhelm Jüngeling formally accepted the offered rifle. The first model Dutch AR-10 was born.

In a coup, production did begin on time after all, with rifles quickly starting to roll off the line, and the parts accumulating for more, just waiting for orders to come in so they could be assembled. The ecstatic Dutch engineers sent a congratulatory cable to Fairchild in Maryland and ArmaLite in California, thanking them for their help, and expressing their faith in the bright future of the enterprise.

## A Photographic Record of Dutch Production

Around the time when the first rifles were being put together, Peter Marcuse, an aspiring photographer, had recently finished his military service, during which he had become acquainted with Friedhelm Jüngeling. He was invited to serve as the official photographer for Artillerie-Inrichtingen's AR-10 pro-duction effort. Over the next four years, his camera would preserve in high-resolution black-and-white the only photographic account of the AR-10's brief shining moment as a mass-produced military rifle. Marcuse, coming from Zaandam, was, as he remains today, very personally invested in the history of the Dutch-produced ArmaLite AR-10.

1 (previous page). Commemorating the coming to frui-on of the partnership between ArmaLite and Artillerie-richtingen, Friedhelm Jüngeling of the Dutch anufacturing concern (left) presents Fairchild president chard Boutelle (right) with AR-10 number 000001 at a remony at Boutelle's Hagerstown, Maryland home on arch 10, 1958.          Smithsonian Institution collection

144 (right). A machinist observes the process of forming the lugs and profile of the AR-10 bolt. A finished bolt stands on the machine at lower center.
Institute of Military Technology

145. A skilled operator examines an AR-10 barrel on a traditional barrel straightening fixture to ensure the absolute conformity of the bore.

photo courtesy Peter Marcuse

146. A technician checks the straightness of an AR-10 barrel by means of a sensitive measuring gauge.

photo courtesy Peter Marcuse

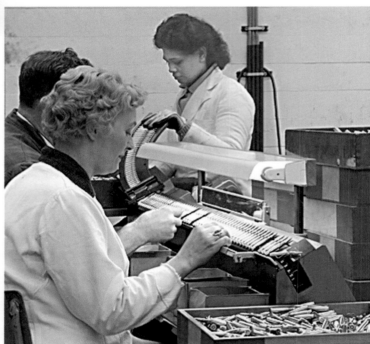

147 (right). Equally important to the success of the Dutch AR-10 program was the manufacture of high-quality 7.62mm NATO ammunition by the A-I small arms ammunition division. Here women inspectors examine newly-produced cartridges to cull out any defective rounds.

photo courtesy Peter Marcuse

# Scoping the Dutch AR-10

Artillerie-Inrichtingen wanted to offer optional telescopic sights, which could be mounted on test rifles for the salesmen to demonstrate to foreign armies. These scopes needed to be not only specially-designed for the AR-10, but rugged and light.

They began by examining and testing several of the standard type of large, powerful scopes seen on sporting and dedicated sniper rifles, which featured long bodies, large adjustment knobs, and wide ocular and objective lenses. After attempting to produce some telescopic sights in-house, however, they quickly reached the conclusion that they needed to outsource production.

After the initial testing phase of scopes designed for use by dedicated snipers, A-I opted to follow the doctrine of low-power scopes for more general use by regular infantry. These would be small, short and compact, with a narrow uniform width which would not only help make the scopes lighter, but stockier and more robust. They settled on a design that offered 3x magnification, and had a 25mm objective lens, a unit that drew much inspiration from the German ZF-4 scope of WWII, and was dwarfed by most conventional telescopic sights of the time, including their own first experimental models.

148. The original 3x25 telescopic sight, serial no. 015, manufactured by DelftOptik for use on Dutch AR-10s with suitably modified carrying handles.

Note as described in the handbook (fig. 150) the adjustments are made by means of concentric rings, with windage at the front and elevation at the rear.

Michael Parker collection

Although they were successful in designing a scope that met their requirements, producing it in-house was determined to be prohibitively expensive; several times the cost of fabricating a working rifle with accessories. They therefore identified the nearby company DelftOptik (located in the city of Delft) which would produce the telescopic sights for $140 each, a figure still considerably higher than even the eventual sales price (including royalties to sales firms and Fairchild) of an entire AR-10; equivalent to around $1,200 today. The price, although exorbitant, was considered acceptable, as telescopic sights are extremely demanding pieces of engineering, and such technology was in its early stages.

In addition to boasting extremely precise optical qualities, a rifle scope must also be able to handle not only the inherent violent recoil of a rifle firing, but also the bumps, dings, moisture, and dirt of a battle-

149 (right). The cover of the small four-page handbook produced by Artillerie-Inrichtingen in August, 1958 to illustrate and describe the operation of the AR-10 rifle scope.                                          editor's collection

# „Artillerie-Inrichtingen"

HEMBRUG — ZAANDAM      THE NETHERLANDS

Description of
Sniper Scope 3 x 25
for ArmaLite AR-10 Rifle

## Sniper Scope 3 x 25
## for ArmaLite AR-10 Rifle

### 1. Technical Data

| | | |
|---|---|---|
| Magnification .... | 3x | |
| Field of View .. | 5° | |
| Dia. Exit Pupil .. | 8,33 mm | ( .328" ) |
| Distance between E.P. and End of Sight ........... | 85 mm | ( 3.4" ) |
| Setting of Distance | 100 up to 700 m | |
| Setting of Windage left and right, maximum ...... | 5⁰/₀₀ | (5 mils) |
| Net Weight ...... | 270 g | (9½ oz) |
| Optics ......... | „coated" | |

### 2. Fastening to Rifle

2.1. Turn down open sight notch as far as possible.

2.2. Clean support faces thoroughly.

2.3. Fasten scope with the aid of its screw and the knurled nut on upper receiver. Screw on this nut tightly, so that the scope be firmly pulled into its seat.

**Note:**

The knob with division 1 - 7 shall be directed **backward.** Be careful that this knob does not get into touch with the rear sight.

### 3. Adjustment

3.1. The rear knurled knob of the sniper scope is used for setting the elevation and for that purpose it has been provided with a graduated ring, bearing the numbers 1 up to and inclusive of 7 (hundreds of metres). One click corresponds to a vertical displacement of the line of sight by .5 mils (i.e. 5 centimetres per 100 metres). For adjustment purposes the graduated ring had been made stiffly revolvable in respect of the knurled knob, and it is turned either by hand or by means of a special wrench, which grips in one of the holes 2 mm diameter in the graduated ring.

3.2. The **front** knurled knob permits the setting of the **windage;** to that end, a graduated ring has been applied on it, with a zero line and divisions unto 5 mils on either side.
One click corresponds to a lateral displacement of the line of sight by .5 mils (i.e. 5 centimetres per 100 metres). The adjustment is effected as indicated in point 3.1, last paragraph.

**Observation**

In view of the fact that the scope is to be found about 8 cm above the barrel, in the case of testfiring at distances shorter than 100 metres the distance knob should be turned in the direction of **greater** distances.
For the usual test-firing distance of 50 metres the distance knob should be set at "2".

151-0858

150. The two pages of instructions from the August, 1958 A-I handbook, covering the use and adjustment of the original 3x25 telescopic sight produced for the AR-10.
editor's collection

field environment. This reliability and ruggedness against environmental factors was stressed in the extreme, and as such, a very expensive scope was deemed not only acceptable, but necessary. In comparison to rifle scopes today, the Artillerie-Inrichtingen-designed unit would be considered mediocre in performance and extremely expensive in price, and while boasting excellent ruggedness by the standards of any era, its lenses were not fog-proof and the unit was not waterproof by contemporary standards. For its time, however, the DelftOptik unit had excellent optical qualities, and was small and robust, offering unmatched ruggedness and reliability.

# The Original Dutch AR-10 Sling

While the Dutch had been making use of surplus U.S. M1 Garand slings for their initial development of the AR-10, they made the conscious decision to produce a distinctive carrying strap for the first series of rifles. In doing so they opted for the more traditional leather sling which was still preferred to stabilize the weapon for accurate, long-range shooting. The piece they designed was both unique and conventional, closely resembling that of the WWII German K98k carbine. It was relatively narrow, measuring just 7/8" in width, with three distinctive fixed leather keepers to hold the loose ends. On one end was a button that fit through a notch, making a closed end, secured with a single keeper. This part was attached to the rear sling mounting point, while the other end with an integral steel buckle was fixed to the front sight base. This front end held the metal adjustment buckle and other two keepers.

151 (right). A photograph from October, 1958 shows one of the first Dutch AR-10s sent to the U.S. for examination being tried out for size by U.S. Army Vice Chief of Staff General Lyman Lemnitzer. The venue appears to be the Fairchild booth at a prestigious East Coast trade show.

Note the rifle is equipped with the rare first type of AR-10 sling, initially made of light tan leather as shown here but later given a darker brown color.

Michael Parker collection

## The Pros and Cons of Naming the Dutch AR-10 Models

The Dutch AR-10 model first represented by serial no. 000001 would come to be known by several names. However, it is the opinion of this author that, where possible, the tradition of naming the production AR-10 models after the countries that adopted them is the easiest to follow, and so this earliest of serially numbered variants should, as we shall see in Chapter Six, correctly be called the "Cuban" model.

There are disadvantages to this system, but identifying AR-10s by models sequentially (first, second, third, etc.), is also problematic, because it is difficult to delineate the exact beginning and end of a distinct model, as the AR-10 went through more than one period of intense evolution, during which certain different components were used interchangeably. This is perhaps the reason why naming production versions of the AR-10 after their country of adoption became the standard in the U.S. collecting and historical community, and as the names for the most famous and heavily-produced models are well established, this is the method that will be followed here.

## Stop the Press: a Roundup of Early U.S.-Made AR-10 Magazines

As we were in the last stages of preparation for printing this book, some material arrived as a result of a fortuitous find on the part of Eric Kincel, a former employee of Knight's Armament who is now the Director of Research and Development for Bravo Company USA Inc.

The photos on the following page, received too late to be assigned figure numbers in the normal sequential series, show four variations of rare early U.S.-made AR-10 magazines. The bodies of these were actually made in two pieces, with separate left and right sides pressed out of sheet aluminum by Tom Tellefson in his garage, formed on dies which he had constructed of epoxy. The two halves were then welded together, front and back.

As can be seen, there are some interesting variations in the form of the components, as well as in the overall length of these hand-made magazines.

The earliest, shown here, were unmarked, while toward the end of U.S. production the floorplates were marked "ARMALITE" in a smaller font than was later used on magazines of Dutch production (fig. 295).

151a. Top view of four early U.S. hand-made AR-10 magazines. Only the second from left has the "smiles" cut into the feed lips to act as stop tabs to prevent over-insertion of the magazine into the upper receiver.

The two leftmost magazines have die-cast followers made of aluminum or zinc, which are connected to the magazine spring.

The two rightmost magazines have machined aluminum followers which are not connected to the magazine spring. They appear to be anodized and dry lube coated.

Eric Kincel collection

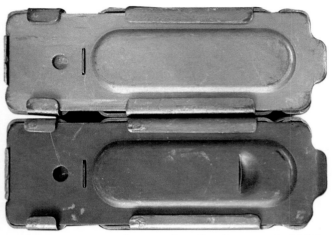

151b. Bottom view of two styles of early unmarked U.S.-made floorplates. Compare with Dutch production, shown in fig. 295.                                    Eric Kincel collection

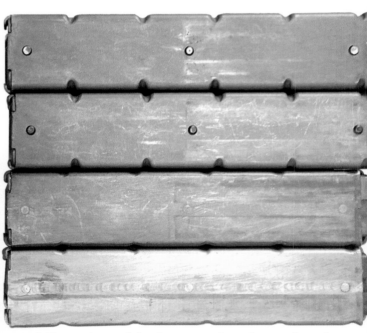

151c. A side view comparison of magazines fitted with the two types of followers shown in fig. 151a.

Note the body at right, with a machined aluminum follower, is approximately 1/8" shorter than the magazine at left, which has a die-cast aluminum or zinc follower.

Eric Kincel collection

151d. Front views of the four magazines shown above in fig. 151a, showing welded and flushed seams. There is another seam up the back, joining the two halves of the magazine body together.

Note the rivets holding the front reinforcing strip, described in the patent (page 58) as preventing the fronts of the cartridges from denting or rupturing the front wall during firing.                                    Eric Kincel collection

*Chapter Six*

# The Grand Tour Begins

**M**eanwhile, in February, 1957, when the brass from the Springfield Armory tests was barely cold on the range floor and the sales and production agreements had not yet been finalized, ArmaLite staff undertook a demonstration tour of Europe to excite interest in military circles for their mysterious, advanced American rifle. Making their way from Austria through Germany, Sweden, Denmark, France, Italy, and finally to Spain in May, ArmaLite's representatives exhibited the wondrous performance of the AR-10 to cadres of infantry and ordnance officers of the new NATO alliance. In a tale of redemption fit for Dostoyevsky, the trusty rifle that accompanied these employees of the plucky American firearms design firm was none other than AR-10B number 1002, replete with its new chromoly barrel and replaced handguard. The same rifle that had experienced a catastrophic failure and nearly cost a young American technician his left hand, all because of the hubris of George Sullivan and the sabotage of Studler and Carten, now tore through courses of fire with round counts that made the Springfield evaluations look like summer days shooting soda cans in the countryside.

These demonstrations would set the stage for more in-depth testing and product improvements that would take the AR-10 not only to its perfected form, but eventually into some of the fiercest combat of the 20th century, under the harshest of conditions. However, it was in oft-overlooked and far less strategically important areas that some of the most fascinating and colorful chapters of the AR-10's history unfolded.

## The Salesmen Start to Work Their Territories Abroad

Later, after Dutch production had begun and the world had been divided into sales regions, which established Jacques Michault of SIDEM, Boutelle's own Fairchild Arms International, and Bobby MacDonald of Cooper-MacDonald as the AR-10's appointed representatives in the developed and industrialized nations of Europe, North America, and the burgeoning markets of Southeast Asia, Sam Cummings's International Armament Corporation (Interarmco) had been selected to represent the AR-10 in some of the most lawless and undeveloped nations on earth. Instead of treating this as a potential hindrance, however, the master arms merchant saw the chaos, disunity, and corrupt authoritarianism in his sales territory as an advantage in the marketing of weapons of war. Cummings's previous experience in that part of the world had convinced him that it was often in the most destitute of countries where the most effective weapons were craved.

Over the next two years, Cummings would prove extraordinarily prolific in aggressively pushing the AR-10 all over Latin America and Africa, and would not only secure the largest orders in the history of the Dutch-produced AR-10, but in so doing would make many new important acquaintances in high places, and experience events that stand out even in the animated history of this most excellent battle rifle.

Cummings had made sure that his company had the impressive credentials to back up his aggressive sales approach. These were described on a further page of the ArmaLite AR-10 brochure shown in figs. 152 and 153 as follows:

> *Interarmco is a fully registered importer and exporter with the U.S. Dept. of State in all categories of armaments. Registered armaments dealer, manufacturer, and Class I special occupational tax payer (automatic weapons) with the U.S. Treasury Department; U.S. Federal Firearms and Manufacturer's License No. 1, District of Maryland.*

From the modern design staff and metallurgical experts of one of the world's leading aircraft builders, FAIRCHILD, has sprung the ultimate in an advanced, lightweight military rifle—the amazing ARMALITE AR-10—"Tomorrow's Rifle Today".

**Tomorrow's RIFLE TODAY!**

LIGHTWEIGHT AUTOMATIC

BASIC INFANTRY WEAPON

**THE ARMALITE AR-10**

LIGHTEST WEIGHT!

FINEST WORKMANSHIP!

MOST VERSATILE DESIGN!

AVAILABLE NOW THROUGH THE GLOBAL FACILITIES OF
INTERARMCO, AMERICA'S LEADING ARMAMENT DEALER

THE SENSATIONAL
FAIRCHILD
ARMALITE AR-10
U.S. CALIBER .30 T-65

**INTERARMCO**
P. O. BOX 3722
WASHINGTON 7, D.C. U.S.A.
CABLE "INTERARMCO"

**INTERARMCO**
P. O. BOX 3722
WASHINGTON 7, D.C. U.S.A.
CABLE "INTERARMCO"

And now, through the exclusive facilities of INTERARMCO, this lightweight weapon is available to all nations under U. S. Department of State export licensing, and at a cost which is amazingly low.

Furthermore, when the sensational Armalite AR-10 is purchased through INTERARMCO —— Interarmco, and *only* Interarmco —— will make a generous trade-in allowance on all old, obsolete, worn-out or useless small arms and ammunition, regardless of age, type or condition.

**BUY FROM THE LEADER!     SELL TO THE LEADER!**

**BUY FROM INTERARMCO—SELL TO INTERARMCO**

152. The first page of ArmaLite's famous "Tomorrow's Rifle Today" brochure, with Sam Cummings's Washington address clearly printed at bottom right.
*Michael Parker collection*

153. The second page of the ArmaLite AR-10 brochure, purpose-printed to clearly stress the advantages of dealing with Cummings. The forceful message "BUY FROM INTERARMCO—SELL TO INTERARMCO" would not have gone unnoticed in countries with obsolete surplus weapons on their hands.          *Michael Parker collection*

The brochure went on to state in no uncertain terms the offer that "INTERARMCO WILL DEMONSTRATE THE AMAZING AR-10 LIGHTWEIGHT RIFLE AT ANY TIME AT ITS OFFICES AND PROV-

ING GROUNDS NEAR WASHINGTON, D.C., OR, UPON INVITATION, ANYWHERE IN THE WORLD!"

## Jumping the Gun

As we have seen, the first serially-numbered Dutch AR-10 was completed during the first few days of January, 1958, although as discussed below series production did not get under way until the Cuban order was manufactured during the middle of 1958, with the 100 ordered rifles not shipped from Cummings's Manchester warehouse to Cuba until that November.

Meanwhile, as recounted below, the eager salesmen's first and second tours of demonstrations in a number of countries around the world were over by the middle of 1958, with the rifles demonstrated perforce being the old faithful original ArmaLite AR-10Bs which the salesmen themselves had purchased from Fairchild, for the simple reason that no new Dutch rifles were as yet available.

## Overview: Three Strikes Already Against the AR-10

Along with the issue of U.S.-supplied surplus weaponry being considered "good enough" (or certainly cheap enough) by a number of countries, and fledgling new domestic designs being championed by others, this sheer unavailability of product was to become the most persistent stumbling block to adoption for many who were lavish in their praise of the AR-10, but simply could not wait for Dutch production to catch up.

# The Cuban AR-10

## Sam Cummings Meets Fulgencio Batista

One of Cummings's first stops in his sales tour was Cuba, in early 1957. President Fulgencio Batista was at the time dealing with a worrisome nationalist revolt, and was particularly keen on giving his government forces an edge over the rebels. Cummings arrived in Havana, and put on an enthralling demonstration of his own personal AR-10, serial no. 1001, at a military range. With the spectacular results that would become a hallmark of Cummings's performances whenever he exhibited the AR-10, he executed different trick shots and showed off the AR-10's controllability by guiding strings of automatic fire effortlessly in straight lines. To conclude, he retrieved a magazine he had loaded with tracer ammunition, and fired it into cans of gasoline he had set up on the range. The resulting fireworks were later described by Cummings as "better than a John Wayne movie".

Batista eagerly purchased one of the demonstration AR-10B models and, after a small amount of firing with it, placed an order with Cummings for 100 rifles with which to equip his élite presidential guard. These rifles were specified to be the type currently being prepared for manufacture in Holland. Confident that larger orders were sure to follow, Cummings was very happy about making this first-ever firm sale.

154. Fulgencio Batista, president of Cuba, whose American friends included the likes of Meyer Lansky, Charles "Lucky" Luciano—and Sam Cummings.
public domain - the Internet

## Castro's Rebels Gain Strength as the Cuban AR-10s are Manufactured

The increasingly well-organized rebel forces hiding in the countryside and mounting ever-more daring and successful raids against Batista's military at the time Cummings made his first visit to Cuba were led by none other than Fidel Castro. Not openly possessed of communist or pro-Soviet leanings at the time, the charismatic nationalist leader attracted to his cause not only the disenfranchised poor of the country, but also idealistic volunteers from places like the United States, keen on fighting for the liberty of an oppressed people.

During the time which elapsed between Batista's order being given to Cummings in early 1957 and the Dutch factory receiving and completing it, the revolutionaries continued to make steady gains all around the island country. It was already the summer of 1958 by the time the 100 AR-10s for Batista's forces were off the line in Zaandam, and the forces still loyal to the unpopular dictator were rapidly losing ground in desperate holding actions while great swaths of the military were switching over to the side of the nationalists.

A favorite business tactic of Cummings's, which he would use to great effect many times in the future, further delayed Batista's badly-needed AR-10s, even as his grip on the country was unraveling completely. For several years Cummings had based a large part of his business in Manchester, England, and one of his main warehouses was located there. As he usually purchased weapons without a designated buyer already lined up, he had the standard practice of first importing any weapons he acquired in that part of the world into England for storage at this facility. This importation into England required the rifles to be inspected, and then classified as being registered in that country.

At some point early in his relationship with Fairchild, Cummings realized that he could use this procedure to great advantage with relation to the sales of AR-10s, which were different from Cummings's usual deals in that when the weapons entered shipment, there was already a buyer. What he decided to do was quite ingenious. Instead of having the AR-10s sent straight to the purchaser, he would instead have them shipped to his Manchester warehouse.

When the rifles reached England and were registered there as imports into the country, Cummings would enter them as inventory and then ship them out again, this time framing the deal as if he were the seller, instead of A-I. This allowed him to not only earn his commission from Fairchild, but to charge an extra commission from the buyer. As he was the only authorized salesman in his "contract" areas, which included Cuba, the purchasing country had to either accept these terms or not receive AR-10s from any source. As he predicted, Batista, yearning for advanced arms to give his forces an edge against Castro's rebels, agreed to Cummings's demand.

## The Changing of the Guard

Finally, after strongarming the strongman, Cummings had the 100 AR-10s loaded onto a freighter bound for the Caribbean in November, 1958. However, by the time the ship docked in Havana harbor, Batista had taken a one-way flight to Spain, and the presidential palace had a new occupant. Only a few weeks into his new job, Fidel Castro received a letter directly from Cummings, informing him that payment had not yet been tendered for the 100 AR-10 rifles of which the new president had recently taken delivery, and asking Castro how he would like to proceed with settling the account.

In response, Castro invited Cummings to return to Cuba to discuss the matter, as well as the new, extremely interesting rifles. The Interarmco founder obliged, and in early 1959, joined Fidel and Raul Castro, who was later to succeed his brother as the Cuban president in 2008, as well as the infamous revolutionary Che Guevara, at the firing range of the Defense Ministry barracks in Havana to test the AR-10. Cummings had stood in just that spot several

155. Fidel Castro, shown shortly after his triumphal entry into Havana after having ousted Fulgencio Batista.

Although he opposed his predecessor's policies vigorously, Castro was just as fascinated by the AR-10.

public domain - the Internet

times, most recently while demonstrating his own AR-10 to the despot who had been hell-bent on wiping out his current hosts. For their part Castro and his party were duly impressed at the apparent combat prowess of the new Dutch AR-10s, and Castro decided on the spot to pay for and keep the rifles. Che remarked that if the revolutionary forces had possessed these 100 rifles in their mountain hideouts,

they could have moved in on and taken the capital at once!

It has been reported, though never confirmed, that both Fidel and Raul Castro were so enamored with the AR-10 that they each kept one of the 100 new rifles as their own personal weapons. The new acquaintances even discussed further orders, as Castro made it plain that he desired further quantities of AR-10s to equip his armed forces.

## Describing the First A-I (Cuban) Model AR-10

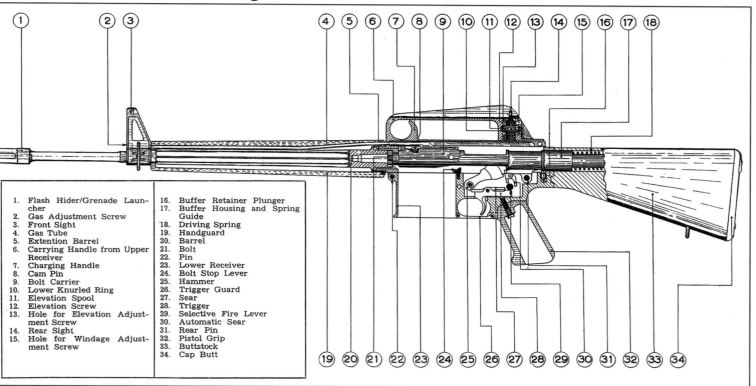

| | |
|---|---|
| 1. Flash Hider/Grenade Launcher | 16. Buffer Retainer Plunger |
| 2. Gas Adjustment Screw | 17. Buffer Housing and Spring Guide |
| 3. Front Sight | 18. Driving Spring |
| 4. Gas Tube | 19. Handguard |
| 5. Extention Barrel | 20. Barrel |
| 6. Carrying Handle from Upper Receiver | 21. Bolt |
| 7. Charging Handle | 22. Pin |
| 8. Cam Pin | 23. Lower Receiver |
| 9. Bolt Carrier | 24. Bolt Stop Lever |
| 10. Lower Knurled Ring | 25. Hammer |
| 11. Elevation Spool | 26. Trigger Guard |
| 12. Elevation Screw | 27. Sear |
| 13. Hole for Elevation Adjustment Screw | 28. Trigger |
| 14. Rear Sight | 29. Selective Fire Lever |
| 15. Hole for Windage Adjustment Screw | 30. Automatic Sear |
| | 31. Rear Pin |
| | 32. Pistol Grip |
| | 33. Buttstock |
| | 34. Cap Butt |

156. A fold-out diagram from the undated Artillerie-Inrichtingen handbook covering the Cuban model AR-10.

Although this first detailed instruction manual was not printed until after the shipment of AR-10s had been sent on its way to Havana, the rifles accepted by Castro were accompanied by a similar large fold-out diagram showing all of the parts and inner workings of this first A-I production model of the AR-10.          Michael Parker collection

Named for its seminal sale to the Republic of Cuba, this first Dutch production model of the AR-10 proved a marvel to the few who had the privilege of using it.

The Cuban AR-10 looked distinctly different from its predecessor, the American-made AR-10B. Most noticeably, instead of the trademark recoil compensator found on AR-10s since the 1955/56 demonstrations, the plain barrel forward of the handguard, extremely thin for its caliber by today's standards, was simply capped at the muzzle end by an open three-prong flash hider, which measured exactly 22

millimeters in diameter and was capable of launching NATO rifle grenades.

The front sight base was also new, although it had actually been used on the original Dutch-produced AR-10B prototype that incorporated the top-mounted gas tube. Everything else externally was the same, and internally the shorter gas tube and gas key on the bolt carrier were present on this model.

The carrying handle was sleeker, and rose from the rear at a somewhat lesser angle, giving it a lighter, more streamlined look than that of the the "humpback," as the AR-10B is sometimes called.

157. Right side view of a typical Cuban model AR-10.
Note the distinctive open three-prong flash hider on the
muzzle end of the thin barrel.

Institute of Military Technology

158. Left side closeup of a typical Cuban model AR-10,
serial no. 000776, shown broken open for field stripping
by simply pushing out the rear assembly pin with the point
of a cartridge. The cocking handle and bolt assembly have
been removed from the upper receiver, and the magazine
disengaged.

One of the features, aside from the light weight, accu-

racy, and controllability, that appealed so immediately to
the Castro brothers was the extreme ease with which the
AR-10 could be disassembled in the field to be cleaned and
maintained.

Note the small section of a Fisher Scientific metric ruler
included as a measuring aid, for many years the trademark
of the Ontario Centre of Forensic Sciences.

photo from Charles Kramer collection

159. Right side view of a late Cuban-pattern AR-10, serial no. 002478, one of the very first Dutch models to be modified to accept the 3x25 telescopic sight (fig. 148).

As shown, the furniture on this rifle was painted green. Different color schemes were experimented with as being of potential interest to foreign buyers.

Dutch Military Museum collection

The streamlined handguard, containing both more insulation and an aluminum heat shield, provided extra comfort when firing this model for long periods. The silver color of the aluminum liner inside the sleeker handguard could be seen along the edges of the vent holes, which were now cut in four rows on the top, bottom, and sides of the handguard to allow air to circulate.

The Cuban model was also slightly heavier than the AR-10B, weighing 7 lbs. 2 oz.

Aside from one example (the first off the line) which had been ceremonially presented to Richard Boutelle (fig. 141), the quantity delivered to West Germany for trials, and a single example provided by ArmaLite for U.S. Army evaluation at the Detroit Arsenal (fig. 167), nearly all the the other examples of this model were sent to Cuba. The further exceptions were a few of these first production models which from time to time were used as test beds for various experiments and trials, such as no. 002478, shown above.

## The U.S. Embargo Shuts Off Further Supply - No More Cuban AR-10s

World affairs would bring down the curtain on this business relationship, however, as before any arrangements could be made for further purchases, the United States imposed embargos on the island nation, and the Dutch were no longer willing to sell rifles to the increasingly authoritarian and Soviet-aligned Castro. Cummings was still able to procure quantities of 7.62x51mm NATO ammunition and modern FN FAL rifles (the AR-10's most able nemesis) for Castro without offending the U.S. government, but the possibility of Castro acquiring any more AR-10s was closed off forever by this turn of events.

Just like every distinct version and shipment of Dutch AR-10s ever received by a military force, the 100 "Cuban-pattern" AR-10s would go on to play an interesting and influential role in the history of the region where they ended up. This story, along with the history of combat service of the AR-10 elsewhere, is a topic for another chapter.

## Playing Both Sides: Cummings Sells Two Central American Rivals on the AR-10

In the course of just one of many seminal deals in his private career as an arms merchant—acquiring and flipping the surplus Panamanian arms stash discussed earlier—Cummings had made some powerful friends. Among them were Panamanian President José Figueres Ferrer, Nicaraguan President Anastasio

Somoza García, and Guatemalan President Jacobo Árbenz. These three leaders all shared two things; a violent relationship with one another, and arms dealings with Cummings before he had ever even heard of ArmaLite or the AR-10.

## Flashback: the Guatemalan Caper

The last mission Sam Cummings performed for the CIA before deciding to go into business for himself was that of acquiring arms for Colonel Carlos Castillo Armas, the leader of the opposition to Árbenz's rule in Guatemala, to help him in a proposed attempt to take power back and oust the feared Soviet-friendly president. This plan was uncovered by President Truman's Secretary of State, however, and cancelled.

Under the Truman Administration, although there was fear that Guatemala would fall under Soviet sway, it was considered repugnant to actually sponsor and even lead the violent overthrow of a friendly government. Instead, the U.S. had imposed an arms embargo on the country, which all nations friendly to the U.S. were obliged to obey. This ban on foreign governments selling to Guatemala provided a possible opportunity for private arms dealers to fill a market need. Cummings explored this possibility, but was unwilling to act without approval from the U.S. government. To this end, however, he lobbied, and at the end of Truman's tenure, he was allowed to make private sales to the Central American nation, buying M1 rifles and Thompson submachine guns from government stores and selling them to the Guatemalans.

## Prelude to the Bay of Pigs - the CIA in Guatemala

Cummings's dealings were allowed to continue through the first months of Eisenhower's presidency in 1954, until the Dulles brothers, Eisenhower's Secretary of State John Foster Dulles and CIA Director Allen Dulles, got wind of the CIA plan for Árbenz's overthrow, and put it into action. This operation, which was carried out in June, 1954, involved Guatemalan exiles under the command and leadership of the Nicaraguan Army, trained, equipped, and funded by the CIA, invading from that neighboring country. Orchestrating the logistics for the incursion was General Anastasio Somoza DeBayle, son of the president of Nicaragua. He had been introduced to Cummings in 1951 by his father, and the two had stayed in touch.

Of course the Guatemalans resisted this invasion, using the M1 rifles which they had purchased from Cummings just a few months before. In the world of international arms deals, though, and particularly in Central America, the major players were unique in accepting the "just-business" role of the weapons brokers. It did not end up hurting Cummings's reputation and standing with his Nicaraguan contacts that guns he had sold were used against them, and in fact, they even went to him to sell the arms they captured from the Guatemalan insurgents.

# The Nicaraguan AR-10 - Close, but No Cigar

In 1956, the junior Somoza was appointed Director of the National Guard of Nicaragua. This organization was very different from its eponymous counterpart in the United States, and actually included the entire armed forces organization of the country, and also operated as the national police force. It had been established by the United States in one of its early-20th-century incursions into Latin America, and had become the private military organization of the rulers of the country. Unlike the territorial forces of many nations like Britain, Australia, and China, the Nicaraguan National Guard was actually an élite military force (for its size and funding), and was primarily used to put down insurrection and dissent. It received funding and training from the U.S., and was the perfect instrument to ensure that Nicaragua's military junta remained in power for most of the 20th century.

In September, 1957, with the Cuban order already on the books but not yet completed, Cummings resumed his sales tour of Latin America and the Caribbean islands in Nicaragua. Sporting his newly-acquired AR-10B, serial no. 1001, he went to work in his trademark style, putting on a show for the assembled general staff of the National Guard. Being a serious enthusiast of weapons himself, General Somoza got in on the act, joining Cummings on the firing range and testing out the rifle himself. Needless to say, he was thoroughly impressed by its performance, and he placed an order for 2,000 rifles on the spot, which was to be followed by another order for an additional 5,000. It was designated to be the official rifle of the National Guard, giving them an even greater firepower advantage over internal dissidents, and also against their neighbors. Somoza reasoned that a quantity of these advanced rifles in the

FORM 3 (FIREARMS)
REV. MARCH 1955

U. S. TREASURY DEPARTMENT - INTERNAL REVENUE SERVICE
**RETURN OF FIREARMS TRANSFERRED OR OTHERWISE DISPOSED OF**
(Chapter 53, Internal Revenue Code - See Instructions at bottom of form)

TO BE FILED WITH THE DIRECTOR, ALCOHOL AND TOBACCO TAX DIVISION, IMMEDIATELY UPON THE TRANSFER OF FIREARMS

1. FIREARMS TRANSFERRED OR OTHERWISE DISPOSED OF *(Date)* 10/4/57

2. DESCRIPTION OF TRANSFER

| AUTHORITY FOR TRANSFER 1/ (a) | PURCHASER OR TO WHOM TRANSFERRED: NAME (b) | ADDRESS (c) | KIND OF FIREARM (d) | MODEL (e) | CALIBER (f) | SERIAL NO. (g) |
|---|---|---|---|---|---|---|
| Refer Form 9 of 9/30/57 | Brigadier General Anastacio Somoza | Jefe Director, G.N. de Nicaragua, Managua, Nicaragua | SMG | AR-10 | NATO-30 | 1035 |
| | SHIPPED | 10/4/57 - Air Way-Bill No. 026-16-049020 | | | | |
| | | Via - Pan American Airlines | | | | |

1/ Indicate date of approval by Director, Alcohol and Tobacco Tax Division, of Forms 4 or 9 (Firearms) or date of application for exemption, Form 5 (Firearms), or in the case of a transferee registered as a special-tax payer (Firearms), the serial number of his special-tax stamp.

I declare under the penalties of perjury that the above statements are true and correct to the best of my knowledge and belief.

3. NAME AND ADDRESS OF SPECIAL-TAX PAYER *(Transferor) (Number and street, city or town, zone, and State or Territory)*
INTERARMCO LIMITED
10 PRINCE STREET
ALEXANDRIA, VIRGINIA

4. SIGNATURE
RICHARD BREED

5. DATE
4 October 1957

6. TITLE OR STATUS *(State whether individual owner, member of firm, or if officer of corporation, give title)*
VICE-PRESIDENT

7. SPECIAL-TAX STAMP NO.
375716

INSTRUCTIONS

1. PERSONS REQUIRED TO FILE RETURNS.—Immediately upon the transfer or other disposition of any firearm, every manufacturer, importer, and dealer (including pawnbroker) shall execute an accurate return on this form, in duplicate, setting forth the information called for in paragraph 2 below. All transactions occurring during a single day may be included in one return filed at the close of that business day. These returns shall be filed with the Director, Alcohol and Tobacco Tax Division, Washington 25, D. C. The duplicate will be retained by the person making the return for a period of 4 years, and be at all times readily accessible for inspection. (Sec. 5842, Internal Revenue Code of 1954.)

2. PERSONS REQUIRED TO KEEP RECORDS OF FIREARMS.—Every manufacturer, importer, and dealer (including pawnbroker) shall make and keep at his place of business a record showing (1) the sale or other disposition of all firearms taxable under chapter 53, Internal Revenue Code of 1954, (2) the date of such sale or other disposition, (3) the serial number, model, caliber, and trade name as well as other marks identifying each firearm, and (4) the name and address of the person to whom any firearm is sold, transferred, or otherwise conveyed. This record must be preserved for a period of at least 4 years from the date of disposition of the firearm, and be at all times readily accessible for inspection. (Sec. 5842, Internal Revenue Code of 1954.)

3. SCOPE OF RECORD AND RETURN PROVISIONS.—The provisions of section 5842 of the Internal Revenue Code of 1954 relating to records and returns are applicable to persons engaged in business within the States of the United States, the Territories of Alaska and Hawaii, and the District of Columbia.

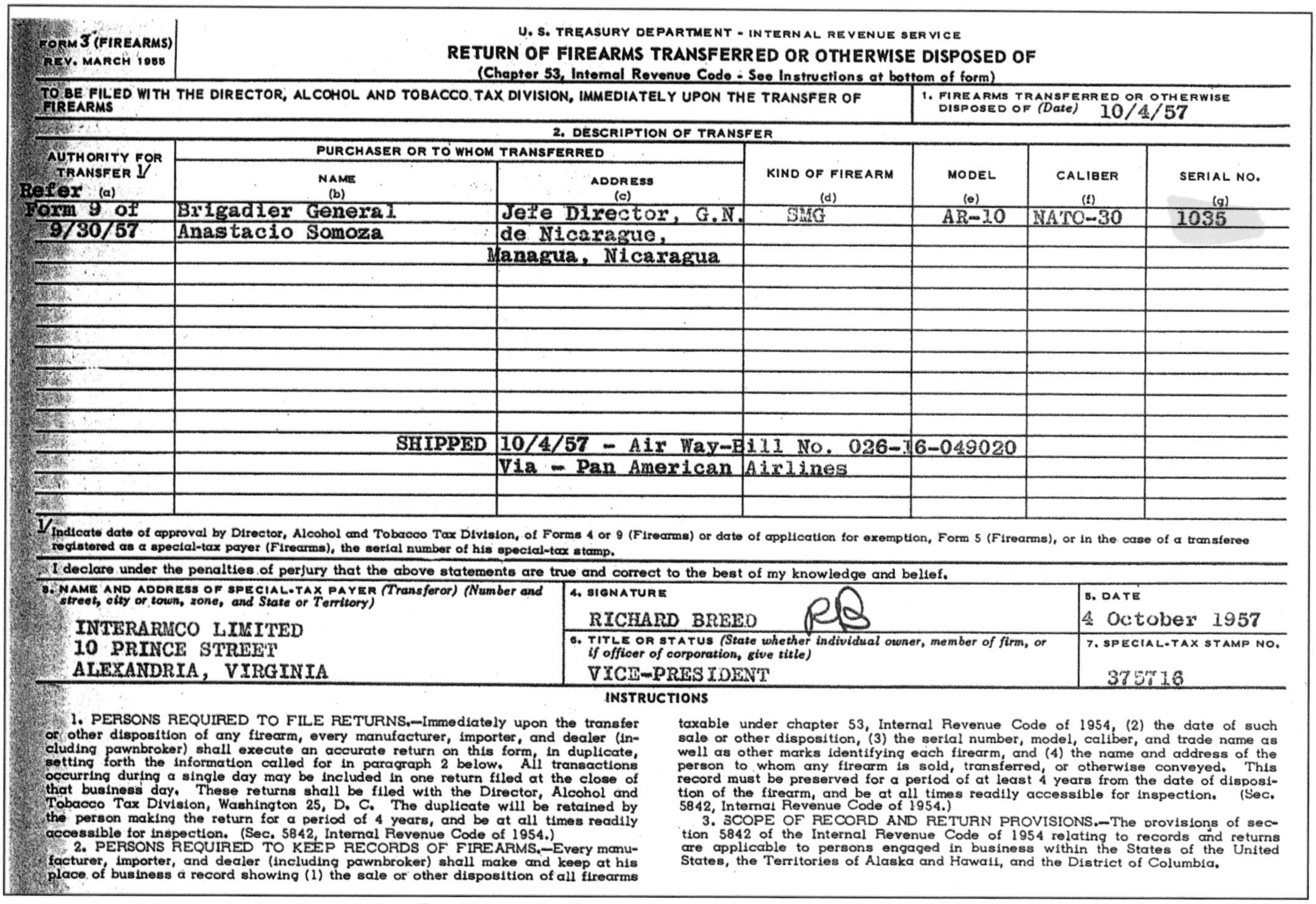

160. The Treasury Department Form 3 authorizing the transfer of ArmaLite AR-10B serial no. 1035 from Interarmco to Brigadier General Anastasio Somoza in Managua, Nicaragua. Michael Parker collection

hands of his élite troops could be a paradigm-shifting asset, making Nicaragua, and by extension his family, the dominant players in Central America.

Somoza did have one stipulation; that a test rifle be supplied to him which would be capable of passing a 7,500-round endurance test with reasonable reliability. On October 4, 1957, AR-10B serial number 1035 was shipped from ArmaLite's machine shop to Nicaragua, where Cummings handed it over to Somoza. Unfortunately, as we shall see, it would have been highly advisable for someone at ArmaLite, or Cummings himself, to have examined and test-fired this sample to make sure it was in perfect order, but inexplicably this was not done.

It appears from later Interarms documentation (fig. 163) that Cummings had also provided another AR-10B to Somoza, serial no. 1030.

# The Guatemalan AR-10 Order

The next stop for the savvy arms dealer was a return to Guatemala. Despite the fact that Cummings had armed the previous president, whom the current government had recently overthrown, the chairman of the ruling military junta, Colonel Óscar Mendoza Azurdia, welcomed the locally-famous weapon merchant with open arms. Cummings demonstrated AR-10 no. 1001 again for the members of the military government, and gained their interest in it. Their enthusiasm was not as great as it had been in Nicaragua, but the officials expressed that they would like very much to adopt the rifle. Instead of asking for a sample rifle for an endurance test, as their neighbor Somoza had done, the general staff was satisfied by the reliability of Cummings's demonstration piece, and placed an order for 450 rifles to be issued to the cadets of their military academy.

The *Escuela Politécnica*, or Polytechnic School, was Guatemala's officer candidate training institution, and was considered the preeminent military academy in Central America. The rifles were to be used in drill, firing practice, and maneuvers, to evaluate whether the AR-10 was suitable for general issue to the Guatemalan Army. As an order as large as 450 rifles could not come from ArmaLite, the academy was content to wait until Dutch manufacture had begun, when a quantity of production-quality rifles could be delivered. In placing this order, the Guatemalans also reserved the option of setting their own requirements for rifle characteristics and functionality, and transmitting these specifications to Artillerie-Inrichtingen when they were ready to begin manufacturing their rifles.

## More Latin American Orders for Small Numbers of AR-10s

Cummings continued to work his way down the isthmus and into South America. He reported back to ArmaLite that the armed forces of all the nations of the continent except for Paraguay and Bolivia invited him to demonstrate the AR-10, and expressed interest in it. Mexico was enthused enough to buy one sample for field testing, as was Panama. Venezuela, which was staking its claim as an aggressive power with territorial ambitions in the region, was very excited, and bought six for extensive testing. A list of all the substantial number of AR-10s supplied by Interarms to various potential customers is recorded in Chapter Ten.

More than any other of his developing-world customers, the Venezuelan Army provided Cummings with useful feedback on the AR-10's performance. Their notes echoed the recommendations that Melvin Johnson had made the previous year (it was 1958 by this time), that the edges of the bolt lugs should be relieved at an angle to provide for primary extraction and make the locking/unlocking process smoother and easier on the working parts. This suggestion was forwarded to Artillerie-Inrichtingen, but was never adopted.

## Fast Friends with Trujillo: Cummings Winds Up his Latin American Tour

The next nation on Cummings's itinerary, and the location of the last of his sales attempts in the Americas, was the Dominican Republic. This island nation was ruled by Rafael Leonidas Trujillo, one of the most heavy-handed, enigmatic, and egomaniacal of all the Latin American dictators. He was also, however, uniquely effective in ruling and developing his country, and was an old friend of Cummings's.

First contacted by Trujillo by way of the Dominican ambassador to the United States in 1954, Cummings had happily accepted the dictator's invitation to visit the capital of the Dominican Republic; a city that was historically known as it is today as Santo Domingo, but which at the time was styled "Ciudad Trujillo" (Trujillo City). This in itself should give the reader an idea of the depth of the Dominican military ruler's megalomania, and of the extreme cult of personality he fostered.

In need of the best possible protection for himself and his family against the seething opposition in the country he ruled, Trujillo had sought a broker in 1954 who could acquire American Thompson submachine guns for his personal guard. Since Cummings had a source for them in England, he arranged for the sale, and met Trujillo in the Dominican Republic to discuss the details. There he also became acquainted with Alexander Kovacs, a Hungarian veteran of the WWII invasion of the Soviet Union, in which his country had been allied with Nazi Germany. When the tide turned and the Soviets were nearing Hungary, the bleak communist future for his part of Europe appeared imminent, and Kovacs emigrated to the Dominican Republic, where his weapons expertise had made a great impression on Trujillo. He was made a general and put in charge of procurement for the Dominican military.

Kovacs and Cummings became fast friends, and after buying the Thompsons, Kovacs often visited Interarmco's Alexandria headquarters, where they discussed new needs for the Dominican dictator's forces. Trujillo also quickly counted on Cummings's expertise, inviting him often to Ciudad Trujillo to ask for his military advice. Their relationship became extremely productive, with Cummings selling huge quantities of small arms to the Dominicans, particularly some U.S. M2 .50 caliber Browning heavy machine guns which he had acquired from the Panamanians.

The transaction that truly brought Cummings into the inner circle of the Trujillo family, however, was his successful sale to the Dominican dictator of twenty-six British Vampire jet fighters, which he had purchased from the Swedish air force. Rómulo Betancourt, the strongman of Venezuela, had previously acquired twenty-five Vampire jets from Britain, so that with this purchase Trujillo had quite literally one-upped his rival, and his twenty-six Vampires gave the Dominican Republic's air force the dominant position in the region.

The AR-10, especially in comparison to the M1 rifles and carbines Cummings had previously supplied, really blew Trujillo and Kovacs away, and they placed a tentative order for 25,000, intending to make the AR-10 the standard infantry rifle of the Dominican armed forces. They were not yet willing to place a firm order, however, and wanted to wait until series production had begun in Holland.

161. Rafael Leonidas Trujillo, president of the Dominican Republic, one of the most heavy-handed, enigmatic, and egomaniacal of all the Latin American dictators.
public domain - the Internet

## Somoza's Test AR-10 is Defective: the Nicaraguan Order is Cancelled

With his sales tour of Latin America deemed a great success, Cummings returned to Alexandria to check on his business there and prepare the paperwork needed in order to process the Nicaraguan and Guatemalan orders. On a Saturday afternoon in early 1958, however, he received a long-distance call from a furious General Somoza, who accused Cummings of trying to kill him. The head of the National Guard himself had been conducting an endurance trial of the uninspected sample AR-10B, serial no. 1035, and while firing just the eleventh magazine of ammunition, the locking lugs sheared off the bolt, some of which gave Somoza Junior a close shave.

162 (right). Anastasio Somoza, president of Nicaragua and head of the élite Nicaraguan National Guard, who narrowly missed death or injury when one of the AR-10Bs Cummings had supplied blew up during testing.
public domain - the Internet

| FORM 6 (FIREARMS)<br>(REV. AUG. 1956) | U. S. TREASURY DEPARTMENT – INTERNAL REVENUE SERVICE<br>**APPLICATION AND PERMIT FOR IMPORTATION OF FIREARMS**<br>*(Chapter 53, Internal Revenue Code. Submit in triplicate)* |
|---|---|

**SECTION I - APPLICATION**

TO Director, Alcohol and Tobacco Tax Division
Internal Revenue Service, Washington 25, D. C.

The undersigned hereby makes application to import the firearm(s) described herein, pursuant to the provisions of section 5845, Internal Revenue Code.

1. NAME AND RETURN ADDRESS OF IMPORTER *(Number and street, city, zone, State or Territory)*

INTERARMCO LIMITED
10 PRINCE STREET
ALEXANDRIA, VIRGINIA

2. INTENDED PORT OR PLACE OF IMPORTATION

ALEXANDRIA, VIRGINIA

3. HAVE YOU EVER BEEN CONVICTED OF A FELONY? *(If "Yes," explain in a separate statement indicating disposition of case and attach same to this application)*
☐ YES ☒ NO

4. DESCRIPTION OF FIREARM(S)

| NAME AND ADDRESS OF MANUFACTURER (a) | TYPE OF FIREARM *(Machine gun, shotgun, rifle, silencer, etc.)* (b) | MODEL (c) | CALIBER OR GAUGE (d) | LENGTH OF BARREL (Inches) (e) | SERIAL NUMBER (f) | OTHER MARKS OF IDENTIFICATION (g) |
|---|---|---|---|---|---|---|
| Armalite Division of Fairchild Aircraft Corp.-6567 Santa Monica Blvd. Los Angeles, Calif. | 2-SMG's | AR-10 | NATO 30 | | 1030<br>1035 | |

5. NAME AND ADDRESS OF SELLER AND/OR CONSIGNOR IN FOREIGN COUNTRY

Brig. Gen. A. Somoza, Jefe Director, G.N. de Nicaragua, Managua, Nicaragua

6. FOR WHAT PURPOSE IS FIREARM TO BE USED?

To be altered and perfected and returned to consignor.

7. HAS ANY ATTEMPT BEEN MADE TO OBTAIN A FIREARM SIMILAR TO THAT DESCRIBED HEREIN IN THE UNITED STATES OR ANY TERRITORY UNDER ITS CONTROL OR JURISDICTION? *(If "Yes," describe fully. Use separate sheet if necessary)*

☒ YES ☐ NO

8. STATE IN DETAIL REASONS WHY THE FIREARM DESCRIBED HEREIN IS UNIQUE OR OF A TYPE WHICH CANNOT BE OBTAINED WITHIN THE UNITED STATES OR ANY TERRITORY UNDER ITS CONTROL OR JURISDICTION. Returned sample: Rifles will be sent by us to Fairchild Aircraft in California for alteration and perfections. SMG's then to be returned to Somoza.

9A. DO YOU HOLD A MANUFACTURER'S (IMPORTER'S) LICENSE UNDER THE FEDERAL FIREARMS ACT? *(If "Yes," give license number)* ☒ YES ☐ NO

9B. LICENSE NO.
See Above.

10A. HAVE YOU REGISTERED AND PAID SPECIAL TAX UNDER NATIONAL FIREARMS ACT AS IMPORTER OR MANUFACTURER FOR CURRENT FISCAL YEAR? *(If "Yes," give special tax stamp number)* ☒ YES ☐ NO

10B. SPECIAL TAX STAMP NO.
375716

I declare under the penalties of perjury that the above statements are true, correct and complete to the best of my knowledge and belief.

11. SIGNATURE OF IMPORTER
RICHARD BREED

12. TITLE OR STATUS OF PERSON SIGNING *(Individual member of firm. If officer of firm, give title)*
VICE-PRESIDENT

13. DATE
February 26, 1958

**SECTION II - PERMIT (For use of Internal Revenue Service. MAKE NO ENTRIES IN THIS SECTION)**

The application has been examined, and the importation of the firearm(s) described herein is:

14. ☐ APPROVED. However should it afterward be shown that your application is improper you may be subject to the penalties prescribed by law.

15. ☐ DISAPPROVED FOR THE FOLLOWING REASON:

16. SIGNATURE OF DIRECTOR, ALCOHOL AND TOBACCO TAX DIVISION

17. DATE

163. The Treasury Department Form 6 authorizing the re-importation of the two ArmaLite AR-10Bs which Cummings had supplied to Somoza. As noted, one of these, serial no. 1035, was defective and blew up during testing, causing the irate dictator to cancel his entire order.

Michael Parker collection

After calming his client down, Cummings told him to ship the rifle back to Alexandria, where a cursory examination showed that the barrel had been installed incorrectly so that the bolt lugs, which were supposed to spin exactly 22.5° into full contact with their locking recesses in the barrel extension, were only locking with a small fraction of their surface, providing very little support for the cartridge in the chamber. The repeated firing had weakened the lugs, and one breaking under the pressure of firing had caused a chain reaction that snapped all of them off, blasting the hardened steel fragments in all directions, even perforating the aluminum upper receiver. Cummings rushed down to Nicaragua to placate General Somoza, and even offered him his personal rifle, serial no. 1001, to complete the test. Although no longer wanting Cummings dead, Somoza turned down Cummings's offer, and then cancelled his en-tire order. With this the AR-10's status as the standard infantry weapon of the armed forces of an entire nation, and one of the largest orders Artillerie-Inrichtingen would have ever received, melted away due to the carelessness of a string of people from ArmaLite, to Interarmco, and the Nicaraguan military.

This extreme bad luck was in some ways not as severe a blow to the AR-10 as that suffered at Springfield Armory, but on the other hand, when one considers that the unsatisfactory results of the U.S. Ordnance evaluation were largely orchestrated and thus pre-ordained, while the catastrophic failure in Nicaragua was due to negligence on the part of the weapon's own creators, it could be said that this very avoidable misfortune was an uncharacteristic low point for ArmaLite.

# Redemption in the Old World

Meanwhile, with a burning desire to prove that the Springfield Armory tests had been a sham, Melvin Johnson, George Sullivan, and Leo Killen were men on a mission. Through their well-connected friend, Jacques Michault, the ArmaLite representatives set up a series of demonstrations for the military forces of several European nations. Starting in February, 1957, just days after the Springfield Armory test report was released, the AR-10 was on the road, making its first stop in Austria.

## Unbridled Teutonic Enthusiasm in Austria

The alpine nation of Austria, standing on the frontier of the Soviet Bloc, faced to its east both the fierce Hungarians, renowned for their warrior spirit and outstanding fighting prowess, and the Czechoslovakians, a nation of unparalleled arms design expertise. National defense was consequently a very high priority in Austria which, although not a member of NATO was one of the most enthusiastic anticommunist nations, and the Austrians were eager to acquire a modern automatic rifle to replace the antiquated straight-pull, packet-loading Steyr Mannlichers with which their forces were still issued. Presented to the army in the Tyrol was none other than AR-10 no. 1002, the ArmaLite rifle that had almost survived the Springfield tests unscathed and with a near-perfect record until its controversial composite barrel had blown up. It had been retrofitted with a new chromoly steel barrel and a new handguard, which contained twice the previous amount of foam plastic liner, providing better heat insulation.

Over the period of just a few days, AR-10 no. 1002 was put through its paces in a harsh-condition endurance test that accounted for 7,500 rounds. When the smoke had cleared, the reborn AR-10 rose from the ashes like a phoenix, having experienced only two stoppages during the entire test, and a single extractor breakage, which actually did not induce a failure. The astonished Austrians informed the ArmaLite team that the test FAL rifles the Belgians had brought to demonstrate had displayed a considerable number of parts breakages, and that until the AR-10 test, the Austrian ordnance officials had simply accepted this state of affairs as inevitable in the brave new world of self-loading battle rifles.

The Austrians were sold on the new rifle, and indicated vociferously a desire to acquire some, and eventually to produce it themselves.

Unlike the small Latin American banana republics, in which the national treasury was nothing more than a dictator's bank account, this plucky anti-Soviet bulwark had to go through the appropriate procurement institutions before buying quantities of new rifles, instead of just cutting a check to the salesman. At this point even the preliminary agreement on series production of the AR-10 in Holland had not been finalized, but the Austrians were so happy with the AR-10's performance that they promised to wait however long it took (within reason) for production to begin, all but writing off adoption of the FN FAL. To tide them over until the Dutch began manufacturing the rifle in earnest, they ordered five U.S.-produced hand-made AR-10Bs for further evaluations.

## The West German *Bundeswehr* Gets a First Glimpse of the AR-10

164. Endurance trials of AR-10B no. 1002 at the West German Proving Grounds at Meppen in May, 1957.

A magazine has just been emptied, and the bolt group is held open, with the rear portion of the side-mounted gas tube extension visible through the ejection port.

Note the handguard is partly obscured by smoke from the hot barrel.          photo from Marc Miller collection

165. An observer watches as the firer changes magazines on ArmaLite AR-10B no. 1002.

Note the expended shell casings covering the ground, offering silent testimony of the great number of rounds that this rifle digested in Germany without a hiccup.

photo from Marc Miller collection

Moving on from their sweet success in the Austrian Alps, the ArmaLite team, now consisting of Johnson with Killen in tow, traveled west to the next appointment, which Jacques Michault had set up for them. Landing in Kiel, West Germany, the team headed to the Meppen Proving Ground to organize a demonstration for the *Bundeswehr*, the newly-created armed forces of the Federal Republic of Germany. Jacques Michault himself personally oversaw the testing, instructing the soldiers in the use of the AR-10.

## A Brief History of German - Spanish Relations

The Germans were no less impressed with the AR-10's performance than the Austrians had been, but there were a number of considerations present in their own arms industry that had not burdened their southeastern neighbors. This had not only to do with the extremely strong arms design tradition of their country, but West Germany's relationship with certain other nations.

Germany was the only major European power at the time that still maintained very close ties with Spain, which had held the unenviable status of low-level pariah ever since WWII. While remaining technically neutral in that conflict, as had Sweden, Ireland, and Portugal, Spain's military dictator, General Francisco Franco, had agitated strongly for the Axis cause, going so far as to offer direct state sponsorship for the Nazi fight against the Soviet Union.

Declaring a "Crusade against Bolshevism," Franco had all but declared war on the Russians, sending huge numbers of "volunteers" to fight on the *Ostfront*, not only incorporated into units of the German, Hungarian, Romanian, and Italian armies, but sometimes actually in independent formations which flew the Spanish flag.

Whereas Germany had been conquered and thoroughly brought to heel by the Allies, Spain was seen by those who had fought against the Axis as the unreformed bastard child of the fascist order. After the war, Spain played host to a number of expatriate Germans who had been working within the wartime German armament establishment. One such man was Ludwig Vorgrimler, whose postwar developments based on the wartime Mauser roller-locked *Gerät* 06H (StG45), produced first by the *Centre d'Etudes d'Armement Mulhouse* (CEAM) in France and later by the *Centro de Estudios Technicales de Materiales Especiales* (CETME) in Madrid, went on to revolutionize Spain's military armament industry and equip this poorest of Western European powers with a first-rate battle rifle.

## Other Strong Contenders: the FAL and the CETME

Due to its low prospective cost and the fact that it had been developed by one of their own, the CETME was quite attractive to the Germans.

The *Bundeswehr* had also given the FAL a very close look, and were much more satisfied with it than their Austrian allies were. During 1955 the Germans had ordered a few thousand of the same model of FAL that Canada had adopted, for issue to their *Bundesgrenzschutz* (BGS) Border Guard units. In the months leading up to AR-10's evaluation at Springfield Armory, the West Germans ordered 100,000 additional FAL rifles, produced to their own specifications, for general issue to the *Bundeswehr*. These, along with consignments of CETME rifles, were put into service in large-scale troop trials to determine a successor to their leftover surplus Mausers.

The AR-10 swooped in, however, and blew the doors off its competition. Everything was put on hold for this radical new rifle. Unlike the ultra-conservative American top military brass who felt that a rifle should be made of wood and steel and bear traditional lines, the Germans, already credited with such unorthodox and radical-looking small arms as the FG42 and StG44, actually saw the strange appearance of the ArmaLite weapon as a selling point. As in Austria, the German AR-10 tests results were equally spectacular, with several thousand rounds being fired with no malfunctions or parts breakages. Although they were rather far along with their evaluation and procurement of FALs and CETMEs, the German Ordnance Department made the decision to include the AR-10 on their short list of rifles for possible adoption. They resolved that once Dutch production began, they would acquire several hundred AR-10s for troop trials, and would not make a final decision on rearmament until these had taken place. In the meantime, Germany also ordered a quantity of five rifles which could be provided by ArmaLite's California workshop, for limited testing.

Also important to the Germans was a topic that had been mentioned in Austria, but not to such a great extent: that of domestic manufacture. The Germans were very serious about eventually being able to produce the rifle they adopted within their own borders. Johnson and Michault assured the Ordnance officials that if a sufficient quantity were procured from the Dutch, they would surely be able to work out a deal in which Artillerie-Inrichtingen would grant limited production rights in exchange for a royalty. The Austrians had also expressed interest in domestic manufacture, but the Germans were very forceful on this requirement, as they had been with the Belgians and Spanish.

## Another Stumble over ArmaLite's Poor Quality Control

Some of the five sample AR-10s that ArmaLite's Hollywood workshop had provided to the Germans did encounter a problem not witnessed during the demonstrations of rifle no. 1002, wherein the bolts suffered freak breakages early in testing. At ArmaLite's advice both nations halted testing temporarily, and Arthur Miller flew in to examine the bolts. He quickly ascertained that the heat treatment process used on the bolts, which had been subcontracted out to a separate machine shop in California, had been done incorrectly, and a few errant tool marks had caused cracks to propagate in the all-important bolt locking lugs. This was a frightening echo of the fiasco of the U.S. Marine Corps tests the previous month, which had combined institutional paranoia and conservatism with grossly indifferent quality control and substandard components. George Sullivan had ten replacement bolts crafted and carefully heat treated in-house, tested, and sent off with Stoner, who hand-delivered them to the Dutch, who relayed them to the Germans and Austrians.

By all accounts the new bolts were satisfactory, as no more reports of malfunctions were heard. Additionally, it can be taken as evidence that these replacement parts made the rifles run well enough that within a few months both interested parties were clamoring for quantities of production rifles for troop trials.

## A Cool Reception in Sweden and Denmark

For their next port of call, the representatives of Fairchild ventured outside SIDEM's sales territory and into that of Interarmco. Being a stickler for ethical behavior, and not wanting to infringe on his colleague's territory, Michault went straight to Denmark where he began preparations for that country's testing. Meanwhile, Cummings accompanied the ArmaLite team, once again including George Sullivan, to the Swedish military proving grounds, where the AR-10 was put through its paces. The single rifle brought for this series of demonstrations across Europe, no. 1002, turned in the type of performance that was becoming characteristic of it.

For a number of reasons, however, the Swedes were not as ready to commit to the AR-10 as the Austrians and Germans were. First, Sweden was neither a member of NATO nor a fierce foe of the Bolsheviks, like Austria, and as such was not on the same type of military footing as most of the rest of Free Europe. They also had their own domestic champion rifle cartridge which they had no intention of abandoning, and not being part of NATO, there was no extra incentive for them to move away from their pet 6.5x55mm round. They had also just in the last decade developed their own self-loading battle rifle, which coincidentally also operated on the direct gas impingement system, the AG-m/42b, in which they took much national pride.

In order to make the adoption of the AR-10 more attractive to the hesitant Swedish, George Sullivan dangled the possibility of domestic production, and even the ability of the AR-10 to be chambered for their favorite cartridge. Although this was probably technically feasible, Stoner and Boutelle were dismayed to learn about this offer, as such a change would actually require considerable effort, time, and resources. Friedhelm Jüngeling at Artillerie-Inrichtingen was also quite perturbed to hear about Sullivan's unilateral offer of production rights, as the Dutch were not only contracting and negotiating for the sole manufacturing license, but had to date already invested considerable resources in gearing up their plant in order to begin production.

A similar outcome was reached in Denmark the next week, where the AR-10 performed extremely well, but failed to make a big splash. The Danes were new and enthusiastic NATO members, and so the ammunition stalemate encountered further north was not an issue. Nevertheless, the powerful Danish Industry Syndicate, Denmark's state weapons concern, was at the time hard at work on its own NATO-chambered battle rifle. Constructed using parts crafted from aluminum alloys, the Madsen Light Automatic Rifle, or "LAR", was a source of great hope to the particularly American-admiring Danes.

Denmark had a first-rate martial tradition, and their Captain Vilhelm Herman Oluf Madsen had spearheaded the development of the first field-ready light machine gun in world history. The Madsen Machine Gun, adopted by the Danish army in 1902, went on to be produced in a dozen calibers and sold as standard infantry equipment to over 30 nations. It was used to great effect in both World Wars and numerous other conflicts, and still remains in second-line use today by the Brazilian Army, after 112 years.

Encouraged by the prospect of their own domestic battle rifle, the Danes, like the Swedes, declined to seriously entertain the idea of purchasing the AR-10. The ArmaLite team was not overly discouraged by these two lukewarm receptions, however, as they were overshadowed by the outstandingly strong interest shown by Germany, and especially Austria. The team accepted the Swedish and Danish conclusions philosophically, and moved on to complete the European testing tour.

# The AR-10 in France

The AR-10 was now trotted out before the French Army, for demonstration to this disgruntled principal member of NATO. The already-contrarian French had been incensed and insulted both by the lack of support they had received during their war in Indochina (later to become the Vietnam War), and the rebuke from the United States that had resulted in their withdrawal from Suez the previous year.

France had developed her own direct-impingement battle rifle which had been issued, at least in limited numbers, since 1940, and for general use since 1949. However, this rifle, the 7.5x54mm MAS-49/56, was not even chambered for the NATO round, and efforts to do so had caused problematic functioning.

Nevertheless, even though the Fourth Republic felt very disillusioned by NATO's apparent weakness in refusing to help them with their territorial and colonial disputes, they were still interested in standardizing their equipment. Despite all of the public rhetoric to the contrary, French political and military leaders admitted in private discussions with their British, American, and West German counterparts that should the Soviet Union take action in Europe, all of their differences would at once be forgotten, and they could count on France as a strong ally. To this end, ammunition and equipment commonality was still a high priority in the French military.

The French AR-10 firing trials, though not as extensive as those performed in Austria, were reported as a complete success, again with no malfunctions experienced by the rifle. As it had done everywhere else, the AR-10 caused great excitement on the part of the officials who saw it in action. At the conclusion of their stay in France, the government made it clear to the ArmaLite team that they were very interested in the AR-10, and once production had started in Holland, they would wish to resume contact and move forward with procurement.

# On to Italy: the Élite COMSUBIN Commandos are Enthusiastic

During the last week of May, 1957, the AR-10 was exhibited to a joint commission of the various branches of the Italian military at a range outside Rome. The Italians, who had continued to issue U.S. M1 Garands since the end of WWII, were delighted with the prospect of the AR-10's firepower, and particularly its light weight and modularity. Expressing particular enthusiasm at the trials was the representative of the Italian Navy's secretive and extremely élite COMSUBIN underwater commando force. This unit took pride in being on the point of the spear, having carried out some of the earliest and most daring raids in the history of modern warfare. In fact, they can be rightly considered the first modern special forces unit, being later emulated by the British and Germans.

At the time the Italian military was relatively satisfied with its surplus U.S. rifles, and while they were searching for a more modern design, the tight budget constraints under which they were working would have made imminent large-scale adoption of the AR-10 very difficult. The COMSUBIN was another story, however, and promised that they would certainly acquire a quantity of AR-10s once standardized production had begun.

# A Final Debut Demonstration, in Spain

On June 8, 1957, Johnson, Sullivan, and Killen proceeded on to Madrid for the final leg of the AR-10's European debut. They sent a telegram back to Richard Boutelle reporting the particularly good reaction they had received from the Austrians and Germans, as well as interesting anecdotes that had been related to them at almost every stop of their tour. The ordnance departments of Austria, Germany, Denmark, France, and Italy had all received correspondence and calls from Sam Cummings representing himself as the sales agent for the AR-10, and offering to set up demonstrations. Jacques Michault had been particularly irritated by this development, as by doing this Cummings was encroaching on SIDEM's sales territory. Although the "standby" territories had not yet been assigned, and indeed the final production deal on the AR-10 had not yet been finalized, it was well recognized that Michault would have domain over non-Scandinavian Europe, and Cummings had been so informed. The telegram contained a curt request to "please remind Cummings to stay in his assigned territory." This pugnacious move by Cummings to flout the rights of a partner would not be a one-off occurrence in the history of the AR-10's promulgation.

The Spanish tests went very similarly to those in Sweden and Denmark: the AR-10 functioned beautifully and was acknowledged as being an excellent rifle, but the host country had its own design into which it had sunk considerable time, resources, and money, and was stoically reluctant to consider an outsider to supplant that which it had labored to produce. Spain was the home of the CETME, and almost simultaneously with the ArmaLite team's arrival in the country, that rifle was officially adopted as standard equipment of the Spanish Army, replacing their outdated Mauser bolt-action rifles and carbines.

## Summing Up the AR-10's First Old World Tour

At the end of the tour the armed forces of several major European powers were seriously considering its adoption, and it had turned heads everywhere it visited. Although the AR-10 had encountered what would become one of its biggest foils—the existence of other domestically-designed and -produced battle rifle designs in Sweden, France, and Spain—its first trip to the Old World could certainly be considered a success. ArmaLite AR-10 no. 1002, the sole example of its type demonstrated to the European powers, fired more than 20,000 rounds in its travels, exhibiting only two recorded malfunctions, both of which happened in Austria and did not deter the officials of that host country in the least.

# Round II in Europe

As the date for the commencement of series production at Artillerie-Inrichtingen drew nearer, and the last of the subcontractors fell into line across northern Europe, the AR-10 returned to the continent on a callback from the Austrians, and also for demonstrations in more countries. This time, ArmaLite was ably represented by Horace Kennedy and Edward Bishop, who took ArmaLite AR-10Bs nos. 1002 and 1004 to some very forbidding locations.

## More Flying Colors in Austria - the AR-10 is "Superior to All Other Weapons"

The first stop was again Austria, where the two rifles were put through endurance trials of 15,000 rounds apiece, and were still rated by the Austrian forces at the end of the examination to be in excellent condition, with no additional parts breakages or recorded malfunctions. It was there in September, 1957, that the Austrian observers declared that the AR-10 was superior to all other weapons they had examined, which included the Belgian FAL, the Spanish CETME, and a dark horse Swiss entry into their rifle adoption contest. The Austrian military resolved then and there to adopt the AR-10, if sufficient quantities could be provided, and supply and manufacturing concerns could be settled with the Dutch factory once the rifle was in regular production.

## Test Rifles for the Italian COMSUBIN

The team then swung by Italy again to drop off the test rifles for the COMSUBIN underwater commandos. This unit, although small, provided special promise for a successful adoption. Unlike most national armies, including that of Italy, small élite forces like COMSUBIN were in charge of their own equipment procurement, and did not have to cut through bureaucratic red tape in order to get the weapons they wanted. They were not bound by the armament decisions of the rest of their nation's armed forces, and were generally a budget priority. While the AR-10 at the time appeared too expensive for general issue to the Italian Army, which was one of the most satisfied of the European nations with its surplus U.S. M1 rifles, COMSUBIN acted with relative autonomy, and its expenditure on individual weapons could be greater than that of the regular army.

# Breaking a Leg in Switzerland

Next, Kennedy, Bishop, and Michault broke new ground by taking the AR-10s before the scrutiny of the Swiss Army. Possessed of one of the most respected fighting forces in Europe, Switzerland had maintained its strict neutrality for over a century not by expert diplomacy and forbidding geography alone, but in large part by making itself a very prickly target for any would-be conqueror. Considered part of the free world because of its civil liberties, democratic system of government and robust capitalist institutions, it was still a place where Soviet diplomats, citizens, and students were welcome in this era of polarization.

Maintaining their neutrality in the age of the superpowers demanded that the Swiss be equipped with the best possible weaponry to repel aggression from either the Eastern Bloc or NATO. This policy of friendship and charity to all nations did not mean, however, that the leadership of Switzerland bore any illusions about which side in the ideological conflict promised a more prosperous and peaceful world, and which possessed the most advanced military technology. Although joining NATO was a political impossibility for the alpine confederation, the renowned Swiss Army nevertheless showed interest in the most effective and ubiquitous 7.62mm NATO cartridge.

Having only two months earlier, at the September callback to Austria, learned of the new Swiss rifle design that was in the final prototype stages, the ArmaLite representatives went prepared into this new national venue for their rifle. As it did everywhere else it was demonstrated, the AR-10 greatly impressed the observers, and generated significant interest. It was reported back to ArmaLite the next month (December, 1957) that the Swiss had been extremely enthusiastic about the AR-10.

All this was to lead nowhere, however, as that same month, the Swiss Confederation announced that it had selected a new rifle to equip its forces. The prototype previously known in the military arms industry as the "S-16" had been adopted as the Fass-57 or Stgw-57 (given two designations due to the polyglot nature of Switzerland), standing for Model 1957 Assault Rifle. Though not actually meeting the definition of an assault rifle, the SIG (Schweitzerische Industrie-Gesellschaft) Stgw-57 was a formidable, if very heavy and unwieldy, weapon, weighing 12 lbs. 9 oz., operated by a retarded blowback roller-locked action.

The SIG Stgw-57 traced its lineage back to the very strong Swiss tradition of arms design which had produced the Schmidt-Rubin series of rifles, which were considered the most effective, fastest firing, and most accurate bolt-action military issue rifles ever devised. The former two traits were mostly due to these weapons being fitted with a "straight-pull" bolt action, which did not necessitate the raising or lowering of the bolt handle to actuate, only requiring a single horizontal back-and-forth motion like the charging of an automatic rifle with a reciprocating charging handle. Its barrel was free-floated in the stock, which made for unparalleled accuracy. Several features of this extremely precocious bolt action rifle were carried over into the new Stgw-57.

The new Swiss "black rifle" also fired a domestically-designed cartridge that had been in military service for over half a century. Very forward-looking for its time, the 7.5x55mm Swiss round not only matched the ballistics of the 7.62x51mm NATO almost identically, but its bullet was exactly the same diameter, meaning that NATO projectiles could be loaded in Swiss ammunition, and vice versa. The Swiss cartridge was four millimeters longer than its American-originated counterpart, but rifles designed for the latter, such as the AR-10, could be easily converted to fire the slightly-extended Swiss round.

The Stgw-57 was a very interesting and innovative battle rifle, possessed of certain similarities to the AR-10, like a straight-line stock and raised sights.

Unlike the AR-10, it was enormously heavy, taking the title of heaviest standard-issue battle rifle in history, with its extra-large 24-round magazine capacity only accentuating this hulking mass.

Designed domestically, it was a source of great national pride, and its adoption was a priority for the Swiss military establishment, which had sunk a great amount of time and resources into its development.

Chambered for a domestic cartridge that provided essentially identical ballistics to the 7.62x51mm NATO round, it helped to foster both Swiss insularity and uniqueness, as well as keeping the industry built around its ammunition manufacture strong.

Ironically, the timing of the first AR-10 demonstration before the Swiss Army coincided almost exactly with announcement of the acceptance of the SIG Stgw-57.

# Renewed Hope in Norway

Fresh from their disappointment in Switzerland, the ArmaLite team moved on to their next testing ground: Norway. Adding injury to insult, Bishop had been accompanying the Swiss infantry on field trials in the rugged Alps, and had lost his footing, breaking his leg in the process. The nagging pain of this injury, accentuated by the coming winter cold that was especially pronounced in Scandinavia, made the already-strenuous travel schedule even worse for this dogged ArmaLite representative.

Norway gave a more interested reception to the AR-10 than Sweden and Denmark had in the summer; being both polite and inquisitive. Norway, with its small population, did not have a very large fighting force to equip. Additionally, its Army had only recently received a large quantity of free U.S. M1 rifles, in exchange for their cooperation in NATO, and permission for the U.S. to establish bases on their soil. The Norwegians had become particularly enamored of their M1s, and had no intention of spending more money to change from what they considered a very acceptable rifle at the time. This issue of choosing between a superior, new weapon that would cost money, and antiquated-but-free surplus equipment, would continue to be a bane not only to the AR-10, but to future ArmaLite rifle designs. The intrigued looks they gave the AR-10, however, would grow into something more serious in the coming years.

# An Extended Greeting in Finland

In Finland, on the other side of Sweden, the AR-10 received a different and quite unexpected type of welcome. In 1939 and 1940, standing up against Stalin's bullying demands for cession of resource-rich parts of their country, the feisty and independent-minded Finns had fought the Soviets to a remarkable standstill. With only a tiny force compared to the Russian juggernaut, Finland had killed and captured five Soviets for every man they similarly lost, destroying 100-fold the number of tanks they lost (they had had very few at the beginning of the war compared to the great masses of Soviet armor), and shooting down between four and ten times the aircraft they lost.

Eventually, broken by the sheer weight of Russian numbers, the ruthless strategic bombing of civilian population centers by the Red Air Force, and the refusal of the international community to come to their aid, they joined the Germans the next year in fighting back against Stalin. Unlike the other smaller allies of the Nazis like Romania, Bulgaria, Hungary and Italy, who had all fought the Soviets, the Finns only committed themselves to taking back the territory that had been stolen from them the year before. With far more justification to hate the Russians than any member of the international coalition that came close to ending the Soviet Union in 1941, the Finns showed great restraint, only fighting to drive the Soviet invaders from their homeland, and refusing to cross into Russia and doing to the enemy what had been done to them.

Finland suffered for its allegiance to the losing side, however, and lost all its newly-regained territory back to the Russians in 1945. Its cooperation in the invasion of the Soviet Union won it Stalin's ire, and the small Scandinavian nation was once again bullied by its massive neighbor to pay reparations and establish something like a tributary status. This Soviet arm-twisting prevented Finland from joining NATO, and required its leaders to play ball with the imperialists in the Kremlin; but the Finns were able to stay out of the real Soviet sphere of control marked by the Iron Curtain surrounding the Eastern Bloc. This contact with the Russians, however, motivated the resourceful and practical Finns to eye with great interest adoption of weapons in Russian calibers.

The Finns were exceptionally delighted at the performance of the AR-10. They had continued to issue their own, much-improved version of the Russian Mosin rifle along with some domestic and British varieties of submachine guns, but due to the strictures placed on them preventing close military contact with the Western powers, they had almost no experience with modern automatic rifles.

# Envisaging a 7.62x39mm "Intermediate" AR-10

The Finnish high command expressed very strong interest in adopting the AR-10, but with a strange twist. They had been provided with some weapons for testing from the Soviet Union, and were of the mindset that the doctrine of the assault rifle, as opposed to that of the long-range battle rifle championed by NATO at the time, was superior. They saw the potential in the AR-10 for a singularly effective assault rifle, and Cummings helped them get in contact with Artillerie-Inrichtingen to set about developing an AR-10 variety chambered for the Soviet 7.62x39mm intermediate cartridge. This trip also saw a personal sacrifice of one of the ArmaLite team members, with Horace Kennedy getting frostbite in his ears after subjecting them uncovered to the polar December winds on the Finnish proving grounds.

## Winding Up the Second European Tour

On the final leg of their journey the representatives swept back again through France and West Germany, conducting a demonstration at Bourges for the French Army from the 17th to the 19th of December, 1957. At the request of the West Germans, two AR-10Bs were delivered to them (down from their original request for five) for further endurance testing. Over the next few days these two weapons, nos. 1011 and 1012, were each put through 10,000-round trials using only the fully-automatic fire mode, and at rates of fire that made the Springfield tests look soft by comparison. Each rifle came through the ordeal with a sterling record, having a significantly lower rate of malfunctions and parts failures than any of the other contenders for German adoption.

This return trip of the post-Springfield AR-10s to Europe provided both success and disappointment to their advocates. The greater interest generated in Austria and Germany was a definite plus, while the bad timing encountered in Switzerland and the intransigence of the Norwegians were letdowns.

Perhaps the most interesting result of this trip was the cooperation that would take place over the next two years between the Finnish Army and Artillerie-Inrichtingen, resulting in the strangest and most radical version of the AR-10 ever envisioned, let alone manufactured.

The most important achievement of the trip, however, to those personally invested in the AR-10's success, was the phenomenal performance that both AR-10s nos. 1002 and 1004 turned in, laying to rest forever the trumped-up concerns spouted by Springfield Armory, and permanently unmasking those tests as a fraud.

## Turkish Delight

In addition to acting in concert with the representatives of Fairchild Arms International, Jacques Michault, acting on his own initiative, staged demonstrations of the AR-10 in some of his contract and standby territories. One of the latter, the Republic of Turkey, presented a unique opportunity. A secular nation of 27 million Muslims, Turkey stood bravely on the border with the Soviet Union, maintaining a state of readiness whereby they were always prepared for the imminent possibility of having to defend their lands from the Red imperialists to the north.

Comprising one of the largest military forces in NATO, and holding the distinction of being the only NATO member state in Asia, modern Turkey had possessed a particularly strong warrior ethos since its founding in 1923 by General Mustafa Kemal, the Hero of Gallipoli. Having gained great renown for the ferocity with which they had fought in Korea, the Turks were nevertheless still equipped with WWI-era German small arms.

Eager to leapfrog the first generation of battle rifles and get in on the ground floor of the new wave in infantry arms, the Turks enthusiastically welcomed Michault, and immediately declared their intent to pursue adoption of the AR-10. They at once placed orders for several examples from Artillerie-Inrichtingen for use in troop trials.

# Success in the Sudan

## Sam Cummings Horns In to Pull Off the First Big Sale of the AR-10

In January, 1958, a chance turn of fortune, for once not detrimental to the AR-10, led to interest in the new battle rifle in one of the most unexpected places on the globe. On January 1, 1956, the previously-known "Anglo-Egyptian Sudan", a British protectorate since the end of the 19th century, had become independent from both British and Egyptian rule, changing its name to the Republic of the Sudan.

Feeling very out-gunned with their British Lee-Enfield bolt-action rifles, and eyeing the growing power of Egypt, replete with Soviet military aid, the Sudanese wanted the most advanced new weapons available with which to equip their military. During the run-up to independence over the preceding few years, the Sudan Defense Force had become completely independent from British command, and so by this time Sudan had an established, albeit inexperienced, underfunded, and small, armed force.

From its very inception as an independent state, the Sudanese had an additional and far-more-pressing reason to be searching for a modern and effective military rifle than did most of their contemporaries. The Sudan was not a unified nation in the modern sense of the word, and was instead a patchwork of culturally disparate regions that had been arbitrarily agglomerated into a single political unit by Great Britain. In August, 1955, just over four months before independence, the Equitoria Corps, the all-infantry division of the newly-independent Sudan Defense Force responsible for the security of the southern provinces, had rebelled against the northern administration.

This insurrection underlined the main cultural flashpoint in the Sudan: the northern two thirds of the country was ethnically predominantly Arab, and practised Islam, while the southern third was populated by sub-Saharan Nubians who followed either Christianity or the native animistic religions of the region. As independence drew nearer and nearer, it became increasingly clear that the Arab north would dominate the political landscape of the new country. When preparing the Sudan for independence, the British had not consulted or given a forum to the leaders of the southern peoples, and as a result, the Muslim north was on track to have a virtual monopoly on power when independence was declared.

When the British or Egyptian intervention on which the rebels were counting in 1955 did not materialize, they called an end to the formal resistance, and agreed to surrender in the city of Torit. When the northern troops arrived, however, the city was empty, and the insurrectionists had retreated into hiding in the vast forests and swamps of the southern Sudanese countryside. Although they did not possess many arms or supplies, the rebels had the support of the local Nubians, and while the new Sudanese armed forces were assured that they were not in a position to imminently threaten northern control of the south, they predicted that these irregular forces, led and organized by former British-trained army officers, could prove a serious threat to them in the future.

To this end, the Sudanese military sent attachés to its embassies all over the world in hopes of acquiring a more advanced infantry rifle. While nations whose main concern was resisting massive Soviet invasions and nuclear strikes were often exclusively focused on the most sophisticated aircraft, tanks, and missile technologies, a military that had as its immediate and main threat a guerilla insurgency by local rebels, saw the infantry rifle as essentially the most useful tool in its prospective arsenal. Therefore, state-of-the-art fighters, bombers, and armored vehicles were not the priority for the Sudanese Army, which put first and foremost the task of equipping their soldiers with the best rifle available with which to combat the southern rebels.

One such general-colonel team attached to the London embassy of the Republic of the Sudan had inquired in early January, 1958 with the U.K. Ministry of Defence about how to obtain quantities of the newly-adopted L1A1 rifle, the British version of the FN FAL. The L1A1 was having its own teething problems in England, and as such Ministry officials informed the disappointed Sudanese that the rifle was not yet in production, so that no timetable for acquisition could be given. One of the officials, however, was familiar with a successful American arms merchant who had offices nearby. These were none other than Interarmco's London headquarters on Piccadilly Row, and Sam Cummings just happened to be working from there at the time. The Sudanese officers were told that he was the sales representative of an impressive new type of rifle that he had been touting all over the world.

Later that day, Sam Cummings was surprised in his office by the two high-ranking Sudanese Army officers, who came calling to learn more about the new rifle of which they had heard. Cummings showed them his personal demonstration model, serial no. 1001, and both were impressed by its light weight. They wanted a firing demonstration, of course, but the AR-10 was chambered for the new 7.62mm NATO caliber ammunition which, ironically, was not yet available outside tight military circles in Europe. Only Sweden, a neutral non-member of the alliance, was producing it on a large scale for sporting and testing use. It happened that Cummings had connections with the Norma Cartridge works in Sweden, near the Norwegian border, so the three chartered a flight to Sweden and flew up in the dead of winter. The bitter cold was especially new to the Sudanese, who had lived their entire lives in the tropical and desert surroundings of North Africa.

Provided for free with all the ammunition he could desire, Cummings showed off no. 1001 in front of the two amazed officers. It then came their turn to fire it, and they were so impressed that they informed Cummings on the spot that the Sudanese government would be ordering AR-10s. Not bound by the same procurement bureaucracy that both served and plagued the major powers, the Sudanese were much more like the Latin American and Caribbean clients with whom Cummings was used to dealing, and could decide to adopt a new weapon with the simple word of a general.

Back at his office after the tests, Cummings received an invitation from none other than the commander-in-chief of the Sudanese armed forces, General Ibrahim Abboud, to demonstrate the weapon for him personally, and finalize the details of the sale. There was just one problem, though; the Sudan, anachronistically called "Anglo-Egyptian Sudan" in the Fairchild sales agreement, was located in the "standby" territory of Jacques Michault's SIDEM International. However, having been sought out personally by the Sudanese military, Cummings felt that he, regardless of considerations of overstepping his territory, deserved to profit from the sale. He was also known to be ruthless with his competitors, and never saw the ArmaLite sales arrangement as anything other than a competition. Not one to pass up an opportunity for profit, Cummings accepted Abboud's offer, and prepared to make his way to Africa.

ArmaLite internal documents admonishing Cummings for representing himself to various European military institutions as the sole salesman of the AR-10 show that his misbehavior in the Sudan was not the first time he encroached on sales territory that was not his. Photographs and documentary evidence showing him demonstrating rifles for the Thai armed forces in 1960 also give compelling proof that it was not his last. The Sudanese order, however, was the only time this ruthlessness paid off, although Michault reportedly never forgave him for it, cutting off contact with Interarms, and not speaking to Cummings again for the rest of his life.

Air transit directly to Khartoum was very limited in the 1950s, and so Cummings decided to take advantage of the stopover in Cairo to ply his trade for the first time in Egypt. Spending a few days in Cairo, he offered to buy any surplus weaponry the Egyptian Army might have on hand. Knowing his reputation, the new Egyptian president, Gamal Abdel Nasser, invited Cummings to his palace to discuss arms, and

166. President John F. Kennedy poses beside General Ibrahim Abboud, commander-in-chief of the Sudanese forces, with Air Force One in the background.
public domain - the Internet

ended up selling Cummings some older British surplus, which Cummings later resold commercially in the U.S. At the time, an American in Egypt dealing directly with the nation's head of state was a rare event indeed, although Nasser, who had recently fought the British, French, and Israelis to a standstill in the Suez Crisis and, with Soviet help, had become a bitter enemy of the NATO alliance, was definitely not in the market for AR-10s.

Arriving in Khartoum, Cummings was picked up directly by General Abboud in his staff car, and taken to his accommodations. Their friendship would end up being fruitful for both, as Abboud bought from and sold to Cummings on several more occasions. Abboud, concerned about the increased activity of the rebels, who were becoming more cohesive and attracting more recruits, wanted the AR-10 rifles particularly for several élite battalions of the Northern Command; particularly the infantry companies of the Eastern Arab Corps. The Northern Command was used to suppress unrest in the south of the country, because its members were ethnically and religiously different from the southerners, and would therefore show little sympathy to the "infidel" Nubians.

Eventually, Abboud expressed to Cummings that he intended to arm almost all of the Northern Command's infantry with the AR-10, but at the present time, he needed 2,500 for the élite companies to be used in what were essentially special-forces operations. Chief among these outstanding units was the Sudanese Airborne Brigade, part of the Northern Command, which was moved to a location just over the border between the southern and northern regions of Sudan, at Al Mijlad, located in the southeastern part of Darfur Province.

These rapid-action units were not intended to do the everyday work of policing the south and suppressing protests and riots, but were meant to be deployed at short notice by air and mechanized transport to surround and wipe out rebel strongpoints or camps hidden in the forests or swamps, as soon as they were located. At this time less than 10% of the rebel troops possessed firearms, so the élite government forces for whom the AR-10 purchase was intended were held back for attacks on very important headquarters sites where organized resistance with semi-modern weapons could be expected.

## Specifications Agreed for 2,508 Sudanese AR-10s

The sale was therefore agreed by the Sudanese commander-in-chief, and the ideal specifications were laid out. The total number of rifles ordered at that time for equipping the élite units was exactly 2,508 pieces. In the price of each rifle was included a sling, bayonet with its internal tools, cleaning kit, and four magazines. At $100 each, the AR-10s were considerably more expensive than the antiquated surplus weapons that most of the Sudan's neighbors were using, but were still one third less than the amount for which the FN FAL was being sold in similar parts of the world.

### Cummings Comes Through Again - with Lances for the Sudanese Camel Corps

While in the Sudan, Cummings, always the student of military history, was taken on a tour by Major General Hassan Beshir of the ground where the Battle of Omdurman had been fought in 1898, in which the last successful large-scale cavalry charge in world history was made by the British against Sudanese Mahdi (an Islamic messianic movement) rebels fighting to drive them from the country. When discussing this battle with Cummings, the general lamented that the remaining mounted brigade of his Camel Corps could not acquire lances for their duties. The Camel Corps, one of the major divisions of the Army that was used to patrol the vast deserts of the south-central Darfur region, had been mechanized in 1956, but one brigade had been kept for ceremonial duties, still mounted on their "ships of the desert." It happened that Cummings had come into several hundred German WWI lances a few years earlier, and presented them to Abboud as a present to commemorate their $2.8 million deal; the largest AR-10 sale up to that time. That price reflected more than just the cost of the sold rifles, and included the scopes specified on eight of the ordered rifles, each of which was more valuable than the rifle itself, plus some other items Cummings sold to the Sudanese that were not connected to the AR-10.

## A Side Trip to Ethiopia - Scotched Again by American Largesse

Before returning to Europe with the lucrative Sudanese order in hand, Cummings first crossed into his own sales territory of Ethiopia, where he demonstrated the AR-10 for that country's central command. The AR-10 made a very favorable impression, but world events once again prevented a sale of the ArmaLite weapon to an interested buyer. The U.S. had recently entered into a cooperative agreement with the government of Emperor Haile Selassie, by which they would be allowed to construct air bases in the country from which to conduct surveillance against Soviet activities in Africa and the Persian Gulf. To sweeten the pot for the East African monarch, the American military mission had in return made available to his forces an unlimited supply of free surplus U.S. equipment. The promise of a limitless supply of M1 Garands, M1 Carbines, and M3 and Thompson submachine guns was simply too attractive to the Ethiopian military for them to spend scarce money on the ArmaLite weapon, regardless of its superiority.

## Modifying the Cuban AR-10 for the Sudanese

Unperturbed by his failure to get the Ethiopians on board, Cummings returned to Europe, delivering the signed Sudanese order personally to the Artillerie-Inrichtingen factory, where the stated requirements of the Sudanese military had to be applied to the blank canvas that was the Cuban AR-10. These modifications are discussed in Chapter Seven.

Meanwhile, more fascinating experiences awaited the AR-10 and its salesmen in far-flung corners of the world.

# A Further Look at the AR-10 by the U.S. Army

## A Cuban AR-10 Shines in Detroit

167. Left side view of the Dutch AR-10 serial no. 000030, sent to the Detroit Arsenal for evaluation by the 507th Ordnance Detachment (Technical Intelligence).
Detroit Arsenal negative no. 56362 dated 11 August, 1958, editor's collection

As the AR-15 was just beginning to turn heads in the U.S. and the first small cracks in the dam of intransigence and inertia supporting the M14 were appearing, the command of U.S. Army's Technical Intelligence Service decided to break with the tyranny of Springfield Armory and show their dissatisfaction with the sketchy adoption of the T44, and evaluate the Dutch-produced AR-10. In August, 1958, the 507th Ordnance Detachment's Technical Intelligence Group received a single Cuban-model AR-10, numbered 000030, from Artillerie-Inrichtingen, and set about putting it through its paces at Detroit Arsenal in Center Line, Michigan.

The rifle was first disassembled and examined, much as was the practice at Springfield Armory. Detailed technical data concerning its apparatus and parts were recorded, and the strong appreciation the Technical Intelligence unit conducting the examination had for the novel piece of weaponry was plainly expressed in the report cataloguing this mechanical information. The innovations of the barrel-locking rotating bolt, economical Stoner-perfected direct gas impingement system, its construction from lightweight, cutting-edge materials, and straight-line stock design were all outlined in detail to clearly identify what distinguished the ArmaLite weapon from its contemporaries.

By all accounts, this first incarnation of the Dutch-produced AR-10 performed exceptionally well in Detroit, being declared serviceable by the 507th. While admitting that the testing staff was not at all familiar with the weapon or its type in general (indeed at that point the AR-10 was truly *sui generis*), the official report praised it highly. In addition to the perfect reliability provided by the Cuban AR-10 in tests at the Detroit Arsenal, a few of its best qualities were singled out by the testers.

The Detroit Arsenal's Preliminary Technical Report on the AR-10, dated September 25, 1958, is excerpted as follows:

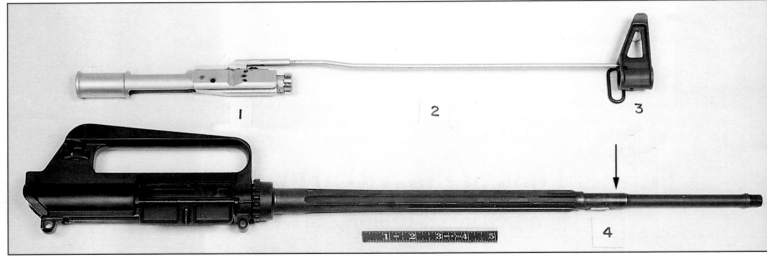

168. Right side view of the bolt assembly (1), the top-mounted gas tube (2) and front sight tower (3) of Dutch AR-10 no. 000030, shown above the barreled receiver.

Preliminary disassembly was carried out with great care, as opposed to the rough handling the prototype AR-10Bs were said to have endured at Springfield the previous year. Special attention was paid to the components of the gas system, as this feature of the AR-10 was particularly intriguing to the Detachment performing the testing.

Note the fluted steel barrel (4), and the location of the gas port (arrow).      Detroit Arsenal negative no. 56360 dated 11 August, 1958, editor's collection

# 507th Ordnance Detachment

## (Technical Intelligence)

### Detroit Arsenal

*Although the AR10 rifle was manufactured in Europe under license agreement by "Artillerie-Inrichtingen", Hembrug-Zaandam, The Netherlands, subject weapon was originally designed, manufactured and patent applied for by the Fairchild Engine and Airplane Corporation, U.S.A. The AR-10 is a cleanly designed, light weight, easy-to-handle, gas operated shoulder weapon employing selective fire. Features include a winter trigger, quick take down for field stripping and a close fitting dust cover which opens automatically upon firing the first round. An extremely light weight (4 oz. empty) box-type magazine of 20 rounds capacity is employed . .*

#### Components with unique features and design:

*Gas system: The gas system of the AR-10 can best be described as being diversified in that it eliminates the use of a piston and operating rod. It utilizes only gas pressure to complete the firing cycle.*

*When the rifle is fired and the bullet passes beyond the gas port . . gas passes through the gas tube . . and into a chamber . . formed by the bolt carrier and the bolt.*

*High pressure gas enters the chamber formed by the bolt and the bolt carrier; the bolt is in the locked position acting as a stationary piston. The entering gas pressure causes the bolt carrier to move.*

*After traveling approximately one-eighth of an inch, the bolt carrier cuts off the supply of gas from the gas tube. Expansion of the gas now trapped in the chamber formed by the bolt and bolt carrier provides the energy necessary to continue the movement of the bolt carrier rearward.*

*As the bolt carrier continues its rearward motion, it rotates the bolt, unlocking it and carrying it rearward. As the bolt assembly travels rearward, the gas is exhausted through a port in the side of the receiver and the cycle is completed through inertia.*

#### Comments:

1. *Normal weapon "climb" appeared to be reduced when subject weapon was fired from the shoulder under full automatic fire. This may be a result of the recoil forces being in a straight line from the barrel through the stock.*
b. *Sight adjustments function smoothly and with ease. Accuracy appeared to be good.*
c. *Weapon tested had a tendency to pull to the right under automatic fire.*
d. *The application of corrosion resistant materiels, simplicity of design and ease of assembly should combine to substantially reduce service and maintenance . .*

A single performance criticism was leveled at the AR-10, however; its tendency to pull to the right when fired in the full-automatic mode. This was not considered a defect of the design, as literally all fully and semi-automatic stocked weapons naturally pull to the right when fired from the shoulder of a right-handed shooter. This is an effect of simple physics, in that a recoiling weapon will always move in the direction of least resistance, which in the case of a stocked rifle fired from the right shoulder, is the right side. The fact that it did not also pull up as its contemporaries were wont to do is a strong testament to the effectiveness of the straight-line stock design.

Although this examination by a U.S. Technical Intelligence unit had no real hope of seeing the AR-10 adopted, even by the armored and mechanized forces for which the 507th generally tested equipment, its excellent performance and the praise it earned flew in the face of the results of the Springfield tests, providing further proof of the unfair nature of those evaluations of 1956-57.

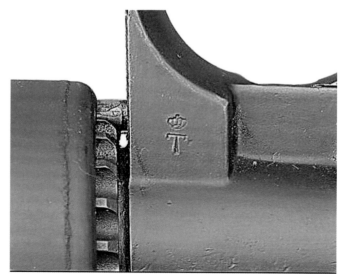

169. Left side closeup of a Cuban model AR-10 upper receiver, showing the crowned 'T' proof mark applied to all weapons produced at the Dutch State Arsenal.

Charles Kramer collection

# Cooper-MacDonald Presents the AR-10 in Asia

Comparatively little is known about the sojourns of Robert "Bobby" MacDonald across the Asian continent and in the new nations of the East Indies, as he left no direct accounts that are available today, and his company has long since ceased to exist. However, the sparse sources that there are paint a vibrant and interesting picture of demonstrations and dealings in some of the most exotic locales on earth.

## The AR-10 in Burma

Demonstrating all over the Pacific Rim, MacDonald caught the particular interest of the governments of two new nations; the Union of Burma and the Republic of Indonesia. Burma, a former British colony, had gained its independence in 1948, and since that time had felt itself very threatened by its rambunctious neighbors. The then-ongoing Chinese Civil War on its doorstep was a constant source of worry; the brutal

170 (right). The first page of another copy of ArmaLite's "Tomorrow's Rifle Today" brochure. Compare with fig. 152: this one is overprinted with the name and address of Cooper-MacDonald, Inc.

The rifle shown is one of the original U.S.-made AR-10Bs from the first series of five rifles.

Marc Miller collection

war the French were fighting in their colonies in Southeast Asia made for a tense situation to its east; and the British counterinsurgency operations in Malaya to its south did not help matters.

Burma, uniquely military-minded of the former British colonies, thus focused heavily on defense matters from very early in its independence, working around a strategy of conventional warfare relying on tanks, mechanized infantry, and aircraft. Though possessing some relatively modern British vehicles, the infantry of the *Tatmadaw*, as Burma's Army is internationally known, was not particularly well-equipped to handle large invading forces. Their regular soldiers were still armed mostly with British Lee-Enfield bolt-action rifles and Sten submachine guns, as well as a smattering of captured Japanese surplus. These weapons did not stand Burma in good stead in this period of rapid evolution of infantry long arms.

This fact was brought home to the new Burmese government in 1949, when a large force of soldiers from the defeated Republic of China first invaded along the border between the two countries, seeking to take control of a section of the dense, mountainous Burmese jungle and use it as a base from which to strike back into China, with the hope of eventually driving the newly-victorious communists from power.

The Nationalist Chinese were in large part equipped with American weapons like the M1 Garand and M1 Carbine, which were a great force multiplier against *Tatmadaw* infantry. Thus, when the Burmese forces confronted this U.S.-supplied and -supported enemy, they were soundly defeated, the Burmese attacks being easily repulsed by the dug-in and battle-hardened Chinese. These conflicts continued for over a decade, always with much the same result.

It was into this atmosphere that Bobby MacDonald arrived in 1957, demonstrating his sample AR-10B to the enthusiastic *Tatmadaw* leadership. Significant interest was evinced, and a potential sale was noted for further pursuit.

## The Mysterious Indonesian AR-10

From Burma, it was just a short hop by air to the capital of Indonesia, newly-renamed Jakarta. This young country presented a unique quandary for a representative of Artillerie-Inrichtingen. Made up of a collection of Pacific islands, Indonesia had only recently been a colonial possession of The Netherlands known as the Dutch East Indies.

The growing nationalist movement had declared the colony an independent state two days after the surrender of the occupying Japanese in 1945, and had then to fight a brutal insurgency campaign against their erstwhile Dutch masters, who were bent on reoccupying the islands which had been stolen from them by the Japanese in 1942. Unlike the Philippines and other areas, Indonesia was never forcibly liberated by U.S. forces, and so the fledgling nationalist leadership there which had collaborated with the occupiers were not denounced and arrested as they were in other parts of Japan's conquest.

Although initially commanding wide international and domestic support (within the islands), framing their reconquest as the removal of a puppet state set up by Japanese collaborators, the atrocities committed by the Dutch forces, and the successes of the independence guerillas, led to outrage in the United Nations, and particularly within the U.S. government, which had been funding Holland's postwar reconstruction under the Marshall Plan. The American threat to cut off aid proved to be the last incentive needed for the Dutch to relinquish control of their territory in the Pacific, with the exception of New Guinea, to the independent state of Indonesia.

This very recent and painful history with the Dutch meant that the new Indonesian government would not even think of buying arms directly from Holland. As MacDonald found, however, this did not preclude their interest in the excellent new rifle that was being demonstrated to them.

To get around the bad blood between the producing country and the potential buyer, MacDonald came up with a rather ingenious plan, which he relayed to Fairchild Arms International's headquarters in Canada. Instead of having the rifles shipped directly from Holland, he proposed that the parts for building the rifles be bought by Fairchild for assembly in the U.S.—thus qualifying the weapons as being American-produced—and then selling the completed rifles to the Indonesian government through Cooper-MacDonald. Artillerie-Inrichtingen was not particularly happy with not being able to make the sale directly, but did not object to the implementation of this plan, should the Indonesians decide to place an order.

# The Hand of the CIA? the Cloak-and-Dagger Nature of the Indonesian AR-10

It turned out that this was not a problem at all, as by December 19, 1957, the Indonesians had already agreed to the arrangement devised by MacDonald, and he had in hand the payment for the first 500 weapons. Little is known about the details of the sale to Indonesia, but the arrangement of sourcing the weapons through the United States has given rise to one of the most persistent and seductive myths within the AR-10 enthusiast community.

As part of the procedure of obscuring the origin of the rifles sold to the Indonesians, the A-I factory records, which identify the exact order quantities and serial numbers of weapons shipped, were either altered to leave out the serial number range for the consignment of rifles sold to Indonesia (1,000 in total), or were never created with reference to those numbers at all.

Adding further to the mystery and allure of these thousand rifles, and making them even riper for conspiracy theory, is the fact that none of the serial numbers in their range have ever turned up (except for perhaps five rifles in private hands) in photographs or on the market. This is particularly strange since from the late 1970s through the middle of the 1980s, most of the countries that had originally purchased and issued AR-10s decommissioned them from military service and sold them on the international commercial firearms market. However, no such sale from Indonesia, or any but a vanishingly small number of examples from the missing serial number range, have ever turned up.

There is a swirl of rumor surrounding this series of the AR-10, with by far the most persistent explanation being that the CIA bought up the entire quantity for some black project, and thus had the records of the factory altered, at the same time convincing all A-I personnel to keep permanently silent about the transaction. That these rifles have never been seen to turn up anywhere only adds credibility to the idea that some secretive organization still has them, and what better bogeyman could there be but the CIA? It is believed likely by those that hold with this theory that the AR-10s in the hands of the world's premiere spy agency were never used in the project for which they were intended, and that they sit crated up in some anonymous government warehouse, much like the Ark of the Covenant in the movie *Indiana Jones*.

However it is the opinion of this author, given what he has learned from the confidential files of both ArmaLite/Fairchild and Artillerie-Inrichtingen, that the purported purchase of 1,000 AR-10s by the CIA is simply a myth, and that the gaps in the records are explained by documentation from Fairchild that discusses the sale to Indonesia and its special arrangements.

Of course, despite it being supported by facts and documentation, this is just the author's opinion, and as stated before, the full compendium of the Dutch factory documents has not been made available for study during the writing of this book. Therefore, it is entirely possible that a "Lost Ark" full of AR-10s is collecting dust somewhere in a secret warehouse in the U.S. It is also possible that another explanation entirely exists for the "missing thousand": for example that their serial number range was reserved for an order that never materialized; or was earmarked for use on a special version of AR-10—like the light machine gun—that never came to fruition, and so simply no rifles so numbered were ever fabricated in the first place.

However it seems by far the most plausible explanation to this author that Indonesia ordered, but did not publicize their possession of, these AR-10s, and unlike other nations, never pulled theirs from service and sold them abroad in later years.

Given the relatively small quantity which they acquired, and their vast performance superiority over either Indonesia's antiquated stock of captured Japanese arms or their eventual choice for a main infantry rifle, it is quite likely that the mystery AR-10s were issued to a unit of *Komando Pasukan Khusus* (the Indonesian Special Forces) that did not march with them in parades, used them only in special operations, and have kept a close watch on their storage and maintenance ever since.

It is the fervent hope of this author that this mystery will someday be definitively solved, but it is feared that even full publication of the files from Artillerie-Inrichtingen pertaining to AR-10 production and sales will not lay these rumors to rest. It is possible that only the actual discovery or disclosure of the location of the rifles will permanently put an end to speculation. Such an event would be a grand day indeed for enthusiasts of the AR-10, but until then, the roundabout sale to Indonesia discussed in intra-office communications within Fairchild offers the most plausible explanation for this enigma.

# A Final Appeal for U.S. AR-10 Production

With the great interest shown in the AR-10 everywhere it was demonstrated, and the triumphant sales made to Cuba, Guatemala, and the Republic of the Sudan, the strong interest shown by Burma, Indonesia, and Portugal, and its imminent testing by Austria and West Germany, ArmaLite felt that its position was excellent, and the time was ripe to produce the now-famous promotional film about the AR-10, touting it as "the most advanced weapon in the world."

Riding this wave, and having moved in October of 1957 to a larger facility at 118 East 16th St. in nearby Costa Mesa, George Sullivan had Richard Boutelle lobby the Fairchild board of directors for funding to set up large-scale production in the United States, which market Fairchild had reserved for itself.

This was to no avail, however, as the parent company was having a down year, and Boutelle himself was catching some heat for his strong advocacy of ArmaLite, which had as of yet not brought in much money for Fairchild.

Late that year, Boutelle was removed from his position as president, and made vice-chairman of the board. Although this was technically a promotion, it was also meant to sideline him and his ideas, which were considered too risky during the downturn Fairchild was experiencing. Therefore the request was denied, and series production of the AR-10 in the U.S.A. would have to wait almost four decades before becoming a reality.

*Chapter Seven*

# Two New "Country" Model AR-10s

## Background to West German Rearmament

In the early years following the surrender of Nazi Germany at the end of WWII, the zones of western Germany occupied by France, Britain, and the United States were allowed to form a country once again as the Federal Republic of Germany, while the Soviets set up the German Democratic Republic (GDR) in their control zone in the east of the former "Thousand-Year Reich".

### Establishing the *Bundesgrenzschutz*

Initially the West Germans were not permitted to establish any form of military force, but in 1951, due to the rising tensions of the Cold War and the Soviet threat from the earlier Berlin Blockade, the Allies allowed the establishment of the *Bundesgrenzschutz*, or Federal Border Protection Force.

Originally armed with surplus U.S. equipment, this new organization was tasked with finding a suitable rifle with which to equip the skeleton force that was to defend the republican West from the Soviets and their puppets to the east. West Germany was still at the time completely demilitarized, relying totally on the French, British, and Americans for their defense, and, given the events of the past thirty years, none of these countries seemed particularly keen on providing arms for full-scale West German rearmament. Spain, however, had no qualms whatsoever about dealing with the country by whom they had stood even as it had descended into madness and inflicted terror on the world during WWII. The fact also that formidable German weapons designers like Werner Heynen and Ludwig Vorgrimler were by this time in the employ of the Spanish, made this bilateral cooperation all the more acceptable.

The West Germans were involved to a greater or lesser degree with each step in the progress of what would become Spain's CETME battle rifle, and indeed their commercial interest was the driving force behind the weapon's upgrade to fire the full-power NATO round, instead of Spain's proprietary, undercharged 7.62x51mm CETME cartridge. This round would have suited the Spanish requirements, but it was realized that when trying to sell the CETME to other nations, having it capable of firing the round that was becoming standard for all nations opposed to Soviet hegemony would be a necessity.

Serious delays were experienced even as the CETME approached nearer and nearer to an initial production stage. A tentative agreement had been reached with the German *Bundesgrenzschutz* for the CETME to become their standard service rifle, but as described in a quote from the Collector Grade title *Full Circle*, Dipl.-Ing. Werner Heynen, the former general manager of the wartime Gustloff-Werke who had in 1950 accepted the position of head of the fledgling CETME development team in Madrid, recalled,

> . . Despite the pressing need [for the CETME], however, the [BGS] order was persistently delayed. The delivery time became increasingly more pressing and shorter, and finally a large number of [FN FAL] weapons was ordered from Belgium because there, as a result of ongoing production, shorter delivery times could be guaranteed, while the Spanish were still not yet even in production.

Therefore, the *Bundesgrenzschutz* cancelled their contract with CETME early in 1956, opting instead for the FN FAL.

## West Germany Joins NATO, with Plans for a New Arms Industry

In 1956, the same year that the FAL was chosen by the West German Border Guards, the members of the new NATO alliance decided against a plan to keep on shouldering the burden of, and responsibility for, the exclusive and complete defense of the Federal Republic of Germany, voting instead to allow the West Germans to enter NATO as a full member with their own regular military force. The French, in particular, who had exhibited the greatest trepidation at allowing Germany to rearm, finally changed their minds when confronted with the realities of the cost and logistical demands of defending another territory almost the size of their own, and acquiesced in voting to allow for German NATO membership, in charge of their own self-defense.

## The *Bundeswehr* Selects a New Rifle

Accordingly, the new German *Bundeswehr*, or "Federal Defense" was formed, and being a real military, instead of what was essentially a paramilitary police agency like the BGS, assumed the initiative for evaluating new weapons.

Initially they intended to follow the BGS decision to keep the FAL as their standard arm, but negotiations with Fabrique Nationale began to stall. The initial run of FALs for the Germans was priced at $110 per weapon, and the new defense force attempted to get that figure down considerably. The $90 then offered by FN was still a bit high for the new army, but the real kicker came with production licensing. The Germans were very concerned with self-sufficiency, being the new first line of defense against the Eastern Bloc, and were particularly insistent about being able to produce their chosen weapon domestically.

While the French and Belgians were grudgingly willing to let the West Germans into their new alliance, the idea of allowing them to set up their own weapons production capability instead of buying arms abroad was another thing entirely. FN, under the policy of the Belgian government, at first flatly rejected German overtures for a licensing agreement but, as further stated in *Full Circle,*

> . . *despite voluble protestations from the FN management that the FAL would never be produced under license in Germany, such an offer was indeed made . .*

The *Bundeswehr* did purchase 100,000 FN FAL rifles of a type known originally as the *Gewehr* DM 1 (*Deutsches Modell* 1), and later simply as the G1 (*Gewehr* 1). However, as stated in a memorandum dated October 15, 1956, signed by German Defense Minister Matthauser, "it is intended to equip the troops with a different model eventually."

## German Trials of Four Rifles: the G1, G2, G3 and G4

The *Bundeswehr* accordingly initiated a testing program to find another rifle to be adopted and produced in Germany. This opened up the opportunity of which Jacques Michault and George Sullivan attempted to take advantage in late 1957 and throughout 1958, and was the cue not only for ArmaLite, but also for the Spanish, and both began aggressively pursuing the German market.

The West German Army set up a classification system to designate all of the rifles it accepted for field trials. The contest was among four rifles; the G1, or FN FAL; the G2, the designation given to the Swiss SIG SG510/PE-57; the G3, the CETME's newly-assigned model number; and the G4, the Cuban model AR-10.

The competition had actually begun in 1957, but just as had happened when the Continental Army Command had ordered Springfield Armory to halt the Lightweight Rifle Program in favor of waiting for the verdict on the AR-10 in the United States, in West Germany everything was put on hold to await the arrival of the Dutch AR-10s.

The FAL (G1) was a known entity, and the CETME (G3) was still not performing well in troop

171. Two of the contenders up for adoption by the West German *Bundeswehr*.

In the foreground is the special 7.62mm NATO caliber Swiss SIG 510 (designated the G2), the heaviest entry by far, fitted with a rubber butt to ease the shock of firing rifle grenades. Behind it is the German version of the Dutch Cuban model AR-10 (the G4), fitted with a bipod and externally threaded flash hider to allow the use of the first type of BFA.                    courtesy Dr. Elmar Heinz

trials, and so tests were set up for the two new entries: the Swiss SIG 510, in a special model chambered in 7.62mm NATO caliber (the G2), and the lightweight ArmaLite AR-10, as produced by A-I in Holland (the G4).

The strong interest shown by the West Germans in the AR-10 led to them being the first national force to actually take delivery of an ordered quantity of the brand-new Dutch-manufactured ArmaLites. (The story of how Sam Cummings had delayed delivery of the Cuban order by insisting that it be shipped first to his warehouse in Manchester is related in Chapter Six). After a demonstration of some Cuban model AR-10s for members of the *Bundeswehr* and *Bundes-grenzschutz* command on February 26, 1958, the renascent German military placed its order on March 22, 1958 for 135 rifles for troop trials. An earlier order for 400 AR-10Bs had been cancelled due to the inability of the new Costa Mesa plant to produce the weapons, but with Artillerie-Inrichtingen now on line, the Germans saw the real potential to take delivery of a perfected, mass-produced version of the new rifle.

Despite the difficulties the Dutch continued to experience, they made their delivery on schedule, with all 135 rifles in German hands by September, 1958. The first 100 of the rifles produced for this contract were of the standard infantry variant, while 35 were fitted with experimental bipods that fit and functioned in exactly the same way as ArmaLite's first experimental bipods. Each "winged" leg of the bipod was contoured the same as the handguard, and even had vent holes that indexed with the ones on the forearm, so when folded they mirrored the look of the area of the weapon they covered.

# Describing the German G4 "Cuban" Model AR-10

172. Left side closeup of a typical German G4 (AR-10), serial no. 000260, fitted with a bipod shown folded over the handguard.

Note the *Bundeswehr* eagle acceptance stamp following the serial number, and the added German initials around the selector switch: 'D' (*Dauerfeuer*) for AUTO, forward; 'S' (*Sicherung*) for SAFE, up; and 'E' (*Einzelfeuer*) for SEMI.

courtesy Dr. Elmar Heinz

The German G4 AR-10 was basically the first A-I production "Cuban" model, produced after, but received before, the consignment which had been sent through Sam Cummings to Batista, and collected by Castro.

The sole variation between the Cuban AR-10s and the German G4 test rifles was in the open-ended three-prong flash hider, which had replaced the original recoil compensator. The flash hider on the G4 rifles had threads cut onto the outside of the ends of the three prongs, in order to accommodate a blank firing adapter. The Dutch had established the need for this accessory from input received from various prospective customers, who had stressed the necessity for the use of blanks when training troops unfamiliar with automatic rifles. All of the operating systems of self-loading rifles require the round fired to develop a certain amount of pressure in order to work the action and, while Stoner's direct impingement design was the most able to handle very low amounts of gas pressure, it still would not cycle reliably when there was no bullet in the cartridge. A blank firing attachment threaded onto the rifle's muz-

173. Right side closeup of the barrel muzzle of a German G4 AR-10, showing the threaded exterior of the Cuban-style open three-prong flash hider, made to accept the first type of BFA, shown in fig. 174.

Charles Kramer collection

zle would restrict some of the flow of gas from the blank cartridge, allowing the pressure to rise much like it would do if a standard round were fired, but with none of the potential danger.

This requirement was very new for a military, as not only did bolt-action weapons have no need for it, but the earlier generation of automatic rifles, like the M1, SVT-40, and Johnson, had no provision for blank firing.

## The Two Types of AR-10 Blank Firing Adapter (BFA)

Artillerie-Inrichtingen found an ingenious way to accomplish both grenade launching and blank firing, first by cutting the mentioned threads onto the outside front portion of the flash hider, and crafting a blank-firing attachment (basically a barrel plug with a threaded cap) that could be screwed onto the flash hider, while standard rifle grenades could also slide over the outside diameter of the same flash hider. Obviously these two functions could not both be performed at the same time, but they added additional functionality that allowed each infantryman to act as a grenadier, and also provided for cheap, safe training and orientation with blank cartridges.

As discussed in *Full Circle*, the German AR-10s were distributed to various units and institutions as they arrived, with 45 (34 standard and 11 bipod models) going to the Infantry School at Hammelburg, 45 (same number of each model as above) to the Armored Corps School at Munster-Lager, 14 (11 standard and 3 bipod) to the Airborne School at Schongau, 14 (10 standard and 4 bipod) to the Mountain and Winter Warfare School, and the remaining 15 (10 standard and 5 bipod) to the central equipment proving grounds at Meppen.

174. A comparison of the two types of blank firing adapters (BFA) developed for the AR-10.

Left: the first type, as used on the German G4 test rifles and Sudanese-contract production. As shown, this featured threads on the inside of the painted aluminum cap, allowing it to be screwed onto the modified flash hiders, which were threaded on the outside ends of their prongs.

Right: the second type, for later-production rifles with flash hiders threaded internally.

Dutch Military Museum collection

## The G4 AR-10 Performs Admirably

The German forces at the several schools ran their AR-10s through various batteries of tests while procurement personnel negotiated with the various firms.

So threatening was the AR-10 that CETME and Fabrique Nationale, unlikely allies but both competitors for the German contract, even colluded to try and stop the ArmaLite rifle. Communication between these firms, memoranda, and pricing data show that at least these two companies were intentionally attempting to undercut the AR-10. The Belgians continued their refusal to budge an inch on price, while the Swiss were amenable to the idea of licensing, but could not get their price down low enough. The Spanish were very open to licensing, and offered an extremely low price of $45 per weapon for the first series made to German specifications, while they tooled up to produce it in-country. This was well below the already astoundingly-cheap $75 per rifle to which the BGS had previously agreed before abrogating the CETME contract in favor of the FAL.

The Germans, however, would not be denied their favorite weapon, and despite the much higher cost, forged ahead in favor of the AR-10.

As testing progressed, it became clear to the troops firing the weapons that the AR-10 was demonstrably superior to all of the other contenders. Not only did it have a staggering advantage in weight, rapid-fire controllability, and accuracy, but it actually experienced fewer malfunctions than any of its competitors.

By the end of 1957 the Swiss rifle had been essentially rejected due to its extremely heavy weight, while the CETME and FAL were both considered at least adequate. The reliability issues which had dogged the initial Spanish entry in 1957 were mostly corrected by the time it went head-to-head with the AR-10, and the FAL displayed excellent dependability, if not the light weight, accuracy, and ease of use of the AR-10.

## The G4 is Done In by Timing

In the end, however, the decision came down to simple timing. Even contemplating the fantastically low price at which the Spanish were offering the CETME, the Germans were still completely willing to pay the extra money to receive the superior rifle, and even to forgo domestic production until Artillerie-Inrichtingen's exclusive license ran out in 1961. The Dutch offered the Germans Cuban model AR-10s (or whatever more advanced model could be developed in cooperation with them) at a price of $110 each; almost three times what the Spanish were asking. What the prospective customers could not stomach, however, was the time it would take before they could begin equipping their infantry with the superior ArmaLite weapon, when staring down the constant possibility of a Soviet/East German invasion.

The prospect of waiting for over three years to be able to produce in-country was difficult enough for the Germans, but the problem that was unique to the AR-10 was that Artillerie-Inrichtingen simply could not crank out rifles fast enough, even at the higher price, to arm enough troops to ensure the viablity of the new German *Bundeswehr* until they could take over and produce the AR-10 for themselves. This stemmed from the fact that, while A-I was majority-owned by the Dutch government, it actually received little to no funding or operating subsidies from the state, and indeed its new Defense Minister, Sim Visser, was no friendlier toward Artillerie-Inrichtingen than Kees Staf, the previous minister, had been, and continued to inconvenience A-I at every turn.

It was thus announced in January, 1959 that the CETME proposal had been accepted, and that a modified version of the G3 trials rifle would be adopted as standard German equipment.

Helping console the *Bundeswehr* at the prospect of not receiving their first choice of rifle was the price of the CETME. Already extremely low due to its brutally simple stamped-steel construction, the amount charged was even further artificially depressed, thanks to the multi-company conspiracy against the AR-10 often lamented in Fairchild's in-house communications.

In addition to receiving their first shipment of CETMEs at essentially cost price, the Germans were permitted not only to produce the rifle, but to modify it, sell it abroad, claim it as their own proprietary design, and even grant production licenses to other countries. The German firm Heckler & Koch, which had been very involved in the Spanish CETME program, received the production rights to what was initially known as the H&K DM3 (CETME), and the now-famous G3 began its evolution.

Nevertheless, the details of the deal between the Spanish and Germans which came out later showed just how much better the Germans thought the ArmaLite weapon was than the CETME, as they were willing to select the AR-10 even at the much higher price, and wait years to produce it domestically.

# The "Mile High Club": the Rare KLM AR-10

Before the larger, more important orders for world militaries were filled, four very distinctive rifles were shipped out to a private entity, for a very unusual purpose. In 1957, Royal Dutch Airlines (KLM; the official airline of Holland and a company with which Fairchild had had many dealings), announced that the next year they would begin flying a new route to Tokyo straight over the North Pole, stopping for fuel in Anchorage, Alaska. Using their new, American-built DC-7C passenger planes, KLM was among the first airlines to offer routing over the pole to save distance when traveling to particularly remote locations.

In preparation for this route, KLM began designing survival kits to protect the crew and passengers against the elements and keep them alive until rescuers could arrive in the event that one of their planes had to make an emergency landing. At first, these kits included cold-weather essentials such as tents, thermal sleeping bags, goggles, axes, saws, knives, fishing gear, hunting traps, stoves, and food. It was quickly realized, however, that patrolling the ice floes of the high arctic and Greenland, as well as the forests and glaciers of Alaska, were fierce and aggressive polar bears. KLM management reasoned that all the survival gear in the world would serve only to keep snacks warm for bears, if no weapons were available to fight them off.

The close connection between Fairchild and the Dutch aviation industry paid off, and the idea of using a modern, advanced weapon that could be easily handled by even small and untrained personnel was successfully pushed to the planners of this route. KLM therefore placed an order for four specially-designed AR-10s, one of which was to be included as part of the survival kit aboard each of the DC-7Cs destined to fly the new route.

Once the characteristics of the weapon were approved by KLM management, the order was placed on September 3, 1958, and rifles were delivered to the airline's corporate headquarters at Amsterdam's Schiphol Airport two days later.

## Describing the KLM AR-10

175. This photograph, part of a press release by both Fairchild and KLM in September, 1958, shows KLM stewardess Johanna Van Duffelen modeling the parka and mittens included in the polar survival kits, along with the distinctive KLM-model AR-10.

Thanks to its plain muzzle, this version of the AR-10 held the title of the lightest full-length AR-10 ever produced, tipping the scales at just six pounds, fifteen ounces.
Smithsonian Institution collection

The rather distinctive KLM version of the AR-10 was essentially a Cuban model but with a plain muzzle, instead of the three-pronged flash hider. The "survival" AR-10 was consequently the lightest variant ever sold in the entire history of the weapon, and with only four examples produced, it stands as the rarest of all AR-10s. Those who appreciate the preservation of historical artifacts can rest assured that at least

three of the four KLM models made are safe and well-cared-for in private hands today.

In addition, according to a KLM pilot who flew the polar route and examined one of the rifles, its lower receiver was also unique in that it was only capable of semi-automatic fire, and the takedown pins were attached by small lanyard chains so they would not be misplaced by the relatively untrained airline crewmembers when disassembling the rifles for stowage.

The four rifles of this order, being manufactured around the same time as those of the Sudanese contract, discussed below, bore the serial numbers 002301 through 002304, and were shipped with one sling and one magazine each. Actual combat was not anticipated, and it was reasoned that twenty rounds of potent 7.62x51mm NATO ammunition would be quite sufficient for any situation involving four-

legged predators; the crew having ample time to reload a partially-expended magazine if the weapon ever did need to be used.

Although none of these four weapons ever had to be put into action to defend a downed airliner, they traveled many miles at very high altitudes, and stand out as the AR-10 variant with the most interesting, unique, and unexpected history. Without a doubt, Ms. Van Duffelen had a very peculiar and fascinating story to tell her colleagues in the airline industry.

# The Sudanese AR-10

## The Cuban AR-10's First Transformation

With orders from two military forces—Guatemala and the Sudan—entered and paid for, it was time for Artillerie-Inrichtingen to make good on its aspirations and promises to Fairchild of being able to quickly and seamlessly adapt the Cuban model AR-10 to the specific requirements of different foreign military customers, who almost by necessity had different mission and supply requirements covering what they wanted their AR-10s to be able to do.

After supplying the German trials rifles and the four KLM survival AR-10s, the next to be manufactured were for the Sudanese contract. General Abboud's representatives transmitted their specifications through Sam Cummings to Artillerie-Inrichtingen, and the Dutch design team, spearheaded by Van Der Jagt, Spanjersberg, and Bakker, set about, under the leadership of Friedhelm Jüngeling, to apply the necessary changes to their first (Cuban) production model. Jüngeling's management style and role within the A-I organization was very similar to that of George Sullivan's at ArmaLite, but

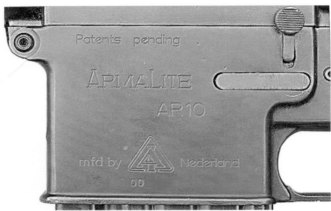

176. Left side closeup of an early prototype of the Sudanese AR-10. No serial number has yet been assigned.

photo from Charles Kramer collection

Jüngeling was himself even more directly responsible for the Dutch AR-10, similar in importance to the role Eugene Stoner had played for ArmaLite.

## Describing the Sudanese AR-10

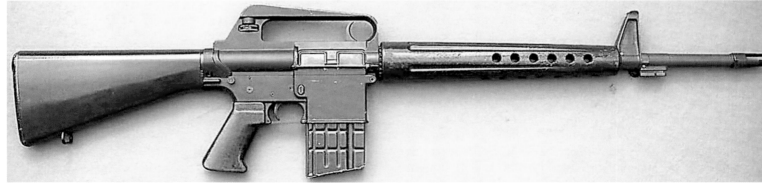

177. Right side view of a typical Sudanese-contract AR-10. The open, externally-threaded three-prong flash hider/grenade launcher was retained, but a solid sleeve with Ger-

man-style bayonet lug was pinned to the barrel ahead of the front sight tower.     Instititute of Military Technology

This model of the Dutch-produced AR-10 is commonly known as the Sudanese, named for its large-scale purchase by the government of the Republic of the Sudan in 1958, the result of the sale orchestrated by Sam Cummings.

The Sudanese AR-10 was the pinnacle of extreme light weight in the line of Dutch AR-10s that saw large-scale production and widespread combat use. It was an updated version of the Cuban model,

weighing 7 lbs. 4 oz., and still employing the same lightweight, fluted steel barrel, but with the front section of the thin barrel profile covered by a barrel shroud, which supported a bayonet lug.

The open-ended three-prong flash hider, threaded on the exterior like that of the German G4s, both allowed the firing of 22mm rifle grenades and the attachment of a first-type blank firing device for use in training with blank ammunition.

## Rudimentary Night Sights, and Eastern Arabic Range Markings

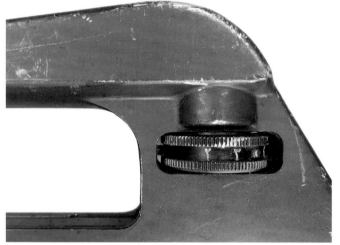

179. Left side closeup of a Sudanese AR-10 receiver, showing the rear sight range gradations marked in Eastern Arabic numerals.          BATF collection

As shown in fig. 178, the AR-10s produced for the Sudanese contract featured rudimentary night sights, for use in nighttime or low-light applications.

In addition, the rear sight, adjustable for elevation as before, differed in that the range gradations were marked in Eastern Arabic numerals, as traditionally used in Arabic- and Farsi-speaking nations.

178 . Top closeups of the rear and front sights on a typical Sudanese AR-10.

Note the small dots of white paint, containing luminous tritium, above the rear sight aperture (left) and below the front sight blade (right).          BATF collection

## The Special Sudanese Barrel Shroud and Bayonet Lug

Not willing to abandon the time-honored martial tradition of bayonet use, both for close-range work and ceremonial duties, the Sudanese requested that their rifles be able to mount a bayonet. Instead of simply welding a bayonet lug to the bottom of the barrel, the Dutch took a different route, seeking to maintain barrel harmonic integrity and minimize accuracy issues that could result from mounting a bayonet, and created a means of insulating the barrel

from contact with the bayonet handle. To this end, they developed a barrel shroud to cover the portion of the barrel forward of the front sight base.

The Sudanese barrel shroud consisted of a simple steel tube pinned in place ahead of the front sight tower, which extended forward to make firm contact with the back of the flash hider. On its bottom, just forward of and underneath the front sight tower, was a lug to which the spring-loaded bayonet catch could

180. Left side closeup of the front end of a Sudanese AR-10, serial no. 000380, showing the special barrel sleeve which provided a long and very rigid attachment for the German-style bayonet lug, so much so that no muzzle ring was required to secure the blade of the bayonet.

BATF collection

be clipped; so securely, it turned out, that there was no need for a ring on the crossguard to stabilize the bayonet.

The Sudanese further requested that their rifles' furniture be of a color that would allow effective camouflage in desert and forest environments, and the original brown of the stocks, handguards, and

pistol grips first designed for the Cuban model ended up being perfectly sufficient for these purposes. The Dutch did, however, have Dynamit A.G. produce a few experimental stocks for the Sudanese-pattern rifle, as they had done for the Cuban model, finished with a coat of thick green paint (fig. 186).

## Brass Stock Medallions for Unit Identification

At the urging of the Sudanese, and possibly the Austrians, A-I began installing brass discs, suitable for stamping unit marking identification, on the left side of the buttstocks of the rifles produced under the first large-volume contract.

As shown here, many were left unmarked; but finding an AR-10 in good condition with one of these medallions on the buttstock—marked or not—is considered a rare treasure among AR-10 collectors.

181. Sideways closeup of a Sudanese buttstock fitted with a brass identification disc, which has been left unmarked. Note the trap in the stamped steel buttplate.

BATF collection

## The Sudanese Bayonets - Subcontracted to Carl Eickhorn of Solingen

In line with earlier plans, Artillerie-Inrichtingen neither designed nor manufactured the bayonets for the Sudanese contract rifles, instead relying on a subcontractor recommended by Sam Cummings. Founded by Carl Eickhorn; the preeminent master bayonet designer and manufacturer, probably in the world,

Eickhorn Waffenfabrik (still in business today as Solingen Knife), in Solingen, West Germany, was given the task of liaising with both the Sudanese to fabricate a bayonet based on their requirements, and A-I to apply the Sudanese specifications to the actual mounting on the AR-10.

182. The Sudanese-contract AR-10 bayonet and sheath,
manufactured by Carl Eickhorn of Solingen, Germany.
Note the lack of a muzzle ring on the crossguard of the
bayonet.                                  Marc Miller collection

183. Left side closeup of the hilt of a new-condition
Sudanese bayonet, showing markings, including the name
of Sam Cummings's Interarmco, the sales agency which
had secured the Sudanese contract.     author's collection

184. The combination tool stored in the grip of the
Sudanese bayonet allowed the infantryman to adjust his
rifle's gas settings, open cans, drive in screws, pry lids off
ammunition boxes, and uncork the copious bottles of
Cabernet Sauvignon the Sudanese Army carried into battle. (Or, perhaps water, or oil, came in corked containers?)
Dutch Military Museum collection,
photo by Marc Miller

This cooperative effort resulted in a very unique
bayonet. The first noticeable trait of the Sudanese
bayonet was that, like the German military bayonets
for the 98 Mauser, it lacked a muzzle ring. A-I reasoned, and Eickhorn concurred, that the long
Mauser-style bayonet lug on the barrel shroud was
so robust that it alone was sufficient to support the
integrity of the bayonet, even in hard use.

The actual knife, fitted with brown plastic panels on the handle, which matched perfectly with the
rifle's furniture, contained within its hollow hilt a
multi-tool device comprising several special-purpose
tools. The strangest of these was a corkscrew, which
seemed extremely out-of-place in a very arid country
where the majority of the population was Muslim,
and thus where bottles of wine certainly never
needed to be uncorked. The most critical component
of this toolkit was a special key, used for adjusting
the gas valve on the front surface of the front sight
base.

## The Rare Sudanese Web Sling

185. The distinctive Sudanese-issue AR-10 sling, made of
light-colored woven canvas.
Note the snaps, steel keeper, and end grommets.
author's collection

The Sudanese had experienced issues with leather rifle slings drying out, cracking, and falling apart with great speed in the very harsh desert environments of much of their country, just as they were prone to rotting in the humid swamps of the south. They therefore commissioned the Dutch to design a new carrying strap for their ArmaLites. The resulting product was made of simple woven canvas, as most modern rifle slings are, but with a very distinctive pattern of snaps, end grommets, and a heavy steel keeper. The sling was also narrower than those found on the American M1 and British Enfield rifles the Dutch forces were currently issuing. These slings are among the rarest of AR-10 accessories today for the collector, and fetch an exorbitant price.

## A Sudanese AR-10 with Green Furniture and British Proofs

## No Bipods on the Sudanese AR-10s

Although both ArmaLite and Artillerie-Inrichtingen had been experimenting with bipod mounting since their very earliest cooperative work on the AR-10 in 1957, there is no evidence that the Sudanese ever sought or acquired any bipods for their rifles.

## Adding 25 Scoped AR-10s to the Sudanese Order

The Sudanese further added to their order 25 scoped models to be issued to designated marksmen in the élite units armed with AR-10s. Because of the high cost of the precision telescopic sights, these particular pieces commanded a price of more than twice that of the standard model rifles, but they were prized for their ability to take extra advantage of the AR-10's naturally excellent long-range accuracy.

Fig. 148 shows one of these earlier, lower-powered 3x25 scopes, mounted to a Cuban model AR-10 but first sold as part of the Sudanese contract. In order to mount the scope, the sides of the carrying handle had to be cut down and a free-rotating, captive nut mounted underneath. The machine screw on the bottom of the scope body was inserted down through the vertical hole in the carrying handle above the nut, and the wide knob of the nut, seen protruding from the side of the handle, was turned to tighten it into place. The scope held its zero by its bottom surface indexing perfectly with a bracket mounted to the shaved top of the carrying handle.

# The Guatemalan AR-10

187. Left and right side views of an A-I AR-10 from the Guatemalan contract, serial no. 003198. Note the telltale vent holes in the barrel shroud.

The Guatemalan model was extremely similar to the Sudanese, and in popular discussions of the AR-10 this has caused it to be almost universally mistaken for its more heavily-produced and -used predecessor.

Dan Shea collection,
photo courtesy Frank Iannamico

36 (previous page). Left side closeup of a further Suda-se AR-10, serial no. 001284, showing a thick coating of een paint on the well-worn handguard.

In addition to the Dutch crowned 'T' proof, this well-aveled rifle is marked on the receiver with British proofs om the Birmingham Proof House, including the Birming-am crossed-scepter Deactivation symbol.

photo from Charles Kramer collection

All of General Abboud's specifications were satisfactorily met in just six months, and on July 30, 1958 Artillerie-Inrichtingen was able to report to Interarmco (Canada) Ltd. that an initial 186 rifles, serially numbered between 000042 and 001136, had already been shipped, and that the remainder of the 2,508 rifles ordered for the Sudanese Army would be completed and ready for shipment by October.

A-I were ready to keep the line running for an additional 450 units after the Sudanese order was completed, as by then the Guatemalan AR-10 specifications were nearing formulation.

Guatemala proved a very easy customer for Artillerie-Inrichtingen to satisfy, as the Guatemalans insisted on only one modification to the Sudanese model. They wanted to increase the surface area of the barrel to facilitate cooling, and were satisfied when the Dutch simply modified the Sudanese-pattern barrel shrouds by drilling a series of vent holes in both sides.

Although it was extremely similar to the Sudanese model, it is justifiable to call the Guatemalan AR-10 a distinct version, owing both to its insignificant-but-noticeable differences from the Sudanese, and the fact that this version was sold to a single national military force, and no one else.

Such an exclusive order by a foreign military has provided us with logical and clear-cut definitions of the distinct models of the Dutch AR-10 we have discussed above, but as we shall see, this was about to change.

## The Mysteriously Mismatched Guatemalan Handguards

188. Underside closeup of a Guatemalan AR-10 handguard, showing cutouts intended to mate with the folded legs of a bipod.

This has caused much confusion among collectors, as no bipods were ordered as part of the Guatemalan contract. The use of these handguards resulted simply from a shortage of standard-pattern handguards when these rifles were being assembled.                    BATF collection

In a move which has long mystified historians and collectors, many of the AR-10s sent to the Polytechnic School under the Guatemalan contract were fitted with handguards contoured for the bipod of the next Dutch model, although none of the delivered examples were actually fitted with bipods. This mismatch came about as a result of an early production surplus of handguards for the rifles being assembled for Dutch military trials, together with a corresponding shortage of standard Cuban/Sudanese handguards for the Guatemalan rifles.

## *Chapter Eight*

# The Transitional AR-10s

## The One-Off "Pre-Transitional" Prototype

189. Right side view of the one-off "Pre-Transitional" AR-10, serial no. 002477, with features as described in the text.

This "Pre-Transitional" is one of many excellent examples of evolutionary stepping stones between different models of AR-10s, and serves as undoubtedly the most profound, clearly demonstrating features from two different well-known and easily-defined type classifications.

Dutch Military Museum collection

The Transitional series which followed the first three "country" model AR-10s—the Cuban, Sudanese and Guatemalan—was preceded by a one-off prototype that bridged the gap between the Guatemalan and the First Transitional models. This is particularly interesting, as the First Transitional itself, by its very name, was simply a stepping-stone. Its name also implies that the type classification actually covers several variants that differ considerably from one another, generally much more so than say, the Guatemalan differed from the Sudanese.

The single rifle that displayed these incremental changes partway between the Guatemalan and First Transitional variants was never given a type designation and, unlike other varieties, has not been named by collectors or been assigned a classification based on its sale to a particular country. Therefore, in the interests of saving time and furthering continuity, this author will refer to this one-off as the "Pre-Transitional."

The Pre-Transitional retained the Sudanese buttstock with steel buttplate and sling attachment, as well as the standard open, externally-threaded three-prong flash hider.

The similarities with its predecessors and differences from the First Transitional ended there, however. The lower receiver had the modified selector switch that designated the SAFE mode as the first (forward) position, discussed below, and the strengthened bolt release lever. It also was fitted with the new, thicker barrel, and had installed on the forward portion of that barrel a shroud that bore a strong resemblance to that found on the First Transitional variant. Although it is not integral with the flash hider, the barrel shroud illustrates perfectly the evolutionary "missing link" between the earlier series and the Transitionals. It also incorporated the new front sight base with thicker rings to protect the gas adjustment screw, and the stronger and thicker extension to the lower receiver for the pivot pin. In fact, this model conclusively answers the question of which version out of the first three (Cuban, Sudanese, and Guatemalan) was chronologically latest. The Pre-Transitional's unique barrel shroud is contoured and ribbed like that of the First Transitional and its successor, but it is also perforated with cooling holes, as seen previously only on the Guatemalan version.

## The "Pre-Transitional" Provides a Provable Chronology of Models

This "fossil record" reinforces the data from Artillerie-Inrichtingen that the Cuban design was their first production model, the Sudanese came next, and for the Guatemalan order, the engineers made some small modifications to the Sudanese, manufacturing and shipping it last.

It is very likely that the Pre-Transitional was put together in order to serve as a test rifle, to make sure that the modifications that had been made to the more-or-less proven Sudanese and Guatemalan designs did not hinder performance or cause any unexpected or unforeseen problems.

Of particular concern at the time was that the barrel be capable of excellent accuracy, since the issues encountered with work stress in the first runs of fluted barrels produced on their new hammer-forging machine were still very fresh in the minds of the A-I personnel. In order to avoid such fiascos as had occurred during the Springfield and Quantico tests, the Dutch spared no expense in ensuring proper functioning, reliability, and safety of any new design they put out, especially for the extremely grueling weapons evaluations that the First Transitional variants would face.

# The "First Transitional" AR-10

190. Right side view of a typical First Transitional AR-10,
serial no. 002477, with features as described in the text.
Dutch Military Museum collection

Before the delivery to the Republic of the Sudan was even in the crates, the next steps in polishing the AR-10 to its pinnacle of design were already under way. By the time the Sudanese rifles were finally all manufactured, their design was no longer the cutting edge of AR-10 innovation. At the time rifle number 000001 rolled off the line, there were only firm orders in Zaandam for weapons for the Sudan (the 2,508 quantity secured *sub rosa* by Cummings), with the 135 for limited field trials by the West Germans soon to follow.

The Dutch military, which had been waiting in the wings, wanted something a bit more substantial for their upcoming field trials, and the rifle designed in late 1958 for this domestic evaluation was formulated with a great deal of direct input, far more than

that exerted by the Sudanese, and featured several modifications to the Sudanese and Guatemalan contract models.

The actual contact with the Dutch military at this time was more informal than anything else, as the Ministry of Defense had not actually announced that the Kingdom of the Netherlands was officially intending to replace the U.S. M1s they were currently using. However, informal contacts with their neighbor Belgium had been going on for at least three years, in which the FN FAL had been demonstrated to certain personnel and evaluated. These same members of the Dutch defense community made the suggestions that saw the further evolution of the AR-10 from its Sudanese configuration to the First Transitional type.

# A New Buttplate and Sling Swivel

One of the first modifications was an attempt to remedy somewhat the discomfort engendered by extended rapid fire. The earlier elimination of the recoil compensator, and then the incorporation of a steel buttplate, while not threatening to the AR-10's status as being in a class by itself in rapid fire controllability, did make the process of achieving this control much harder on the shoulder of the rifleman. Although it meant eliminating the feature of stowing a small cleaning apparatus in the stock with easy access through a hinged trap in the buttplate, it was elected that the steel buttplate should be replaced with one made of thick, forgiving rubber.

Also on the buttstock, the rear sling attachment was changed. The original Cuban, Sudanese, and Guatemalan configurations all had a rectangular metal retaining ring with rounded edges that was an integral part of the buttstock, located underneath it. This First Transitional model kept the same placement of the attachment, but instead of a fixed ring, added a swivel joint to which the same type of ring attached, and hung down lower. This allowed the sling's rear attachment point to move more freely for access and tension, and put less stress on the fiberglass shell of the buttstock.

# Altering the Fire Selector Positions

191. A closeup comparison of the two types of Dutch AR-10 selector switch markings.

Left: from the handbook for the Cuban model, showing the original settings, in which "SAFE" was in the up position between "AUTO" (forward) and "SEMI" (rearward).

Right: from the Second Transitional handbook, showing the redesigned selector settings, with "SAFE" forward, "SEMI" up, and "AUTO" rearward.

handbooks from Charles Kramer collection

Two items on the lower receiver were upgraded from their original form. Since Stoner's very first AR-10 prototype, the M-8, the selector switch, which was used to dial in the appropriate fire mode, had the "SAFE" position central or upward, with "SEMI" being forward and "AUTO" to the rear. The Dutch considered it cumbersome to have to go through two selections to change from semi-automatic to fully automatic, or vice versa.

To this end, the "SAFE" position of the selector lever was moved to the forward location, so that when taking the weapon off safe, one first selected "SEMI", the default fire mode for all but Soviet and Chinese infantry doctrine, and then by turning the

lever one more stop, "AUTO," making the change from semi-automatic to the fully-automatic fire setting quicker. As noted above, this configuration had first appeared on the one-off "pre-Transitional" rifle.

The other change made on the lower receiver was to replace the original bolt release lever with a wider, larger, and more robust component, as also seen on the pre-Transitional. Although this change was standardized as part of the first of the Transitional rifles, the first few examples of this model still possessed the older "lollipop-type" bolt release (fig. 176), due to the factory using up a production overrun of Sudanese lower receivers.

# Upper Receiver and Barrel Modifications

The more cosmetically obvious of the two important modifications to the upper receiver of the First Transitional AR-10 was a natural next step in the evolution of the front end of the AR-10 barrel since the Cuban model was first introduced. As noted above, the area forward of the front sight base on the first (Cuban) version of the A-I AR-10 consisted of a plain barrel of very thin diameter, with a simple, open-type, three-prong flash hider screwed onto the threaded muzzle. The German G4s, as well as the Sudanese and Guatemalan variants, still utilized a screw-on flash hider, now threaded externally for use with the first type of AR-10 blank firing attachment (fig. 174), but the latter two models had also introduced a barrel shroud that indexed with the front sight and was held in place by a pin that went through it and intersected the edge of the barrel. The main original purpose of this shroud was to provide a place for a bayonet lug.

For the version that was to be tested by the Dutch Army, the A-I engineers combined the shroud and flash hider into one integral unit, improving both at the same time. In addition to the obvious fact that one single part would be more cost-efficient and provide less potential for failure, these two functions being combined into one unit that had only one attachment apparatus, instead of the two (pinned shroud and screwed-on flash hider) of the previous versions, would also improve modularity, ease of cleaning and assembly, parts replacement, and even accuracy, due to the more predictable barrel harmonic interaction of one part as opposed to two.

The actual shroud and flash hider was modified from the original types, and brought a distinct new look to the front end of the AR-10. It had a long surface leading up to the flash hider component, with rings ribbing it at three roughly-equidistant points along its surface. The first of these was wider, providing a thick, beefy housing for the transverse pin that locked the shroud to the barrel. The front end merged seamlessly into a flash hider that departed considerably in appearance from, and added additional functionality to, the earlier design.

Unlike that found on the earlier models, the First Transitional flash hider was closed at the end, making it the type of muzzle device known colloquially and collectively as the "birdcage." The function of the closed front end was to prevent the prongs of the flash hider from catching on brush or other foreign objects when the rifle was being carried through rugged terrain.

The closed front of the flash hider also had two other positive effects, albeit small ones, on the function of the rifle; one which was probably intended, while the other was likely a pleasant side effect. The full 360° of metal at the front end of the muzzle device provided a better area for internal threading to accommodate a new and final version of the blank firing adapter, which was threaded on its exterior, so it could screw into the new flash hider.

The strange and doubtless-welcome latent function of this configuration was the elimination of an irritating calling card of the earlier Dutch-produced AR-10s, and indeed to a greater or lesser extent of all rifles with open prong-type flash hiders. The shape of the previous trident flash hiders was much like that of a tuning fork, and when sufficient vibration was imparted to it, it would give off a high-pitched ringing or humming. Due to this acoustic principle, firing a Cuban, Sudanese, or Guatemalan rifle would invariably leave a ringing in the air for a good few seconds after the last shot was fired. This ringing could also be produced when reloading the rifle and slapping the bolt release to close the bolt. If anything, the noise is even more noticeable in these situations, at least in this author's experience, because there is no residual report of the fired round to mask its shrill buzzing.

The new barrel shroud/flash hider of this model did not feature a bayonet lug, as the Dutch did not require a bayonet for their initial testing. However, it had been discussed from the start, and particularly with later input from the Portuguese Air Force, that this shroud would be an ideal place to mount a bayonet, in that not only would it be sturdy, but with the lug connected to the shroud tube and not directly to the barrel, the presence of a bayonet would not affect barrel harmonics to the degree that a directly-mounted bayonet would, thus not impeding accuracy. How well this theory worked in practice has never been sufficiently researched, but additional studies over the past half-century have shown that the majority of the point-of-impact changes associated with mounting a bayonet come from shock wave deflection, with the reflection of the muzzle blast off the bayonet blade actually pushing the bullet off course.

# Beefing Up the First Transitional AR-10

As had consistently been the case since the Dutch began reworking ArmaLite AR-10B design, durability and parts longevity was the order of the day in this update.

One area that the A-I engineers decided to make a bit more robust was the front sight base itself. On the Cuban, Finnish (discussed below), Sudanese, and Guatemalan rifles, the bottom area of the front sight base that formed the ring that fit around the barrel, as well as the gas adjustment screw housing that extended up from it to form the bottom of the hollow portion of the front sight tower, was rather thin. On the earlier rifles, there was a distinct ring around the barrel, with the screw housing extending up from it being much thinner and coming to a sharp ridge from an angled cut on either side. It was decided that this thin, angular ridge did not do enough to protect this critical area and reduce overheating, so necessary to the correct gas modulation of the action. Therefore, the ring-and-ridge was replaced by two thicker rings that connected in the middle, so that they shared one another's structural strength. This eliminated breakable sharp edges, and provided considerably more metal to both shield and increase heat dissipation from the gas port area.

Further, along these lines, the area surrounding the front pivot pin that held the upper and lower receiver halves together during disassembly was beefed up. The forward-extending portion of the lower receiver that contained this pivot pin on the earlier models was quite thin, and its overall size was increased, with a much more gradual angle and curve introduced to eliminate the thin, weak point just behind the cutout for the pin, where breakage was seen as being a possibility. When viewing the lower receivers of a Cuban, Sudanese, or Guatemalan rifle side-by-side with that of the First Transitional update, the pivot pin extension seems much sharper and thinner on the earlier models as opposed to the gradual slope and thicker extension of its successor.

# Eliminating the External Barrel Flutes

Finally, the most integral, yet hardest-to-notice change to the AR-10 that brought it into the Transitional generation took place under the handguard. Ever since the disaster at the Springfield range in February, 1957, all the remaining AR-10s produced by ArmaLite (serial numbers 1006 through 1047) featured thin, fluted alloy steel barrels. The fluting ridges, also found on the early A-I production rifles, helped greatly to decrease the weight and added stiffness to dampen barrel harmonics, making the Cuban, Sudanese, and Guatemalan rifles quite accurate. However, there were concerns that the milled flutes provided weak stress points where cracking could occur. Moreover, the extreme light weight of the barrel that necessitated the fluting, more than the process itself, made the part vulnerable to breakage and overheating. Since absolute minimum weight was no longer as important to the future potential customers of the AR-10 as durability, a new barrel type was put together for the Transitional model.

The design of the new barrel profile was simplicity in itself; the barrels that came off of the vertical cold hammer-forging machine were now simply left as-is and stress-relieved, with no fluting cuts milled into them. This resulted in a thicker barrel that took longer to heat up and further improved accuracy, but did, of course, add considerable weight. By this time, however, the low weight promises of so many other rifle designs had fallen by the wayside, and the AR-10 was so secure in its place as the standard of light weight to which all other battle rifles were compared, that abandoning the original seven-pound weight limit was not considered a problem by any of the interested military forces.

# Summing Up the First Transitional AR-10

This first variant of what is collectively known as the Transitional model of the AR-10 was developed concurrently with the Sudanese- and Guatemalan-contract production weapons, in order to supply the troop trial needs of the Dutch Army.

Updated in several areas from the Sudanese model, its weight was increased to 8 lbs. 4 oz.

A thick rubber buttplate replaced the earlier stamped-steel model, in order to help control recoil and aid in shooter comfort.

The unitary barrel shroud/flash hider produced more uniform barrel harmonics, and allowed a potential bayonet mounting site, although no lug was actually attached to this variant. The flash hider itself had a closed front end to eliminate the problem of the open-type three-prong flash hider catching on foliage.

The gas adjustment valve in the front sight base was retained, and although it sported the same furniture, including the one-piece handguard, it was fitted with a new, heavier, non-fluted barrel.

The "Transitional" series, so named after-the-fact, provided the link between the extremely light-weight Sudanese and Guatemalan versions and the even heavier and more robust Portuguese/NATO model, the pinnacle of AR-10 design, that was to follow.

# First Transitional AR-10s in Dutch Evaluations

In August, 1958, as the lot of 135 Cuban-model weapons was on its way to Meppen for the *Bundeswehr* trials, the Dutch government formally announced that they were beginning a program to select a new infantry rifle. It stressed the official position of the military that the current foreign surplus weapons used by the Dutch forces (mostly U.S. and British) were outdated, and in need of replacement.

On October 9, 1958, a working group was formed to supervise and coordinate the testing and adoption. At the first meeting on October 30, 1958, the chosen representatives of the Royal Dutch Army, Air Force and Navy suggested an overall plan for submission of rifles to a technical test, after which the three best rifles would be selected for further troop testing. The rifles chosen for initial examination were the U.S. M14, the ArmaLite AR-10, the Spanish CETME, the FN FAL, the Swiss SIG, and the French MAS A58.

Holland was quite literally the only foreign country to give serious consideration to the adoption of the Springfield T44 design, and this reflected the strong relationship between the two countries' military institutions, as well as the deep respect the Dutch had for their U.S.-surplus M1 Garands. Before they were reequipped after they had been liberated by the Allies in 1945, the parsimonious Dutch had been relying on perhaps the most antiquated main infantry weapon in regular use at that time - the packet-loading Model 1895 Mannlicher rifle, and several later carbine versions of the same design, all chambered for the rimmed Dutch 6.5x54R military cartridge. Therefore, the world of difference they perceived with their Garands led them to take a hard look at the M14, which the Interservice Committee initially described in a meeting report dated December 16, 1959 as "A modernized Garand. A reliable rifle that keeps functioning well under harsh conditions. Firing rifle grenades was not possible because a grenade launcher is lacking."

The mentioned French A58 rifle was apparently never produced in France: certainly examples of it were never sent to the Dutch, and so this rifle disappeared from the roster without being examined or tested.

## Specifications Required in the Dutch Trials Rifles

The working group quickly promulgated a list of standard requirements that each rifle entered for testing ought to possess. These included the ability to launch rifle grenades, and at least the possibility of mounting a special sight for them; a bipod that would be readily removable and foldable; and a flash hider. A desired, but not required, feature was the incorporation of a day/night sighting system, which should be attached to one of the weapons submitted for the tests. The ability to attach this device to standard-issue rifles was not required, and it was expected that the weapon that was so equipped would perforce be modified to a considerable degree.

This announcement put the respective firms on notice as to what characteristics their weapons should have, and also admonished them to make sure that the chambering tolerances on the rifles they submitted for testing should be somewhat forgiving, as ammunition for use in the tests was being procured from a plethora of sources, both in the United States and Europe, and so insensitivity to ammunition variations was considered of great importance to success in the evaluations. The specification called for submission of a total of four rifles of each type (including the optional night-sighted piece); 96 magazines (24 per rifle); accessories like slings and cleaning kits; and chamber and bolt gauges to measure the tolerances for the technical evaluation that would be the prelude to the firing tests.

The testing timetable was forthwith released to the manufacturers. It called for all rifles to be submitted no later than January 1, 1959. The technical tests would then take place, terminating on August 1, 1959, at which point the firing trials would begin. These would last until the middle of 1960, and January 1, 1961 was decided as the date on which the selection of the new rifle would be officially announced.

Though there was technically some experience in living memory of the Dutch military's running selection trials of a new rifle, the 63 years since the Model 1895 Mannlicher had been adopted were not spanned by anyone's career in the armament industry. While the technical tests were performed on the submissions, Major Van Benthem, Army Director of Procurement, headed a sub-working group including personnel from the Marines, Army, Infantry Command, and Technical Command. It was reflected in their reports how they often felt confused and overwhelmed at the task of evaluating rifles that were so much more advanced than the antiquated bolt-action pieces that had last been tested by their forces. This inter-service working group thus looked very closely at how the American, British, Canadian, and Spanish had gone about comparing examples of this very new technology.

## Contrasting FN's History of Cooperation with A-I's Aloof Hubris

The lobbying of Belgium's Fabrique Nationale had been very effective in the years leading up to the testing, and had put the FN factory in a uniquely influential position by having worked closely with the Dutch armed forces, advising them on developments in small arms, providing parts and technical assistance, and reiterating their public support for the Dutch in rebuilding their armed forces. On the other hand, Artillerie-Inrichtingen could be said to have really dropped the ball. In the opinion of the armed forces, the privately-managed, state-owned Dutch weapons concern acted with extreme hubris in dealing with the working group, apparently not taking seriously what the military considered a real and important process.

Whereas FN had always been on the spot with advice and specifications, frequently reiterating the seriousness with which they viewed the Dutch military evaluations, in their presentation to the working group the A-I management informed their hosts that they "would like to offer a contribution of the AR-10 to the Dutch Army, if it did not cause their contractual obligations to other countries to suffer." The effect of this *prima donna* attitude on the testing staff cannot be underestimated, as it gave the ultimate arbiters of military selection a very bad first impression of the AR-10's manufacturer.

For their part, A-I felt justified in making this rather arrogant statement for two reasons. The first was that they were the Dutch military arsenal, and had been the exclusive manufacturer of weapons for their country's defense forces since shortly after its independence in 1830. They did not see any real possibility of this status quo changing, and did not take the other firms, FN in particular, for the threat that they were to what A-I saw as the AR-10's inevitable adoption.

Their other chief reason was their confidence in the ArmaLite weapon. In their presentation to the working group, the factory's management also stated that "this weapon [the AR-10] is the best in the world, and that serious consideration of any other designs is superfluous."

Jüngeling's staff essentially told the procurement officials of his own country, whose adoption of the AR-10 he had personally promised to Boutelle, that their military evaluations of different rifles was a farce, and that they should simply cut out the theatrics and select at once the superior weapon built in their country. The fact that the Artillerie-Inrichtingen personnel were completely justified in considering the AR-10 to far surpass its competition did not excuse their arrogance on this point. It is entirely possible that this grievous miscalculation of their own military's earnestness cost the AR-10 its chance to be adopted by Holland, and indeed cost the Dutch armed forces the opportunity to be equipped with what was far-and-away the best battle rifle in the world, in the ubiquitous opinions of not only the A-I engineers but the brave élite troops who were soon to carry it into combat elsewhere.

## The AR-10 and the FAL Lead the Pack in the Technical Tests

The technical tests were completed on schedule, with the AR-10 and FAL both receiving the highest marks. The technical staff particularly appreciated the simple trigger mechanism of the ArmaLite rifle, which had been carried over almost exactly from Stoner's first prototype. They also noted as excellent the light weight and the ease with which the rifle was adapted to modern manufacturing processes and

featuring completely interchangeable parts, requiring no hand fitting.

# The Firing Trials Begin

192. Testing the First Transitional AR-10 at Hardewijk,
with the operator firing prone. A bipod is fitted, but is
folded, and the rifle is supported only by the man's two
arms.                              Dutch Defense Ministry collection

193. The same soldier is now aiming a scoped First Transitional AR-10, with the bipod deployed.

Long-range accuracy was a high priority for the Dutch
forces, and as such scoped examples were ordered in large
quantities for these first rounds of testing.

The First Transitional models still utilized the original
3x scope with the mounting done by machine screw,
rotating nut, and bracket.

Dutch Defense Ministry collection

194. The Royal Dutch Marines also made their own evaluations of the First Transitional AR-10s at their Proving Grounds at Harskamp.
Dutch Defense Ministry collection

On schedule, the firing trials began at Harderwijk, under the supervision of Major Phillibert of the Infantry. The AR-10 impressed there as well, turning in the best performance in accuracy, rapid fire controllability, and ease of handling of any of the weapons. The FAL also performed well, and was considered satisfactory. The Swiss SIG SG510 was rejected at this point due to its excessive weight.

In the testing at Harderwijk, special attention was paid to the accuracy potential of the different entrants. For this purpose, the Army requested rifles with integral folding bipods, which would aid in prone firing precision. Interestingly, this design was clearly in the works when the Guatemalan contract rifles were being produced, as some of those pieces were shipped with handguards cut for these Transitional bipods (which would be the standard model used throughout the rest of the AR-10's life at Zaandam), although no bipods or attachment apparati were present on any of the Guatemalan rifles.

Making a particularly poor showing at Hardewijk was the U.S. M14, which not only experienced defects that the Springfield Armory representatives could not overcome in time, but its production capability was judged insufficient to meet even the needs of the armed forces of so small a country as The Netherlands. More than two years after the official adoption of the T44 as the M14, its production was still stalled in the U.S., with the assurances Springfield had made about its ease of manufacture using mostly existing M1 tooling turning out to be illusory. As stated in the final notation in the Interservice Committee's December 16, 1959 interim report concerning the M14, "The chairman states that the M14 can only be acceptable for further testing if these weapons can be delivered in time for simultaneous testing of the three chosen rifles."

More details about the dismal failure of the M14 in sharp contrast to the expectations and flowery predictions of Springfield Armory will be discussed later in an event that was a particular vindication of the AR-10, but suffice it to say that within just a few short years of the M14's adoption, the U.S. military had already standardized a rifle that looked remarkably like the AR-10.

# Results of the Firing Trials

The firing portion of the evaluations did point out several areas in which the AR-10 could be improved. Of particular note was the relative fragility of the plastic buttstock when launching rifle grenades. The method of testing this capability was by placing the buttstock of the weapon against a steel railroad rail, and firing a large number of grenades. The ideal weapon would come through this test without damage to the buttstock. In this test the AR-10 with its lightweight polymer stock did not fare as well as its competitors, which boasted wood and metal furniture.

The locking lugs of the bolt were also criticized as being a potential weak point, but since this concern was theoretical, it was not counted as a defect.

That did not stop the A-I engineers from taking this observation to heart, however, and it led to a later modification that would be critical to the great success of the final production version of the AR-10.

It was generally found during the first rounds of testing that while the modern battle rifle was a formidable weapon, possessed of capabilities far exceeding older designs, it still could not effectively mimic the close-range, hip-fire potential of a submachine gun, and although adaptable to the light machine gun role, it was not as easy to use prone in the squad support role as the far heavier, dedicated light machine guns of the past. This capability was considered satisfactory, however, provided bipods could be included on some of the rifles.

# The Crucial Dutch Field Trials - Stressing Reliability Above All

The Dutch field trial evaluations, slated to begin on July 1, 1960, were to take place in Dutch New Guinea, one of the harshest, muddiest, and most humidly corrosive environments on the planet. They would consist of various "torture tests," with one type of weapon assigned to each of two battalions, who would run their respective rifles through the course of evaluation for two months.

In the sub-working group set up to plan and administer the adverse condition testing, the hand of Minister Sim Visser's hostile Defense Department could be seen, along with the justifiable resentment felt by the Dutch Army at Artillerie-Inrichtingen's high-handed approach to the tests. The testing schedule included mud, rain, dust, and rust tests, and would stress reliability as being paramount over all other concerns.

A-I objected to this regime of making all other considerations take a backseat to reliability. Although the AR-10, particularly in its newest forms,

was considered a very reliable weapon, and had not experienced any issues in the tests that took place in Holland, it had also not previously been subjected to humid jungle environments, and the factory management worried that it would not live up to the FAL's capabilities in that regard.

One must bear in mind that the FAL that would be facing the AR-10 in the jungles of New Guinea was not the T48 of 1953 that had almost given up the ghost in Alaska, but a formidable, if heavy and old-fashioned, weapon, the recipient of the largesse of a magnificently well-funded weapons design bureau which had spent over a decade ironing out the kinks of its design and making it into a robust and reliable, if increasingly heavy, weapon. The weight added to the AR-10 as it was being beefed up for the toughest endurance testing was not unique, as all of the rifle designs adopted by a major military force routinely added about 10% of their prototype weight when put into actual production.

# New Specifications for the Dutch Adverse-Condition Trials

On April 12, 1960, the sub-working group in charge of adverse condition testing released the list of requirements for the weapons in these tests, as well as the number to be acquired. Aside from the characteristics like grenade-launching and general light

weight, which had already been laid out and were also common knowledge for the industry when seeking government contracts, a few more detailed specifications were brought to the attention of Artillerie-Inrichtingen.

### Scopes and Night Vision Equipment Should Fit All Rifles

The scope and infra-red night sighting device requested by the Army should, if adopted, be able to be

fitted to all the rifles without many special provisions, such as a heavily-modified upper receiver.

Since this was as yet not an absolute requirement, it was noted by A-I for later development.

The fire control selector switch layout was also requested to be in the form used on the First Transitionals, instead of how they appeared on the Sudanese contract rifles.

It was also specified as a requirement that the rifle be able to move a few degrees around its bore axis when deployed on its bipod, to facilitate rotating it to the side for easy removal of empty magazines and insertion of full ones. This had not been the case on either the earlier experimental bipod models, or on the several submitted to the first round of testing at Harderwijk.

### Loading Magazines from Charger Clips

A request was made for a device that would allow the quick loading of empty magazines with ammunition packed on five-round charger clips. Both FN and A-I had already formulated simple stamped metal magazine loaders that would fit over the magazine and hold the clip still while the rounds were pressed down into the magazine (fig. 133).

### Removable Handguards

A stipulation of particular importance to the appearance of the next version of the AR-10, was that the rifle should have handguards that could be quickly and easily removed to allow for their cleaning, and the cleaning of the barrel and gas tube/piston assembly. Though quite an efficient and robust design, the one-piece handguards found on all previous models of the Dutch-production AR-10 were only removable after the front sight assembly and any muzzle device had been removed, which was not something that the infantry soldier was trained to do, or had the tools and facilities to accomplish.

# The Second Transitional Model AR-10

The A-I solution to the problem, implemented in early 1959 while examples of the First Transitional variant were still undergoing testing by the Dutch in the first round of technical trials, was ingenious, and actually led to a new variant of the AR-10 called the Second Transitional.

Although the Second Transitional variant appeared strikingly different from the First Transitional pattern, the only real difference between the two was the new four-part handguard assembly. Incorporated to satisfy the informal requests of the Dutch military for ease of disassembly, this later became a formal requirement when the rifles were being prepared for the second and final round of Dutch evaluations.

## Describing the New Handguard and Liner Assembly

The handguard assembly devised for the Second Transitional model was the most successful handguard design of any AR-10 version, and was retained throughout all subsequent production. It can be rightfully viewed as the progenitor of the handguards found on most modern military rifles today, chief among them being the AR-10's diminutive successor, the AR-15.

It was decided that the one-piece handguard should be split in half, so that it could be taken off laterally. Instead of trying to incorporate the heat shield into each half of the split-type handguard, the A-I engineers elected to make the two parts into separate units.

The new forearm was thus designed as a four-part assembly, composed of two interior halves and two exterior halves. The interior halves were configured as perforated, top-and-bottom stamped aluminum liners (fig. 195), contoured to cradle the barrel and gas tube and protect the integrity of both parts, with riveted-on flanges to accept the fronts of the shorter two handguard halves. The left- and right-sided external halves, initially made of walnut and later of Bakelite, fit over the rear two-thirds of the perforated metal liners, providing a comfortable surface for the rifleman to grip. When combined with the metal liners, these retained the weight of the previous model at 8 lbs. 4 oz.

The new detachable split handguard liners were held in place at the front by a retaining cap, which indexed with the front sight base. To install, the front ends of the upper and lower liners would

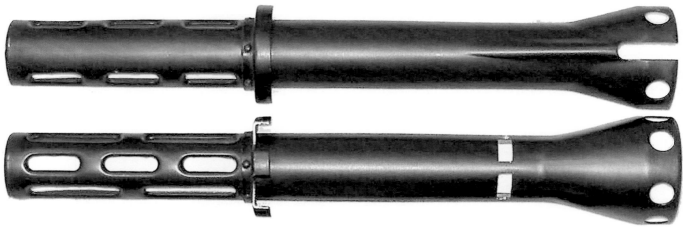

195. The two aluminum handguard liners developed for the Second Transitional AR-10, showing the riveted-on brackets for attaching the fronts of the (first wood, later Bakelite) handguard shells.

On the upper liner (above), note the provision at right for the receiver end of the gas tube.

photo from Charles Kramer collection

be inserted into this cap, and the left and right handguards would be slipped into the flanges on the liners. Holding the rear portion in place was a handguard locking ring; basically a circular metal ring with an interior flange, threaded internally to attach it to the barrel nut. To complete the installation of the handguard assembly, the locking ring was screwed back as far as possible over the barrel nut, the handguard components were swung in at the rear, and the retaining ring was then screwed out until it enclosed the rear circumference of the handguards, holding the entire assembly in place.

Once assembled, the sheet-aluminum liners formed a tube inside the outer grip components, which acted as both a heat sink, becoming very hot from conducted barrel heat, and an air-cooling device; with many perforations designed to draw air in to be heated by the barrel and then expelled, cooling the weapon more rapidly, while the outer wooden portions of the handguard would insulate the rifleman's hand from the extreme temperatures and the rapid heat exchange taking place between the barrel and the interior handguard liners.

## An Initial Experimental "Throwback"

With their usual inventive and creative spirit, Artillerie-Inrichtingen's designers first experimented by incorporating features of several previous models into the initial example of the Second Transitional rifle. These retro features, illustrated and described in fig. 196, were soon abandoned, but this extremely mismatched rifle provides one of the best examples of the modular nature of the AR-10, even in exchanging parts between distinct model types.

## Second Transitional Variations - With and Without Bayonet Lugs

197. Left side view of a Second Transitional AR-10, serial no. 004224.
Note the bayonet lug is the early German style, as used on the Sudanese contract rifles.
Institute of Military Technology

198. Left side view of a Second Transitional AR-10, serial no. 004497, fitted with a bayonet lug designed for the new A-I-produced bayonet under the grenade launcher. As shown in other illustrations, this lug could be located on top of the launcher if desired.
The scope is still the original 3-power unit, mounted by means of a machine screw, rotating nut, and bracket.
Institute of Military Technology

...6 (previous page). Right side view of the initial experi-...ental Second Transitional AR-10, serial no. 002308, ...hich in addition to the new handguard design incorpo-...ted features from several previous models.
The vented barrel shroud, externally-threaded open ...sh hider, and the bayonet-mounting capability, were all ...ken from the Guatemalan model (fig. 187).
The special "winged" bipod design, with holes posi-...oned to match the handguard vents, had been developed ... ArmaLite for the AR-10B and then employed on the ...uban model rifles sent to Germany for testing as the G4 ...gs. 171 and 172), while the rear sling attachment point ...own was as used on the Cuban through the Guatemalan ...riants. Dutch Military Museum collection

Several different very slight variations of the Second Transitional existed, all concerning the presence and location of a bayonet lug. The versions sent for testing by the Dutch, as well as by certain other forces, included robust German-style bayonet lugs on the bottom of the barrel shroud. An order sent to Portugal for troop trials possessed a shorter, differently-configured bayonet lug, mounted on top of the barrel shroud. A slight variant of this had the bayonet lug mounted on the bottom.

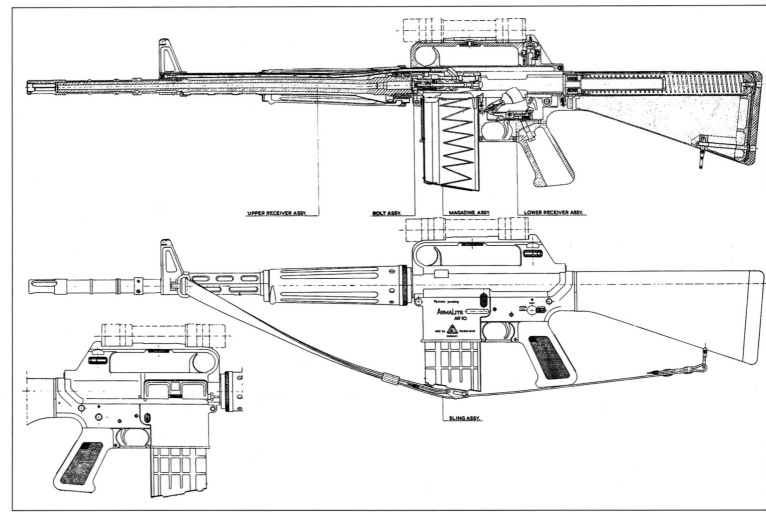

199. Drawings of the scoped Second Transitional rifle, from the September, 1959 A-I instructional handbook.

The selector setting for "SAFE" is still shown in the forward position, and the scope is still the original 3-power unit.

Note the absence of a bayonet lug.

Michael Parker collection

A final variation simply possessed a plain ringed barrel shroud with no bayonet lug, just as was seen on the First Transitional variant.

## A New Bayonet Design

For this newest, heaviest-to-date version of the AR-10, a new bayonet was designed, changing the original policy of Artillerie-Inrichtingen to outsource their production of bayonets. For the second round of Dutch testing they designed a simple knife-type bayonet with a muzzle ring, that could be mounted to the top or bottom of the new integrated flash hider/barrel shroud. The bayonets were carried in a polymer scabbard with a steel ring at the top, which was attached to a cloth belt loop and snap-closed retention strap that held the handle against the body during sheathed carry.

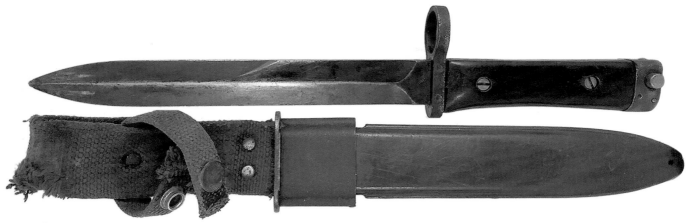

200. The AR-10 "Portuguese" bayonet and sheath, so named because as discussed in Chapter Ten, it was ordered along with the later Portuguese/NATO AR-10 rifles.

The metal of the bayonet was finished in dull black phosphate, to reduce its visibility in reflected light. The one-piece, wraparound walnut grip was secured in place with two set screws, and the pommel held the spring-loaded locking apparatus, while a muzzle ring secured it tightly to the front of the flash hider.

The sheath resembles a standard U.S. bayonet sheath, with a coarsely woven canvas frog and snap fastener.
author's collection

201 (right). Left side closeup of the above bayonet, showing the stylized Artillerie-Inrichtingen logo stamped at the base of the blade.                    author's collection

## The Second Transitionals Ready for Trials Ahead of Schedule

The required 240 A-I weapons were prepared and submitted for the tests ahead of the deadline. They each were sent with sling, bayonet (when applicable to the model), and four spare magazines (five in total per rifle). They consisted of 112 standard rifles, plus 107 equipped with a new type of bipod, five infrared-capable sniper units, which also possessed bipods, with the remaining 16 demonstrating rare and non-standard varieties that never reached the large-scale production stage.

## An In-Depth Second Transitional Parts List

Part of the package prepared by Artillerie-Inrichtingen to support the Dutch military trials was a complete parts list of the Second Transitional AR-10, shown here along with the several pages of nomenclature listing the numbers and names of all the parts.

202 (right). The cover of the Illustrated Parts List for the AR-10 rifle and bipod, produced by A-I to support the Second Transitional AR-10s in the Dutch trials and elsewhere.                    Charles Kramer collection

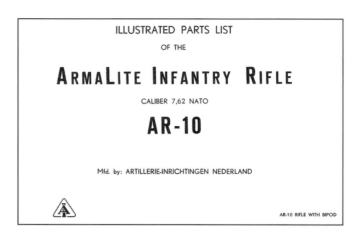

ILLUSTRATED PARTS LIST
OF THE

**ARMALITE INFANTRY RIFLE**

CALIBER 7,62 NATO

**AR-10**

Mfd. by: ARTILLERIE-INRICHTINGEN NEDERLAND

AR-10 RIFLE WITH BIPOD

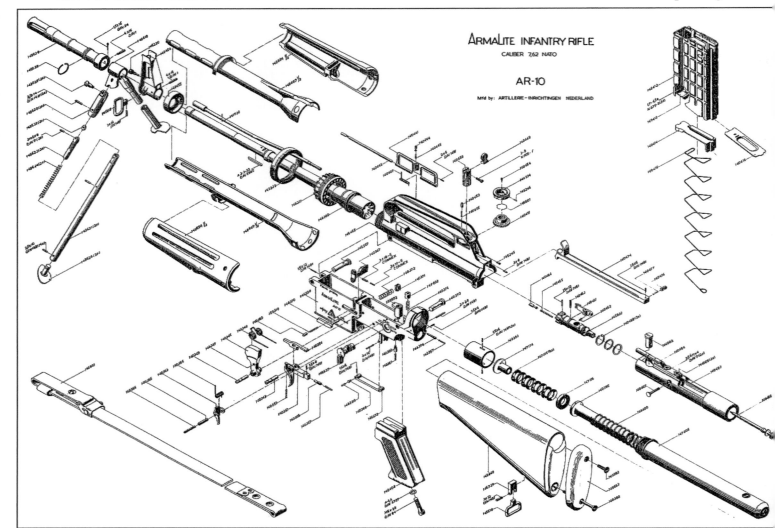

203. Copies of a booklet containing this exploded parts diagram were provided by Artillerie-Inrichtingen to facilitate training and the quick and correct replacement of parts during the testing of the Second Transitional AR-10s.

As shown below, each component was described by name and part number with reference to the diagram.

Charles Kramer collection

204. An actual example of a Second Transitional AR-10, serial no. 004386, with no bayonet lug.

Note the rear sling attachment point, relocated to the side of the buttstock to facilitate "carbine carry."

Institute of Military Technology

**AR-10 Rifle with bipod**

- 1 -

| Description | Part No | Description | Part No |
|---|---|---|---|
| Receiver, upper | 146465 | Roll pin | DIN 1481-1,5x6 |
| Stop pin sight spool | 145245 | Assembly, barrel extension | 146389 |
| ☒ Assembly, dust cover | 140470 | Nut, barrel | 149231 |
| Cover, dust | 140441 | Assembly tube, gas | 141730 |
| Latch, cover | 140444 | Assembly, regulator, gas | 146230 |
| Spring, cover latch | 140445 | Locking ring, handguard | 149232 |
| Roll pin | DIN 1481-2 x 6 | Locating ring, handguard, front | 149455 |
| Pin, cover | 140446 | Front sight | 149454 |
| Spring, cover | 140447 | Key | DIN 6885-A-3-3-32 |
| Spring, rear sight | 140455 | Taper pin | DIN 1-4 x 16 |
| Detent, rear sight | 140453 | Flashhider-grenade launcher | 149638 |
| ☒ Assembly rear sight | 146600 | Taper pin | DIN 1-4 x 16 |
| Aperture rear sight | 140449 | Cotter pin | DIN 94-1,5x15 |
| Screw rear sight elevation | 140450 | Detent ring, grenade | 146699 |
| Nut, rear sight | 140451 | Swivel, sling | 145619 |
| Spool, elevation rear sight | 144334 | Roll pin | DIN 1481-3 x 12 |
| Stop pin, sight spool | 145246 | Liner, handguard in 2/2 | 149887 |
| Screw, aperture clamp | N 1105-F 3 x 8 | Handguard in 2/2 | 149694 |
| Screw, lock rear sight | 140454 | Receiver, lower | 146228 |
| Spring, ring rear sight | 146601 | Spring, buffer retaining | 140376 |
| Roll pin | DIN 1481-3 x 8 | Pin, buffer retaining | 147803 |
| ☒ Assembly, handle charging | 141608 | Assembly, extension receiver | 147805 |
| Handle charging | 140474 | Roll pin | DIN 1481-2 x 20 |
| Detent, charging handle | 140476 | Spring action | 140400 |
| Spring, charging handle detent | 140477 | ☒ Assembly, buffer | 147113 |

☒ not shown on the photograph

205. Page 1 of the nomenclature list for the Second Transitional AR-10. The part numbers refer to those on the exploded diagram (fig. 203).    Charles Kramer collection

- 2 -

| Description | Part No | | Description | Part No |
|---|---|---|---|---|
| Buffer . . . . . . . . | 140393 | | Roll pin . . . . . . . | DIN 1481-1,5x8 |
| Guide, action spring . . . . | 140395 | | Pin, hammer, trigger . . . . | 140349 |
| Guide, buffer discs . . . . | 147114 | | Sear . . . . . . . . . | 140354 |
| Disc, buffer . . . . . 8x | 147115 | ✕ | Assembly, hammer . . . . . | 140344 |
| Locking ring, buffer . . . . | 147116 | | Hammer . . . . . . . . | 140345 |
| Roll pin . . . . . 2x | DIN 1481-1,5x6 | | Retainer, hammer pin . . . . | 140347 |
| Magazine catch . . . . . . | 140370 | | Spring, hammer . . . . . . | 140348 |
| Spring, magazine catch . . . . | 140372 | | Pin, hammer, trigger . . . . | 140349 |
| Button, magazine catch . . . . | 140371 | ✕ | Assembly, automatic sear . . . . | 140362 |
| Spring, bolt catch . . . . . | 150318 | | Sear, automatic . . . . | 140363 |
| Plunger, bolt catch . . . . . | 140369 | | Spring, automatic sear . . . | 140364 |
| Bolt catch . . . . . . . | 140367 | | Bushing, automatic sear . . . | 140365 |
| Roll pin "Connex" . . . . . . | S 3x10 | | Pin, automatic sear . . . | 140366 |
| Roll pin "Connex" . . . . . . | S 2x10 | ✕ | Assembly, trigger guard . . . | 146689 |
| Safety . . . . . . . . | 146637 | | Trigger guard . . . . . | 146573 |
| Plunger, safety detent . . . . | 140360 | | Plunger, trigger guard . . . | 140388 |
| Spring, safety detent . . . . | 140361 | | Spring, trigger guard . . . | 140389 |
| Pistol grip . . . . . . . | 140408 | | Roll pin . . . . . . . | DIN 1481-1,5x6 |
| Lock, washer . . . . . . | DIN 6797-A-6,4 | | Roll pin . . . . . . . | DIN 1481-3x16 |
| Screw . . . . . . . . | DIN 84-M 6x28 | | Hinge pin . . . . . . . | 146557 |
| ✕ Assembly, trigger . . . . . | 140350 | | Roll pin . . . . . . . | DIN 1481-1,5x12 |
| Trigger . . . . . . . | 140351 | | Pin, take down . . . . . | 140373 |
| Spring, trigger . . . . . | 140353 | | Plunger, take down detent . . . | 140374 |
| Plunger, sear . . . . . . | 140356 | | Spring, sear, take down pin . . . | 140357 |
| Spring, sear . . . . . . | 140357 | | Roll pin . . . . . . . | DIN 1481-1,5x6 |

✕ not shown on the photograph

206. Page 2 of the nomenclature list for the Second Transitional AR-10. The part numbers refer to those on the exploded diagram (fig. 203).     Charles Kramer collection

208 (following page). Left side view of the cutaway demonstration Second Transitional AR-10 fabricated by Artillerie-Inrichtingen, serial no. 004903.

Demonstrated for the Dutch military as well as the South Africans, it helped the engineers and salesmen explain the technical genius of the AR-10 in an easy-to-grasp way.     Dutch Military Museum collection

- 3 -

| Description | Part No | | Description | Part No |
|---|---|---|---|---|
| ✻ Assembly stock with recoil pad | 146847 | | Swivel head, bipod | 149619 |
| Assembly stock | 146848 | | Head, bipod leg | 149620 |
| Block, rear sling swivel | 146737 | | Leg | 149621 |
| Recoil pad | 149663 | | Plunger, bipod leg | 149622 |
| Screw, long butt | 140382 | | Roll | 149623 |
| Swivel sling | 145619 | | Compression spring | 149624 |
| Roll pin | DIN 1481-3 x 12 | | Lock screw, bipod leg | 149625 |
| Screw, short recoil pad | 140383 | | Foot, bipod | 149626 |
| ✻ Assembly, bolt | 146709 | | Roll pin | DIN 1481-2,5x10 |
| Carrier, bolt | 140483 | | Roll pin | DIN 1481-2,5x14 |
| Key, bolt carrier | 140484 | | Pin, roll | DIN 7-3 m 6x8 |
| Socket head cap screw | 2x 144005 | | ✻ Assembly, magazine | 140411 |
| Lock pin, socket screw | 2x DIN 7-2,5 h 11x4 | | Box, magazine | 140412 |
| Bolt | 145933 | | Retainer, magazine floor plate | 140413 |
| Extractor | 140461 | | Rivet | 3x N 675-0-1,7-4,5a |
| Spring extractor | 140462 | | Follower, magazine | 140414 |
| Pin extractor | 140463 | | Spring, magazine | 140415 |
| Ejector, bolt | 140464 | | Plate, magazine floor | 140416 |
| Spring, ejector | 140465 | | Sling | 146801 |
| Roll pin | DIN 1481-1,5x12 | | | |
| Ring, bolt seal | 3x 140468 | | | |
| Pin, cam | 144002 | | | |
| Pin, firing | 140466 | | | |
| Pin, firing pin retainer | 140467 | | | |
| ✻ Assembly, bipod | 149618 | | | |

Hembrug-Zaandam, March 1960

Staatsbedrijf
Artillerie-Inrichtingen

✻ not shown on the photograph

207. Page 3 of the nomenclature list for the Second Transitional AR-10. The part numbers refer to those on the exploded diagram (fig. 203).    Charles Kramer collection

# An A-I Instructional Cutaway Second Transitional

209. Left side closeup of the Second Transitional cutaway, serial no. 004903, showing the key moving parts painted in various bright colors, the better to follow the series of events in the firing cycle.

Actual cutaways and drawings of the action components were an important part of training and familiarization in the dawning era of the automatic infantry rifle, when the inner workings of such weapons were not well-understood.                    photo from Charles Kramer collection

Further aiding in the training of Dutch unit armorers, and facilitating demonstrations of the AR-10 to interested foreign parties, the Dutch prepared an instructional Second Transitional rifle with portions of the receiver and buttstock neatly cut away and the interior parts brightly painted to show how they interacted, in order to demonstrate the revolutionary inner workings of the ArmaLite rifle. In an age before the ubiquity of ultra-high-speed photography, the only way to visualize in real time the way the internal components of a rifle interacted was to simply cut one open and observe their movements as the action was manually actuated.

# Two Types of A-I Magazine-Fed Squad Support AR-10s

210. Left side view of the first type of Squad Support AR-10, as fabricated for the Royal Dutch Navy's New Guinea tests, serial no. 004412, with fixed heavy barrel.

The standard handguard assembly has been replaced by a perforated steel cylinder, housing the very heavy barrel. A folding bipod which would allow the rifle to rotate for the changing of magazines was attached forward of the front sight base, and an additional carrying handle was present forward of the one on the upper receiver, to compensate for the new center of balance due to the extremely heavy barrel, handguard, and bipod.

Institute of Military Technology

211. Left side view of the second type of the A-I Squad Support AR-10, serial no. 004534, with detachable barrel.

Coming a bit later in time, judging from its slightly later serial number, this featured some improvements such as the quick-change barrel and vertical foregrip.

Institute of Military Technology

Leading up to the New Guinea tests, the Dutch Navy had expressed interest in an AR-10 capable of being used as a light machine gun, but one which fed from magazines. This led a parallel program that had been progressing slowly at Artillerie-Inrichtingen to be kicked into high gear, and resulted in a wildly divergent AR-10 variant that is not well known to the collector or historical community, and which has not been given any informal type designation. We will refer to this model, of which there are two variants, as the "Squad Support" AR-10.

Both of the main squad support variants were adaptations of the Transitional series of rifles, fitted with heavier, more robust barrels. They both utilized standard detachable 20-round magazines, but were also fitted with special perforated metal handguards to allow for more effective barrel cooling, and an extra carrying handle to further isolate the hand of the user from the extreme heat generated by the sheer volume of fire for which these weapons were designed.

The earlier of these variants was the one formulated for the Royal Dutch Navy, and did not diverge very far from the standard rifle configuration, while the later version, with its quick-change barrel, scope-mounting carrying handle, vertical foregrip, and extended foldable bipod, was much better adapted to the role of the light machine gun.

# Publicizing the Artillerie-Inrichtingen AR-10 "Family"

212. A popular staged action photograph of the AR-10 "family", taken in the fall of 1957 as part of Artillerie-Inrichtingen's advertising brochure titled "Tomorrow's Rifle Today."

In the foreground it illustrates the factory's earliest attempt to fabricate a squad support model (a Cuban-model AR-10 fitted with a "winged" bipod), which can be seen in the hands of a Royal Dutch Marine who is lying prone and about to launch a grenade.

Behind him we see three advancing riflemen, with standard Cuban AR-10 rifles, plus a kneeling man at center with a scope-mounted Cuban "sniper" model.

This photo was taken by another eminent Dutch military photographer named Jan Schiet (actually translating to "shoot" in English), who was the predecessor to the later Peter Marcuse. Michael Parker collection

# The Joint ArmaLite/A-I Transitional AR-10 Carbine

213. Left side view of a Second Transitional carbine, serial no. 004263, produced for the Dutch New Guinea military trials. This experimental version was only 35" long overall, compared to the 41" length of the standard rifle.

Note the shortened 16.1" barrel, with special grenade launcher/flash hider (fig. 214), and the handguards, which spanned the space beteeen the front sight tower and the receiver.                         Institute of Military Technology

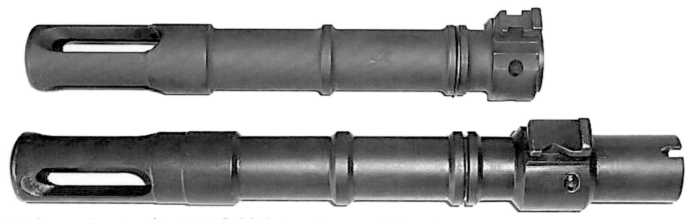

214. A comparison view of two AR-10 flash hider/grenade launcher assemblies.

Above: the shortened carbine version, with bayonet lug dimensioned to accept the rare M4 bayonet design used with the carbine (fig. 215).

Below: the standard rifle version, with bayonet lug dimensioned to accept the standard "Portuguese" bayonet (fig. 200).                         Charles Kramer collection

With the squad support type accounting for twelve of the total number of rifles allotted for the Dutch military tests, the remaining four were representative of a very interesting and innovative AR-10 version which, while never being successfully sold to a military buyer, allowed the AR-10 to fill yet another role quite capably. Noting the dissatisfaction of the testing staff at Hardewijk with the battle rifle in general when applied as a submachine gun, Artillerie-Inrichtingen took the initiative, working especially closely with ArmaLite back in California, to develop a carbine version of their Second Transitional model.

A shorter, handier companion to the Second Transitional rifle version, the AR-10 carbine achieved the signature power and controllability of the original design in a shorter, lighter package, with only a small diminution of accuracy and rapid-fire controllability. Weighing 7 lbs. 15 oz., the carbine measured a mere 35" overall, compared to the 41" length of the standard rifle.

Sporting a 16.1" barrel with modified barrel shroud and flash hider, the carbine model was still able to launch rifle grenades, utilize a blank firing adapter, and even mount a bayonet.

The gas system was shortened somewhat to optimize the pressure curve and provide for reliable functioning, given the different pressure and velocity characteristics of the shorter barrel.

Many decades ahead of its time, the Transitional Carbine represented one of the first successful attempts to make a shortened version of a standard service rifle that still used the same magazines and ammunition, as well as many of the same component parts.

The construction of the carbine version was quite simple, and really involved only modifying the barrel and incorporating a slightly different barrel shroud. The two outer shell components of the new type of handguard assembly were used by themselves without the liners on the carbine, and took up the entire space between the upper receiver and the front sight base on the shorter barrel. The area forward of the front sight base had to be shortened as well to achieve the desired reduction in overall length, and so the flash hider/barrel shroud device from the Transitional models was simply cut down in length to allow it to fit the shorter barrel.

Even with the different pressure curve of the shorter barrel and gas system, the standard receiver extension tube, buffer, buffer spring, and stock were able to be employed.

# The Rare Kiffe M4 AR-10 Carbine Bayonet

215. Left side view of the rarest of all AR-10 bayonets, the Kiffe-contracted, Japanese-produced modified copy of the U.S. M4 bayonet, fitted with a dog-leg muzzle ring set back to clear the flash hider vents on the shortened carbine version of the AR-10.

Dutch Military Museum collection

The shortened barrel of the carbine necessitating the truncated barrel shroud caused an unforeseen issue for Artillerie-Inrichtingen; one which Fairchild and its U.S. industrial connections solved in short order. The standard AR-10 bayonet as fabricated at the Dutch plant would not fit the new shorter muzzle device properly, as its muzzle ring protruded too far and overshot the end of the flash suppressor. ArmaLite quickly deduced that the small bayonet that had been developed for the U.S. M1 Carbine was well suited for shortened rifles, and a surplus example showed the Dutch that it was a good starting point. However, the muzzle ring was still a bit too far forward, sitting right in the middle of the flash hider vents and interfering with the proper functioning of the device, and becoming exceedingly dirty from burnt powder fouling.

Fairchild quickly contracted with the Kiffe Knife Co. of New York to produce a modified M1 Carbine bayonet (Bayonet Model M4), on which the muzzle ring was set back further. The very small consignment of these specialized pieces were produced for Kiffe at the Howa Machine Works in Japan, and are without a doubt the scarcest, if not the most intricate or inventive, bayonet model ever devised for the AR-10.

# Summing Up the Non-Standard Dutch AR-10 Rifles

The attempts made by Artillerie-Inrichtingen at the aggressive urging of Fairchild Arms International to adapt the AR-10 to various roles—the squad support role in particular—did bear fruit, but unlike their incorporation of sniping scopes on certain rifles, these variants were never commercially successful. While the squad support model was from all reports relatively advanced and robust, setting up an AR-10 to fill the role of a true belt-fed machine gun was never marketed successfully. The sniper version and its night-sighted variant stand as the only nonstandard versions of the AR-10 to be selected and issued to any national armed force, with the machine gun and squad support weapons being relegated to the status of museum pieces.

# Maintaining the AR-10 in the Dutch Trials

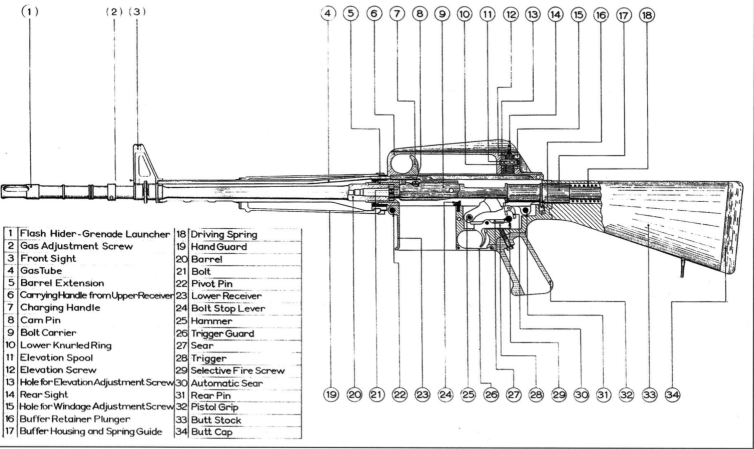

| | | | |
|---|---|---|---|
| 1 | Flash Hider - Grenade Launcher | 18 | Driving Spring |
| 2 | Gas Adjustment Screw | 19 | Hand Guard |
| 3 | Front Sight | 20 | Barrel |
| 4 | Gas Tube | 21 | Bolt |
| 5 | Barrel Extension | 22 | Pivot Pin |
| 6 | Carrying Handle from Upper Receiver | 23 | Lower Receiver |
| 7 | Charging Handle | 24 | Bolt Stop Lever |
| 8 | Cam Pin | 25 | Hammer |
| 9 | Bolt Carrier | 26 | Trigger Guard |
| 10 | Lower Knurled Ring | 27 | Sear |
| 11 | Elevation Spool | 28 | Trigger |
| 12 | Elevation Screw | 29 | Selective Fire Screw |
| 13 | Hole for Elevation Adjustment Screw | 30 | Automatic Sear |
| 14 | Rear Sight | 31 | Rear Pin |
| 15 | Hole for Windage Adjustment Screw | 32 | Pistol Grip |
| 16 | Buffer Retainer Plunger | 33 | Butt Stock |
| 17 | Buffer Housing and Spring Guide | 34 | Butt Cap |

216. A sectioned drawing of the Second Transitional AR-10, with components numbered and named.
These are the part numbers referred to in the following armorers' instructions.       Michael Parker collection

A four-page typescript set of armorers' instructions dated January, 1960 was prepared by A-I for use by military armorers tasked with supporting and maintaining the AR-10 during the Dutch New Guinea trials. This illustrated document, from the Michael Parker collection, is excerpted below, with the component numbers mentioned corresponding to the numbers shown on the section drawing of the Second Transitional AR-10 shown in fig. 216.

## *Instruction for Fitting the Barrel of the AR-10 Rifle and for the Use of Inspection Gauges*

### 1. Fitting new barrel

*The barrel (20) is slipped into the upper receiver (6) [fig. 217], after which the upper receiver is placed in the barrel assembly- and removal fixture No. 147300. The barrel nut is screwed onto the barrel as far as possible by hand [fig. 218].*

*By means of torque wrench No. 147330, the barrel nut is tightened with a torque of 8 - 12 kgm*

*[fig. 219] in such a manner that a slot in the barrel nut lines up with the gas tube hole in the upper receiver, so that the gas tube (4) can easily pass [fig. 220]. Care should be taken that the gas tube and barrel are paralleled in such a way that their respective gas-ports are opposite each other without the gas tube having to be bent. Next the clamping ring is screwed to the barrel nut; in doing so,*

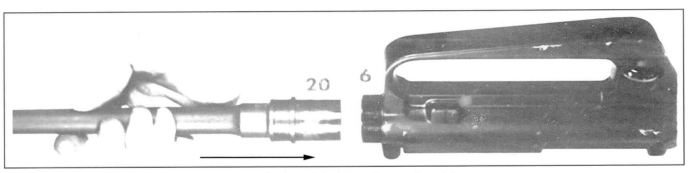

217. Inserting the barrel (20) into the receiver (6).
Michael Parker collection

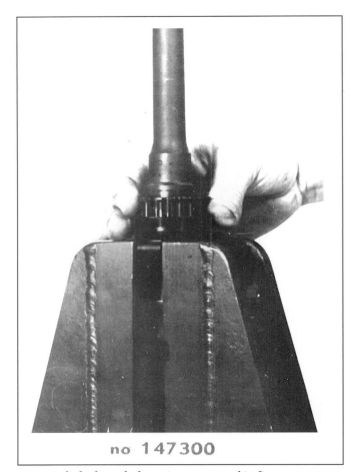

no 147300

218. With the barreled receiver mounted in fixture 147300, the barrel nut is screwed on as tightly as possible by hand.
Michael Parker collection

no 147330

219. The barrel nut is tightened by means of torque wrench No. 147330, with a torque of 8 - 12 kgm in such a manner that a slot in the barrel nut lines up with the gas tube hole in the upper receiver.          Michael Parker collection

one should take care that the knurled edge is to the rear [fig. 221].

Thereafter the [handguard] liner retainer is pushed on the barrel (hollow side backwards) as far as it will go; then the key is pressed into its groove in the bottom of the barrel and the front sight (3) is slipped over the barrel [fig. 222]; hereafter the gas tube is fitted [fig. 223]. The tapered hole in the front sight should be aligned with the correspond-

ing grooves in the barrel and gas tube. For this use the special aligning tool (taper pin with handle).

Next the taper pin groove, which is still cylindrical, is sized and adequately tapered (conicity 1 : 50) by means of the taper reamer from right to left. This being done, the tapered pin is driven in [fig. 224] with the use of a hollow-pointed drift punch.

If the rifle is equipped with a bipod, this is now placed on the barrel; heed should be paid to the correct position of the key groove in the swivel head with respect to the key in the barrel [fig. 225].

Then the grenade launcher is placed on the barrel [fig. 226] in such a manner that the tapered hole of same is aligned with the appropriate groove in the barrel  .  .

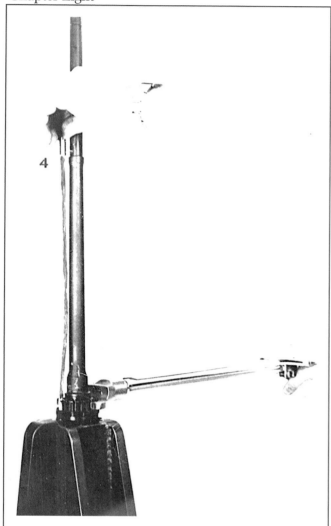

220. Ensuring that the gas tube (4) and barrel are paralleled so that their respective gas-ports are opposite each other without the gas tube having to be bent.

Michael Parker collection

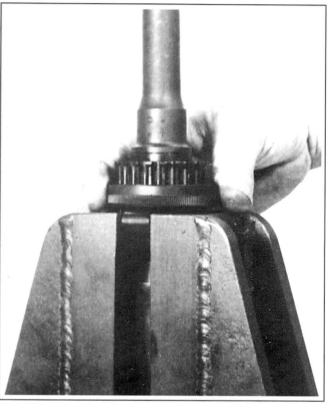

221. Screwing the clamping ring onto the barrel nut, with the knurled edge to the rear.    Michael Parker collection

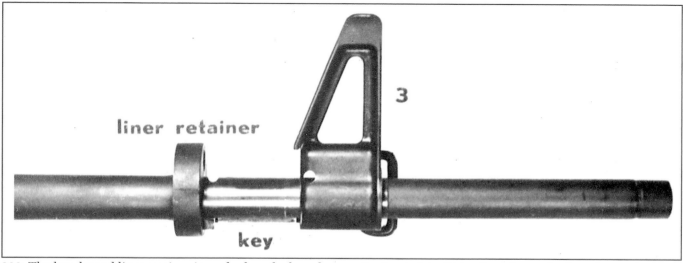

222. The handguard liner retainer is pushed on the barrel (hollow side backwards) as far as it will go; then the key is pressed into its groove in the bottom of the barrel, and the front sight (3) is slipped over the barrel.

Michael Parker collection

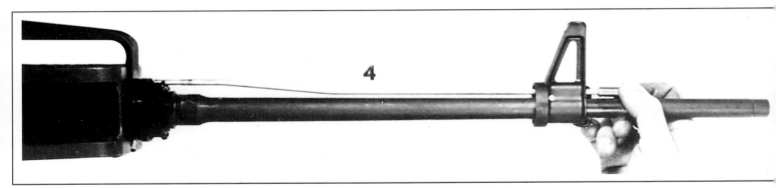

223. Fitting the gas tube (4).
Michael Parker collection

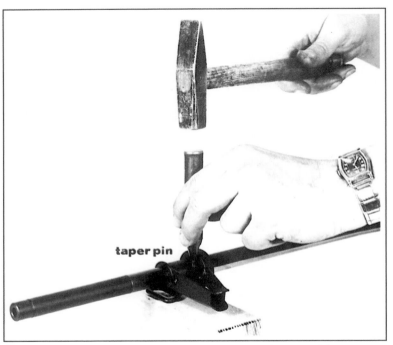

224. Securing the gas tube and front sight assembly.
Michael Parker collection

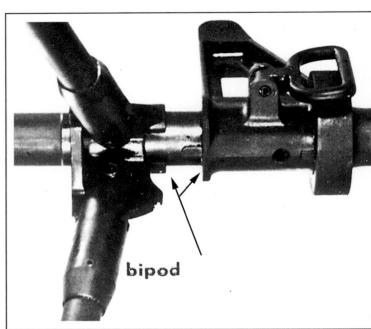

225. Ensuring the correct position of the key groove in the swivel head of the bipod, with respect to the key in the barrel.                              Michael Parker collection

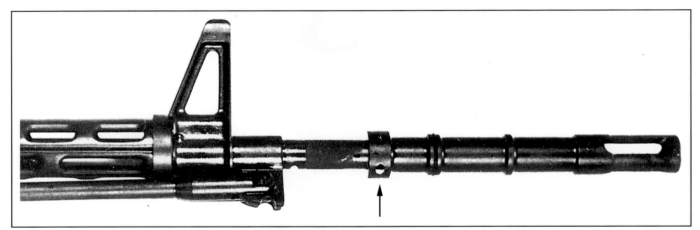

226. Placing the grenade launcher on the barrel, with the tapered hole (arrow) aligned with the appropriate groove in the barrel.                    Michael Parker collection

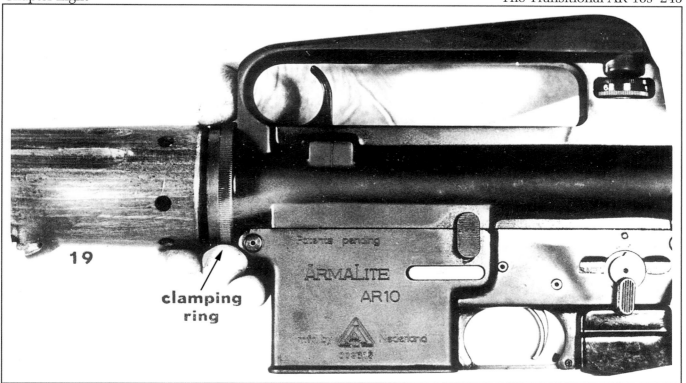

227. In preparation for disassembly of the handguard (19),
the clamping ring is screwed entirely rearward, so that it
no longer contacts the handguard.

Michael Parker collection

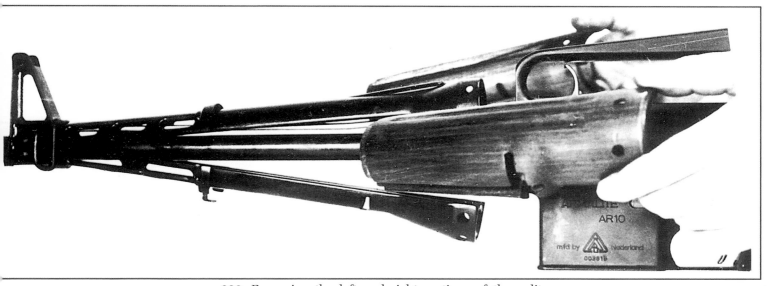

228. Removing the left and right sections of the split
handguard, after which the upper and lower halves of the
liner are taken off.          Michael Parker collection

*Then the barrel groove (still cylindrical) must be
sized and adequately tapered (conicity 1 : 50) by
means of a taper reamer from right to left. Hereafter
the tapered pin is fitted; here too, the hollow-
pointed drift punch is used.*

*Finally the cotter-pin is placed into the appro-
priate hole of the grenade launcher from the top*

*downward, after which its legs are spread, so that
the tapered pin is prevented from falling out  . .*

**Note**

*1. In new barrels the manufacturer has made
the holes for the tapered pins cylindrical and
undersize. After assembly they shall be sized*

and adequately tapered, by means of the reamer and wrench included in the supply.

2. If, after the holes have been reamed conical, it should appear that the original tapered pin can be driven in too deep and that the head might get <u>under</u> the surface of the front sight or grenade launcher, use shall be made of an oversize tapered pin, which forms part of the supply.

<u>Disassembly of the barrel</u> is accomplished by reversing the above operations except that a different wrench must be used to unscrew the barrel nut.

### 2. Disassembly and assembly of the handguard

The clamping ring is screwed entirely rearward, so that it no longer contacts the handguard (19) [fig. 227].

Then the left and right sections of the split handguard are removed, after which the upper and lower halves of the liner are taken off [fig. 228].

For the purpose of assembling, one should proceed in reverse, paying heed to it that the half of the liner that has a recess in its back for the gas tube, be placed at the top. The handguard sections are also recessed for the gas tube  . .

### 4. Checking of headspace

After the rifle has been opened, the steel headspace gauge is fixed to the bolt (21). The locking system is pressed forward by hand; in doing so, one should

229. A complete set of AR-10 headspace gauges, numbering 4133, 4138, 4143, 4148, 4153, 4158, and 4163, previously owned by Henk Visser of NWM.

photo from Charles Kramer collection

headspace gauge

231. Bolt locked on minimum headspace gauge.
Michael Parker collection

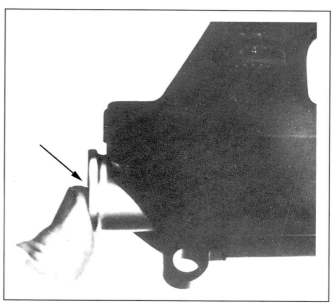

232. Bolt unable to lock on maximum headspace gauge.
Michael Parker collection

*take care that the gauge gets into the chamber [fig.
230].*

*When the minimum gauge (1.633 inches) is
applied, it should be possible to lock the bolt [fig.
231]; if, on the other hand, the maximum gauge*

*(1.637 inches) is employed, locking of the bolt of a
new weapon should be impossible [fig. 232].*

*If the bolt of a used rifle closes with a 1.641
inches field-reject headspace gauge, it is necessary
to change the bolt of this weapon  .  .*

## No Substitutions Allowed in the Dutch New Guinea Trials

As the submission date neared for the commencement of the Dutch adverse-condition trials in New Guinea, Artillerie-Inrichtingen had actually completed the design of what was to be the final and most advanced AR-10 ever, and was tooled up to produce this model. As discussed in Chapter Nine, this further-improved version is known as the Portuguese/NATO model AR-10.

A-I requested that the Dutch Army allow them to substitute the newly-redesigned weapons for the Transitionals they had already submitted, but this request was denied. Therefore, the AR-10 was forced to compete in the jungles of New Guinea in a less capable format than was available at the time.

The tests began ahead of schedule, on June 17, 1960, and were slated to be completed in September, due to the elimination of the Italian BM-59 (the 7.62x51mm M1 Garand as modified by Beretta) from competition.

However the results of the first two weeks of firing did not bode well for the Transitional AR-10. Although it had performed with flawless reliability in the tests to which it had been previously applied, the jungle mud, humidity, salt, and corrosion proved very harmful, and it malfunctioned at a rate considerably higher than the FN FAL.

The trials report, much to the dismay of Artillerie-Inrichtingen, focused almost completely on the 96 rifles assigned to the adverse-condition exposure tests, in which the rifles were abused, left exposed and intentionally dragged through mud and sand before being tested for reliable functioning. In these evaluations, the AR-10 experienced an average of 96 jams per weapon, still a relatively low malfunction rate but one which paled in comparison to that of the FAL, with 24 malfunctions. The results from the other weapons assigned to performance testing, in which the AR-10 shot circles around its competition (as well as the control M1 Garands) in accuracy, modularity, handling, and rapid fire controllability, were almost completely ignored by the Inter-Service Working Group.

230 (previous page). Inserting the bolt group with headspace gauge aligned to enter the chamber.
Michael Parker collection

## The Dutch Special Forces Argue in Vain for the AR-10

Echoing the concerns of the manufacturers, the representative of the *Korps Commandotroepen* (Commando Corps; the élite Dutch Special Forces), loudly protested this emphasis in the testing report, pointing out that the FAL was a much more mature weapon, and in very short order, perhaps even at that time, the AR-10 could be made to surpass its competitor even in the arena of adverse-condition functioning. He highlighted the inescapable superiority of the accuracy, modularity, ease of maintenance, parts interchangeability, handling, rapid fire control, and light weight of the AR-10 over any other existing rifle design. However, his advice was not heeded, and the report was allowed to stand.

# The Dutch Adopt the FN FAL

233. Left side view of a Dutch FAL rifle, adopted by the Royal Netherlands Army in 1961.

As discussed in the Collector Grade titles *The FAL Rifle* and *More on the Fabled FAL*, the Dutch FAL had several unique features, including the distinctive grenade launcher, tunnel foresight, fixed rear sight (the same as used on the folding-stock Para FAL), and the fixed rear sling loop on the buttplate.                    editor's collection

The conclusion of the trials report that the FAL was better suited to the Dutch forces under adverse conditions added new weight to the omnipresent skepticism with which a rifle not made exclusively of wood and steel was still viewed in military organizations at the time. As a result of this the Dutch Defense Department, already possessed of great personal enmity toward A-I director Friedhelm Jüngeling, and the working group, who had been treated so flippantly by the Artillerie-Inrichtingen representatives at the commencement of their task, opted in 1961 to adopt the FN FAL as the new standard rifle of the Netherlands armed forces.

The final conclusions regarding the AR-10, which were influenced by an October, 1956 statement from a "U.S. weapons expert" at Springfield Armory that the AR-10 "would take five years or more to take it through tests to adoption," were as follows:

*This rifle is far from being [fully] developed and designed. Further development to come to an acceptable weapon will require a lot of extra time, and there is no assurance that either the materials used, or the gas- and locking systems, will be acceptable for use.*

As for the FAL, the subcommittee had already reported to the Headquarters General Staff on October 8, 1960 that

*. . In the light of technical tests, and compared to the other tested weapons, it is suggested that the FAL be accepted as the new rifle of the Dutch armed forces . .*

# A Royal Gift is Offered, but Not Accepted

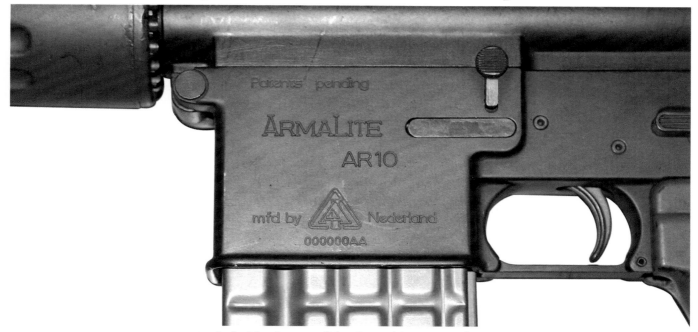

234. Left side closeup of the Second Transitional AR-10 specially prepared by Artillerie-Inrichtingen for presentation to Prince Bernhard, serial no. 000000AA.
Dutch Military Museum collection

235. Right side closeup of the special AR-10 prepared as a gift for Prince Bernhard, which he declined to accept. Note the green-painted furniture, and the crowned 'B' engraved on the magazine well, shown enlarged in the inset at right.          Dutch Military Museum collection

Around this time Crown Prince Bernhard (fig. 97), the husband of Holland's Queen Juliana, who reigned from 1948 to 1980, was offered a specially-marked A-I AR-10 rifle as a gift. However, due to the political situation engendered by the Dutch rejection of the AR-10 and adoption of the FN FAL, he declined to accept it. He was later offered an A-I AR-10 carbine, which he did accept. Both these weapons are now in the Dutch National Military Museum.

# The Short-Lived 7.62mm NATO Caliber Garand Conversion Project

Meanwhile, following on the heels of the ill-fated AR-10 trials in New Guinea, Major Van Benthem, chairman of the Small Arms Selection Committee, paid a vist to A-I's Hembrug - Zaandam works with more unwelcome news. It was believed at the time, with perhaps only Friedhelm Jüngeling appreciating the depth of political spite motivating the Defense Ministry, that the New Guinea tests had not been conclusive, and that since the NATO AR-10 now being shipped to the Portuguese had soundly outperformed the FAL even in reliability in that country's evaluations, that it would still be given a chance by the Dutch military. The A-I staff were thus extremely eager to learn details of the expected test regimen to which their perfected AR-10s would be put, expecting this to be the purpose of Van Benthem's visit. However there was only one topic which the major wanted to discuss, and that was not it.

From the beginning the Dutch working group had been liasing with the Italian Beretta factory regarding the BM-59—their inexpensive and efficient conversion of the M1 Garand to fire the 7.62mm NATO cartridge from a detachable box magazine—and had decided at the urging of the Defense Ministry that Holland should make its own attempt at this goal. This came as a nasty surprise to A-I, as both the FAL and the AR-10 had been conclusively deemed superior to the BM-59 by the Dutch armed forces, to the point where that weapon had been eliminated from consideration without having been formally tested. Van Benthem explained that while the BM-59 had not been deemed satisfactory, it was after all based on the M1 Garand, which was popular with the Dutch military, and thus a suitably converted M1 deserved consideration.

The chairman of the committee then struck a lower and even more painful blow to the plucky A-I staff, informing them that Defense Minister Visser had just in the last few days adopted as policy that the choice of weapons up for adoption was now only between the FAL and the U.S. M14. Unlike the BM-59, the M14 had been evaluated side-by-side along with the AR-10s, FALs, and other trials submissions at Harskamp and Harderwijk, and had performed pitifully next to the current frontrunners, not even matching the Swiss entry in capability. It became crystal clear at that point that the vendetta against the AR-10, which had begun with the deliberate obstruction of attempts to acquire machinery and continued throughout the heavily biased nature of the military evaluations, would result in the AR-10 never being chosen by the Dutch forces, regardless of by how much of a margin it outstripped its competition.

Resigned to their fate, on January 9, 1961 the Artillerie-Inrichtingen board agreed to cooperate with Beretta in working up a conversion of the Garand resembling the BM-59, which could be performed domestically on the existing M1 rifles in Dutch service and additional surplus M1s acquired from the U.S.

A meeting to compare notes was held the next month in Italy, where it was determined that two examples of the converted weapon should be worked up to test the feasibility of the plan, and to explore the technical challenges inherent in the conversion. A-I was represented at this research symposium by Jüngeling, executive secretary Van de Dijk, and technical engineer Dongo. Although bitter about the likely fate of the AR-10 in the race for a new standard infantry weapon for their country, and skeptical that an M1 Garand conversion would be able to equal the performance of even the contemptible M14, the delegation confirmed that such a project posed no real technical challenges to their organization. However, Jüngeling in particular declared himself loath to embark on yet another rifle development program for a weapon that might well not be given serious consideration by the armed services.

Upon returning to Zaandam on the 22nd, just five days after the commencement of negotiations with the Italians, the A-I works were paid a visit by none other than the grim reaper himself: Defense Minister Sim Visser. The news he brought was by

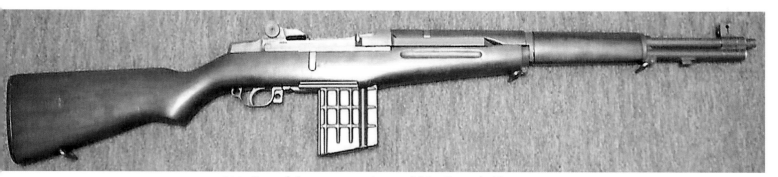

236. One of the two experimental M1 Garand rifles converted to 7.62mm NATO caliber and fitted with a detachable AR-10 magazine by Artillerie-Inrichtingen in 1961.
courtesy Vic Tuff

this time expected: the AR-10 had been finally and permanently rejected for consideration for Dutch adoption. The gleeful minister, contemplating the defeat of his political rival Jüngeling, did offer one small olive branch, declaring that he had not yet determined conclusively whether the AR-10 was unfit to be sold abroad. It was obvious to all concerned that this salve to their wounds was just another step towards the inevitable scuttling of the ArmaLite program in its entirety, as after all it was not within the purview of the Defense Minister of the Kingdom of The Netherlands to decide whether the AR-10 was suitable for adoption by the militaries of other nations.

Nevertheless, the construction of two Garand conversions commenced at A-I in short order. There was hope among some that if their nation would not adopt their primary and best-performing product, they at least might be able to begin a valuable program of converting and supplying NATO-caliber M1s. It was obvious right from the outset that while the basic idea of reworking surplus U.S. rifles to fire the new NATO cartridge from a detachable box magazine was in most general terms one of Italian origin, the Dutch effort would be almost completely independent and would reach it own solution, and not simply copy the Beretta conversion.

To this end, the characteristics of these "Dutch BM-59s", as they were called, were formulated by Spanjersberg and Bakker. The barrels of the two prototypes were made to different lengths, with the first keeping to the original 24" Garand specification, while the second rifle received a 505mm (20") barrel, made to duplicate exactly this characteristic of the AR-10. The stocks were recontoured to fit the new barrels, and to allow for the insertion of a modern battle rifle magazine.

The magazine chosen was another point of divergence from the Beretta design, and was an obvious choice for the Dutch. In inventory at A-I were huge quantities of AR-10 magazines, produced by Nederlandsche Metallfabrik, and the unique way in which these were mounted straight up and detached with the push of a button was incorporated in the converted Garands. This was a major improvement over the BM-59, which used the method of "rock and lock" magazine change omnipresent on most other contemporary rifles, from the Kalashnikov and FAL to the M14 and CETME.

The cost saving in modifying existing rifles rather than adopting new ones was one of the most attractive features of this program, and it was calculated that A-I could convert Garands at a price of $40 per weapon, provided a quantity of 120,000 rifles (the number of rifles the military needed) were ordered. Factoring in the cost of acquiring the additional 65,000 surplus rifles from the Americans, it was estimated that the total program could be completed for a cost of $5.45 million, compared to the $14.4 million that Defense Minister Visser was offering for M14s, in contravention of the wishes of the armed forces.

With news of the increased U.S. interest in the AR-15 finding its way across the Atlantic, there were also rumblings in the Dutch defense community at that time that the 7.62mm NATO cartridge itself might be on its last legs. This eventually further favored the M1 Garand conversion idea, because converting them to fire the smaller 5.56mm cartridge was seen as eminently practicable.

Not long after, however, in the same fell swoop that was to put an end to the entire AR-10 project in Holland, the Dutch Defense Ministry also cancelled the fledgling Garand conversion initiative, with both experimental prototypes of the "Dutch BM-59" being sold to private collectors, in whose caring hands they remain today.

## Better Late than Never: the Dutch Adopt an AR-10 Descendant

As a fitting finale to the story of the AR-10 in Dutch service, collector Marcel Braak offers the following:

*Perhaps it is nice to mention that in the early nineties the Dutch Armed Forces accepted the Diemaco C7 [rifle] and C8 [carbine] as their new service weapons. By doing so, they were equipped with a descendant of the AR-10 after all.*

# Moving On - A-I Eyes the Civilian Market

Back in 1956 and 1957, Melvin Johnson had written at length to both Richard Boutelle and George Sullivan concerning the imminent commercial promise of the AR-10 as a sporting rifle for the American civilian market. This advice went completely unheeded on the ArmaLite end, where all hopes had initially been pinned on military adoption by the U.S.

When that failed to materialize, the order of the day became mass production in Holland for sale to foreign militaries. However, in 1959, with the completion of the Second Transitional AR-10, the A-I management decided to attempt to market the AR-10 commercially in a more "civilian friendly" form.

## The Feds Shoot Down the First Semi-Auto AR-10

237. Left side view of the single prototype Dutch semi-auto-only sporting rifle marked "AR-10", serial no. S0072.
This sole attempt to get a rifle designated "AR-10" into the American civilian arms market was turned away by the U.S. Customs Service for several reasons.
Institute of Military Technology

In their attempt to enter the U.S. civilian market, A-I submitted a sample of their first semi-auto-only variant to the U.S. Customs Service for approval. Initially, the designation of the weapons was not changed from "AR-10," and although the receivers were manufactured so that they would not accept full-automatic fire control parts, the selector switch markings still included "AUTO" as a holdover from military production. In addition, although the rifle was not fitted with a flash hider, it was shipped with a protector cap over the same threaded muzzle, meaning that it was still possible to mount a standard flash hider/grenade launcher. The magazine was modified to a 10-round capacity, but the magazine wells were standard,

meaning that a standard 20-round magazine would fit and function.

U.S. Customs refused to certify this initial entry for import. The only feature of the rifle that actually violated American law was the lower receiver, which although modified was that of an "AR-10," a type of weapon that was registered as a machine gun under the U.S. National Firearms Act. American law treats any weapon with a receiver that has been, or even shares a manufacturing designation with, one that is capable of fully-automatic fire as a machine gun. The federal authorities also objected to the threaded muzzle and the ability to accept standard 20-round magazines, although the law did not proscribe such features.

# The "AR-102" Purpose-Built Sporter
## Back to the Drawing Board - Designing a New Lower Receiver

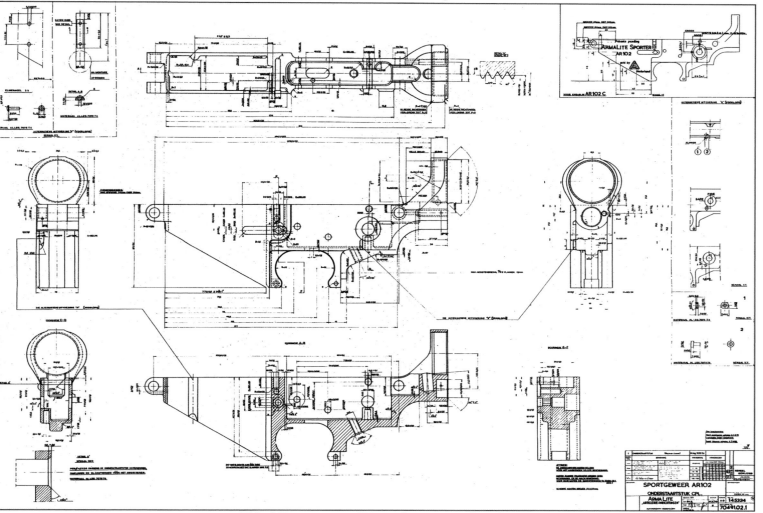

238. Titled "Sportgeweer AR-102," a new set of machining diagrams were drawn up for a lower receiver that completely eschewed any parts from the fully automatic trigger mechanism. These receivers were manufactured completely separately from those for the standard AR-10, so that the AR-102 could truly be classified as being a completely different type of rifle.

Note the curved metal ridge inside the magazine well, shown in the drawing at top center, intended to prevent the insertion of a standard 20-round magazine.

Charles Kramer collection

To address these concerns, A-I came up with the new designation "AR-102," the receivers of which did not include the "AUTO" receiver marking, even though their initial offering had been incapable of fully-automatic fire. Achieving both the goals of a distinct model of lower receiver and limiting the magazine capacity, the Dutch engineers gave the magazine wells on the new lower receivers a distinctive slant cut, allowing easier insertion and removal of the shorter magazine.

# The Proprietary 10-Round Magazine

The new slanted magazine well also included an interior metal ridge which indexed with a deep indentation in the right side of a unique new 10-round magazine. This ridge prevented the insertion of standard magazines, and only allowed those with the indentation, which were only produced in the reduced capacity. They then eliminated the barrel threads, leaving the barrel with a plain muzzle.

The 10-round magazines were produced by simply cutting a standard body shell in half, and fitting the standard follower and floorplate with a half-length spring. The proprietary cut on them was made at the factory, although standard magazines could be easily modified to fit in the AR-102's magazine well with a simple file cut on the feed lip portion of the crease on the right side of the magazine. By the same token, an end user could just as easily have filed down the blocking ridge inside the lower receiver to allow the use of regular-capacity 20-round magazines. However, these modifications were enough to comply with the original import requirements.

239. Left front three-quarter view of the proprietary AR-102 10-round magazine. The extension of the vertical crease through the right side of the magazine allowed only these magazines to be inserted into the well of an Artillerie-Inrichtingen AR-102 sporter.

Dutch Military Museum collection

# The AR-102 is Made in Small Numbers Only

240. Left side view of a completed Dutch AR-102 Sporter, serial no. 000002. Note the plain muzzle, and the slanted bottom on the magazine well.

Designed expressly to be civilian-legal, the purpose-built, semi-auto-only AR-102 was provisionally approved for import by the U.S. Customs Service but was ultimately unsuccessful.                    Institute of Military Technology

242 (following page). A rare example of a cased AR-1( found at a recent meeting of Dutch AR-10 collectors.

Note the flash hider, incapable of launching grenade the coiled leather sling, and the two ten-round magazine

photo by Marc Mill

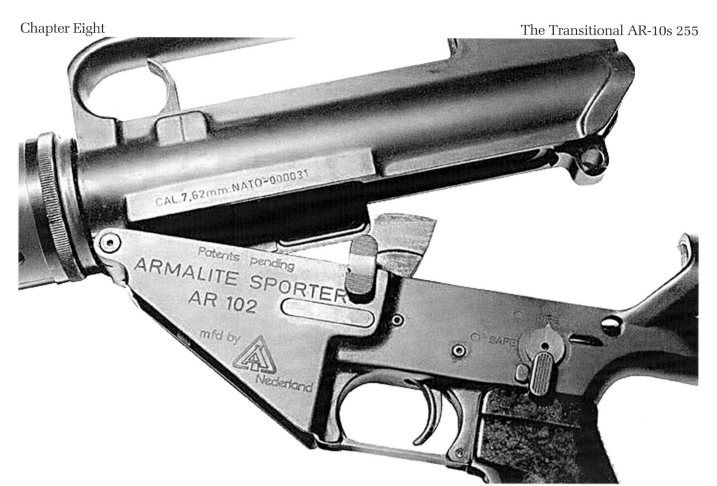

241. Left side closeup of AR-102 serial no. 000031, with the standard AR-10 caliber marking and serial number on the upper receiver, and "ARMALITE SPORTER, AR 102," on the slanted magazine well of the lower, instead of the "ARMALITE AR10." Note the selector stop, and the new markings reading only "SAFE" and "FIRE".

Despite the changes, the AR-102 offered the same modularity and extreme ease of disassembly for cleaning and maintenance as found on the standard AR-10.

photo from Charles Kramer collection

## Casing the AR-102 - to No Avail

It was intended that each AR-102 would be shipped in a beautifully-finished wooden chest with a stylized red-and-gold "AI" logo emblazoned on it. Along with the rifle and two 10-round magazines, a distinct version of the original Cuban-type dark brown leather sling was also included, which measured 1.25" in width.

As shown in fig. 242, A-I also fabricated several examples of the AR-102 featuring a flash hider reminiscent of that on the Second Transitional variant, but which was incapable of launching rifle grenades.

## A Cutaway Instructional AR-102

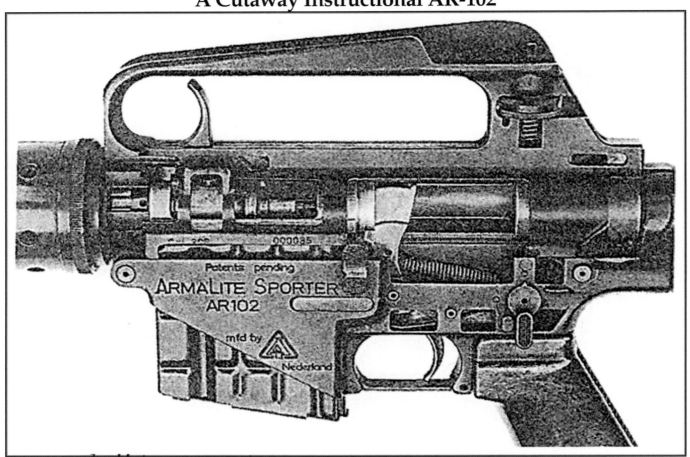

243. Left side closeup of the first of four steps illustrating the charging and firing cycle featuring the instructional cutaway made from AR-102 serial no. 000035, labeled "Bolt forward and locked, hammer resting on firing pin."
photo from Charles Kramer collection

A cutaway example of the AR-102 was produced for demonstration purposes to the arms import agencies of various potential foreign markets. A montage of the charging and firing process is illustrated in four steps in figs. 243 - 246.

245 (following page). Left side closeup of the third of four steps illustrating the charging and firing cycle featuring the instructional cutaway made from AR-102 serial no. 000035, labeled "Recoiling parts held in rearmost position, hammer forced down below point of sear contact by bolt carrier." photo from Charles Kramer collection

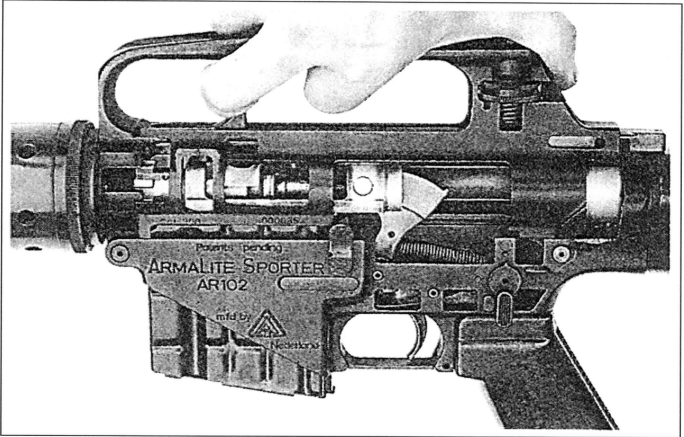

244. Left side closeup of the second of four steps illustrating the charging and firing cycle, featuring the instructional cutaway made from AR-102 serial no. 000035, labeled "Cocking handle moved back 3cm - bolt unlocked, hammer being forced back."

photo from Charles Kramer collection

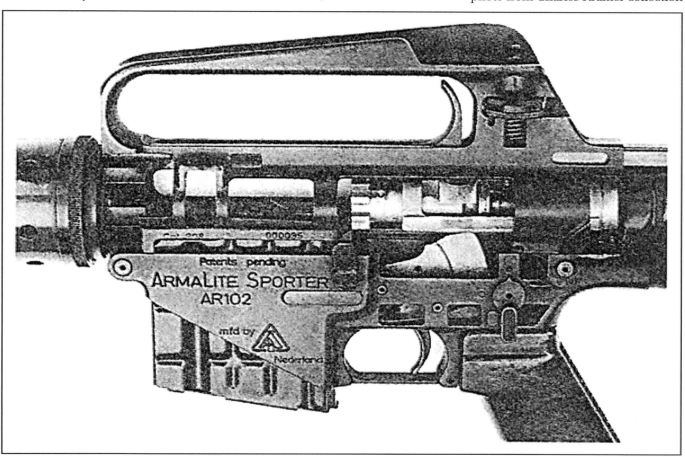

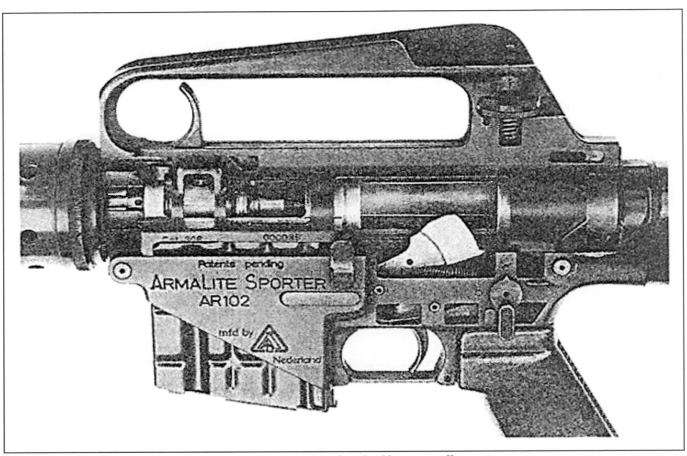

246. Left side closeup of the fourth of four steps illustrating the charging and firing cycle featuring the instructional cutaway made from AR-102 serial no. 000035, labeled "Bolt forward and locked, hammer cocked, held on sear."
photo from Charles Kramer collection

## Envisioning a "Family" of Sporters

As shown in fig. 247, in addition to the full-length standard AR-102 rifle variant envisioned for sale on the commercial market, Artillerie-Inrichtingen also had on the drawing board plans to produce and market a scoped version, and a short-barreled carbine called the AR-102C.

Over 20 years later, Colt's successfully released commercial variants of the AR-15 based on this paradigm, but the original concept was pioneered by the Dutch.

## The AR-102 is Still Denied Access to the Commercial Market

Even then this venture did not meet with commercial success. Despite having faithfully adhered to each of the U.S. Customs Service's demands, the AR-102s were still denied entry to the American market. The arms import officials in the U.S., Britain, and other nations where there was a good consumer base for such products simply found the rifles to be far too radical and "dangerous-looking." This ignorant misconception that a rifle that looks particularly modern must be inherently more prone to criminal use than one with a conventional wood stock is one that continues to dog the AR-10 and its descendants to this day.

Thus this final semi-auto-only Transitional type was never even shipped abroad, as by the time Customs turned thumbs down on the AR-102 A-I had abandoned the entire project as being unprofitable.

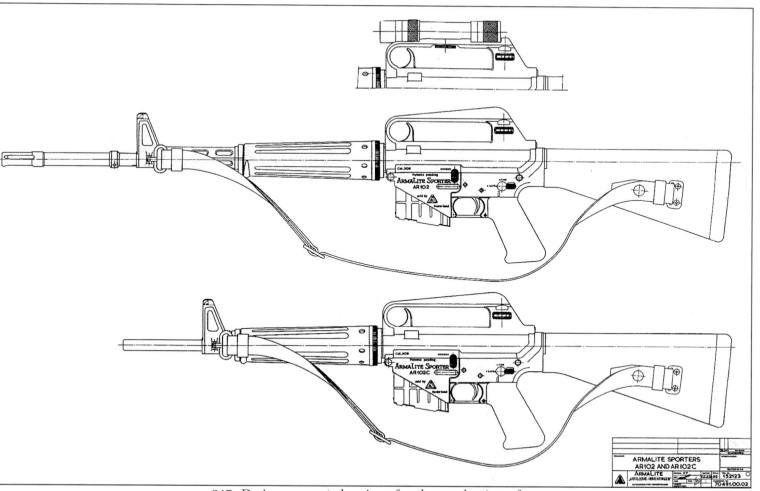

247. Design concept drawings for the production of a scoped version of the AR-102, plus a carbine-length model dubbed the AR-102C, were finalized at Hembrug-Zaandam. Had the U.S. authorities been more open to the import of these firearms, an entire "family" of civilian-legal, semi-automatic sporting AR-10s might have been seen.                drawing from Charles Kramer collection

# Competing for South African Adoption

Sam Cummings, who had set up yet another Interarmco office in Pretoria and fostered close relations with many in the South African defense community, arranged for the AR-10 to be an entrant in that controversial African nation's ongoing rifle adoption program. Marginalized more and more since WWII for its policies of white minority rule and racial discrimination, the Union of South Africa saw as its best chance for survival and success as complete a state of military self-reliance as could possibly be achieved. Therefore, the contract to equip their armed forces, due in large part to the size of the

country, promised to be one of the largest and most lucrative of any at the time.

The government intended to make an initial purchase of 25,000 rifles of the type it selected, and then to seek a licensing agreement for massive domestic production. As many of the nations who were interested in selling arms to South Africa had governments and populations who were increasingly hostile to continued dealings with the rogue African regime, the military leadership were endeavoring to conclude their selection process and begin producing and issuing the rifles in very short order, so that future sanctions could not preclude them from arming their infantry with the best possible equipment.

From November, 1959 to January, 1960, the Danish Madsen, the Belgian FN FAL, the Spanish CETME, the German G3 (modified to sufficiently differentiate it from its Spanish predecessor), and the Swiss SIG SG510 competed alongside the Second Transitional model AR-10 in firing trials for the prize of the South African contract.

When all was said and done, the AR-10 turned in a performance not unlike that which it was about to repeat in the Dutch trials in New Guinea, discussed above. The AR-10 surpassed by leaps and bounds its competition in every area except for adverse condition functioning, in which it still came in a close second to the FN FAL, beating in that arena the Swiss, Spanish, and German entries, which had already developed reputations for excellent reliability.

## Unwilling to Gamble - South Africa Adopts the FN FAL

Once again, the FAL, which had been engineered for far longer and had received much better funding and support, won the day on its reliability, in addition to the impressive momentum it had built up by being adopted by so many other countries. It seemed to the South Africans to be the "safe choice," not only being more comforting to their institutional conservatism (being composed of wood and ordnance steel), but because Belgium, Canada, Great Britain and numerous other countries had already chosen it. Just as had happened many times before (and was soon to happen in the very country of the AR-10's production), the South African Army leadership was not willing

to gamble on the less-proven AR-10, despite its superior capabilities.

The result of the South African tests was one of the main driving forces behind Artillerie-Inrichtingen pushing the Dutch military so hard to allow them to submit their newest variant of the AR-10 for adverse condition testing, which they claimed would outperform even the FAL in reliability. It would turn out that this assertion was amply justified but, as discussed above, the unwillingness of the Dutch Inter-Service Working Group to upgrade from the model the South Africans had already tested doomed the AR-10 to coming in once again a close second to the FN FAL in New Guinea.

## Further Pursuit of the Multi-Role Weapon - the Belt-Fed A-I AR-10

The successful development and fielding of a carbine, sniper, and even a squad support variant of the AR-10 rekindled interest in attaining a reliable belt-fed machine gun version.

Working directly from the second type of magazine-fed squad support weapon (fig. 211), A-I went right back to basics and ordered a new, dedicated upper receiver forging (fig. 248) for use in this new model, which allowed plenty of material to accommodate a belt-feed device. A cam slot was also machined in the side of the bolt carrier, so that it would operate the mechanism to index the belt of linked ammunition with each reciprocation.

The result was a weapon that bore essentially all of the features from the latter of the two A-I squad automatic/heavy barreled AR-10 variants, but with an open slot on the upper receiver. The quick-change heavy barrel and vertical foregrip were still present, as was the perforated steel heat shield in place of the handguard, with the front sight and bipod integrally mounted to it. Strangely, this type, unlike the squad support weapon from which it was devised, did not have the secondary carrying handle above the barrel,

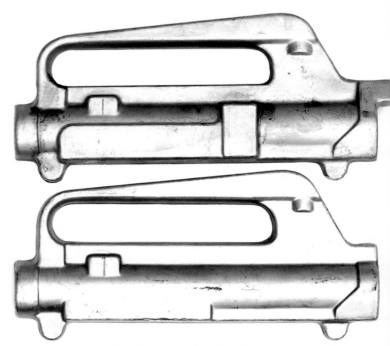

248. A comparison between the aluminum upper receiver forging as produced for the belt-fed model, above, and the standard AR-10.                    Institute of Military Technology

249. Left side view of a Dutch belt-fed AR-10, with the belt feed mechanism installed.

This updated belt feed device, which was detachable to allow the weapon to fire from standard magazines, mounted securely to the left side of the weapon, and only necessitated cuts in the upper receiver, instead of both the upper and lower as seen on the 1957 ArmaLite prototype.
Institute of Military Technology

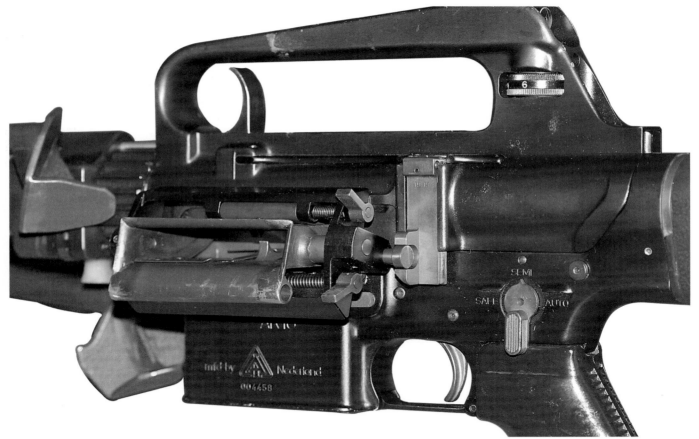

250. Left side closeup of Dutch belt-fed AR-10 serial no. 004458, showing details of the belt feed mechanism.
In use, the vertical reciprocator at rear would be activated by the bolt carrier to index the belt with every shot.
photo from Charles Kramer collection

although it still would have had the same balance characteristics.

The belt-feed mechanism could be installed and removed easily by simply folding the upper and lower receiver halves apart. Taking it out converted the LMG AR-10 variant essentially into a heavy-barreled squad support weapon, able to make use of standard magazines.

251. Left side view of a further Dutch belt-fed AR-10, serial no. 004219, with the belt feed mechanism removed to show the openings in the upper receiver only.

This model appears almost identical to the second type of squad support weapon developed for the Dutch rifle trials in New Guinea (fig. 211), and in this configuration it is indeed functionally the same.

Institute of Military Technology

## Summing Up the "Very Capable" Dutch Belt-Fed AR-10

Unlike the squad support, carbine, sniper, and standard versions, the AR-10 belt-fed machine gun did not gain the attention of prospective customers, and was never adopted or produced in any quantity. The few that were made were demonstrated to the Portuguese, but did not generate any orders.

Despite this lack of commercial success, the Dutch attempt at sustained automatic fire capability did, by all accounts, result in a very robust and serviceable weapon. ArmaLite's 1957 belt-fed prototype (fig. 121) was reported by those who worked on and tested it (it was never officially put through trials by any military force) to be quite finicky, unreliable, and prone to parts breakage. The Dutch model, on the other hand, being heavier and having the advantage of the numerous engineering advances made during the two-plus years of further development it had enjoyed at Artillerie-Inrichtingen, could apparently handle adverse conditions and long strings of fire quite well.

The accounts both of the engineers in Zaandam and the collector who owns and has made extensive experimental use of both of these variants (1957 ArmaLite and Artillerie-Inrichtingen), reports that unlike its California-produced predecessor, the Dutch LMG was, and continues to be, a very capable automatic weapon.

## A Pyrrhic Victory for the Transitional AR-10 at Aberdeen Proving Ground

Putting into perspective the relative issues the AR-10 had experienced with reliability in adverse conditions, a light automatic rifle test was performed by the U.S. Army Ordnance Corps at Aberdeen Proving Ground in Maryland from August 30 to October 20, 1960.

The tests were ordered by the authority of the Chief of Ordnance in a directive letter dated June 23, 1960, excerpted as follows:

### SUBJECT: *Test of Fairchild AR-10 7.62-mm Rifle*

1. *Six Fairchild AR-10 7.62-mm Rifles, together with accessories, equipment and descriptive literature will be shipped in the near future to the Proving Ground, Attention: D&PS (Mr. L. F. Moore). It is requested that three of the six rifles be selected at random and subjected to the appropriate standard engineering test. Adequacy of the accessories and equipment should be assessed.*
2. *The test report will cover the Fairchild materiel only and will be classified "For Official Use Only" . .*
4. *It is requested that this office be advised when the above mentioned materiel has been received at the Proving Ground, and of the date established for initiation of the test. A representative of the Fairchild Engine and Airplane Corporation will be present to observe portions of the test.*

Accordingly, through the late summer and fall of that year, three of the six Second Transitional AR-10s submitted by Artillerie-Inrichtingen were put through essentially the same course of endurance, adverse-condition, and performance tests to which the U.S.-made AR-10B had been subjected less than three years previously at Springfield Armory. At that time the AR-10 had soundly defeated the T44, producing a considerably lower (and almost nonexistent) malfunction rate, and even beating its rival in the dust and extreme-cold evaluations. The AR-10s even averaged, with iron sights, groups smaller than one minute-of-angle, far surpassing the possible performance from any standard-issue military rifle of the day, and even most specially-modified M14s and M1s used for sniping and competition. Then, as recounted above, the composite barrel of AR-10 no. 1002 had suddenly burst during the endurance test, thus engendering the damning, if foregone, conclusion by the Carten-Studler R&D cabal that the AR-10 was not suitable for military service.

The Aberdeen test report included no conclusions or recommendations, but the Abstract from that document, Report no. DPS-101 dated November, 1960, reproduced as follows, makes interesting, not to say astounding, reading:

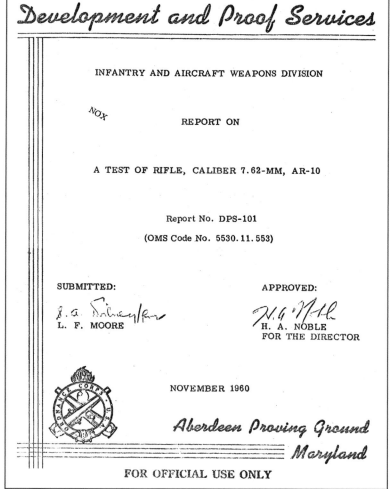

252. The title page of Aberdeen Proving Ground Report No. DPS-101, titled "A Test of Rifle, Caliber 7.62-MM, AR-10," dated November, 1960.

Michael Parker collection

### Report On
# A Test of Rifle, Caliber 7.62-MM, AR-10

### Abstract

*Three rifles [serial nos. 004219, 004412 and 004534] were subjected to the light automatic rifle test, and two rifles were subjected to additional accuracy tests. The AR-10 rifle weighs 10.11 pounds when fully loaded with the flash hider - grenade launcher assembled, and is 41.2 inches long, over all. The rifles tested were chambered for the 7.62mm NATO round. The average mean radius for 10-round targets fired from a bench rest at 100 yards was 1.0 inch. The average number of shots fired semiautomatically in one minute was 57.7, and the average number of hits on the "E" target at 100 yards was 49.0. When fired automatically the average number of rounds fired was 97.3, and the average number of hits was 21.0. With the rifle held normally the average malfunction rate was 0.22 per hundred rounds. The average velocity loss in firing 6,000 rounds was 29 feet per second. The average mean radius for 10-shot bench-rest targets fired at 100 yards before the endurance test was 1.0 inch and after the test it was 1.2 inches. No part was broken in any rifle during the endurance test. About normal functioning was obtained in the unlubricated, extreme-cold and dust tests. In the mud test the performance varied from 5 rounds fired with 13 stoppages to 40 rounds fired with 1 stoppage. Cartridge case failures, caused by excessive pressure arising from water in the bore [see below], resulted in damage to two rifles in the rain test. The damaged parts were replaced, and the*

**ABERDEEN PROVING GROUND**
S18-001-2723-1005-59-3P/ORD-60        30 August 1960

Project TS1-2/265. Rifle, Caliber, 7.62mm, AR-10, Figure 1.
Top and right side views.

253. APG photo no. S18-001-2723-1005-59-3P/ORD-60, titled "Rifle, Caliber, 7.62mm, AR-10, Figure 1. Top and Right Side Views," dated August 30, 1960.

This was one of the three A-I Second Transitional rifles submitted for trial, serial nos. 004219, 004414, and 004534.
Michael Parker collection

*rifles completed the test. A cook-off occurred after firing 220 rounds in 1 minute, 9 seconds, but no cook-off resulted after firing 200 rounds in 56 seconds.*

## Comments on Test X (Rain Test)

The damage reported during the 600-round rain test occurred when the cartridge cases ruptured and damaged the rifle. The damage was confined to magazines, extractors and firing pins. When these parts were replaced, the rifles functioned normally.

An investigation into the causes of the damage concluded that it was "possibly due to a faulty test fixture", which prompted the following explanatory note in the official report:

*. . The rain test facility was inspected after the test was completed, and it was found that the AR-10 rifle was elevated slightly when positioned in the firing slide. The flash hider has three equally-spaced slots. The part is oriented in such a manner that one slot is at the top. This permits the drops of water to pass through the top slot and impact on the bottom of the hider. The drops had sufficient impact to splash some water into the bore. With the bore at a slight elevation it is expected that a considerable amount of water would accumulate in the bore during a 10-minute exposure period. Extremely high chamber pressure was developed when the rifle was fired under this condition . . Only an exceptionally strong mechanism could withstand this excessive pressure with no damage to major components.*

These three A-I AR-10s would remain in American hands, and over the next three decades would be used in secret development projects to attempt to arrive at a weapon accurate enough for use as a sniper rifle, but still being capable of producing a high volume of semi-automatic or fully-automatic fire. A few reports and even photographs have filtered out over the years of these Second Transitionals, heavily modified, bearing scopes, and even fitted with many pieces of furniture taken from M16s, being tested at Aberdeen and at Picatinny Arsenal in New Jersey. These weapons would not be forgotten, and the work done on them by the U.S. Army would come in handy one day for the very man who had invented them.

# 7.62x39mm AR-10 Assault Rifles in Finland

Meanwhile, through connections provided by Sam Cummings, from the summer of 1958 until late 1960 Artillerie-Inrichtingen had embarked on a unique and fascinating project in cooperation with the Finnish Army to adapt the AR-10's design to meet the special needs of this small Scandinavian nation. Because of the unique position Finland occupied—being both a democratically-ruled market economy friendly to the West, and a friendly neighbor to the Soviet Union, with which it also maintained close ties (partly from Russia's postwar bullying in response to Finland's WWII participation in the German campaign against the Soviets)—Finland desired to adopt a rifle that fired the standard cartridge of its communist eastern neighbor, instead of that of the newly-formed NATO alliance, of which it was not a member.

The Finns, therefore, who had been intrigued by the demonstration that had been put on for them in the dead of winter in early 1958, saw the obvious potential of the AR-10, and very quickly began working with a small subset of A-I engineers to develop a prototype of the AR-10 chambered for the Soviet 7.62x39mm assault rifle round, for evaluation in, and possible adoption by, Finland.

## A Windfall for Western Analysts

In a peculiar twist, even more than a decade after the adoption of the Kalashnikov assault rifle and Simonov carbine, next to nothing was known within the NATO alliance about the 7.62x39mm round, on which the Soviet military had almost completely standardized for their infantry needs. In order to ascertain the specifications for the Finnish AR-10's chamber, bolt, mainspring weight, and gas system, the Dutch requested and received extremely detailed load data concerning the Russian cartridge, and even a considerable quantity of loaded ammunition. These data piqued the interest of the military intelligence organs of many NATO nations, as well as the American CIA, who first learned the intricate details and capabilities of Soviet infantry weapons, all thanks to the AR-10's dazzling performance for the Finns.

Due to the protracted nature of this project, which as noted lasted from the summer of 1958 until late 1960, the various prototypes built for the Finns followed the ongoing developmental status of the standard AR-10 itself, with versions chambered for the Soviet intermediate cartridge being built directly off the plans for rifles in the Cuban and Second Transitional series, before a final proprietary carbine was fabricated.

## The First (Post-Cuban Model) Finnish Prototype (1958)

The shorter length and extreme taper of the 7.62x39mm cartridge had traditionally necessitated magazines that deviated significantly from those of rifles that fired cartridges with straighter bodies, like the 7.62x51mm NATO. The initial Dutch solution to this dimensional issue was quite ingenious, and did not even require changing the lower receiver of the weapon. The inside rear of the standard magazine body was blocked by a spacer bracket, and the follower was shortened to accommodate the shorter cartridges.

In addition to the magazine, a few other modifications had to be made to convert the AR-10 to function with the new cartridge. A different bore diameter was required, because the Soviets measured their 7.62mm bores differently, giving them an actual diameter of .311 - .312", as opposed to the .308-inch NATO barrels. The chambers for this dis-

254. Top view of the magazine from the first 7.62x39mm Finnish prototype AR-10 (fig. 255), showing how the Artillerie-Inrichtingen engineers took a minimalist approach by simply inserting a spacer in the back of a standard magazine to suit the shorter, tapered cartridge.
Institute of Military Technology

tinctive cartridge case also had to be cut shorter and dimensioned differently. A bolt-and-carrier combination was made up from an unfinished set intended

255. Right side view of a post-Cuban AR-10, serial no. 001271, converted by the Dutch to 7.62x39mm for the initial Finnish evaluations in 1958.

This first Finnish prototype looks almost identical to the Cuban, except for the threaded flash hider of the Sudanese model, and the distinctive black-finished, non-contoured bolt carrier. Without examining the interior of the magazine or chamber, one might well overlook this rifle as being simply a late Cuban model, as this author almost did.                         Institute of Military Technology

for a standard-caliber rifle. The cartridge seat in the bolt face was cut smaller to suit the narrower rim of the Soviet round, and the bolt carrier was given less of a round contour. Instead of being chrome-plated like the standard pieces, this combination was given a dull military parkerized finish.

The buffer springs were updated slightly to allow positive functioning with the different recoil impulse and pressure characteristics of the smaller assault rifle round when used with the standard AR-10 gas system. The rear sight adjustment range was changed to take into account the steeper ballistic path of the slower, chunkier Russian projectile.

Other than these relatively minor modifications, the first (and second) Finnish prototypes used standard receivers, buffers, furniture, front sight bases, gas tubes, and flash hiders. In fact, one is hard-pressed when viewing either the first or second weapon type fabricated for the Finns to tell them apart from standard versions of either the Cuban or Second Transitional models.

A number of between one and eight of these specially-modified Cuban-type rifles were supplied by A-I for tests in Finland in 1958.

## The Second (Second Transitional) Finnish Prototype (1959)

Running concurrently with the updates made to the AR-10's design based on other customer demands, the next Finnish AR-10s were visually identical (with the ejection port door closed) to those of the Second Transitional variants formulated for the Dutch New Guinea tests. These were fitted with the same modified barrel and chamber, bolt carrier group, and straight magazine with a spacer to accommodate the 7.62x39mm cartridge, but were still very modular, with most of the other parts interchangeable with those of the standard production rifle. Again, a total of from one to eight were supplied by A-I for Finnish tests in 1959.

## The Purpose-Built Third Finnish Trials AR-10 Assault Rifle (1960)

Finally, in 1960, Artillerie-Inrichtingen made up a radically divergent prototype for the looming Finnish evaluations, still based on the Second Transitional AR-10 variant, which took advantage of many of the idiosyncrasies of the Soviet cartridge. To satisfy a Finnish desire to have a higher-capacity magazine, the lower receiver was given an angled cut similar to that found on the AR-102 sporting rifle, and a guide, reminiscent of the Kalashnikov series magazine re-lease, was housed in the magazine well, taking up some of the space and allowing a taller, narrower (front-to-back) magazine to be inserted. This did not actually function to hold the magazine in place or release it, but provided a channel into which the magazines could be inserted, where they were engaged by the standard magazine release lever found on all Dutch-production AR-10s. The new 30-round

256. Left side closeup of the final 7.62x39mm Finnish prototype, produced for trials in 1960, showing the unique features that it and only it embodied. This one-off was produced from a standard lower receiver with a modified magazine well, cut through the middle of the existing standard manufacturer's markings and retrofitted with a magazine guide behind the new, curved magazine.

The charging handle extension lever was reversible from the right to the left side, based on shooter preference.

Note the flat-topped carrying handle, rounded at the rear to protect the rear sight, fitted with a range scale modified to fit the trajectory of the ballistically inefficient Soviet round.

All of the remaining parts of the rifle, except for the chamber and actual bore diameter, were identical to those of the standard Second Transitional carbine.

photo by Marc Miller

magazine was distinctly curved, closely resembling the standard Kalashnikov magazine.

As the 7.62x39mm round was designed to burn all of its powder in about sixteen inches of barrel travel, the barrel was therefore optimized for this cartridge by being shortened to this length, which also reduced the overall weight.

This final Finnish prototype incorporated the same handguard design, barrel length, and shortened barrel shroud/flash hider as the A-I carbine, even down to the Kiffe bayonet which mounted on the set-back muzzle ring below the barrel.

The carrying handle on the upper receiver also displayed several features with which the Dutch were experimenting at the time of the Transitional

rifles. For charging while wearing thick winter mittens, the right side of the cocking handle was fitted with a spring-loaded extension lever that folded along the side of the carrying handle so it would not get in the way during firing, and could be adapted for right- or left-handed users. The carrying handle also had a different shape than any production AR-10 model, with the high sides that traditionally ran down the entire length of the handle, protecting the rear sight, only being present directly on either side of the rear sight. The remaining area of the handle was flat on top, perhaps to save weight and material, or perhaps to allow for the modular mounting of a scope should the weapon be adopted by the Finns.

## The AR-10 Wins - but Discretion Favors the AK-47

As desired by the Finns, a total of ten of these final prototypes were completed by early 1960, and sent to Finland for tests against a version of the Danish Madsen—also chambered for the Soviet cartridge—and some examples of the actual Russian AK-47.

Despite the AR-10 winning the tests, the Finnish government made the decision that in furtherance of better relations with their bully of an eastern neighbor, and for reasons of parts and logistical commonality, they would simply adopt the Russian weapon. The Finns therefore designated a slightly-modified version of the AK-47 known as the Rk-60, which was updated two years later to become the Rk-62, a weapon that is still in regular service with the Finnish armed forces.

Although this weapon was essentially a Kalashnikov, it was manufactured domestically, and to such incredibly high quality standards that it was capable of performance that the rough Soviet rifles could not hope to match. Just as the Finns had done previously in adopting and domestically manufacturing an improved version of the standard Russian Mosin-Nagant bolt-action rifle to immeasurably higher quality standards, they saved money and still got an excellent weapon for their troops. Even while both are basically the same weapon, comparing the Finnish Rk-62 to a Russian AK-47 or AKM is like comparing a Rolls Royce to a Yugo.

It was thanks to these Finnish prototypes, chambered for the intermediate 7.62x39mm cartridge, that the ArmaLite AR-10, which normally belongs in the category of a main battle rifle, made its sole, brief appearance as a true assault rifle. Due to its appearance and capabilities, the AR-10 has been and continues to be described as an assault rifle by the ignorant and misinformed (indeed some poor souls even believe that the "AR" stands for "assault rifle"), but in all of its standard versions, right down to the present day, it is a battle rifle.

# Wrapping Up Abroad

## The Results of AR-10 Demonstrations in Various Sales Territories

As discussed in Chapter Six, a number of evaluations using various versions of the Dutch-manufactured AR-10 had been performed in a great number of nations, especially in Europe, where it had generated serious interest.

Though of lesser importance to Artillerie-Inrichtingen (arrogant presentation notwithstanding) than the Dutch and South African trials, these tests had by this time been completed, and the results of each are summarized below.

### Sweden

The Swedes had never really given the AR-10 much of a chance, and they would end up sticking with their domestically-designed AG-m/42b until adopting a slightly modified version of the H&K G3 as the Ak4 in 1965, most of which were manufactured under license at the state-owned Karl Gustafs Stads Gevärsfaktori in Eskilstuna.

### France

The French had been eyeing the AR-10 very seriously, but with the dissolution of the Fourth Republic in 1958 and the threat of military insurrection prompted by the ongoing colonial wars, combined with the attendant focus on self-reliance and insularity, France finally opted to stick with its own domestically-designed and -produced MAS-49/56 battle rifle, in its proprietary 7.5x54mm chambering.

### Norway

Norway, on the other hand, which had been keeping the adoption of a new rifle on the back burner, decided in 1960 to move forward with trials to replace their vintage M1s. They ordered 10 Second Transitional AR-10s that year, which soundly defeated the German G3s and Danish Madsen battle rifles against which they were pitted in proving ground tests and limited field trials. The ArmaLite weapon, particularly in its advanced stage of design, impressed the Norwegians with its ruggedness and ability to function in extremely cold conditions.

However the unfortunate events that were befalling Artillerie-Inrichtingen at this time, discussed below, precluded the possibility of supply to the small Norwegian military, and they perforce adopted their second choice; the G3, known in Norway as the AG 3, which was produced under license by the Kongsberg Våpenfabrikk.

# Latin America and the Caribbean

Aside from Cuba, Nicaragua and Guatemala, whose stories have already been recounted, demonstrations in other Latin American and Caribbean countries did not result in any orders for equipping troops. As discussed in Chapter Ten, the intrepid Sam Cum-

mings had promoted the rifle almost everywhere in the region, and while Venezuela and the Dominican Republic had expressed serious interest in adoption and large purchases, things did not turn out that way.

# Venezuela

Venezuela had also been testing the FAL, and was given a special incentive by Fabrique Nationale to adopt it. Since the beginning of the century, the Venezuelan military had been particularly infatuated with the 7x57mm Mauser round; one of the first truly modern rifle cartridges which, when loaded with a boat-tailed "spitzer" or pointed-tip bullet, was effective at very long ranges yet produced relatively mild recoil, especially when compared to the larger .30-caliber U.S. and 8mm German rounds.

One of the last designs that had been proposed during the postwar search for the ideal military rifle round, which ended in the adoption of the U.S.-promoted 7.62x51mm NATO cartridge, was the

7x49.15mm "Second Optimum", described in *The FAL Rifle* as a joint Anglo-Canadian venture [which] was continued and marketed by FN in Venezuela as late as 1957 as the "7mm Liviano." FN agreed to assist the Venezuelans in keeping the ballistic performance they had learned to love in this shortened 7mm cartridge, which would fit and function in the FAL. This incentive was too good to pass up, and on November 30, 1954 Venezuela had ordered 5,000 examples of a special variant of the FAL in the "7mm Liviano" (7x49.15mm) chambering. A few years later, they ordered more FALs in 7.62mm, along with a sufficient quantity of new barrels to retrofit all their existing 7mm FALs to the 7.62mm NATO caliber.

# The Dominican Republic

Trujillo had shown great interest in the AR-10, and despite a violent event that might well have changed his mind (discussed in Chapter Six), is reported to have placed a large preliminary order. However, during the time period when AR-10 production was commencing in Holland, events were coming to a head in the Dominican Republic. Not only did the dictatorial rule of Trujillo grow more tyrannical as 1958 and 1959 passed, but his people became less willing to tolerate it.

His agitation against the Venezuelan government began to make him a political liability to his allies in the area, and in 1960 he ordered an assassination attempt on the president of that country. This angered the other Latin American and Caribbean states, and caused the United States to lose patience with his regime. He started to be viewed as a new

Batista, and American public opinion as well as government policy saw it as best for him to be relieved of his position. This sentiment made it increasingly difficult for Sam Cummings to deal effectively with him, and the Dutch government was particularly disinclined to sell arms to a rogue state led by a butcher, as Trujillo was increasingly seen.

The possibility of an AR-10 sale to the Dominican Republic ended with the finality of an artillery strike on May 30, 1961, when Trujillo himself was assassinated by a disaffected faction of his military, who were enraged at his bungling and increasingly corrupt leadership. Although his family attempted to hold onto power, the country was thrown into a state of chaos, making the prospect of ordering foreign arms an impossibility.

# Austria

The Austrian military forces were among the most enthusiastic towards the AR-10, and after their first tests of Dutch-produced AR-10s in 1958, which kept soundly defeating the FAL, they advised A-I that they were essentially committed to its adoption. The lack of tropical coastal environments in that alpine country shows that even the earliest AR-10 prototypes were extremely reliable in temperate, hot-and-dry, and cold conditions.

However, when the Austrians expressed a desire to order quantities of the rifle, Dutch production was only just beginning, and they were told that it would be two years before the quantity they required could be manufactured and shipped. This timeframe was just too long to sit well with the Austrians, however, and Fabrique Nationale pounced on this opening, supplying a test quantity of FALs for comparative troop trials, which the AR-10 once again won handily.

A final attempt was made to get Artillerie-Inrichtingen to either promise to speed up production, or to grant a license to allow domestic production of the AR-10 in Austria. The former was impossible, and the latter was something A-I was not in the least inclined to do: as the sole production licensee they had paid good money for this exclusive right, and furthermore, any weapons produced outside their own factory would earn them no royalties. Thus, Austria, just like West Germany, while preferring the AR-10 but unable to accept the long delay before major production could begin, adopted the FAL. In their specifications, however, they made several modifications reminiscent of features of the AR-10, such as a polymer buttstock. As a result, the *Sturmgewehr* 58 (Model 58 Assault Rifle) produced under license by Steyr, is still considered essentially the most capable of the standard FAL variants.

# Italy

Italy, which had been enthusiastically considering the AR-10 for two years, eventually went a less costly route, choosing to adopt the far cheaper BM59, a modified M1 Garand reworked by Beretta.

The development and successful sale and use of the BM59 stood in especially stark contrast to the M14, the cost-conscious Italians being remarkably successful in adding a few inexpensive tweaks to their surplus U.S. M1 Garands, resulting in a rifle that weighed in at 10 lbs. 1 oz., and was functionally the same as the M14. The incorporation of a flash hider capable of launching grenades, a selective-fire trigger group, a bipod, a rubber buttpad, and a detachable 20-round magazine made the antiquated Garands, if not the equal of weapons like the FAL or AR-10, into modern, capable battle rifles.

When the BM59 was standardized in 1959 and accepted by the Italian Army, they were selling on the world market to foreign militaries for the paltry sum of $60 each, a price even higher than that offered domestically to the government. The BM59 gave excellent service to the Italian Army into the 1990s, when they were finally given up in favor of a more modern assault rifle.

# Indonesia

The Italian BM59 conversion also saw much success abroad, offering to less developed countries the same effectiveness as the rifle the U.S. was using. In addition to becoming the standard rifle of the Italian Army for a staggering 38 years, with some units reportedly still being equipped with it today, the BM59 was adopted by several foreign countries, including Ethiopia, Somalia, and Indonesia.

In the latter country, the BM59's cheap price and a subsequent agreement for its domestic production under license proved too good to refuse, and so the AR-10 was never made the standard rifle of all Indonesian forces.

The success of the BM59 may have precluded large-scale adoption possibilities for the AR-10 in some countries, but its underselling of more modern pieces did not end up closing out certain other opportunities for the ArmaLite rifle. The mystery of what has become of the 1,000 AR-10s the Indonesians bought (500 of which were confirmed and paid for before Dutch production even began) remains a mystery, although perhaps at some point in the future these Indonesian AR-10s will be declared surplus and sold commercially.

# Italy, Again

Just as the BM59's success did not stop Indonesia from purchasing a sizeable quantity of AR-10s (a large quantity compared to total AR-10 production, but paling in comparison to the order sizes of other rifles by other countries), it did not close the door on Italian interest either.

Coming to the rescue for ArmaLite's hopes in that new NATO member state, as was the case in many places where the AR-10 met with success, were the élite unit of the Italian Navy, the COMSUBIN. Their purchase and use of the AR-10 would be kept a closely-guarded secret, with only rumors of it escaping, making the likelihood of the Indonesian *Kopassus* (a shortened name usually given to the special forces command) being able to keep a quantity of AR-10s away from prying eyes on an isolated, militarized island in the Pacific quite plausible. The variant adopted by the Italian Naval Special Forces ended up being the final and most advanced version of the AR-10, the details of which are discussed below.

# Turkey

Holding out longer even than the Austrians had, the stalwart Turks were still intent on standardizing on the AR-10 in early 1961, having tested quantities of these rifles extensively and been greatly impressed with their performance. Other weapons that had been evaluated alongside the AR-10 were the BM-59 and the FAL, with the G3 having been eliminated.

The Turkish military continued to express its preference for the AR-10 right up to the bitter end, when any possibility of acquiring them was finally cut off once and for all by the politically-motivated forced cessation of AR-10 production in Holland. In exasperation, Turkey finally accepted the all-too-common consolation prize of a dirt-cheap production licence from the West German MoD for the G3, its fourth-choice rifle, which was manufactured domestically in the G3A3 (fixed butt) and G3A4 (retractable butt) versions by the Silahsan weapons factory, a division of the state-owned Makina ve Kimya Endüstrisi (MKE) Kurumu in Ankara.

According to the entry in *Full Circle*, quantities of FN FALs were also in Turkish service, among which were many ex-*Bundeswehr* FN G1s.

# Burma (Myanmar)

In Southeast Asia, Bobby MacDonald's demonstration for the fledgling, embattled nation of Burma did indeed pay off, with their purchase of a quantity of AR-10s in 1959. As the specific records on this sale were not available at the time of writing of this book, little is known about this order. It is accepted that the quantity was small, somewhat less than 500 pieces, and was either designated for troop trials or for issue to the fledgling rapid reaction force that would two decades later become a very capable, well-organized special operations organization. Given the fact that the current Myanmar government is not willing to speak about the purchase and use of these rifles, it would appear likely that one of the two possibilities discussed above are the case. Given the year of purchase, one can also assume that the type of rifles purchased were certainly one of the Transitional variants, probably the second.

Whether the Myanmar military junta ever intended to equip all their armed forces with the AR-10 is unknown, and possibly unknowable, but what is recorded is that by 1961 they had succumbed to the attractive cost and adequate performance of the German G3 adaptation of the Spanish CETME, along with the West German government's propensity for easily and cheaply granting manufacturing licenses and providing technical support. The 7.62mm G3 would continue as the standard Burmese infantry weapon until the 1990s, when many of them were converted to 5.56mm.

*Chapter Nine*

# The Pinnacle of Performance

# The AR-10 in Portugal

Notwithstanding the disappointments in other countries, discussed above, there remained the enticing possibility of receiving a quantity order for AR-10s from the Portuguese Republic.

## Some Historical Background to Salazar's *Estado Novo*

Not a representative democracy like most European nations at the time, Portugal was what could be described as a somewhat benevolent dictatorship, akin to Dubcek's Prague Spring in Czechoslovakia.

Unlike the by-then-defunct totalitarian regimes of fascist Italy and Nazi Germany, Portugal, ruled since before WWII by the brilliant, devoted, and ascetic Prime Minister António de Oliveira Salazar, had allowed a wide array of personal and economic freedoms that would have been unthinkable in more traditional fascist states. In just one example of Portugal's unique status among contemporary nations, it was the first in human history to set about abolishing the death penalty, and one of the first to completely accomplish this goal.

Although modeled on a corporate system and closer in political doctrine to the authoritarian governments of its neighbors like Italy, Spain, and Germany, Salazar and his government had been particularly disgusted with the racist ideology of the Nazis, and in their revulsion kept themselves distant from any German influence. In response to the infamous Nuremberg Laws of the 1930s, Salazar had notified the Nazi government that all Portuguese were considered citizens of Portugal regardless of their race or religion, and no discrimination against Portuguese of the Jewish faith in Germany would be tolerated by his regime.

Portugal's neighbor Spain, however, was well and truly under the sway of Hitler by the time WWII had begun and, as a result, became somewhat of an international pariah after the conclusion of the war, both in spite of, and because of, its neutrality. Spain

257. António de Oliveira Salazar (1889 - 1970), Prime Minister of Portugal from 1932 to 1968.
public domain - the Internet

did not actively declare war on any of the Allies, but the fact that they were neutral meant that, unlike in the former belligerent states of Italy and Germany, armed forces of the Allies were not able to overthrow the extreme fascist regime there and bring it into the new Europe.

Portugal, on the other hand, had aligned closely with the Allies at the start of the war, with Salazar consulting with the British and resolving to remain outwardly neutral, partly as a means of actually helping the Allied cause.

Portugal and Britain had formed a military alliance, and when Britain declared war on Germany, Portugal announced that it stood ready to honor the alliance and fight by England's side if summoned. It was worked out that a neutral Portugal served as a strong bulwark against the Axis, by keeping Spain out of the war. It was predicted by Salazar, who also correctly prophesized in 1939 the exact final disposition of the war (a prediction that no other European

leader shared), that if Portugal were to declare war on Germany, Spain would likely feel threatened and in turn declare war on Portugal, Britain, and France. With Spain being much larger, Portugal would have almost immediately been crushed, and Spain would then have helped Germany consume France, the defeat of which was not seen as likely at the time.

Because of Portugal's aid to the Allied cause during the war, and Salazar's almost-miraculously successful modernization of his country, the United States and the rest of Free Europe were extremely friendly toward the small Iberian nation in the post-war years. Having been staunchly anti-communist "before it was cool," Salazar's Portugal fit right into the new Cold War mindset of Europe, and was a founding member of NATO.

# Portugal Eyes the AR-10

## Michault Comes Calling - with AR-10B No. 1003

Once he had purchased his own personal demonstration AR-10B, serial no. 1003, Jacques Michault set about demonstrating the rifle on his own throughout his assigned sales territory. One nation that showed unique opportunity was the small and unassuming Republic of Portugal.

In 1958, the confluences of events discussed above, as well as several other characteristics peculiar to the Portuguese Republic, made for a uniquely welcoming environment for the AR-10.

## Portuguese Headaches in the *Ultramar*

The first of these factors was the unique political situation Portugal occupied with respect to its extensive overseas territories. Although quite a bit of Africa was still under the control of various European powers in the late 1950s, only Portugal, France, Belgium, and Spain seemed determined to hold onto their colonial territories. France and Belgium were by the mid-1950s deeply embroiled in fighting counter-insurgencies in their African and Asian colonies. These conflicts would lead to the bloody and bitter independences of Algeria and the Congo, and to the later Vietnam War in what was then French Indochina.

Portugal's situation in 1958, however, was in stark contrast to that of France. Throughout the *Estado Novo*, or "New State" regime of Dr. Salazar, the Portuguese government had taken a very progressive approach to racial harmony and integration of what were once colonies into the framework of statehood. By the end of WWII, the principal possessions of Portugal, called the *Ultramar*, or "Overseas;" including Angola, Mozambique and Guinea in Africa;

Goa on the Indian subcontinent; and Timor in the Pacific Java islands; were no longer treated as colonies, but as provinces, equal in political and economic status to the home European territory. This equality was universal in theory, but existed practically to a varying degree among locales.

The Portuguese government's strong stand against racism and in favor of pluralism could not only be seen in the absolute protection given to Portuguese Jews during WWII, but also in subsidies to ethnically white Portuguese engaged in economically beneficial activities in Africa, including even greater amounts paid to those who married native Africans. A favorite saying of Salazar's was that any indigenous person living in Portuguese-controlled territory was eligible for all offices of state, including President of the Republic. Especially in the 1950s, during a time when the horrendous racism of the Jim Crow South was the *status quo* in the United States, there were many African natives who occupied high positions in all areas of the provincial administration of the Portuguese *Ultramar*, even the military.

However, behind all of the good will and stability that was seen in the former colonies, the Portuguese leadership were under no illusion that their country would be immune from the anticolonial, anti-Western, sectarian, and political strife that was embroiling so many African nations which had either recently become independent or were just on the cusp of gaining their independence. Causing particular trepidation to the colorblind Portuguese was the invidious hand of the Soviet Union in fomenting unrest and sowing racial discord in the newly-independent parts of the world. Being particularly loathe to the idea of communism, the military and political leadership of this plucky, enthusiastic NATO member state saw a strong defense against Soviet-sponsored guerilla insurgencies in the overseas provinces, particularly those in Africa, as a major national defense priority.

## The Appeal of Portugal as a Blank Slate

The other unique characteristic of the Portuguese Republic in 1958 that drew Michault to it like a moth to a flame, was that it was undergoing a large-scale military reform and reorganization, but unlike almost all of the other continental powers, it had neither progressed close to the stage of adopting a foreign battle rifle design, nor developed one of its own. As we have seen above, one or the other of these rationales had kept otherwise-impressed militaries like those of Switzerland, Spain, France, Denmark, Sweden, and Norway from giving serious consideration to the AR-10.

At this time, the Portuguese infantry were still armed with bolt-action German Mauser K98ks, but the military leadership was only just beginning to contemplate the selection of a new automatic rifle to replace them. In addition to the overall need for a new standard infantry rifle, an attitude of greater acceptance towards advanced weapons like the AR-10 was propagated by the newly-formed "Battalion of Hunter-Parachutists", more commonly known in English as paratroopers. As had been the case in the U.S. military and that of WWII Germany, the new Portuguese élite airborne units had adopted a very forward-thinking approach to new military technology and equipment, and were looking for the most advanced, effective, and lightweight weapons available. This stemmed not only from the fact that weight is at an absolute premium for soldiers who have to jump from airplanes carrying all their supplies with them, but due also to the extreme vulnerability of paratroopers by the very nature of their operations. While conventional infantry usually advance along a front line and have rear units established to supply and support them, the airborne troops are, by definition, alone and surrounded in hostile territory, and thus have an extremely strong desire for force multipliers in the form of advanced weapons. At the same time these units are far less concerned with commonality with the regular forces, because they rarely if ever fight in concert with them.

The Portuguese Ministry of Defense had, in 1955, greatly accelerated their country's military reorganization by finally giving in to the relentless advocacy and campaigning of then-Secretary of State for Aeronautics (the Portuguese equivalent of the American Secretary of the Air Force) Kaúlza de Oliveira de Arriaga. As well as being a charismatic and influential political figure, Arriaga, who eventually became a general and commander-in-chief of the Portuguese armed forces, had been a tireless reformer since his appointment to the Portuguese Institute of Military Studies for modernization of the military, particularly in the creation of rapid-reaction air-mobile units.

## Forming the Parachute Hunter Battalion

On December 26, 1955, under Service Ordinance 15671, the *Batalhão de Caçadores Pára-quedistas* (Parachute Hunter Battalion) was created, with funding allotted for training, research, and equipment. This newly-commissioned unit would be the first-ever Portuguese formation to be issued camouflage clothing and berets, which were green. This led to their most common nickname, which is still used for the unit today: *Boinas Verdes*, or "Green Berets."

Research at the Institute for Military Studies, led by Arriaga, had investigated the characteristics and command structure the new battalion was to have. Studying the models of the Americans, French, and Germans, the Portuguese high command decided to establish the airborne units as a part of the Air Force, following the French and German example, instead of making them an Army unit as in the U.S.

The Portuguese Air Force itself had been a very recent creation, being organized on May 27, 1952, and had a more all-encompassing role in air operations than was the case in other militaries. In the U.S. and Soviet command structures, for example, although there was a separate air force, the army and navy both had their own aircraft for operations, particularly supply and transport. Since these roles were also to be filled by the newest branch of the Portuguese armed services, it was reasoned that the paratroopers should be assigned to them. It was envisioned very early on that in addition to jumping from transport planes, helicopter insertions would be a critical component of the evolving world of modern warfare, and the Air Force's trailblazing on this front was predicted to be important to airborne operations.

This forward-thinking and progressive attitude with respect to rapid-response special forces dove-tailed well into the evolving reorganizational strategy of putting counterinsurgency and counterterrorism operations first, with a large-scale, conventional war-fare capability taking a back seat. Although possessed of a long and storied warrior tradition and what was, man-for-man, an excellent and modernizing fighting force, it was clear that in a large-scale war with the Eastern Bloc, Portugal's small army would not likely be a key player in repelling the forces of tyranny from Europe. Not only were there far larger militaries like those of the U.S., Britain, France, West Germany, and Italy to handle this task, but Portugal's geography, being far from the probable theater of operations and not of great strategic importance, made the likelihood of Portuguese forces having to defend their territory from massed Soviet advances very slim.

## Training with the French and Spanish - Critiquing their Issued Arms

In order to form the initial corps of officers and NCOs who would make up the professional backbone of the new para-infantry, a cadre of specially-selected, élite infantry officers (both commissioned and non-commissioned) was sent abroad for specialized instruction in parachuting. Having modeled their airborne command structure most closely after the French, the first group of the 192 to receive foreign training was sent to the town of Pau, in the Pyrenees, close to the French-Spanish border and also very proximately-located to Portugal, where the French airborne training school was located. The cadre then took the regime of courses at the Spanish Military Parachuting School. Although Spain was still somewhat of an international pariah state, the Portuguese government, in their wise WWII neutrality, had maintained excellent relations with their neighbor.

The task of finding and acquiring the best weapons to equip these new light para-infantrymen was always at the forefront of the training and organization of the unit's command structure. One constant observed by the Portuguese students at the two airborne schools was the lack of effective arms in the hands of their host forces. The French troops, in the usual maverick nature of airborne units, were equipped with American M1 Carbines, as opposed to the increasingly-common MAS-49/56 autoloaders or older MAS-36 bolt-action rifles. The extreme light weight of the U.S. weapon, itself the favorite arm of American paratroopers, superseded any concerns that the French command structure might have had about weapons commonality or standardization, and

Carbines were issued both to those air force units, as well as to the foreign legion that participated in airborne operations and close-range jungle infiltration fighting in the war raging in Indochina.

True to its initial design parameters, the French found the M1 Carbine to be a nexus of huge compromises and sacrifices made in the name of saving weight and size. The .30 Carbine cartridge was so small and anemic that it was far closer in power to, and is even considered by many experts today to be, a handgun round instead of a true rifle caliber. Even in the day of the intermediate-power rifle round, either of the reduced power full-caliber or small-caliber high-velocity school, it is a stretch to consider the .30 Carbine cartridge fired by the M1 Carbine a true rifle round. To make matters worse, the round-tipped, flat-base construction of the projectile gave it the worst possible ballistic efficiency, meaning that it lost velocity and power extremely quickly and arced significantly even at short ranges. It also made the round very ineffective at stopping an enemy soldier in his tracks. However, the Carbine was light, maneuverable, had little recoil, and was capable of an excellent rate of fire due to its detachable magazines that were, for their day, considered quite high-capacity. Therefore, the French paratroopers made do with them, despite their glaring shortcomings.

The Spanish were even worse off than their French counterparts, being outfitted with their standard infantry rifle, which was simply their earlier-issue German Mauser 98 variant shortened, given a flash suppressor, and rechambered for first the Spanish variant of the 7.62x51mm NATO round (a less powerful load with a lighter bullet and less powder, intended to tame recoil), and then for the NATO cartridge itself. Although in 1957 the Spanish Legion and the Airborne Light Infantry Brigade were the first units to receive the new-but-heavy CETME battle rifle, when the Portuguese were studying under them during the period from 1955 through 1957, the paratroopers of Portugal's eastern neighbor still carried the long-antiquated Mauser, with its slow rate of fire and five-round magazine, on their backs when jumping from airplanes.

# Developing the Portuguese/NATO AR-10

It just so happened that Colonel Arriaga (his rank at the time of his first contact with Fairchild's representatives), who was spearheading the airborne project, caught wind of the new and fantastically-light rifle that had just entered production in Holland. Through his superiors at the Ministry of Defense, whom he had thoroughly impressed into granting him the funding for the airborne battalion, this rising star of the Portuguese defense community was brought into contact with Jacques Michault of SIDEM International.

As we have seen, since its inception and throughout Dutch production in 1958 and 1959, the AR-10 had evolved several times, and always displayed uncommonly good performance in almost all respects. However, the absolute preeminence of the AR-10 as being "in a class by itself" among battle rifles is based on this final variant, designed and produced by the Dutch with considerable direct input from the Portuguese.

For once, a representative of the AR-10 had made it into an important market in the first wave, not having to play catch-up with Fabrique Nationale, CETME, or SIG.

The Portuguese military in general was profoundly taken with the AR-10 in demonstrations, and immediately began working closely with the engineers at Artillerie-Inrichtingen to craft their desired configuration. Since their very first field trials, the Portuguese had preferred the AR-10 to the FAL and CETME, but were somewhat critical of the extreme light weight of the early models with which they were provided, stating that they would rather have some-

thing a little heavier that would hold up longer and require even less maintenance.

The Portuguese were not in the same huge hurry to reequip that had preoccupied the nations nervously sitting on the border of the Eastern Bloc, like Germany and Austria, or those facing real or perceived enemies on all sides, as were Burma and Indonesia. The Portuguese had, throughout the time of the Artillerie-Inrichtingen countdown to production and fulfillment of early orders, bought small quantities of most competing weapons, including the FAL, CETME, and later the German G3. They conducted field trials with all of these weapons, comparing them first to small quantities of Sudanese-type AR-10s, then Transitionals, and had throughout this period provided the Dutch factory with the most useful input and feedback of any foreign customer.

Thanks to their friendliness, openness, and unhurried nature, which this author has observed as unfailingly characteristic of the Portuguese, the AR-10 reached its final, and one might well say perfected, form, which would earn it the highest of praise from some of the bravest and most élite warriors on the planet, and may well have been a driving force behind the recent renaissance of this, the most superlative of battle rifles.

The close cooperation with the Portuguese would lead ArmaLite and Artillerie-Inrichtingen to the very pinnacle of AR-10 development, and what was without a doubt the most advanced, rugged, and reliable version of the AR-10 ever built and issued to an armed force in history, including the present day.

## Improving the Second Transitional AR-10

The Transitionals tested were considered an improvement over the first A-I submissions, but the Portuguese noted some issues that were soon to be experienced in adverse condition testing elsewhere, particularly when the rifle was exposed to dirt, mud, and salty humidity. This was not, in the eyes of the Airborne, an indictment of the AR-10's design or

capabilities, but more an indication of an area where improvement was possible. The FALs they tested were similarly critiqued, and the CETMEs and G3s were actually defeated by the Transitional AR-10s in the same adverse condition trials, just as happened in South Africa.

With concrete suggestions for improvement coming in from the Portuguese, the Italians—whose naval special forces were also evaluating the AR-10 and thought highly of it—and the Venezuelans, who had put their test models through particularly de-manding endurance trials, Artillerie-Inrichtingen set about to add a series of further modifications to the Second Transitional variant, in order to arrive at a weapon that would stand up to the harshest of torture tests.

## But First - Another Evolutionary "Missing Link"

258. Left and right side views of a Second Transitional variant, serial no. 003420, obviously a "stepping stone" to the final Portuguese/NATO model, which features a unique blend of old and new features.

From left, note the grenade sight bracket on the face of the front sight tower; the finalized gas adjustment dial that could be manipulated with the point of a cartridge; the Bakelite handguard shells, ventilated with both slots *and* holes; and the side-mounted "carbine carry" rear sling attachment point.

However the bolt holdopen remains the early "lolli-pop"style, and the selector markings still have "SAFE" in the original up position.

Dan Shea collection, photos by Frank Iannamico

Fig. 258 illustrates yet another example of what can be considered an evolutionary "missing link" rifle, along with the "Pre-Transitional" (fig. 189), and to a lesser extent the First Finnish and the German test rifles. It is still a Second Transitional variant, but as shown, it is fitted with a combination of old and new componentry, the latter of which are depicted and discussed throughout this chapter.

## Modifications on Later Models Introduced Over Time

As shown above and in a number of the following illustrations, there was often an overlap of old and new componentry in later-production versions of the Dutch AR-10. It is important to keep in mind that when a new component was introduced as a result of an improvement in design or material, the stocks of old parts were not automatically scrapped, and so the new parts were not necessarily featured in all subsequent production. Thus, some later models manufactured or assembled after the supposed date of introduction of a new or modified part were still fitted with an earlier version of that component. This is explained with reference to the normal procedure used in series production, where new parts are manu-factured and then dumped into parts boxes, to be used by assemblers as the rifles were made up. Older parts from the bottom of the box may thus not be used right away, but they would be used if needed to keep production moving.

# Larger Bolt Lugs

259. Closeup of a Sudanese bolt which, while bearing the Dutch Arsenal firing proof, failed after extended use, with one lug shearing off.

Charles Kramer collection

The first modification was one that had been suggested earlier by Melvin Johnson, the necessity for which had been shown by the breakage of some bolts in the Venezuelan and other endurance tests. The weak point on the earlier bolt design was determined to be the seven lugs that protruded from the bolt face and locked into the barrel extension. Though an extremely sound design, round counts in the tens of thousands caused the lugs on some bolts to crack and fail. The solution to this problem was to increase the width of the lugs on the bolt, and modify the

260. Face-on views of two AR-10 bolts.
Left; as used in all previous production.
Right: as introduced in the Portuguese/NATO rifles. Note the larger, more robust locking lugs.

Charles Kramer collection

barrel extension to match. Similar problems with the locking lugs on the barrel extension had not been observed, and it was reasoned that since there was so much more mass to the extension into which the bolt locked, this component could sacrifice a bit of material and theoretical strength to make room for the beefed-up bolt lugs.

# A Stronger Extractor

The weakest point—the part most prone to failure in any automatic rifle design—is the extractor. In keeping with the stronger bolt lugs, the extractor on this series, made possible by the wider locking slots of the barrel extension, was also increased in size and sturdiness. This completely solved the extractor breakage issues that had dogged the earlier AR-10 variants (though not necessarily any more so than other contemporary battle rifle designs), and caused

the end-users of this final version to report a lower incidence of extractor failure than even the G3s and FALs with which it concurrently served.

The result of this development was a smashing success, and test-firing tens of thousands of rounds at Artillerie-Inrichtingen proved that the bolt lug breakage problems were solved, with no increase in wear on, or damage to, the locking recesses in the barrel extension.

# Heavier Bolt Carrier

The rather lengthy body of the bolt carrier was also strengthened, with the positive side effect of slightly reducing the felt recoil of the weapon. As shown in fig. 261 the rear surface area, which had been relieved on previous carriers, was left a uniform diameter

throughout its entire length. This slightly heavier uniform-diameter bolt carrier, first seen in a few of the last-produced examples of the Second Transitional variants, became standard equipment in the Portuguese model.

# Increasing the Chamber Shoulder Angle

Along with the new bolt and barrel extension, another of the suggestions made by Melvin Johnson long ago regarding primary extraction was actually incorporated in this pinnacle of AR-10 development,

albeit tangentially. It was determined that no actual primary extraction cuts needed to be made on the bolt lugs or the corresponding locking recesses, but the chambers in barrels of the final AR-10 model

261. Right three-quarter view of the new bolt carrier, with no reduced area on its rearmost section.

Compare with fig. 260: it appears that the bolt lugs shown here are still the original size.

Charles Kramer collection

deviated slightly from the standard 7.62x51mm NATO dimensions, with the shoulder angle changed from 40° to 42°. In order to check that headspace was correct in these rifles, a special headspace chamber gauge was correspondingly developed by Artillerie-Inrichtingen, because standard ones with the 40° shoulder would not give proper readings in these modified barrels.

## The Experimental Franchi "Dual-Hammer" Rate Reducer

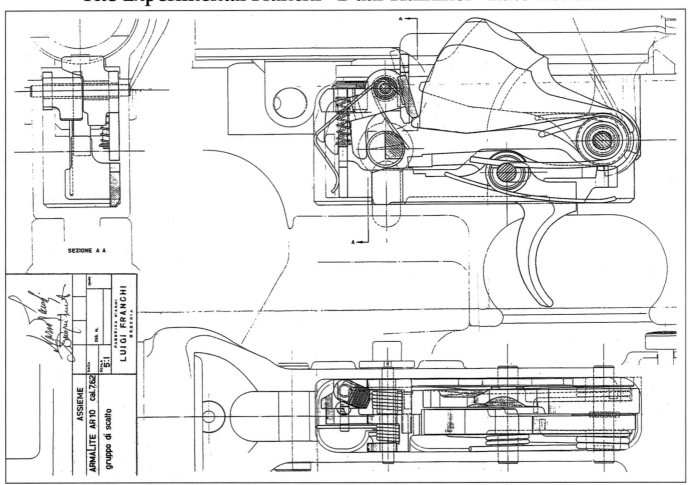

The Portuguese were responsible for the next modification, which also aided the reliability, longevity, and accuracy of the weapon. Concerned with the tendency of the standard trigger assembly to be contaminated by dirt, as well as for the springs to lose power and give poor return strength and hammer force, they suggested a new type of hammer. This sentiment had already been echoed by the Italians, but for the reason that they wanted to reduce the rate

of fire when the weapon was set in the full-automatic mode.

Tests were made using lower receivers fitted with an experimental "dual-hammer" trigger group invented by Attilio Franchi of Brescia, Italy, in which a first hammer that was too short to reach the firing pin would fall and then trip the second hammer to fall and strike the pin, thereby lengthening the time the round was in the chamber waiting for the hammer to fall.

## A New Coil Hammer Spring and Plunger

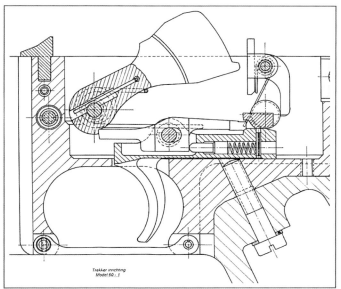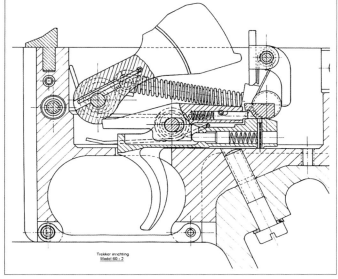

263. A diagram from the final A-I AR-10 handbook, dated June, 1961, shows the two types of hammer springs.
Left: the orginal torsion spring.
Right: the more positive coil spring and strut assembly, which provided a more powerful hammer strike.
Charles Kramer collection

Although the Italian "dual-hammer" design did reduce the rate of fire, it was more complicated, and did not accomplish the requested function of making the trigger more resilient to dirt and environmental contamination, and it was consequently not adopted.

To solve the problem of dirt in the action, the torsion hammer spring was replaced with a strong coil spring wrapped around a strut, much like the hammer spring assembly utilized in other military automatic rifles such as the FAL and the G3. This

strong coil spring provided constant positive pressure on the hammer, and unlike the torsion spring, standard on all the earlier models of the AR-10, the mechanism would still function even after becoming quite dirty, partly due to the raw power packed in the new coil spring design.

Designer L. James Sullivan later commented on this new design as follows:

*That . . trigger variation is quite clever. It reduces the original Dutch AR-10 trigger pull force two ways, first by applying the hammer spring force about 45° from the direction of motion, which reduces the force on the sear. Second, it adds the spring plunger, so only the resistance of that rela-*

262 (previous page). A set of signed drawings from the Fabbrica d'Armi Luigi Franchi of Brescia, Italy, illustrating the proposed Franchi "dual-hammer" rate-of-fire reducer. This was considered too complicated, and was not adopted. courtesy the late William B. Edwards

*tively light force must be overcome by the trigger return spring force.*

This would eventually become the standard spring configuration used in modern AR-15 and AR-10 rifles. It increased the reliability of the weapon, not only in nullifying the effects of dirt in the lower receiver, but also by providing a heavier, more positive hammer strike, which was more likely to activate excessively hard military cartridge primers, even when the mechanism was dirty. The author

can personally attest to the usefulness of this innovation, as once when he was in a competition the lower receiver of his modern-production AR-10 became packed with dust and gravel, but the fire control group continued to function, despite the audible and tactile grinding of dirt in the working parts.

Accuracy was also improved by giving the Portuguese/NATO AR-10 a smoother, lighter trigger pull, due to the more refined surfaces of the fire control group's working parts.

# Explaining the Component Interaction in Semi-Automatic Fire

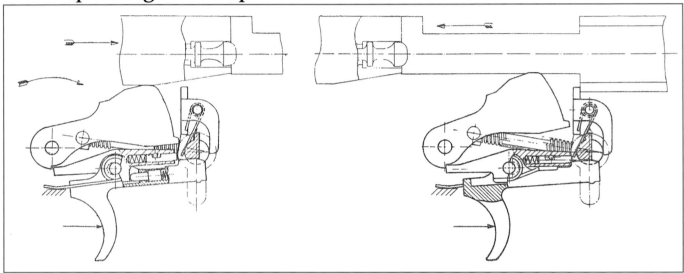

264. Cutaway drawings from the June, 1961 A-I handbook illustrate the action of the mechanism with the fire selector set for "SEMI."                                    editor's collection

An excerpt from the June, 1961 A-I AR-10 handbook explains the actions shown above as follows:

### Semi-automatic fire

*When the safety is in the "Semi" position, the automatic sear is kept out of the way of the hammer and does not partake in the functioning of the mechanism.*

*When the mechanism is pressed down by the recoiling bolt carrier, it is caught by the front tip of the sear and held down when the bolt carrier has returned to its foremost position [fig. 264a]. Owing to the construction of the spring loaded plunger in the rear tip of the sear, the trigger must be released before another round can be fired [fig. 264b].*

# Explaining the Component Interaction in Full-Automatic Fire

A further excerpt explains the actions shown in fig. 265 as follows:

### Automatic fire

*When the safety is in its "Auto" position, the automatic sear can turn backwards under the tension of its spring, which causes the lower lug of the automatic sear to get in the way of the hammer. At the same time the trigger can be pulled farther*

*backwards owing to a milled cut in the body of the safety just over the rear part of the trigger. When the trigger is now pulled, the front tip of the sear is kept fully free from the hammer [fig, 265a], whenever this part is pressed down by the bolt carrier after firing a cartridge. However the hammer is now caught by a lower lug of the automatic sear and kept in this position [fig. 265b] until the bolt*

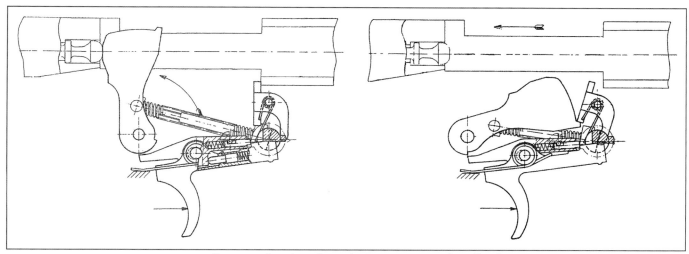

265. Cutaway drawings from the June, 1961 A-I handbook
illustrate the action of the mechanism with the fire selector
set for "AUTO."                                    editor's collection

*carrier had almost ended its forward travel, the bolt
then being in the locked position ready to be fired.*

*At this moment the lower rear edge of the bolt
carrier strikes the upper lug of the automatic sear,
causing it to turn forward and to free the hammer;
the hammer then again strikes the firing pin, firing
the cartridge in the chamber. This action continues
as long as the trigger is being pulled. Releasing the
trigger lifts the front tip of the sear and brings it in
the way of the hammer, keeping it in the cocked
position.*

## The Patented Jüngeling Trigger Group

A-I director Friedhelm Jüngeling would go on to file
for patent protection on his own design of fire control
group in the United States two years later, being
granted U.S. patent no. 3,167,877 on February 2,
1965.

While making no mention of the new coil hammer spring, the patent drawings retaining the torsion
spring of before, this redesign was particularly concerned with ensuring a positive return to a readiness
condition when the trigger is released. In the words
of the patent, ". . the drawback [of the previous
trigger mechanism] is encountered that when the
sear adheres to the projection on the trigger arm due
to dirt in the mechanism, the trigger does not disengage the sear so that the operation is disturbed."

266 (right). Figs. 1 through 3 from U.S. patent no.
3,167,877 titled "Trigger Mechanism for a Firearm",
awarded to F. G. Jüngeling on February 2, 1965.
U.S. Patent Office

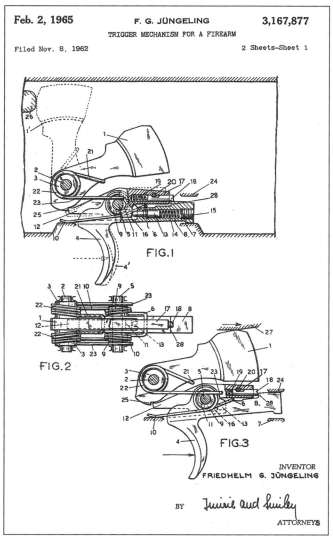

# New Bakelite Handguard Shells

267. Left side view of the improved handguard shells used on the Portuguese/NATO version of the AR-10, made from hard Bakelite, which has a characteristic marbled appearance. This variation has round ventilation holes spaced around the rear ends.

Most of these came in a light tan color (fig. 268), but due to the vagaries of the European plastics industry some were very dark, almost black, as shown here.

BATF collection

In coordination with Tom Tellefson back at ArmaLite, and relying on the greatly-increased capabilities of their plastics contractor Dynamit Nobel A.G., Artillerie-Inrichtingen devised an improved replacement set of outer handguard shells to fit over the standard aluminum liners. Completely eschewing the previous method of impregnating fiberglass cloth with graphite epoxy and molding it with resin, the new handguards were simply injection-molded Bakelite; one of the first types of hard plastic. It provided a better gripping surface than the finished walnut wood found on the Second Transitional variants, and was lighter, with better heat-insulation properties. The injection molding process used to construct the Bakelite handguard shells resulted in an attractive, mottled grain structure, not unlike marble.

In addition to versions of these shells being configured for rifles with and without bipods, three minor variations of Bakelite handguard shells may be seen in the illustrations: those with small round ventilation holes at the rear (fig. 267); those with longitudinal slots front and back (fig. 268); and a combination of both (fig. 258).

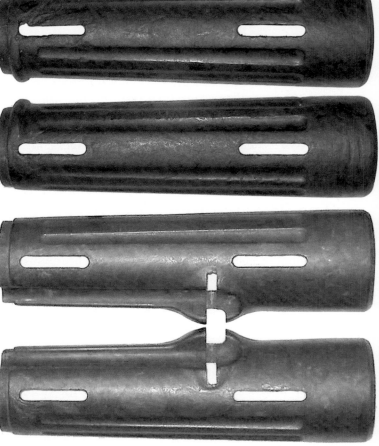

268. Two sets of left and right handguard shells, made from tan-colored Bakelite. Produced by Dynamit Nobel, these improved upon the wooden assemblies they replaced, providing lower weight, better heat insulation, and greater strength.

Above: for standard barrels.

Below: for barrels fitted with bipods.

Charles Kramer collection

## New Pistol Grip and Buttstock

The pistol grip was slightly modified by the elimination of the horizontal serrations that had been present on the rear surface of all earlier models. The Dutch military had complained that this aggressive texture on the backstrap had a tendency to give riflemen blisters in the webs of their hands.

The buttstock was also made of a newer material; a reinforced type of polycarbonate plastic, which provided a stronger and more rigid platform for firing rifle grenades. The new rubber buttplate pioneered on the First Transitional variant was also kept and incorporated onto this new, stronger buttstock.

## New, More Positive Gas Adjustment

269. Left side closeup of the front sight base and bayonet lug of a Portuguese AR-10, fitted with a unique experimental "coin-operated" gas adjustment valve.

Note the grenade launching sight bracket on the face of the front sight tower.      Artillerie-Inrichtingen collection

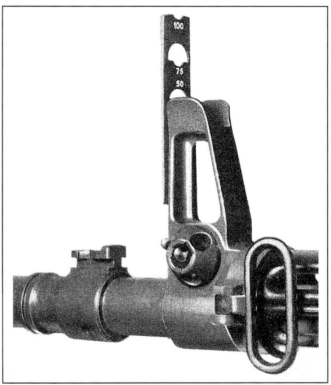

270. A left side view of the finalized front sight base as used on the Portuguese/NATO AR-10, showing the correctly-mounted grenade sight and the perfected, cartridge-actuated gas adjustment valve.

photo from the 1961 A-I handbook,
editor's collection

The Dutch military had complained in the first round of testing about the difficulty soldiers had in manipulating the gas adjustment screw, which since the outset of Dutch production had been accomplished by the use of a special two-pronged tool, latterly part of the tool kit in the handle of the Sudanese bayonet. On some of the Squad Support model variants experiments had been made with a much larger gas regulator dial, that could be manipulated by hand. This was not found to be the best solution, however, because it could be accidentally changed either by bumping the front sight base against a foreign object, or simply by the blast from firing. It was also determined that the front sight base, through which the hot propellant gases passed on their way back to the upper receiver to activate the bolt, would quickly

grow so hot as to scald the hand of a rifleman who attempted to adjust the gas setting on his rifle.

An ideal solution was realized in late 1959, when a large dial on the left side of the front sight base was introduced to replace the small, tool-actuated threaded screw in the front of the sight tower. The first type of this more easily-accessible and positive gas adjustment apparatus featured a large curved slot into which a coin, cartridge rim or knife blade could be inserted to act as a lever to turn the dial. In rifles that had been fired, cleaned, and generally "broken in," this process could be accomplished by simply inserting one's thumbnail into the slot and twisting.

The possibility of this part being too difficult, or too hot, to move, caused the Dutch and Portuguese to seek an alternative means of actuating the dial.

The eventual solution, which satisfied all parties, was a dial with indentations that allowed the gas regulator to be set using the point of a cartridge. Once this was rotated to the desired setting, strong detents locked the dial positively in place. This not only solved the problem of requiring a specialized tool, but essentially eliminated the possibility of the regulator losing its setting from the violent blast of firing; something to which the earlier gas adjustment screw was vulnerable.

## Front Sight Tower with Grenade Sight Bracket

The front sight tower was provided with a small ridged bracket along its forward edge, for use in mounting a detachable ladder sight that would allow the accurate aiming of an anti-tank or anti-personnel grenade out to a range of one hundred meters. This accessory grenade sight, of the type also issued with other rifles capable of launching these large explosive projectiles, is utilized by lining up the tip of the mounted grenade with the appropriate numbered range marking on the sight, and then launching the grenade by firing a special grenade blank cartridge.

271 (left). The rifleman's view of a Portuguese AR-10 with the grenade sight clipped onto its bracket, showing range markings for 50, 75 and 100 meters.
photo from the 1961 A-I handbook,
editor's collection

## Adding a "Carbine Carry" Capability

By request of the Portuguese, and following their own field testing, the A-I engineers elected to move the rear sling attachment point to the left side of the buttstock, thus allowing the weapon not only to be carried flat against the body when slung over the back, but essentially worn on the front with the sling area near the buttstock over the right shoulder, so that the weapon could be carried hands-free, but brought into action almost instantaneously.

## The "Automatic Fire Stop"

An armorer-installable accessory device that the Portuguese in particular requested for the rifles they were evaluating was marketed by A-I as the "automatic fire stop." This small apparatus could be screwed into the rear of the lower receiver's fire control group and tightened with a hex wrench to hold the automatic sear stationary, and block the selector switch from being moved beyond the semi-automatic position to the full-automatic mode.

The Portuguese had found that rapid, well-disciplined semi-automatic fire, in which every shot was aimed, was far more effective than bursts of fully-automatic fire. They reasoned that soldiers would run out of ammunition far too quickly if they sprayed rounds at every possible threat, and the particularly élite and well-trained men of their new airborne units found that they simply had no need for automatic fire most of the time. For tight-quarter infiltration missions, in which close-range rapid fire would be critical, the soldiers were allowed to remove this device themselves.

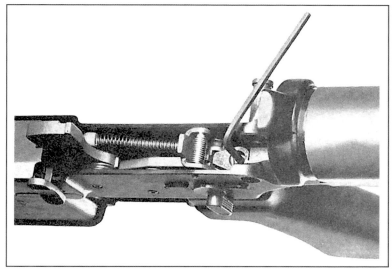

272. This photograph from the instruction manual developed with input from Portuguese Air Force Command demonstrated for soldiers the correct way to remove their automatic fire stops when ordered to do so for special missions in which more ammunition would be carried and automatic fire would be needed.

photo from the 1961 A-I handbook, editor's collection

## New Telescoping Cocking Handle with Positive Bolt Closure

A strange request from the Portuguese military, that would soon be raised by the U.S. Army with reference to the AR-15, was a capability to allow the rifleman to manually force the bolt forward. This was implicit in any weapon with a reciprocating charging handle, but since the AR-10's top-mounted cocking handle did not move back and forth with the bolt, it could only be used to pull the bolt carrier group back and released to charge the weapon, but could not push the carrier and bolt forward to force a round into the chamber.

This was a function against which Eugene Stoner would argue strongly when it was forced on the AR-15 by the U.S. Army a few years later; a position with which most firearms experts today would agree. Stoner's rationale was that if a cartridge did not fully chamber by the force of the mainspring alone, something was wrong, either with the rifle or the cartridge, and the best solution would be to perform the standard military "immediate action" drill, in which the offending cartridge is smartly ejected and an attempt is made to chamber a fresh round. Forcing the bolt forward on a reluctant-to-chamber round could result in a difficult-to-clear jam, or cause the weapon to appear ready to fire when the bolt was not locked completely, causing a possible cartridge burst and catastrophic failure. This is exactly what happened to General Somoza when he was testing the AR-10 in Nicaragua, and Stoner

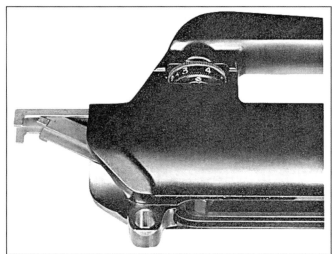

273. Right side closeup of a Portuguese AR-10 upper receiver showing the first experimental model of the two-part telescoping charging handle being removed. The one photographed here for the handbook did not actually have the forward assist function.

photo from the 1961 A-I handbook, editor's collection

thought it would be lunacy to give a shooter the means of possibly endangering himself and the weapon in this way.

Notwithstanding the concerns of the engineers, a new charging handle type was developed, which did bring some real benefits to the weapon, in addition to the perceived benefit of a "forward assist capability," by which the bolt can be forced forward using the handle.

The new cocking handle rail assembly was composed of two parts, the front portion of which telescoped over the rearmost part. The actual protruding handle now rested on a spring-loaded plunger, which when depressed took hold of the bolt carrier, and permitted the shooter to force it forward.

The other benefit of the two-piece charging handle was that it did not protrude from the rear of the weapon, and the back of the upper receiver could thus be closed off.

Removal of the new two-piece telescoping charging handle was a bit more involved than the simple act of pushing the apparatus down through its slot once the bolt carrier group had been removed, as was the case on the previous type. For the new piece, the handle was retracted until the inner telescoping component lined up with a notch in the upper receiver, through which it could be tilted down and removed, as shown in fig. 273. Once it was removed it no longer held the upper portion in, which

could be pulled straight down and removed for service and cleaning.

The closing of the cocking handle channel in the rear of the upper receiver eliminated the possibility of the ingress of dirt and contamination at that point, and also precluded the possibility of either excess gas from the operation of the action, or the blast of flaming powder that could result from the pierced primer of a poorly-made round, from reaching the face and eyes of the rifleman. It did add a very small amount of weight to the weapon, but the greatly-strengthened cocking latch with spring-loaded plunger made the charging handle on this final iteration of the Dutch AR-10 considerably sturdier.

The first few experimental models assembled with the closed receivers and telescoping charging handles actually did not feature the forward assist function, and it was this model that was photographed by Artillerie-Inrichtingen for use in the handbook issued with the final AR-10 variant. However, the actual rifles that were shipped to the purchasers of this model did have this capability.

## Another Late "Transitional Transitional"

274. Right side view of yet another evolutionary link between the Second Transitional model and the final, perfected Portuguese/NATO version of the AR-10 - truly a "transitional Transitional" - serial no. 006038.

While retaining a number of early characteristics, such as the early gas adjustment screw in the front of the front sight tower, no grenade sight bracket, and wood handguard shells (configured to accept the folded bipod legs), this was the first to boast a closed upper receiver and an early version of the two-piece, telescoping charging handle, without the spring-loaded plunger that could force the bolt forward.                    photo from editor's collection, courtesy Masami Tokoi

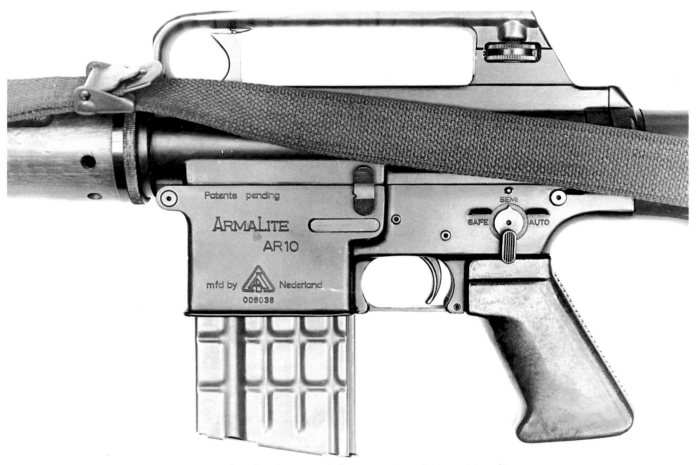

275. Left side closeup of the "transitional Transitional" AR-10, serial no. 006038.

The handguard shells are walnut, the markings are the Transitional style, the bolt holdopen is the stronger, later type, described below; the change lever markings are the later type with "SAFE" forward; while the pistol grip still has the ridges on its rear surface.

photo from editor's collection,
courtesy Masami Tokoi

## The Stronger Bolt Holdopen

A new bolt catch assembly was also incorporated in this model. There had been some complaints in different tests that the bolt catch did not hold the bolt open with complete certainty after the last round of a magazine had been fired. A much thicker and stronger part was accordingly designed and incorporated, and it did the job perfectly; ensuring that regardless of how dirty or hot it got, the bolt would lock back positively after the last shot, even when shooting different or substandard lots of ammunition which were not manufactured to correct specifications and/or provided lower pressures. It further imparted an almost infinite service life to this component.

## Chrome-Lining the Chamber

Finally, one of the most beneficial modifications to the AR-10's design, and certainly the most prophetic, was introduced. One of the complaints from Spring-field Armory was the tendency for cartridges left chambered in the rifle overnight or for longer periods to stick, particularly in adverse environments or when the weapon was dirty. Similar observations had been made by others who had tested the AR-10 in humid, sandy, and tropical environments. Working on their own initiative, the A-I engineers strived to find a way to prevent this issue, while generally making chambering and extraction more reliable. For the solution, they actually looked to an alteration that was made very late in the development of the U.S. M14, and lined the chamber with hard, slick, chrome plating.

## Adverse-Condition Problems Eliminated

When tested by the Portuguese, and then subsequently by the Dutch factory staff themselves, it was found that the chrome-plated chamber, combined with the reinforced bolt locking lugs and extractor, essentially eliminated performance issues connected with adverse conditions. After the Dutch trials in New Guinea were over, A-I subjected several examples of this newest version of the AR-10 to the same tests as described in the test report, and they actually came through them with no malfunctions whatsoever. These results were corroborated in similar evaluations performed by the Portuguese Air Force and Italian Navy, and with this all concerned knew that they had a winner.

## Chrome-Lining the Bore as Well in Later Examples

The chrome-lined chamber, introduced in late 1959 and early 1960 as part of the run of modifications described above, was intended to improve reliability and functioning in extreme environments where regular cleaning was not always a practical possibility.

The AR-10 barrel at this point was still made of the same hammer-forged alloy steel used since the start of Dutch production, with a plain, unplated bore. As production got underway on this variant, the Dutch continued to experiment with lining the entire bore of the barrel with chrome. This offered an increase in service life, it taking the friction of bullets and the heat of propellant gases longer to wear away hard chrome than steel; made the bore more resistant to rust and pitting; and made cleaning easier, as dirt, copper, and carbon fouling have a much harder time adhering to hard, slippery chrome than to plain steel, which was the reason the AR-10 bolts and bolt carriers already had chrome-plated surfaces. Some of the later examples of this model were manufactured with barrels that had their entire interior chrome-lined. This was also a characteristic common to individual barrels sent as replacement parts to the militaries who ordered this final version.

## Adding a Scope-Sighted Capability to All Rifles

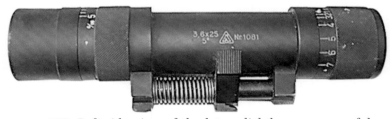

276. Left side view of the later, slightly more powerful 3.6-power DelftOptik scope sight, with redesigned spring-loaded base that required no modifications to the AR-10.
Institute of Miltary Technology

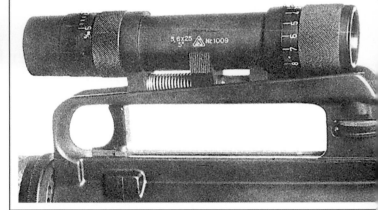

277. Fig. 13 from the June, 1961 A-I handbook, showing a left side closeup of a Portuguese/NATO AR-10 with 3.6x DelftOptik scope sight mounted.        editor's collection

The basic means of mounting a telescopic sight to a slot cut in the carrying handle was carried over into the new Portuguese/NATO variant, but with the exact mounting apparatus and the actual scope used changed somewhat. On earlier models, the scopes could only be attached to dedicated rifles on which the sides of the carrying handle had been shaved down. In conformance with the preference stated by the Dutch Army ahead of the adverse-condition trials that scopes and night vision equipment should fit all rifles, however, discussed beginning on page 226, the carrying handles were now left completely intact, with only a milling cut of very precise contours made within its interior channel. The new scope, cosmetically almost identical to the previous model but of slightly greater magnification (3.6-power vs. 3-power), had integral to its base a spring-loaded locking mechanism that was contoured to fit in the slot in the carrying handle, whereupon the spring would force a locking tab forward, the pressure keeping it in place so sturdily that it did not lose its zero even under heavy use.

# Summing Up the Portuguese AR-10

While the heaviest model of the rifle, this last entry in the history of Artillerie-Inrichtingen AR-10 production added to the extreme accuracy, ease of handling, modularity, corrosion resistance, simplicity of maintenance and rapid-fire controllability of the previous models the most rock-solid reliability, giving it a verifiable reputation of unsurpassed dependability in some of the harshest combat conditions on earth. All this was achieved while still keeping the AR-10 considerably lighter than its competitors, and further improving its already stellar performance in areas like accuracy and controllability.

## Selling the NATO Model AR-10 to Portugal and Italy

278. Left side view of a typical Portuguese/NATO AR-10, with blank firing adapter installed.

This was the final and most formidable model of the AR-10, embodying all the modifications and improvements discussed in this chapter.

Institute of Military Technology

The final and most formidable model of the AR-10, seen here with its BFA installed, is generally designated the "NATO" or "Portuguese" version, due to its sale to Portugal and Italy, both NATO member-states. The quantity sold to Portugal was considerably larger, and the extensive and highly successful use to which the Portuguese put their AR-10s, discussed in Chapter Eleven, made it not only more famous in their hands than those of the Italian Navy, but certainly the most well-known of all the AR-10 models.

Although being so similar in outward appearance to the Second Transitional variant as to cause common confusion between the two, the Portuguese edition featured numerous internal modifications that vastly improved its performance and reliability.

In addition to the previously-discussed new-model two-piece charging handle and gas adjustment valve, which could be actuated by the point of a round of ammunition, the NATO AR-10, weighing 8 lbs. 15 oz., was equipped with a bayonet lug on top of its barrel shroud, giving it a unique appearance when bayonets were fixed, usually for ceremonial purposes.

The new Bakelite handguard and fiberglass stock and pistol grip were sometimes painted green by the Italians, and when they would break or deteriorate through normal wear and tear in Portuguese service, were replaced with different types of makeshift wood and aluminum furniture, giving a varied and interesting character to many of these rifles.

As discussed in Chapter Eleven, along with the Sudanese AR-10 this was the only variant to see extensive combat use, and was certainly the model to see the harshest of that use. This was due partly to the fact that fewer of them were in Portuguese hands than the Sudanese possessed, and also because the Portuguese put them into action in extremely hard fighting against a well-armed and organized opponent in what is undoubtedly the most detrimental climate to the functioning and longevity of firearms.

Recognizing the awesome capabilities of the NATO version of the AR-10, the Italian COMSUBIN special force was finally convinced, and acquired a quantity of these rifles for field use. The small force made it their standard battle rifle and, judging by the sparse accounts that have filtered out of that secretive organization, it served them well.

As the factory data on the original sale made to the Italian government has not been made available for study by this author, the quantity involved must be estimated. Given the variant acquired by COMSUBIN, the sale had to take place in either 1960 or 1961, and was likely for a relatively small number, probably not exceeding 500 examples. Some of the rifles were fitted with the bipod, developed for both the Transitional and Squad Support models, which allowed the rifle to be rotated around its barrel axis for quick reloading of magazines from the prone position. Additionally, the new Bakelite handguards for these weapons were cut specifically to provide a place for the bipod legs to be stowed when not deployed.

The extra weight of this variant, combined with the very sturdy and modular bipod, allowed the Italians and Portuguese to make very effective use of their NATO-model AR-10s in the squad support role. It is not clear whether the Italians initially ordered any of the specialized scoped versions, either with the standard DelftOptik-produced telescopic sights or M3 infra-red night sighting systems, but it is likely that they at least purchased some conventional scoped models for their designated marksmen; a role particularly important in special operations.

*Chapter Ten*

# The Swan Song of the AR-10

## The Portuguese Model 961 Assault Rifle

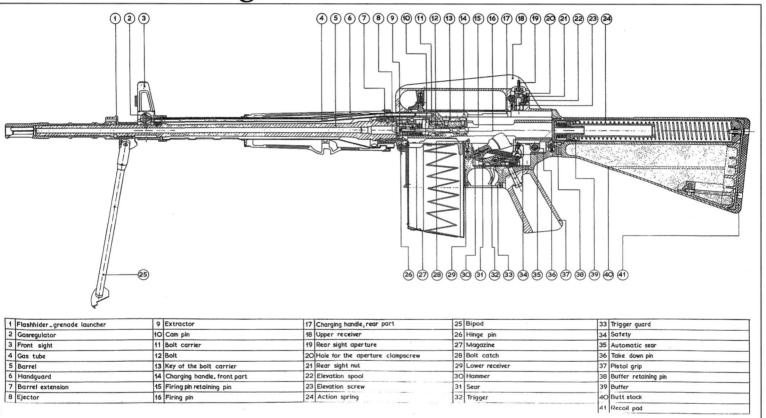

| 1 | Flashhider _ grenade launcher | 9 | Extractor | 17 | Charging handle, rear part | 25 | Bipod | 33 | Trigger guard |
|---|---|---|---|---|---|---|---|---|---|
| 2 | Gasregulator | 10 | Cam pin | 18 | Upper receiver | 26 | Hinge pin | 34 | Safety |
| 3 | Front sight | 11 | Bolt carrier | 19 | Rear sight aperture | 27 | Magazine | 35 | Automatic sear |
| 4 | Gas tube | 12 | Bolt | 20 | Hole for the aperture clampscrew | 28 | Bolt catch | 36 | Take down pin |
| 5 | Barrel | 13 | Key of the bolt carrier | 21 | Rear sight nut | 29 | Lower receiver | 37 | Pistol grip |
| 6 | Handguard | 14 | Charging handle, front part | 22 | Elevation spool | 30 | Hammer | 38 | Buffer retaining pin |
| 7 | Barrel extension | 15 | Firing pin retaining pin | 23 | Elevation screw | 31 | Sear | 39 | Buffer |
| 8 | Ejector | 16 | Firing pin | 24 | Action spring | 32 | Trigger | 40 | Butt stock |
| | | | | | | | | 41 | Recoil pad |

279. From the last A-I handbook, dated June, 1961 (fig. 280), a left side sectioned view of the Portuguese/NATO AR-10, with components numbered and labeled.

editor's collection

**A**lthough not the largest consignment of AR-10s manufactured during the period of A-I production, the sale of the NATO or Portuguese model to Portugal would become the one for which the Dutch production facility is best known today. Starting in 1960, the Portuguese Ministry of Defense accepted the AR-10 as standard equipment, designating it *Espingarda de Assalto* 7.62mm m961 AR-10 (Assault Rifle, 7.62mm Model 961, AR-10). While the

G3s and FALs that had been bought by the Portuguese and given similar designations around the same time were issued to units of the Army for comparison purposes, the AR-10 was adopted outright as the standard rifle of the Air Force's paratrooper regiment, the unit having subsequently grown from battalion size.

The units which were issued the AR-10 turned in glowing reports of its effectiveness and ease of use,

280. The blue plastic cover of the last A-I handbook for the AR-10, dated at Zaandam, June, 1961.  editor's collection

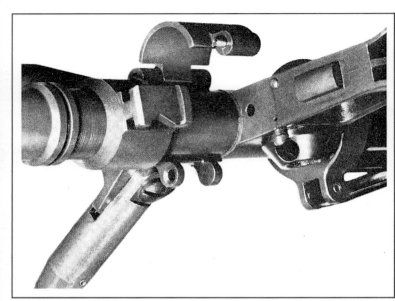

281. Fig. 9 from the last A-I AR-10 handbook, illustrating the correct way of attaching the bipod by pushing in the spring-loaded detent in the top portion of the hinge with the point of a cartridge before locking the hinge around the barrel.                              editor's collection

compared to other weapons they tested, and the government went on to purchase more consignments of AR-10s from A-I. Although a much larger order and eventual domestic production had been contemplated, it ended up being the case that only 1,556 rifles in total were procured by Portugal from the middle of 1960 through July 4, 1961; the fourth anniversary of the signing of the production license contract between Artillerie-Inrichtingen and Fairchild Arms International. While the Portuguese Ministry of Defense was in charge of the first purchases, the Department of the Air Force contracted directly with the Dutch for the later quantities.

Considering all of the modifications that were made to reach the rifle with which the Portuguese

were so pleased, the final model did cost slightly more than the prices paid by the Germans for their Cuban-model weapons, the Sudanese for their eponymous rifles, and the Dutch for their Transitional trials models. The Portuguese rifles ended up costing the Air Force $119 each. This price for each rifle included (at least at the time of the final July, 1961 order) a bayonet, bipod (if that model had the handguard cut for one), Dutch-type M1 rifle sling (the type of sling used for the later AR-10s, while the Sudanese and Guatemalan contract guns had been issued with their own proprietary slings), automatic fire stop, blank firing adapter, and eleven magazines. This magazine count was remarkably high, considering not only that the Portuguese government ordered other consignments of just spare parts and magazines, but that the Sudanese contract price for delivered rifles had included only five magazines each.

## Portuguese Sniper AR-10s

The relatively large-scale production of optical sights for the AR-10s, combined with very sizeable orders from the Dutch government for almost-identical scopes for their newly-standardized FN FAL rifles, meant that the unit costs were severely reduced, and the Portuguese ended up paying only $163 for the

entire kit, including scope and an adjustment tool for zeroing the scope.

The Portuguese sniper rifles, in addition to having their carrying handles milled out to accept the spring-loaded scope mount, also differed in two ways from the standard-issue weapons. First, barrels in-

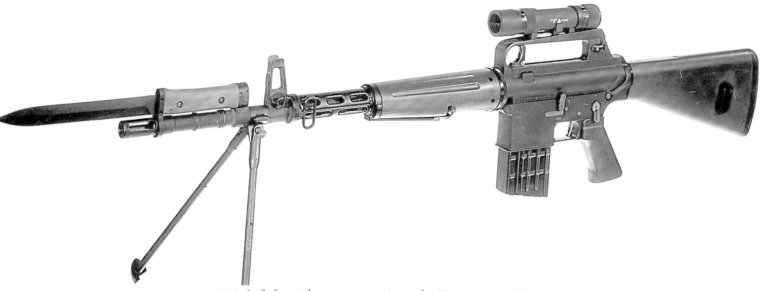

282. Left front three-quarter view of a Portuguese sniper AR-10, with bayonet and bipod fitted.
      The scope is the later 3.6-power DelftOptik unit, with the improved spring-loaded attachment system (fig. 276).
                                    courtesy Artillerie-Inrichtingen

tended for these rifles were given extra attention after having been specially selected for unusual dimensional uniformity and accuracy potential from lots of production barrels. Second, the lower receivers that were meant for assembly into sniper rifles were

manufactured without the fully-automatic fire capability. Just as on the AR-102 commercial model, the hole for the auto sear pin was never drilled, and the standard trigger group, minus this part, was installed.

## A Late Squad Support AR-10, with Rear Cocking Handle

283. Left side view of a late A-I magazine-fed squad automatic version of the AR-10 with detachable barrel, bipod, folding auxiliary front handgrip, forward-mounted carrying handle, and 3.6x scope sight.
      Note the rear-mounted cocking handle, accessible from the right side only.         Institute of Military Technology

During their field trials, the Portuguse even ordered a few of Artillerie-Inrichtingen's experimental belt-fed light machine gun models of the AR-10, and at least one scoped, magazine-fed version, which was fitted with the A-I version of the rear-mounted cock-

ing handle. Neither type was ever adopted, but it was the closest A-I ever came to selling these final heavy-barrel squad support AR-10 versions, which had been originally envisioned by ArmaLite back in 1957.

# Emulating the WWII German *Vampir* Infra-Red Night Sight

284. Right side view of a Portuguese AR-10 fitted with the active infra-red night sight system. The ponderous size and weight of these early units was augmented by the lead-acid battery pack which powered them.

Note that while the standard 3.6x scopes were by this time mountable on any rifle, the night sight system required a dedicated rifle with a modified carrying handle.
Institute of Military Technology

Unlike the AR-10s themselves, "active" night vision looks hopelessly antiquated compared to today's "passive" electronic and optical technology. First pioneered by the Germans during WWII, active infra-red uses an actual light source—essentially a large, heavy spotlight fitted with a bulb that gave off weak infra-red light, invisible to the naked eye, but easily picked up by very low-power infra-red-detection optics. This technology has been replaced by the starlight scopes of later generations and today, which magnify ambient light sources to allow soldiers to see in almost total darkness; or, for low-light visibility, passive infra-red sighting devices that rely on the infra-red radiation being emitted at different levels by objects, terrain features, animals, and people.

As with the vehicle- and weapon-mounted WWII German *Vampir* night sights, the rifleman so equipped for night patrols had to wear an extremely heavy lead-acid battery pack on his back, to which

were connected long cords, running through a specially-designed pistol grip up over the upper receiver to both the infra-red light source and the infra-red-detection optical sight. The unit used on the AR-10s, while descended from the German Vampire unit, was a very close copy of the American M3 infra-red sniper scope which had been mounted to special versions of the U.S. Carbine, designated the M3, during the Korean War.

The "active" infra-red sighting apparatus developed for the late AR-10 was ingenious, but still required some light modifications to the rifle. The night scope system could be attached to the carrying handle of any AR-10 that had been modified with special cuts machined into it to allow the mounting of a standard telescopic sight. However, to allow for the extra clearance required by the extremely bulky apparatus, the normally-high sides of the standard carrying handle were shaved down to make the top of the handle flat. The unit was bolted in place with a set screw that ran through a hole pre-drilled in the modified carrying handle and indexed with the locking bracket clamped around the sides of the top portion of the handle. This held the scope steady and aligned it with the rifle's barrel. This mounting method also allowed the sighting system to be removed from the rifle and reinstalled without the loss of the point of zero.

## Late Experiments I: a Folding Cocking Handle Extension

285. Right side view of a late Portuguese/NATO AR-10, with bipod folded.
Note the folding cocking handle extension, like that found on the late Finnish trials rifles of 1960 (fig. 256).
courtesy Artillerie-Inrichtingen

Even as the end drew near, the intrepid A-I staff did not stop their experiments, pushing forward with a plan to add further functionality to the charging handle on the Portuguese rifles.

A few late Portuguese AR-10s were fitted with a folding cocking handle extension, such as had appeared on the final Finnish trials model of 1960 (fig. 256). Little is known of how this feature was accepted, if indeed it was actually delivered on any of the rifles sent to the Portuguese or Italians, and it does not feature in the handbook.

## Late Experiments II: Pictographic Selector Markings

286. Left side view of a late Portuguese AR-10, serial no. 006646, with accessories.
    From left: bayonet; BFA; bipod; grenade sight; bayonet scabbard; magazine; auto fire stop (behind the pistol grip).

This rifle is fitted with the folding cocking handle extension and, as shown in the closeup view (fig. 287), has pictographic selector markings.
courtesy Artillerie-Inrichtingen

287. Left side closeup of the rifle shown in fig. 286, serial
no. 006646, showing the experimental folding cocking
handle extension and pictographic selector markings.
courtesy Artillerie-Inrichtingen

Along with the cocking handle extension, at least one late-production rifle was marked with pictographic dots to indicate the selector switch positions.

These last examples to come off the line also sported a contoured buttpad with a curve in it (fig. 286), designed to keep the rifle in position on the shooter's shoulder.

# The End of Dutch Production

## The Dutch Government Restricts Arms Shipments to Portugal

A leading voice among those opposed to the Portuguese policy of holding onto their overseas provinces came, ironically, from none other than Holland. The Dutch government could be said by observers at the time to be in no position whatsoever to criticize another nation's strong and sometimes-ruthless attempts to defend its overseas possessions against nationalist rebels. The government of Dutch Prime Minister Jan De Quay, however, had apparently never heard the old epithet about the inadvisability

of those who live in glass houses throwing stones. From a certain point of view, this position was understandable, as Holland had been forced by international pressure to relinquish its long-held possessions in the East Indies, and therefore the Dutch were not inclined to stand idly by and watch another European power attempt to hold onto its overseas possessions when they themselves had not been allowed to do so.

In a move that is hard to accept as a coincidence, considering the level of animosity that still existed between the Dutch Defense Department and Friedhelm Jüngeling, one of the first official expressions of this opposition to the Portuguese attempts to hold onto Angola was the imposition of restrictions on the export of Dutch arms to Portugal. Although it is entirely possible that this was just a natural first step to indicate which of the Portuguese actions were disapproved by the Dutch government, it just so happened that the only Dutch firm in the defense industry with commercial ties to the Portuguese military at the time was Artillerie-Inrichtingen.

Whether it was intentional or not (a fact that may never be proven to the satisfaction of historians), this decision by the De Quay administration had the practical effect of dealing a death blow to Artillerie-Inrichtingen's efforts to produce and market the AR-10. The particularly stinging irony of this development is that finally, after four years of the hardest work in the face of a litany of disappointments and misfortunes, A-I finally had a very robust, perfected product; the capacity for its efficient, high-quality, large-scale production; and a willing customer with large-volume needs for rifles, who had evaluated several designs and chosen theirs. Reportedly so impressed were the Portuguese with their AR-10s, in comparison with the test lots of G3s and FALs they had purchased, that they were not only intending to equip all four battalions of their airborne infantry with it, but gradually to standardize it as the official rifle of all their armed forces.

Thus the Dutch government's restrictions on arms shipments to Portugal curtailed both AR-10 production at A-I and the Portuguese aspirations to equip their forces with a rifle that was superior to all others in the world, and resulted in only two battalions being outfitted with AR-10s. The Portuguese were still able for the next few months to acquire accessories for their rifles, buying up all that they could in the way of spare parts, in order to keep their rifles running for as long as possible, so that their airborne infantry could keep their edge over any potential opponent.

## Makeshift "Jungle" Furniture

When their sparse stock of spare parts finally wore out, the Portuguese took to producing their own makeshift replacement parts and accessories. As a result, a great diversity of replacement furniture came to adorn the rifles in Portuguese service, from fine walnut to crude, painted aluminum.

The wood forearms produced by Portugal as replacement parts generally resembled much more closely the earlier types of handguards issued with the Cuban through the Guatemalan model rifles, in that they were full-length. They were, however, made in two halves, and as such could be removed just as the original-equipment Bakelite shells and aluminum liners could be by backing off the threaded clamping ring around the barrel nut. The aluminum handguards, by comparison, only replaced the outer wood or Bakelite shells of the forearm, leaving the inner aluminum liner in place. The aluminum was often painted brown for camouflage, but on occasion was left bare.

The wood furniture was generally preferred, due to its lower weight and heat insulation properties, while the aluminum pieces would heat up quickly when exposed to strong sunlight or sustained firing. The metal furniture was preferred for some specialized tasks, however, as paratroopers on special missions against enemy fortifications, particularly in Mozambique, found that only the aluminum

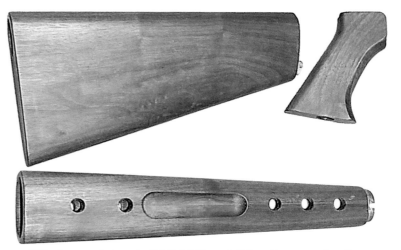

288. A complete set of AR-10 "jungle" furniture, made of walnut.

The handguard is the earlier full-length type, but split vertically into two halves for ease of disassembly.

Marc Miller collection

stocks held up to the recoil of sustained use of rifle grenades. This capability of the AR-10 was greatly appreciated for destroying bunkers and sandbagged machine gun positions.

As shown in fig. 291, two variations of pistol grips were seen in the retrofitted Portuguese aluminum "jungle" furniture, as it was called. The first was

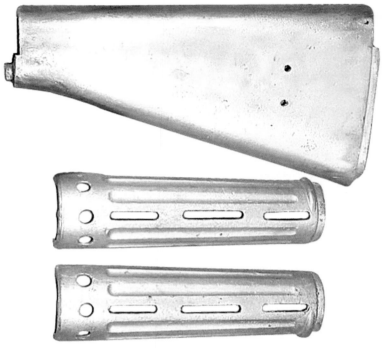

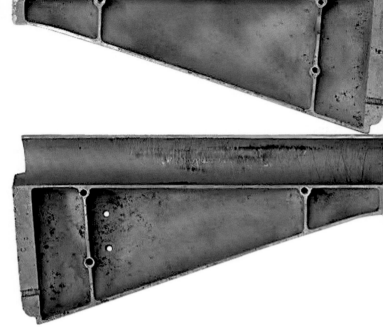

289. A set of replacement AR10 furniture (buttstock and handguard shells), made of aluminum.

Marc Miller collection

290. The makeshift aluminum buttstock fabricated by the Portuguese to aid in the launching of rifle grenades was composed of two hollow castings, which were bolted together.                              Charles Kramer collection

smooth and replicated exactly the contours of the original equipment, while another had crude grid lines cut into it to provide a more aggressive grip texture.

291 (left). A pair of aluminum replacement pistol grips, one painted and coarsely checkered (left) and the other left plain, and showing vestiges of paint.

Charles Kramer collection

## Sauve Qui Peut

The imminent enactment of restrictions on the arms trade coincided with Defense Minister Visser's swooping in for the kill with the announcement on May 25, 1961 that A-I would no longer be authorized to make expenditures on the "ArmaLite Rifle". This prodded the factory to scramble to deliver as many completed units as they could to their Portuguese customers. The plan for the initial orders made by the Portuguese government had been to first acquire a quantity sufficient to equip all of their airborne units, in addition to another élite special forces unit recommissioned in 1961 to combat the Angolan insurgency, with enough spare weapons and parts to keep the rifles running indefinitely, and then to formulate plans for wider-scale purchases to equip the regular forces of the Army. In accord with this somewhat-longer timetable (slated to take at least until the end of 1961), the factory had been producing parts to meet a larger order, meaning that when the news of the imminent policy shift reached them, the parts that they had finished were largely somewhat

292. Left side closeup of a Portuguese AR-10 no. 007026.
Compare with fig, 287: the cocking handle has the built-in bolt closure device but no horizontal extension, while the selector markings are the standard type.
BATF collection

mismatched, consisting of incomplete parts sets for different types of rifles.

At this point the emphasis became one of simply assembling as many complete rifles as possible, using the parts and components already on hand, and shipping them out. In the hodgepodge of parts, for example, there were a considerably larger number of semi-automatic lower receivers and specially-tuned barrels intended for sniper rifles than there were either scopes or upper receivers with the requisite

milling cuts in the carrying handles for scope mounting. Therefore, a number of rifles (between ten and one hundred) were assembled in the standard configuration but with particularly accurate barrels, and/or the lack of a full-auto fire setting. Although most of these rifles produced from mixed parts were in either one or the other of these nonstandard variations, reports show that at least some displayed both.

## The Italian COMSUBIN Scrambles for the Leftovers

Despite the near-superhuman efforts of the A-I staff to get the maximum number of rifles possible assembled and off to Portugal before the politically-motivated sanctions went into effect, the final *bon voyage*

of 300 NATO-model AR-10s for Lisbon on July 4, 1961 left behind the unfinished skeletons of still more rifles. Although they had been prohibited from pursuing further development or production of the

293. A copy of the invoice dated January 19, 1962 covering the final order from the Italian COMSUBIN for 60 leftover AR-10s plus accessories, including 6 3-power scopes, totaling US$10,446.10.        courtesy Artillerie-Inrichtingen

AR-10, the May 25 order had said nothing about *assembling* rifles from existing parts.

Needless to say, in addition to the jilted Portuguese, who had intended on standardizing the AR-10 for all their armed forces, other militaries still salivated for more ArmaLites. The Italian Navy's élite COMSUBIN unit, being the only other recipient of the perfected final model, was particularly delighted with their purchase, and longed to receive as many more examples as they possibly could. Having been bitterly disappointed when further production had been disallowed, the Italians were very happy to learn that there was still the possibility of receiving a few more AR-10s, which would go a long way in such a small unit.

After settling the details with Rome, A-I set about assembling every last rifle they could from what was left over of their inventory of in-house- and subcontractor-produced components. They were able to turn out 60 units, which were shipped on January 19, 1962; nearly a full year after further production had been banned.

Six of the remaining upper receivers used to build these "consolation prize" AR-10s were of the type cut for scope mounting, and so the Italians requested telescopic sights to go along with them. Demonstrating how hard the bottom of the barrel was being scraped on behalf of COMSUBIN, A-I reported that none of the later 3.6-power scopes were left in inventory, and they were not permitted to order more from DelftOptik. That company was by this time hard at work churning out scopes for the Dutch FAL, which happened to be almost a carbon copy of the model they had originally developed for the AR-10.

However, a small quantity of the earlier model 3-power scopes, as originally sold along with the Cuban model AR-10 (with an additional 25 being ordered by the Sudanese), were still available, brand new in their boxes. There were also several of the newer spring-loaded apparati for mounting to the Portuguese carrying handle, and the Italians enthusiastically agreed that the 3x25 scopes were just fine for their purposes. Having the same exterior dimensions as the later scopes, these old models fit the mounting units perfectly, and six of these assemblies were made up and shipped along with the rest of the order.

Thus the Italian Navy received their sixty extra rifles, along with a full complement of accessories such as blank firing adapters, bipods, bayonets, grenade sights, automatic fire stops, and 600 spare magazines. As shown in fig. 293, the total bill for the shipment came to just over US$10,000.

Taking delivery of these treasures early in 1962, COMSUBIN gained the honor, no doubt much appreciated by them at the time, of being the very last customer to receive AR-10s from Artillerie-Inrichtingen.

# Summing Up - the Best of the Best

The standard of accuracy displayed by the NATO model AR-10 was not only peerless among military rifles of its day (and would still be considered so today), but was the best of all the AR-10s, meaning that when used with iron sights, necessitated by the shortage of scopes and upper receivers cut for them, the soldiers who were issued rifles with these higher-grade barrels usually never noticed any practical difference. The lack of a full-auto fire capability did catch the attention of the members of the unit which received standard rifles built on sniper-type lower receivers, but most of the fighting in that theater of operations was done at longer ranges anyway, making semi-automatic fire more effective.

With the last and most promising markets for their perfected rifles now made politically unavailable to Artillerie-Inrichtingen, it was time to close up

shop. On July 6, 1961, two days after sending their last order of 300 rifles and accompanying accessories to the Portuguese Air Force, the board of directors of the state-owned arms factory finally made the decision to close down all further research, development, and production of the AR-10. They then, on the same day, sent a telegram to Fairchild Arms International, reporting that they would not be exercising their option to extend the production licensing contract for a further four years.

Thus ended the promising development program of the greatest and most under-appreciated battle rifle ever devised. All the years of late nights, innovation, industriousness, and efficiency, surpassing the quality of work of all other arms institutions, had been brought to an end by political enmity and international events beyond their control.

## A Roundup of Dutch-Made AR-10 Magazines

The magazines issued with the Portuguese and Italian orders were slightly different in appearance from the earlier type (page 172). They were finished now in a dark black coating, making them less visible than the previous gray (and occasionally green) anodized

aluminum magazines sold to the Cubans, Sudanese, Guatemalans, Indonesians, and Burmese. Artillerie-Inrichtingen had been experimenting with a new type of floorplate, and in their final shipments the Portuguese received rifles with magazines fitted with

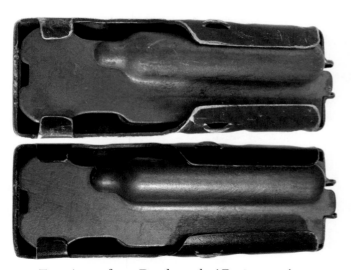

294. Top views of two Dutch-made AR-10 magazines.

Left: original type with grey cast follower as shipped up through the Sudanese and Guatemalan orders.

Right: later Portuguese type with stamped steel follower, as produced first for the Dutch field tests, and then in larger quantities for the Portuguese contract. Differing from the light gray coating of the earlier magazines, these were a much darker color, which not only was less likely to catch the eye and give a hidden infantryman away, but which also matched the color of the AR-10's receiver finish. The follower was given the same dark coating as the magazine's body.

Note the semi-circular "smiles" or stop tabs cut into each of the feed lips, designed to prevent over-insertion of the magazine into the receiver when the bolt assembly was locked in the open position.

Compare with fig. 151a: these stop tabs were introduced during U.S. production, and on the early U.S.-made magazines they were pressed in, rounded, and machined flat, whereas in Dutch production they are left as a simple half-shear.                                        author's collection

this improved floorplate (bottom, fig. 295). It sported two notches at the front which abutted against the inside front of the magazine body, so the floorplate had first to be pried up in order to remove it.

Matching the new exterior color, the traditional cast followers in these new series of magazines were replaced with stamped followers which were also much darker than their predecessors.

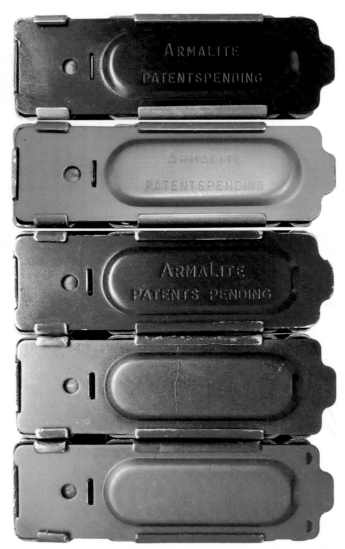

295. Bottom views of five AR-10 magazines, showing the chronological progression of different floorplate types incorporated into the Dutch-produced magazines.

Top: a prime example of the misprint on the first individual production series, with the "S" between "PATENT" and "PENDING" being added after the mistake was realized.                                        author's collection

Second from top: another misprinted floorplate anodized red, fitted to a magazine for use when firing blanks. The cast follower is also dyed red.   Eric Kincel collection

Center: by far the most common, still bearing the ArmaLite mark. This type was standard through almost all of the magazine production at Nederlandsche Metaalfabrik, and was shipped with the Sudanese order right through to the Portuguese.                                        author's collection

Second from bottom: early, unmarked production from when the Dutch began producing additional magazines without notifying ArmaLite, leaving the floorplates conspicuously blank.                                        author's collection

Bottom: final Dutch production, also left blank, with two tabs notched out at the front of the floorplate. These tabs were indented so that they stuck up into the magazine, locking the floorplate against the body.

author's collection

# An Interarms Retrospective

## Introducing Michael Parker, Interarms' "Unofficial Archivist"

Retired Virginia attorney Michael Parker has kindly supplied the following interesting material from the files and archives of Interarms. Mr. Parker describes his long-standing connection with the firm as follows:

*. . I was vice president and counsel at Interarms from 1980 to 1996, and a consultant to the firm both before and after that. When Interarms folded [in 1998] I salvaged many of the records, and am now its Unofficial Archivist.*

The legal firm name was always "International Armament Corporation" (IAC), initially shortened to "Interarmco", which was also the firm's cable address. This was further shortened to "Interarms" in 1967 after the Armco Steel Corporation, a completely separate entity, had objected to the similarity in names and threatened suit. Nevertheless Cummings himself favored the name "Interarmco", and the firm was best known by that name.

Fortunately for us, unlike SIDEM International and especially Cooper-MacDonald, whose dealings were kept close to the vest and are little known today, Interarms kept excellent records, many of which still survive. These include a great deal of the paperwork associated with Cummings's role as by far the most successful of the three star salesmen for the AR-10, and these documents will be explored here, with a special note of gratitude to Michael Parker.

296 (right). A compilation of various Interarmco office and warehouse addresses as used by Sam Cummings over the years to further his extensive business dealings.

Michael Parker collection

**INTERARMCO**
P. O. BOX 3722
WASHINGTON 7, D.C. U.S.A.
CABLE "INTERARMCO"
U. S. Continental Division Offices and Headquarters,
U. S. Warehouse Group II, 10 Prince Street,
Alexandria, Virginia. U.S.A.

**INTERARMCO**
LIMITED
10 PRINCE STREET
ALEXANDRIA, VIRGINIA

**INTERARMCO (CANADA) LTD**
WASHINGTON      LONDON      GENEVA

INTERARMCO (Canada) LIMITED
16, rue Crespin
tél. 35 48 35
cable : Interarmco
GENEVA, SWITZERLAND

INTERARMCO A/S
Østerbrogade 9 D
tel: Tria 369
cable: Interarmco
COPENHAGEN, DENMARK

**INTERARMCO**
**Industrial Group**
**"Le Bermuda" Bloc C**
**29 Avenue Hector Otto**
**Monte Carlo, Monaco**
Telephone: 30-58-60
Cable: "INTERARMCO" Monaco

# Notes on the Interarms Empire in Alexandria, Virginia

297. Sam Cummings himself, holding an early ArmaLite AR-18, takes over vice-president Dick Winter's office-cum-conference room at Interarms' U.S. headquarters, located within "Building No. 1" at 10 Prince St. in Old Town Alexandria, for a photo shoot.

In front of the desk are a German G1 FAL, an FN MAG on its bipod, an M16 rifle, and the *pièce de résistance*, a U.S. two-pounder (1.65") Hotchkiss Mountain Gun, serial no. 363. Undated, this was manufactured by the American Ordnance Company, which had offices in Washington, D.C. and shops in Bridgeport, Connecticut. The steel-rimmed wheels are marked on the hubs "Archibald Wheel Co., Lawrence, Massachusetts." This rig was designed to be dismantled and packed on four mules.

photo credit *The New York Times*,
Michael Parker collection

In addition to the huge, multi-storied warehouse in Manchester, England, and offices in various locales around the world, at one time or another Interarms owned or maintained a total of eleven warehouses on or near the Potomac River waterfront in Old Town Alexandria, Virginia.

Building no. 1, located at 10 Prince St., served as the firm's headquarters. In the earliest years, circa 1958, part of the first floor was used as a warehouse. This area also had a door fronting on S. Union St., and this address was sometimes used back then to create the illusion of multiple locations, although they were both in the same building.

Building no. 2, at 2 Prince St., was used mostly as a dealer showroom for buyers of surplus.

Building no. 3, at 1010 Duke St., was a bonded warehouse until the mid-1980s. In the early 1990s it was converted into a three-level parking garage.

Buildings nos. 4 and 5, siamesed with passageways, were located at 204 and 206 S. Union St. No. 4 was used for storage, and later housed the gunsmiths' shops, a test range, and the shipping and receiving departments.

298. A view of one floor of Interarms' warehouse no. 5, located at 206 S. Union St. in Alexandria, showing some stocks of surplus ammunition awaiting purchasers.

The stack in the right foreground contains banded crates of Canadian-made 7.92mm ammunition. The card label reads "Mauser ammo cal. 7.92mm (8mm) - 1,144 rounds per case."                    Michael Parker collection

Buildings nos. 6 and 7 were at the south end of the same block, another pair of quite large, siamesed warehouses. The rear half of no. 6 was fenced off as a bonded area, and until the early 1970s no. 7 contained a large Manurhin machine which was used for reloading military ammunition with soft-point bullets.

Buildings no. 8 and 9, at the northwest corner of Prince and S. Union Sts., kitty-corner from 10 Prince St., were temporarily leased to store the large shipments of Spanish and Yugoslav surplus in the 1960s.

Buildings 10 and 11 were large corrugated steel sheds located on the corner of S. Union and Franklin Sts., used for storage only. Eventually they were leased to a boatyard, but Interarms continued to occupy part of them for dead storage until the end. They were torn down about 10 years ago, and townhouses now occupy the site.

All of the above were leased by International Armament Corporation from Cummings Investement Associates, which owned the real estate (except for buildings 8 and 9). They have all since been sold for redevelopment.

# "Hunter's Lodge", and "Ye Olde Hunter"

"Hunter's Lodge" was a corporate shell owned by Interarms to handle mail order sales. "Ye Olde Hunter" was a cartoon character used in its ads, but then this name was adopted as the name or dba of a corporation owned by Clarence Robinson, who owned Robinson Terminal next door to the Interarms headquarters at 10 Prince St., where in the early days incoming cargo was offloaded from ocean-going ships for transfer to one or other of the Interarms warehouses. Later, when shipping was containerized and handled out of the much larger port of Baltimore, further up the Potomac, Robinson Terminal was acquired by John Richards, who ran a gun store (a former crab house restaurant) over the water. The last shipment handled through the Alexandria terminal was the large consignment of surplus arms from Spain.

In the earliest years there was a cozy intermingling of these businesses, with surplus being shunted to Ye Olde Hunter on consignment, etc., but that soon ended. Everybody still thinks Ye Olde Hunter was Interarms' "retail outlet"—which impression neither business made any effort to correct—but IAC had no financial interest in it.

299. All that remains of "Ye Olde Hunter", a former crab restaurant on the Alexandria waterfront.

Michael Parker collection

# Interarms Sales of ArmaLite and A-I AR-10s, 1957 - 1996

Much of the following information comes from the NFA Log Book maintained by Alois Hermann, a former WWII *Zollmeister* (paymaster) in the *Wehrmacht* who had served on the Eastern Front, been captured by the Russians, and did not make it back to Germany until the end of the 1940s.

After emigrating to the United States, Herr Hermann worked for many years as Interarms' Import Manager, where he was well appreciated by all for the meticulous records he kept of every transaction which required NFA documentation.

Herr Hermann's log book entries are augmented by copies of relevant BATF documents; mainly Form 2 (recording the receipt and registration of guns imported by Interarmco), and Form 3 (record of tax-free transfers between NFA licensees), all provided by Michael Parker.

### ArmaLite AR-10s

| Serial No. | Sold To | Date |
|---|---|---|
| 1001 | Interarms (Samuel Cummings's personal demonstrator) | April 19, 1957 |
| | sold to Knight's Armament | June 19, 1996 |
| 1029 | ArmaLite Division | March 19, 1958 |
| 1030 | shipped to Brig. Gen. Anastasio Somoza, Jefe Director, G. N. de Nicaragua, Managua, Nicaragua | September 20, 1957 |
| | returned "to be altered and perfected" by ArmaLite Division | November 26, 1957 |
| 1034 | deactivated and donated to the NRA Museum by Fairchild or by Richard Boutelle personally | |
| 1035 | shipped to Brig. Gen. Anastasio Somoza, Jefe Director, G. N. de Nicaragua, Managua, Nicaragua | October 4, 1957 |
| | returned "to be altered and perfected" by ArmaLite Division | November 26, 1957 |

| Serial No. | Sold To | Date |
|---|---|---|
| 1038 | Daniel Rewers* | December 2, 1970 |
| 1039 | from Cummings collection, sold to Knight's Armament | June 19, 1996 |
| 1040 | Daniel Rewers | November 20, 1970 |
| 1041 | Dominican Republic | December 17, 1957 |
| 1042 | Dominican Republic | December 17, 1957 |
| | returned - in bond - sold to Knight's Armament | June 19, 1996 |

## Artillerie-Inrichtingen AR-10s

| Serial No. | Sold To | Date |
|---|---|---|
| 000060 | Carl Schlieper, Solingen, Germany | October 31, 1958 |
| 000064 | Chile | June 20, 1958 |
| 000067 | Cmdr. Robt. S. Curtis for Office of Secretary of Defense, Pentagon | December 16, 1958 |
| | subsequently sold to Lorne David Matheson | June 14, 1962 |
| 000071 | Ministerio De Defensa Nacional, San Salvador, El Salvador | July 10, 1958 |
| 000089 | President of the Republic of Guatemala, Casa Presidencial, Guatemala, C.A. | July 10, 1958 |
| 000092 | O Departamento de Producao E Obras, Ministerio de Guerra, Rio de Janeiro, Brazil | June 11, 1958 |
| | returned - shipped to Artillerie-Inrichtingen for technical examination and not to be reimported | March 11, 1959 |
| 000097 | Carl Schlieper, Solingen, Germany | October 31, 1958 |
| 000122 | Nederlandsche Wapen En Munitiefabriek, 's Hertogenbosch, Holland | November 17, 1958 |
| 000124 | Colonel Carlos Arosemena, Guardia Nacional, Panama, R. P. | August 29, 1958 |
| 000126 | Colonel Carlos Arosemena, Guardia Nacional, Panama, R. P. | August 29, 1956 |
| 000130 | Ministerio de Guerra, Estado Mayor General de Ejercito, Buenos Aires, Argentina | June 26, 1958 |
| 000134 | Henry D. Lopez-Penha, Coronel, E. N. Dirección General Servicios Tecnologicos de la Secretaria de Estado de las Fuerzas Armadas, Ciudad Trujillo, Dominican Republic | June 30, 1958 |
| 000135 | Nederlandsche Wapen En Munitiefabriek, 's Hertogenbosch, Holland | November 17, 1958 |
| 000137 | O Departamento de Producao E Obras, Ministerio de Guerra, Rio de Janeiro, Brazil | June 11, 1958 |
| | returned - shipped to Artillerie-Inrichtingen for technical examination and not to be reimported | March 11, 1959 |
| 000140 | Cogswell & Harrison, Acton, W 3, London, England | January 20, 1959 |
| 000141 | from Cummings collection, sold to Knight's Armament | June 19, 1996 |
| 000143 | Cogswell & Harrison, Acton, W 3, London, England | January 20, 1959 |
| 000145 | taken by Samuel Cummings for demonstration purposes on a tour of South American countries | June 27, 1958 |
| | subsequently sold to Service Armament Corporation, Bogota, N.J. | July 23, 1958 |
| 000146 | Chilean Air Force, Santiago, Chile | June 20, 1958 |
| 000156 | Cogswell & Harrison, Acton, W 3, London, England | January 20, 1959 |
| 000158 | Société de Vente Hispano Suiza, S.A., Geneva, Switzerland | August 29, 1958 |
| 000165 | Major G. C. Brownlee, O.N.C., Ordnance Depot, Accra, Ghana | August 29, 1958 |
| 000173 | Gabriel Tudela, Lima, Peru | July 7, 1958 |
| 000180 | Major G. C. Brownlee, O.N.C., Ordnance Depot, Accra, Ghana | August 29, 1958 |
| 000186 | Cogswell & Harrison, Acton, W 3, London, England | January 20, 1959 |
| 000190 | Erma Werke, Munchendachau, West Germany | January 9, 1959 |
| 000193 | Erma Werke, Munchendachau, West Germany | January 9, 1959 |
| 000195 | Carl Schlieper, Solingen, Germany | June 11, 1958 |
| 000203 | Ministerio de la Defensa, Servicio de Armamentos, Caracas, Venezuela | August 7, 1958 |
| 000214 | Service Armament Corporation, Bogota, N.J. | January 6, 1961 |

SPRINGFIELD ARMORY- ORDNANCE CORPS

**Neg:** 19-058-20/ORD-61  **Date:** 10 Jan 61  **Proj:**

RIFLE, ASSAULT, AR10, 7.62-MM, ARMALITE

Donated to Springfield Armory

Right Side View

300. Right side view of A-I Second Transitional AR-10 serial no. 003769, which as noted in the list below was donated to Springfield Armory on December 6, 1960.

Note the top of the carrying handle, modified to accept the mount of an infra-red night sight (fig. 284).

courtesy the late Edward C. Ezell

| Serial No. | Sold To | Date |
|---|---|---|
| 000228 | Alexander Holst, Christiana Portland Cement Company, Oslo, Norway | April 22, 1959 |
| 000239 | R. F. Keys, Industrial Engineering Division, Frankford Arsenal, Philadelphia, Pennsylvania - for testing and demonstration purposes only | November 13, 1959 |
| | returned - shipped to Nederlandsche Wapen En Munitiefabriek, 's Hertogenbosch, Holland | January 5, 1960 |
| 000246 | Dansk Industri Syndikat, Copenhagen, Frihavnen, Denmark | March 28, 1960 |
| 000278 | Argentina | June 29, 1959 |
| 000280 | Argentina | June 29, 1959 |
| 000303 | Dominican Republic | May 8, 1959 |
| 000323 | Ministerio de la Defensa, Servicios de Armamentos, Caracas, Venezuela | August 7, 1958 |
| 000326 | Dominican Republic | May 8, 1959 |
| 000666 | Albion Arms, Peterborough, Ontario, Canada | April 24, 1959 |
| 000776 | Albion Arms, Peterborough, Ontario, Canada | April 24, 1959 |
| 001271 | Fairchild | May 4, 1960 |
| 001364 | Fairchild | May 4, 1960 |
| 002847 | Browne Industria E Comercio Limited, Rio de Janeiro, Brazil | November 17, 1959 |
| 002871 | Browne Industria E Comercio Limited, Rio de Janeiro, Brazil | November 17, 1959 |
| 003051 | ArmaLite, Inc., Costa Mesa, California | October 26, 1962 |
| 003085 | (scoped) American Firearms and Ammunition Corp., Long Island City, N.Y. | December 2, 1960 |
| 003331 | (scoped) Service Armament Corporation (scoped by D. E. Ott) | January 6, 1961 |
| 003334 | (scoped) Jugoimport, Beograd, Jugoslavia | February 21, 1961 |
| 003414 | (scoped) American Firearms and Ammunition Corp., Long Island City, N.Y. | December 2, 1960 |
| 003698 | American Firearms and Ammunition Corp., Long Island City, N.Y. | December 2, 1960 |
| 003769 | (with infra-red mount) donated to Springfield Armory Museum | December 6, 1960 |
| 003892 | (carbine) Daniel Rewers | November 20, 1970 |
| 003963 | (with infra-red mount) from Cummings collection | June 19, 1961 |
| | sold to Knight's Armament | June 19, 1996 |

| Serial No. | Sold To | Date |
|---|---|---|
| 004102 | (carbine) Daniel Rewers | November 20, 1970 |
| 004197 | Nihon Chuo Sangyo, K.K., Nihonbashi, Tokyo, Japan | November 7, 1961 |
| 004212 | Paul N. Van Hee | June 8, 1961 |
| 004281 | American Firearms and Ammunition Corp., Long Island City, N.Y. (for testing by CIA) | November 21, 1961 |
| 004326 | R. B. Alexander | August 15, 1961 |
| 004389 | Charlie's Gun Shop | May 1, 1961 |
| 004394 | His Excellency Sheikh Mubarak Abdulla Al-Jabir Al-Sabah, Deputy Commander in Chief of the Kuwait Army, Kuwait, Arabia | November 16, 1961 |
| 004452 | G. G. Simmons, Jr. | October 23, 1961 |
| 004526 | David E. Cumberland, Bangkok, Thailand | April 3, 1961 |
| 004530 | Potomac Arms | August 16, 1961 |
| 004540 | Aldrich Museum | August 30, 1961 |
| 004753 | (LMG with heavy-duty barrel) from Cummings collection, sold to Knight's Armament | June 19, 1996 |
| 004979 | (with heavy-duty barrel) Daniel Rewers | November 20, 1970 |
| 006722 | (with infra-red mount) from Cummings collection, sold to Knight's Armament | June 19, 1996 |
| 006966 | (scoped) David E. Kern Ott | May 8, 1969 |
| 006967 | (scoped) Daniel Rewers | November 20, 1970 |
| 006968 | (scoped) from Cummings collection, sold to Knight's Armament | June 19, 1996 |
| 006969 | (scoped) from Cummings collection, sold to Knight's Armament | June 19, 1996 |
| 007372 | (with QC barrels) Daniel Rewers | November 20, 1970 |
| 007413 | Daniel Rewers | November 20, 1970 |
| 007474 | Daniel Rewers | November 20, 1970 |
| 007507 | Daniel Rewers | November 20, 1970 |
| 007564 | Daniel Rewers | November 20, 1970 |
| 007571 | (scoped) from Cummings collection, sold to Knight's Armament | June 19, 1996 |
| 007583 | (LMG with QC barrels) Daniel Rewers | November 20, 1970 |
| 007607 | Daniel Rewers | November 20, 1970 |
| 007669 | J. Curtis Earl | March 5, 1973 |
| 007671 | Daniel Rewers | November 20, 1970 |
| 007704 | Law Enforcement Ordnance Co. | March 6, 1973 |
| 007725 | Daniel Rewers | November 20, 1970 |
| 007726 | Daniel Rewers | November 20, 1970 |

* Daniel E. Rewers was a sheriff's deputy in Cook County, Illinois (Chicago), and later in Wyoming. He was a Vietnam vet, ex-Army EOD, and a reserve CID warrant officer. He was also a gun collector, and partnered up with R. J. Perry of Chicago to buy one lot of machine guns from Interarms back in the 1970s. As shown above, these included AR-10s serial nos. 1038 and 1040 (ArmaLite), and nos. 003892, 004102, 004979, 006967, 007372, 007413, 007474, 007507, 007564, 007607, 007671, 007725, and 007726 (A-I).

## Excerpts from the Samuel Cummings Obituary

Samuel Cummings died on April 29, 1998, at his home in Monaco. Excerpts from his obituary, written by J. Y. Smith and carried as a Special to *The Washington Post* on Saturday May 2, 1998, read as follows:

### Arms Dealer Samuel Cummings Dies

*Samuel Cummings, 71, a former Central Intelligence Agency employee who became one of the world's leading arms merchants and who founded Interarms, a dealership based in Alexandria [Virginia], died April 29 at his residence in Monaco.*

. . *Mr. Cummings began Interarms in 1953 and soon was playing an important role in the global arms trade. Interarms had a number of subsidiaries and maintains extensive warehouse facilities along the Alexandria waterfront in Old Town and in Manchester, England. At one time in the late 1960s, it had an estimated 700,000 weapons in storage in Alexandria.*

*For years, the company had estimated annual sales to foreign governments totaling $80 million, mainly of relatively low-tech items such as rifles, pistols, machine guns and hand grenades. Customers ranged from the regime of Fulgencio Batista, the president of Cuba who was overthrown by the Castro revolution in 1959, to the government of the Philippines, which bought arms to fight communist insurgents.*

*Interarms also supplies rifles, shotguns and pistols to sportsmen in the United States and other countries.*

*Unlike many operators in the often-shadowy world of arms deals, Mr. Cummings loved to talk about his business, and he became something of a media celebrity . .*

*The guiding principle of Mr. Cummings's business strategy was the idea that "the military market is based on human folly - not normal market precepts. Human folly goes up and down, but it always exists - and its depths have never been plumbed . ."*

## Hope Springs Eternal Department

| Caliber | .223 |
|---|---|
| Muzzle Velocity | 3,300 F.P.S. |
| Overall Length | 38.875 |
| Weight | 5.9 lbs. |
| Rate of Fire | 750 R.P.M. |

Manufactured By

**Colt's Patent Firearms Manufacturing Co., Inc.**

| Caliber | 7.62 MM (NATO) |
|---|---|
| Barrel Length | 21 inches |
| Overall Length | 41.25 inches |
| Weight | 6.85 lbs. |
| Rate of Fire | 650 R.P.M. |

Manufactured By

**Artillerie Inrichtingen**

**Air Force Survival Rifle MA-1**

| Caliber | 22 Hornet |
|---|---|
| Length Pack in Stock | 14 inches |
| Length Assembled | 30.5 inches |
| Weight | 2.50 lbs. |

301. Front and back pages of a small ArmaLite brochure which is undated but must have been issued circa 1960, after ArmaLite had sold the rights to the AR-15 to Colt's in 1959 (discussed below), but before Artillerie-Inrichtingen AR-10 production was terminated in 1961.

The early version of the AR-15 depicted, with no flash hider, cylindrical handguard and top-mounted cocking handle, is as shown in fig. 94.

Charles Kramer collection

The brochure shown in fig. 301 was sent out in response to an inquiry dated May 5, 1960 from an Alexander Montgomery, the honorary curator of the Ordnance Museum of the Valley Forge Military Academy ("a quasi-military installation"), concerning the "availability and price of an AR-10 light machine gun and an assault rifle to be possibly acquired for display in our museum."

Along with the brochure, ArmaLite Sales Manager Don E. Carmichael sent Mr. Montgomery a letter dated May 24, 1960, reproduced in fig. 302.

302 (right). The letter, dated May 24, 1960, sent by ArmaLite Sales Manager Don E. Carmichael in response to the inquiry quoted above, stating in no uncertain terms that "ArmaLite AR-10 rifles are <u>not</u> available for individual purchase at this time."          Charles Kramer collection

**ARMALITE®**
DIVISION OF FAIRCHILD ENGINE AND AIRPLANE CORPORATION
118 EAST 16th STREET, COSTA MESA, CALIFORNIA
LIberty 8-7701  •  CABLE ADDRESS "ARMALITE"

May 24, 1960

Mr. R. Alexander Montgomery
123 S. Broad Street
Philadelphia 9, Pennsylvania

Dear Sir:

This is in reply to your letter concerning our AR-10 military rifles.

The Armalite AR-10 rifles are <u>not</u> available for individual purchase at this time. However, we sincerely hope in the future to fill your requirements for a modern semi-automatic rifle.

In answer to many of your questions, we enclose a brochure covering the Armalite AR-10.

Thanking you for your interest in our products, we remain

Sincerely yours,

ARMALITE DIVISION
Fairchild Engine and Airplane Corp.

*Don E. Carmichael*

Don E. Carmichael
Sales Manager

DEC:gw
Encl.

# A Phoenix from the Ashes

Meanwhile, in Holland, the government's draconian edict prohibiting A-I from further manufacture and sales was not the end by any means for the AR-10. As discussed below, in the following months, loyal A-I engineers were to embark on a new, secret but short-lived follow-on project and, as further discussed in Chapter Eleven, the Portuguese/NATO model AR-10 went on to provide the most outstanding combat service imaginable to a brave and élite fraternity of special forces soldiers.

Further, as discussed in Chapter Twelve, the arrival of surplus AR-10s from Dutch production would come to excite the passion and interest of a small group of historians and collectors in the United States and other countries.

Finally, as discussed in Chapter Thirteen, the AR-10 was destined to come back to life in several guises as modern production rifles, adopted for sniper use by the U.S., Canadian, British and other militaries, and becoming very popular with civilian shooters, this author included.

## Cloak and Dagger in Zaandam

The final *coup de grâce* of the Visser Defense Ministry on May 25, 1961 not only disallowed further production of the AR-10, just as the arms export restrictions cut off their last customer, but forbade Artillerie-Inrichtingen from carrying out any future arms production at all, with a threat that any attempts to carry on

weapons manufacture of any type would result in funding for the facility being cut off.

However Friedhelm Jüngeling, not one to be easily dissuaded, discovered a loophole in the government edict. While rifle *production* was clearly no longer permitted, the order said nothing about small arms *research*. Therefore, he decided that performing

303. Right side view of the non-firing aluminum mockup of the Artillerie-Inrichtingen MG09, shown with bipod extended.

The rudimentary butt tube is reminiscent of Stoner's original M-8 (the X-01, fig. 12), while the vertical forward handle repeats the same concept as found on the LMG versions of the AR-10.

Dutch Military Museum collection,
photo courtesy Marcel Braak

further experiments based on the AR-10 would not violate at least the letter of his new directive.

Not more than a few weeks after the last order had been crated up and shipped off to Portugal, a series of secretive meetings began between the director and three of the original and most venerable engineers who had been attached to the AR-10 project from the beginning; Van Der Jagt, Spanjersberg, and Bakker. The regular staff at the plant, and even most of the engineering staff, were not only not privy to what was discussed behind the closed doors of the boardroom, but often did not even know that the conferences were taking place at all, due to the fact that they were generally held at night, long after working hours, when the building was otherwise all but deserted.

The plan formulated in these secretive rendez-vous was to take what had been learned during the development of the AR-10, and set about designing a new modular weapons system.

Particularly disturbing to the lead design engineers had been the failure to interest foreign militaries in a robust, working model of the ArmaLite weapon in the light machine gun configuration. The development order was thus shifted, so that first a light machine gun would be developed, and then a rifle variant would be evolved from that. The rationale was to start with the most challenging type of arm, and then simply scale it down achieve a reliable rifle

version, which was seen at the time as the simpler of the two to fabricate.

Their drawing boards soon produced the first prototype plan, which called for the cobbling together of many slightly-modified AR-10 parts into a belt-fed machine gun, but with a barrel profile and length closer to that of the earlier AR-10 carbine.

Modifying an unused upper receiver, the design team replaced the direct impingement gas system, each group setting to work on one of nine components of the new gas system. Within a year they had two designs in the works, giving them the designations MG3 and MG4 (machine gun 3 and 4). They built a wooden model of the next envisioned prototype, the MG05, as well as an aluminum mockup of the proposed MG09, although apparently no working prototypes of these were ever made.

An interesting feature of these very compact prototype designs was that the ammunition belt feed was on the bottom, instead of through the left side as was the norm on such weapons. This allowed for a detachable modular container that would hold the coils of belted cartridges balanced under the weapon, keeping the ammunition clean and preventing the belt from binding or getting in the shooter's way.

Four years later, developments taking place across the Atlantic brought to the attention of the clandestine working group the growing necessity of adapting their design to the new U.S. 5.56x45mm small-caliber high-velocity assault rifle cartridge.

None other than Eugene Stoner, by this time working as a consultant for the Cadillac Gage Company, a private arms design firm with facilities in Detroit, Michigan and, not coincidentally, in Costa Mesa, California, had developed two very similar modular weapons systems: the 7.62mm Stoner 62, and the 5.56mm Stoner 63. In these Stoner had also started with a light machine gun, and had with minimal modifications developed a rifle version of the original design. At the beginning of the clandestine A-I "post-AR-10" project, the possibility of incorporating the new .223 caliber cartridge was explored, but this was as yet in its infancy and the accepted standard NATO cartridge was still the 7.62x51mm.

## Cancelling the Short-Lived "Post-AR-10" Project

Sadly, by 1965, Friedhelm Jüngeling had fallen seriously ill, and he stepped down as director of A-I that summer. The newly appointed director was informed of the secret project by its ringleaders, and after some consideration and an examination of the mockups and other material, he decided that, considering the government's prohibition against small arms manufacture, and thus the poor likelihood of imminent commercial success, the project should be abandoned.

Although no working weapon system ever resulted from this effort, the enthusiasm and ferocity with which it was pursued is a testament to the extreme devotion that the AR-10 mustered in those who had poured their hearts and souls into its perfection.

## In Memoriam - R. H. G. ("Kick") Koster

The mere existence of this initiative and the living knowledge of its details are thanks to documents and stories revealed by the chief design engineers late in their lives, in interviews and friendly meetings with the late Rudy "Kick" Koster, a true giant of AR-10 research. Without his tireless researching, collecting prototypes and cataloguing documents, much of the accumulated knowledge of the history of the Dutch development, production, and sale of the AR-10 would have been lost. His collections of files and writings, some of which have been made available to this author for use in the production of this work, are imminently illuminating, and paint a picture of an industrial development project that clearly meant much more to those involved than a mere paycheck.

# Contrasting Two Development Projects: the T44 and the AR-10

As the final demise of Artillerie-Inrichtingen's AR-10 program has at last been recounted, it seems fitting and instructive to compare and contrast the program that resulted in the Portuguese AR-10 with that which brought forth the M14.

It is easy after reading the record of Artillerie-Inrichtingen's efforts to pronounce it a flawed initiative, but when one views it in the context of other arms development programs, it actually comes across as extremely efficient with regard to cost, time, and innovation.

The trial-and-error process of developing a mechanical design is already a necessarily unruly one, but when one adds the political considerations, capacity for human error, and whims of fortune, any such endeavor seems predestined to be chaotic at best. This brings to mind a favorite maxim of Otto von Bismarck, who said that if one enjoys military rifles, one should not observe too closely how they are made.

It should be of some comfort to those who respect the AR-10 that from the time the Dutch first received their licence to manufacture and began converting the AR-10B design drawings to the metric system to when the last shipment of rifles was shipped off to Lisbon in July, 1961, the M14, an already mature and well-funded design, had not yet been produced in any serious numbers, although it had been more than four years since the formal announcement of the adoption of the T44 as the M14. Production at Winchester, the first of three eventual civilian M14 contractors, did not even begin until April, 1961.

Growing outrage at the costly Lightweight Rifle Program's failure to deliver a quality product was mounting in the U.S. armed forces, with news of the scandal slowly seeping out to the general public.

Within two years, the M14 would be on life support, soon to be steamrollered into obscurity by the flechette- and grenade-firing SPIW (Special Purpose Individual Weapon) program, which confidently predicted a new weapon that would be "Standard 'A' by June of 1965". The futuristic SPIW turned out to be yet another costly fiasco that was never perfected, leaving the field by default to the AR-10's little brother, the 5.56mm AR-15.

Exposés appeared in the popular press: one titled "The U.S. Army's Blunderbuss Bungle that Fattened Your Taxes," published in *True* magazine in April, 1963, and another titled "The M14: Boon or Blunder?", published in *Gun World* magazine in the same month, laid bare the monumental blunders of Springfield Armory's adoption program. These articles contained interviews with retired generals and other important players in the military firearms community such as Charles Dorchester and Melvin Johnson of ArmaLite, and quoted Defense Secretary Robert McNamara himself calling the M14 project "a disgrace."

By the time of publication, it was reported that the M14's development and production attempts had cost the American taxpayer in excess of $100,000,000, taking almost two decades.

On January 23, 1963, the Secretary of Defense ordered a halt to all M14 production, and by the end of 1963, quantities of AR-15s—85,000 for the Army and 19,000 for the Air Force—had been purchased in the famous "one-time buy."

The M16, without the controversial bolt-closure device insisted upon by the Army, was adopted by the U.S. Air Force in 1964, and the XM16E1, fitted with the bolt closure device, was standardized as "U.S. Rifle, 5.56mm, M16A1" on February 28, 1967. Although reliability problems dogged the M16 throughout its early years of issue, its prowess has long since been proven and attested by the fact that after more than fifty years, the M16 remains as the longest-serving rifle in U.S. history.

All told, compared to the hundred million dollars spent by the U.S. government on the T44, the Dutch only required an investment of three and a half million dollars, and after that point funded their operations with sales of weapons. For not even four percent the budget and in less than half the time, Artillerie-Inrichtingen had perfected a robust production version of the AR-10 that outclassed its Springfield competition in every conceivable way.

On the issue of weapon weight, the 1962 Aberdeen Proving Ground test of M14 rifles for the *Ad Hoc* Committee stated that "With fully-loaded magazine, the average weight of the M14 rifle would be 10.36 pounds as compared with 10.07 for the fully-loaded M1."

A section of this report titled "Undesirable Characteristics of the M14 Rifle" concluded with the devastating understatement that "Factors which limit the accuracy capability of the M14 rifle are Design and Workmanship."

From the time that the M-8 X-01 was only a glint in Eugene Stoner's eye to the departure of the last AR-10s from the Zaandam plant, a period of only eight years had passed, while from the beginning of the Lightweight Rifle Program to when M14s were finally being produced economically by its three civilian contractors (just before their production contracts were abrogated), eighteen years had elapsed.

# Fairchild Bails Out

## The ArmaLite Gas System Patent Rights Sold to Colt's

It was of the bitterest irony for ArmaLite that in 1959, just as Portuguese AR-10 orders had begun to materialize, and interest in the proprietary ArmaLite AR-15 was becoming stronger in U.S. military circles, the Fairchild board, in dire need of some quick cash for operating expenses, decided to sell the rights to the gas system patent for the AR-10 and AR-15 to the (ironically equally cash-strapped) Colt's Patent Firearms Manufacturing Company of Hartford, Connecticut.

Some salient excerpts from the Collector Grade title *The Black Rifle* read as follows:

*. . As Fairchild's need to salvage some of its ArmaLite investment grew keener, Colt's veered ever closer to actual bankruptcy. The sobering fact was that the manufacturing plant had not been updated in virtually a century. The firm was at length purchased by a New York financier for merely the value of the inventory of firearms already manufactured but still unassembled.*

*. . Interestingly, Colt's paid [the newly-formed] Fairchild Stratos Corporation a lump sum of only $75,000 plus a royalty of 4½% on all future production for the rights to the AR-15, while for its good offices in putting the deal together Colt's paid Cooper-MacDonald $250,000 plus a royalty of 1% on future production. Salient portions of the Colt/Cooper MacDonald arrangement, as later read into the record of the Ichord subcommittee hearings, are as follows:*

*Cooper-MacDonald was instrumental in bringing about agreements between Fairchild Engine and Airplane Corporation (now known as Fairchild Stratos Corporation) and Fairchild Arms International, Limited pursuant to which agreements, as*

*thereafter modified and supplemented, Colt received the exclusive license to manufacture and sell a lightweight automatic rifle sometimes known as the ArmaLite AR-15 and any variation thereof which incorporates the ArmaLite gas system as defined and elaborated by U.S. Patent no. 2,951,424, issued on September 6, 1960, entitled "Gas-Operated Bolt and Carrier System."*

*A further broader definition of "ArmaLite military weapons" was later agreed upon, as meaning "the ArmaLite AR-10 and the Armalite AR-15 and any variations (e.g. bipod, tripod, heavy barrel, light machine gun, etc.) thereof incorporating the ArmaLite gas system."*

## Rumblings of Trans-Atlantic Strife

In what was generally a very cordial industrial relationship, a sore spot developed in late 1959 that engendered some ill feelings between Artillerie-Inrichtingen and ArmaLite. The Dutch, who had sunk considerable capital into tooling up for AR-10 production, were startled and angered to learn about the sale of the patent rights to the AR-10 and AR-15 to Colt's. Although their contractual manufacturing rights were not affected, and they still retained the option to renew their production license with the new owners of the AR-10, Jüngeling and his staff felt betrayed by the partner with which they had worked so closely.

Although documentary evidence and firsthand reports of this period are sketchy and often contradictory, the Dutch began to withhold some information regarding their ongoing progress from their American partners. For example, the new trigger design, with its more positive coil hammer spring

(fig. 263), and the sear release improvements patented by Jüngeling (fig. 266), first incorporated in the perfected Portuguese/NATO AR-10, were never disclosed to ArmaLite. The Dutch factory and its subcontractors also ceased the practice of diligently acquiring Fairchild's authorization for shipments.

Also, it has come to light that Artillerie-Inrichtingen acted on its own to authorize Nederlandsche Metaalfabrik to produce additional magazines for the Portuguese forces, which were manufactured both before and after the end of their contractual license rights. As a way of thumbing their noses at ArmaLite, the Rheinmetaal-fabricated magazines from these final, unauthorized production runs omitted the "ARMALITE PATENTS PENDING" marking stamped on the floorplate, and simply left that part blank (fig. 295). These unmarked magazines are among the rarest of all those made for the AR-10 during its three-plus years of production in Holland.

# The AR-10A of 1959- the Final ArmaLite AR-10

Just before the sale to Colt's took place, however, ArmaLite on its own initiative took what had been learned by the Dutch in arriving at the Transitional model and combined it with what they had discov-

ered during their development of the AR-15, and came up with a final, one-off version of the AR-10 that was as intriguing as it was mysterious.

## Scaling *Up* the AR-15

As L. James Sullivan later recalled, "On the AR-10A, Gene Stoner told Bob Fremont and myself to scale up the AR-15 for 7.62 NATO (a kind of reversal from

scaling down the final Dutch AR-10 to the .223 AR-15 . .)"

304. Left and right side views of the last American-designed AR-10, the ArmaLite AR-10A, serial no. 1048.

This rifle displays a combination of features both from Dutch developments, as well as design elements learned from work on the AR-15, such as the rear-mounted cocking handle.

The open-pronged flash hider is clearly of the Cuban type, and the angled proprietary magazine can be identified by its different waffle crease pattern.

The plastic furniture on this rifle and on ArmaLite AR-15 no. 000004 were both painted with the same brown paint, as these two guns were to be used by Colt's as samples.                    Institute of Military Technology

As Sullivan related it, the AR-15 had represented a step forward in evolution from the original AR-10, with literally hundreds of small but significant innovations being incorporated in its design, and it was this improved platform that Stoner wanted to use to form the basis for the very last ArmaLite AR-10.

Somewhat confusingly, the new rifle was named the "AR-10A", repeating the name that had already been bestowed back in early 1956 on Stoner's prototype Model X-03 (figs. 48 and 49) which, as discussed in Chapter Two, had been the first AR-10 to be given a full-scale publicity tour, gaining unprecedented approval from all the U.S. military personnel who had witnessed its demonstrations.

On the new AR-10A the ArmaLite team replaced the original muzzle device with an open-pronged flash hider, as used on the Cuban production series of the Dutch AR-10s.

They had also discovered that the slot in the top of the upper receiver in which the original charging handle rode was a source of premature breakages when test-firing the smaller, faster, higher-pressure .223 cartridge in the AR-15. Therefore the charging handle that was grasped and pulled from the rear,

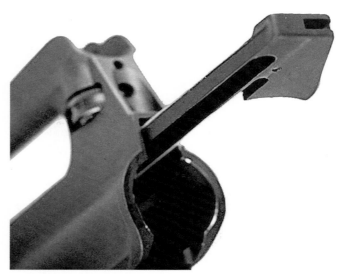

305. Left rear closeup of the rear-mounted charging handle on AR-10A serial no. 1048.

Although this early version was accessible from the right side only, it ushered in the preferred and now long-established cocking handle configuration on both the AR-10 and AR-15.                    *Small Arms Review* Volume I No. 4, courtesy Dan Shea and Robert Segel

originally used experimentally on one of their early AR-10 LMG variants, and which is now near-universal in modern AR-15 and AR-10 production, was reintroduced. As shown, this was accessible from the right side only. This also solved another problem about which some of the purchasers of the Dutch AR-10s were complaining; that the standard top-mounted charging handle, given its close proximity to the gas tube and contact with the gas key, became extremely hot when the rifle was fired extensively. The innovation of moving the cocking handle to the rear, as suggested by Jim Sullivan, solved both of these issues, and as shown earlier, the Dutch had fitted at least two of their experimental magazine-fed LMG versions with this updated assembly (figs. 211 and 283).

The Dutch had hoped to reverse-engineer one of these ArmaLite AR-10As to learn from its improvements, but while it was demonstrated to a visiting group of engineers from Colt's, only one example was ever made, and this unique rifle was not provided by ArmaLite to the Dutch as requested. Artillerie-Inrichtingen therefore fabricated their own rear-charging upper receiver from scratch, with only minimal input from Stoner and L. James Sullivan.

Finally, the magazine well in the lower receiver of the AR-10A was altered to incorporate a distinctive 4° forward cant, which slightly altered the position

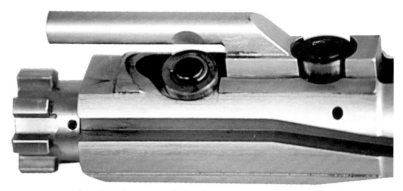

306. Left side closeup of the bolt assembly fitted to ArmaLite AR-10A when these photographs were taken.

Note the rollers incorporated in the bolt cam and gas key, a throwback to Mel Johnson's patented designs (figs. 4 and 5), and the cam cut for actuating a belt feed mechanism. James Sullivan has commented on both these features, stating that this bolt assembly was not original to the AR-10A. *Small Arms Review* Volume I No. 4, courtesy Dan Shea and Robert Segel

of the cartridges in the magazine and improved feed reliability under adverse conditions. As a result of this the AR-10A uses a proprietary magazine, otherwise almost identical to the standard-production magazine but identified by its different waffle-crease pattern.

Only one magazine and rifle of this later AR-10A pattern exist in the world.

## The Colt NATO-Caliber AR Project Falls by the Wayside

Both the AR-10A and the AR-15 utilized the same direct gas system, and the original plan was for Colt's to further develop them both. In an additional except from *The Black Rifle*, Bobby MacDonald relates his recollections of what transpired after the rights deal had been concluded:

*. . In the process of getting started to sell the AR-10, [Fairchild] developed the AR-15. It seems to me that General Wyman, who was head of CONARC at that time, liked the idea of the small caliber, high velocity rifle, and . . the ArmaLite division of Fairchild scaled down the AR-10 to fit the 5.56, or .223 as that [cartridge] was originally called.*

*Mr. Boutelle, at that time president of Fairchild, asked me to come up to Hagerstown to see the rifle. I went up there and fell in love with it. So he was about to take off to go to [Fort] Benning with Stoner for the original Army tests of the rifle when I decided that the smart thing to do would be to take both rifles, the bigger caliber and the smaller cali-*

*ber, and take [them] around the world and shoot [them] under all kinds of conditions.*

*. . So [in the spring of 1959 Gene Stoner and] I went to [Southeast Asia] and we shot [AR-15 serial no. 000004] in the Philippines, Malaya, Indonesia, Thailand, Burma, India . . and then Italy . . the more I shot it, the better I liked it. And it seems to be that is the way most everybody in those days felt about the rifle. The fact of the matter was having both calibers, I ended up giving away 6,000 rounds of 7.62 in the Philippines, because nobody wanted to shoot [the AR-10]. Everybody wanted to shoot the AR-15 . .*

*Well, when I got back to Singapore I cabled Colt who had already spent $100,000 tooling up for the AR-10. I cabled them to stop it and go full out on the AR-15 . .*

Therefore, it was Colt's and not ArmaLite who ultimately benefitted from the unprecedented success of the AR-15, and any future production of rifles like the AR-10 based on Stoner's proprietary gas system was precluded for decades.

Meanwhile, however, it appears that, with the manufacturing agreement with Artillerie-Inrichtingen still in effect, Colt's never intended to build the AR-10, but instead were authorized to manufacture both the AR-15 and the new AR-10A. However, Bobby MacDonald's telegram, discussed above, caused them to concentrate their efforts solely on the AR-15.

MacDonald's comment that Colt "had already spent $100,000 tooling up for the AR-10" seems fanciful, given Colt's severely straitened financial circumstances at the time, and this figure may have included the money spent on preparing the AR-15 for manufacture. It seems that ArmaLite never prepared a complete set of manufacturing blueprints for the AR-10A, and perhaps the Colt engineers could have been spending some money to finalize these drawings. However, the rights deal was signed in January, 1959 and MacDonald's trip was in the early spring, so they did not have a lot of time to spend $100,000.

# The Closeout

With Colt's decision not to pursue the further development of an AR-10-type weapon, the solitary specimen of what was designated the "AR-10A" languished in storage for three decades, before being finally purchased by C. Reed Knight Jr., a collector who appreciates its rarity and historical significance and gives it the care it deserves.

As ArmaLite, and by extension Fairchild, were no longer able to benefit financially from the universal acceptance eventually accorded the Colt AR-15, the board voted to cut ties with ArmaLite in 1961, selling it to its principals, Charles Dorchester and George Sullivan. Not being able to pursue any more designs based on the Stoner gas system patent that had made their designs so successful, Sullivan and Dorchester continued to develop other new arms, both with Stoner and after his departure to start his own company. These met with some commercial success, but never in a big way as they had originally hoped.

The ownership of ArmaLite changed hands several times over the years, and the original partners went their separate ways. Michael Parker, Interarms' "Unofficial Archivist," learned from John Ugarte, later the owner of the ArmaLite name and trademark, that the original ArmaLite Corporation was dissolved in 1988.

# Part III: An AR-10 Retrospective

## *Chapter Eleven*

# The AR-10 in Combat

The tale of how the AR-10 came to reach its final form is only part of its story, and as intriguing as it is to observe how an idea gifted with the spark of genius can come to fruition in the real world, all of the design work was after all simply a means to the end of supplying a friendly force with a weapon that performed well in combat. Thus, even considering all that went into the AR-10's invention, design, development, and sale, it is undoubtedly true that its use in war exceeds its use in peace "as far as day does night; it's spritely, waking, audible, and full of vent," and yet this crucial part of the story has hardly ever been mentioned before.

### Hidden Theaters and Operations

The very nature of both the AR-10's extreme effectiveness and the low numbers which were produced meant that those that were manufactured went to some of the most élite special forces units of their time; organizations that are famously good at keeping secrets. Unlike those of regular armies, the exploits of special-mission units are usually not covered in the media or even known to the general public, often for decades, if at all. This fact makes information about the AR-10's usage in the hands of many of its purchasers almost a complete mystery, with only the occasional, tantalizing detail emerging.

Most of the military forces that adopted the AR-10 have not let out as much as a peep about the uses to which it was put. However, some information has seeped out, as well as details of operations in which the AR-10 was involved that were far beyond the scope imagined by either the seller or purchaser at the time the deals were made. While little to no useful information about the AR-10 has leaked out of many of the nations that bought them, the one truly enduring mystery concerns those that were probably sent to, and are still jealously guarded by, the Indonesian *Kopassus*.

## Slaughter in the Dark: the Sudanese Civil War

Of the military forces that bought quantities of the AR-10, the Sudanese stand out not only as the purchasers of the single largest consignment of rifles, but by the almost complete shroud of secrecy they imposed on the conflict wherein they put their rifles to use.

It is known that the Sudanese used the AR-10 to equip élite infantry in their Eastern Arab Corps and their Airborne Brigade from late 1958, and that the rifles were finally decommissioned out of military service in 1985, appearing that year as surplus on the international civilian arms market, where they were acquired by private individuals. Between these dates they were used throughout the First Sudanese Civil War, and into the first years of the Second. No photographs have yet been found of this weapon even in the hands of Sudanese troops in that conflict, let alone of it actually being fired in anger; indeed, almost no news at all of these conflicts ever made it out of the undeveloped Sudan. Journalists were essentially nonexistent, and the industrialized world had far more pressing and dangerous military threats about which to worry to spend any time paying attention to a tribal and ethnic conflict deep in the interior of darkest Africa.

In suppressing the rebellion fomented by the ethnically and religiously distinct population in the South, the Sudanese government forces were often given to horrendous brutality and reprisals against the civilian populace. Interviews in modern times with survivors of those horrors often contain references to "light, advanced, automatic American rifles" in the hands of particular élite units used to assault the harder targets among rebel camps. Despite considerable effort, this author was not able to locate any veterans still living who were issued the distinct and light Sudanese variant of the AR-10.

# Training and Drill: the Dull Life of the AR-10 in Guatemala

With the demise of Artillerie-Inrichtingen as a small arms producer, any plans that the Guatemalan military might have had about standardizing the AR-10 for their infantry, just as the Nicaraguans were on the cusp of doing before that fateful accident in 1958, were ended. Therefore, the 450 already in their hands were permanently relegated to parade ground duties at *La Escuela Politécnica*, the Guatemalan Army officers' school. These rifles performed that duty until the late 1970s. In the early 1980s the government, in an effort to raise some money, sold the entire quantity to importers in the United States, who paid a princely sum indeed for these exceedingly scarce and extremely light Guatemalan model AR-10s.

With the disposal of their ArmaLites, the cadets of *La Escuela Politécnica* resumed their parade duties with surplus U.S. M1 Garands. For field exercises involving live fire (a relatively new development for this institution), M16s provided by their northern ally are now employed, but the important duty of drill rifles is still reserved for their venerable, classic M1 battle rifles.

It is known with relative certainty that not only were the AR-10s in the hands of the Guatemalan officer candidates extremely well-looked-after and cleaned regularly, but that they were seldom, if ever, fired. Purchasers of these rifles in the United States report that the internal parts are all in essentially new condition, with the only wear present on the weapons being observed on the buttplates, the result of tapping the rifle on the ground during drills, and on the charging handles, which were racked over and over again as another part of parade ground demonstrations.

# In Silent Service with the Italian COMSUBIN

Unlike the other institutional buyers of the AR-10, and even the airline that acquired four, there has not been noted any wide-scale decommissioning and attendant foreign sales of the NATO-model rifles bought by the Italian Navy in 1960 or 1961. A few such rifles have turned up in the hands of private collectors in Italy, and it is even possible that a few exist in the United States, as well as the other countries to which most of the surplus AR-10s went in the period of mass sell-offs, but as it stands, it is very likely that a quantity of AR-10s still sit in an undisclosed military facility in Italy, and even may still be in use by the heirs to the COMSUBIN operators who were so impressed by them over half a century ago.

307 (right). The only photo we have seen of an Italian naval COMSUBIN operator, here shown carrying his NATO/Portuguese AR-10 through the bush.

public domain - the Internet

# The AR-10 as an Unwitting Instrument of Castro's Communist Revolution

If Sam Cummings had been asked when he traveled to Havana in early 1959 to demonstrate the AR-10 to the new oligarchs of the island nation, whether he thought it likely that those rifles would be used by men of a different nation in an attempt to overthrow their government, he would have probably said no. If asked whether he cared, he almost certainly would have said no as well.

With the fostering of close ties with the Soviet Union, Fidel Castro's government quickly became a natural antagonist of the many right-wing authoritarian regimes in the Caribbean-Latin American world. One such particularly-outspoken rival was none other than Cummings's friend Rafael Leonidas Trujillo, of the nearby Dominican Republic.

In 1947, while still a student, Castro himself had joined in an attempted invasion of the Dominican Republic from Cuba, the plan for which was scuttled at the last minute. Therefore, it did not take a lot of deep thought for Castro to come up with a plan, which he was likely already formulating when he met with Cummings, to make use of disaffected Dominican exiles living in Cuba to mount an invasion force and drive Trujillo from power. Within just a few months of his paying for the weapons, the majority of the Cuban AR-10s were already in the hands of Dominican paramilitaries in their training camps on Cuban territory. A number of the quantity of FN FALs he had acquired before arms embargoes went into effect were used to arm part of the planned invasion force destined for the Dominican beaches, while the forces who were considered the élite—those trained for an airborne assault on the island—were issued the AR-10s.

The operation commenced on June 14, 1959. Troop ships set out from Cuba bound for the Dominican coastal towns of Maimon and Estero Hondo, while the improvised, pseudo-air-mobile unit took off in an American Curtis C46 transport. The landing craft were caught out at sea by the Dominican Navy and the Vampire fighter-bombers Cummings had sold to Trujillo, which were always on the alert and patrolling in case of such an incursion. The Dominicans had had advance warning through their intelligence networks, so their air and sea forces were particularly well-prepared to meet the small exile force, massacring them at sea and on the beaches as they attempted to land.

The C-46, piloted by a Venezuelan volunteer and carrying a crew of fighters including several Cubans and even one American, landed at the airport in the town of Constanza, where they disembarked and commenced their plan to fight their way into the mountainous jungles from which to foster peasant support and launch an insurrection, the way Castro and his band had done in Cuba. After a successful landing, however, the rebels were trapped by a large response force that had arrived in the area, and were wiped out to the last man. The strange rifles picked up off the bodies of these fighters caught the interest of the Dominican troops sent to crush the rebellion, and soon enough officers who had been briefed about the weapon that Cummings was trying to sell, and which they had evaluated, identified them correctly as AR-10s.

This news, along with the rifles themselves, made their way quickly to Trujillo, who angrily confronted Cummings, who was in *Ciudad Trujillo* at the time, demanding an answer as to why the rifles he was trying to sell had just been used by Cuban-backed rebels to try and overthrow him. Cummings by this time was an expert cooler, and was able to talk the Dominican strongman down from his ranting, telling him truthfully that he had indeed sold the weapons to Castro, but that he had no idea the Cubans intended to "do anything so rude with them as to invade the Dominican Republic." Since not only was Cummings an excellent source of armament for Trujillo—one of the few who were still dealing with his government at the time—but it was the Vampire jets Cummings had secured for him that had carried the day, by playing a decisive role in detecting and defeating the amphibious portion of the invasion.

Not only were the captured AR-10s not destroyed, but they were arsenaled by the Dominicans with the intention of issuing them to élite forces should the need arise for a special operation of some kind; perhaps a retaliatory raid against Cuba. This was not to be, as Trujillo was assassinated by members of his own army within two years and, within the same time frame, Castro would be on the receiving end of another very similar invasion of armed exiles, which he would stop in much the same way Trujillo had done earlier. The CIA-sponsored Bay of Pigs fiasco was a very similar operation to that mounted on June 14, 1959 (known thereafter in

Cuban and revolutionary Dominican circles as the "June Fourteenth Movement"), and the Americans would have done well to heed how mistaken the

Cubans had been in assuming that a civil uprising would naturally follow such an action.

# Dominican Army Captain Ilio Capozzi and His Cuban AR-10

Before the Springfield Armory tests of the AR-10 had begun in late 1956, a very interesting individual by the name of Ilio Capozzi arrived in the Dominican Republic to begin a job he had just accepted as a military adviser and training officer to the local forces, particularly helping them set up a naval special forces unit. Capozzi was an enigmatic fellow, apparently speaking five languages, including German, Spanish, Italian, and English. A veteran of WWII, he sometimes claimed to have served with the German Gestapo, and at other times to simply have fought in the Italian Army. This war experience may have been simple résumé-padding, but he was apparently half-German and half-Italian, making his claims about dual wartime service possible.

What is not contested is that he was a founding member and instructor of the Italian Navy's underwater demolition and special forces team, codenamed the Frogmen: none other than the COMSUBIN organization which bought some of the last of the AR-10s. Captain Capozzi had resigned from the Italian Navy before their purchase of the ArmaLite rifles, but apparently kept in contact with his old unit, for in 1959 he was immediately attracted to the Cuban-model AR-10s newly-captured by the Dominicans. He kept his position through all of the turmoil that surrounded the Trujillo assassination, and earned his pay in the capital, renamed Santo Domingo after Trujillo's death.

In 1963 the first democratically-elected government in the nation's history came to power, and immediately set about enacting reforms. Being distinctly liberal in what was a very conservative part of the world, the administration of President Juan Bosch set about instituting policies of social welfare and land reform, giving power and economic opportunity to the disenfranchised majority of Dominicans. These people had not seen their lot improve under Trujillo's government, which had progressively become a kleptocracy. Of particular import for the new government was the abolition of the *Latifundium* system, based closely on Classical Roman industrial agriculture, which parceled arable land into huge estates run by wealthy landowners, tying the laborers to the land as virtual slaves.

These moves naturally drew the ire of members of the old order and the landed establishment, who

308. Captain Ilio Capozzi keeps his Cuban AR-10 close at hand as he converses with a group of armed Dominican irregulars.

The man at left is carrying a .30 Carbine caliber Cristobal Model 2 SMG, a Beretta look-alike designed by the Hungarian Pal Kiraly, and manufactured at the Dominican *Armeria* in San Cristobal.          photo credit: *LIFE* magazine

had done quite well under Trujillo's corruption. An attempt to introduce a new constitution that enshrined these values, as well as the allowance of civil divorce and the legalization of homosexuality, also caused opposition to the new government from the Catholic Church, a very powerful institution in the Dominican Republic.

This reactionary sentiment led to a military coup that ousted Bosch on September 25, 1963—just seven months after his assuming office—replacing him with a council of three civilian strongmen known as "The Triumvirate," after the group that ruled the late Roman Republic (it seems that antiquity was a major political theme in the Dominican Republic). Bosch fled to Puerto Rico, while the junta attempted to consolidate power. This effort failed miserably, as the populace generally did not accept the legitimacy of their rule, and by 1965 the time was right for an uprising. On April 24th of that year, the Dominican Army mutinied, led by Colonel Caamaño, and ousted the Triumvirate, calling for Bosch's return. A civil war began the next day, as General Wessin of the military academy launched counter-strikes, with the help of some military units that were loyal to him, as well as the right-wing Military Intelligence Service.

The pro-Bosch faction, known as the Constitutionalists, were not dislodged from the positions into which they had entrenched themselves in the capital, despite the ferocity of the attacks by the new rebels, calling themselves the Loyalists. Fighting intensified in the streets of Santo Domingo, and the increasingly brutal violence of angry mobs of civilians against the police forces, which generally remained on the Loyalist side, caused great uneasiness in the United States. Fearing that they were witnessing another communist insurrection, the U.S. sent in the Marines and the Army, instituting a peacekeeping force that attempted to separate the warring camps. This worked only moderately well, and the fighting continued, but with the Constitutionalists becoming more and more marginalized.

The Americans, although officially there to restore order and seek a political solution, became increasingly convinced that the Constitutionalists were Cuban-sponsored, and eventually sided with the Loyalists and helped install General Imbert as the new president of the country. The party loyal to Bosch was eventually defeated and authoritarian military rule was reinstituted in the Dominican Republic, where it was to last for another twenty-three years.

Joining Colonel Caamaño in fighting for the restoration of the Constitution, now thoroughly indoctrinated in their cause, was Ilio Capozzi. Photographed by the intrepid *LIFE* Magazine photographer Bill Eppridge, the Italian volunteer dug in with the pro-Bosch faction in their camp in the capital, advising them in the planning of raids and actions against both the domestic opposition and the American

309. Even in moments of relaxation, Ilio Capozzi kept his Cuban AR-10 by his side. Here it plays the role of an impromptu guitar.                photo credit: *LIFE* magazine

forces. Never leaving his shoulder as he paced, talked, ate, and even relaxed with his fellow revolutionaries, was a Cuban AR-10. He was observed at different times standing with his AR-10 and sharing a joke with captured U.S. Marines, and even handling their weapons, which included the earliest U.S. military variants of the AR-15.

This author would encourage readers to search the *LIFE* Magazine Image Collection on Google for the Santo Domingo Revolt, in order to see more of these iconic images of the rarest of standard-production AR-10s.

Devoted to his cause to the last, Capozzi died in action on May 19, 1965, leading a daring attempt by the Constitutionalists to retake the national palace. His Cuban-model rifle was in his hands as he fell, and although this use of the AR-10 was not the largest, most effective, or historically significant, it is one that characterized all of the personal drama and tragedy that permeated that sad and bloody period in Dominican history.

# The Portuguese AR-10 in Combat

Although the previously-recounted tales of the AR-10's usage are interesting and sometimes of real historical importance, they all pale in comparison to the legendary service Eugene Stoner's creation gave to the men (and sometimes women) of the Portuguese Airborne.

## *A Guerra Ultramar*: the Angolan Conflict Erupts

Portugal had made their adoption decision and placed their first order for AR-10s just in time, because in January, 1961 a labor strike in Angola turned violent, and by February two nationalist insurgent groups were in open, armed insurrection against the provincial authorities, demanding that Angola leave the Portuguese Republic and become an independent state. The entire province of Angola was soon embroiled in rebellion, and Portugal scrambled to buy as many AR-10s as they could in order to equip all of their airborne battalions, seeing special air-mobile operations by élite forces capable of helicopter and parachute insertion as critical to the defeat of the rebels.

International factors came into play for both Portugal and the AR-10 very early in what would become known as *A Guerra Ultramar*, or "the Overseas War." The initial response to the violence, usually against civilians who were either of white ethnicity or viewed by the rebels as not being supportive of their aims, was often heavy-handed, with the Portuguese military seeing the fighters not only as rebels, but as both racist and communist; two things that were particularly loathed by Dr. Salazar's *Estado Novo* administration.

International criticism quickly began to mount concerning the arbitrary arrest and detention policies employed by the provincial government, as well as the harsh tactics with which the military was combatting the rebels, usually killing out of hand any they encountered. These particular methods were insti-

tuted by special units of the Army that were already in-country at the time, in response to the campaign started by the UPA (Union of Angolan Peoples) in March, 1961, in which insurgents swept through villages massacring entire populations, including men, women, children, and babies regardless of race or ethnicity, though saving their cruelest treatment for white and mixed-race Angolans. The rebels attempted to make propaganda use of the way in which the *Caçadores Especiais* (Special Hunter Companies) units responded; i.e., giving rebels no quarter and mercilessly pursuing them in mobile sweeping operations through the jungles. This backfired, however, as the strong response against them was seen in the international community as rightful action against combatants who were massacring civilians. Some units in these early days even adopted the practice of cutting off the heads of dead insurgents and impaling them on stakes as a warning to prospective rebels that joining the insurgency and massacring civilians would not be tolerated, and that any who made such a foolish decision stood to pay dearly for it.

The year 1961 started and ended very badly for Portugal, with the insurrection in Angola commencing in January, as well as the hijacking of a Portuguese passenger ship by anti-Salazar, pro-provincial independence rebels. It ended with the Indian Army invading the Portuguese overseas province of Goa, which it had administered as Portuguese India for four hundred and fifty years.

## Success in Angola

As the Army's special forces, the Special Hunters, fought hard and made great progress against the rebel groups in Angola, Battalion 21 of the Airborne soon arrived in-country. This unit of the Air Force, which would take control of air-mobile operations in Angola for the entire war, was created in May, 1961 from the temporary command already in place in the territory, being so designated as the first parachute infantry unit of the Second Air Command Region (2nd Region, 1st Unit; 21st Battalion). At the beginning of

hostilities, although the Army's special units were able to make amazing progress across very rough terrain, there were huge logistical problems due to the difficulty with maintaining ground lines of supply and communication in such a tropical country with a rainy season. This difficulty was compounded greatly by insurgent raids against roads and railways. Therefore, the heaviest burden of taking the fight to the enemy and supporting the Army fell squarely on the shoulders of the Airborne in these early months of the war.

In those days the élite-but-small units that were already on the ground fought valiantly against extreme odds, with junior officers often taking the initiative and earning high honors. However, bravery can only sustain a military operation for so long, and at the end of the day, the Special Hunters needed resupply and support. This they got from the Air Force, with fighter-bombers initiating bombing and strafing missions against enemy concentrations, and transport aircraft dropping much-needed ammunition, food, and medicine to the forward echelons that had pushed into the jungle without an open line of communication.

The paratroopers of the 21st Battalion soon took to the skies, to be dropped by parachute and inserted by helicopter to efficiently encircle enemy formations and wipe them out with accurate and sustained fire from their AR-10s. The Portuguese who carried the AR-10s loved them so much that when the spare parts supply dried up and they could not get any replacements made in Portugal's armament factories, they took to improvising whatever spares were needed. Fortunately, according to veteran accounts, the occurrence of parts breakage in that final model of the Dutch-production AR-10 was vanishingly small, and whereas any rifle will eventually lose accuracy as its barrel becomes worn from high numbers of rounds fired, there were confirmed instances of individual rifles firing in excess of 60,000 rounds without any parts replacements, and still maintaining one minute-of-angle shot groups; a level of accuracy and performance no standard version of the FAL or G3 has ever come close to achieving, and one

310. The Portuguese military celebrated national holidays with military parades in the capitals of their overseas provinces. Here, a military formation marches through the main thoroughfare of Luanda on "Portugal Day," June 10.
public domain, courtesy Joaquim Coelho

which requires expert tuning and expensive parts-fitting to be reached by an M14 or an M1 Garand.

So successful were the Airborne's early operations that soon the UPA rebels were in full retreat, heading north for the border with Zaire. By the beginning of October this especially brutal rebel group had been expelled from Angola, and was in exile inside what had recently been the Belgian Congo. This set the pattern for what would often follow: rebel incursions from bases in neighboring countries. In Angola, though, the Portuguese forces did an especially effective job of stamping out the insurgent forces in this province.

## The Portuguese Light Infantry Doctrine of "Vertical Envelopment"

Being one of the first forces to fight a counterinsurgency campaign in the era of modern, air-mobile warfare, the Portuguese were the true pioneers of some of the most effective and advanced infantry practices yet developed. It was still assumed in Europe and the United States that future conflicts would likely come in the form of grand battles fought between massive mechanized forces on open ground, in which armor and vehicle-born infantry would be decisive. However, as the Americans, Australians, and South Koreans were later to learn in Vietnam, the best-equipped and trained military force prepared for this type of conventional warfare can face astounding upsets at the hands of light, mobile infantry that make good use of terrain, infil-

tration, and ambush tactics. The world had been given ample forewarning of this phenomenon when an enormously large and highly mechanized Soviet invasion force was humiliated by a tiny group of lightly-armed Finns during the "winter war" of 1939 - 1940, and it seemed that the Portuguese were the only European power to pay attention to this lesson.

Starting before the overseas conflicts broke out, and honed quickly to a fine point in Angola, the Portuguese forged the operational concept of vertical envelopment, the basis of which is the rapid deployment of infantry forces from the air, (hence vertical), to surround an enemy force and prevent their escape, so that they could be completely defeated. The first attempts at this were made by dropping infantry by

311. Portuguese paratroopers disembark from a French Alouette III helicopter and move to blocking positions on a vertical envelopment mission. Pilots on insertion missions would single out an area that was on high ground with brush or trees nearby to conceal the troopers in their very vulnerable disembarkation stage.

As shown here, when the terrain was uneven (as it often was in the southern African bush) or there was enemy fire incoming, the Alouette helicopters would sometimes not even touch down, but hover over the drop zone and allow the men to jump down to the ground.

courtesy Portuguese Air Force

parachute away from an enemy concentration, and then having the units form up and encircle the position. The rebels in the African provinces, just like the Finns in their northern forests or the Vietcong in the South Vietnamese jungles, would usually attack a Portuguese formation or civilian locale, and then instead of staying to fight it out, would disengage and fade back into the jungle. The Portuguese endeavored to respond not bluntly with massed artillery and bombing, but with subtlety and stealth; adopting the mobile tactics of their enemies, and beating them at their own game. The emphasis in this type of warfare was not to smash a large enemy unit in a protracted action, but to ensure that the opposing forces could not escape.

Actual parachute drops were comparatively rare, with the only ones performed by the 21st Battalion taking place in the first two years of the war. The tactic quickly evolved, especially later in the war after they had received numbers of new French Alouette helicopters, when massed parachute drops were replaced by extremely fast and fluid helicopter insertions, landing men along the routes where the enemy might escape. These troops were brought in with five men per aircraft. Flights of six transport helicopters each would generally insert the light detachment of around 30 men, accompanied by one other Alouette mounted with a small-caliber, rapid-fire artillery piece that would hover and provide ground support during the mission. The airborne

312. An Air Force airborne unit prepares to board their Alouette helicopter for an insertion mission. Not being able to rely on logistical support, the paratroopers had to carry all equipment they would need for the operation on their backs, making the light weight of the AR-10 and its magazines particularly well-suited to this type of use.
courtesy Portuguese Air Force

313. This photograph demonstrates the multi-ethnic character of the men who fought in the Portuguese Overseas Provinces in Africa. The racial integration of even élite units was a key component to the successes of the Portuguese mission, both military and civil.

The photo also shows the changes in equipment that the Paraquedistas underwent during their 13-year service. Initially, the soldiers were equipped with American surplus gear, including BAR belts and chest rigs, made to hold a large number of M1918 LMG magazines, and wearing M1C helmets. The fast-moving and light nature of paratroop operations led them to change from the heavy steel helmet to a light, French-style Bigeard cap for better hot-weather comfort and hearing. The American magazine pouches were by late in the war replaced with Portuguese-manufactured copies.        courtesy Portuguese Air Force

infantry would then have four or five units form blocking positions while one or two moved in towards the enemy, rotating which unit moved and which dug in.

In this way the 30 men would quickly tighten the circle on the enemy, and once contact was made, would drive the guerillas into the other formations which were set up to intercept and block their retreat. Once the operation was concluded, the transports would return (after refueling) to pick up the paratroopers and collect any wounded and prisoners from the central area where the action had ended. These raids could take up to several days, but were more regularly concluded within just a few hours of the first boots hitting the ground.

Although the Portuguese military was not a large force and developed such tactics partly to make the most of their relatively small number of men, these methods of fighting forged in the wilds of tropical Africa have proven not to be simply a crutch for asymmetrical warfare or a means of combatting insurgencies; they have been studied and adopted since by major military forces for their effectiveness in all types of conflict. The Portuguese did not lack armor or heavy weaponry, but the terrain of their overseas provinces made these types of equipment of limited usefulness. So successful and versatile were the Portuguese tactics of vertical envelopment

314. Military order and discipline were kept up by frequent parade demonstrations. Interestingly, one of the only times that the Paraquedistas actually wore their green berets was in such circumstances.      courtesy Portuguese Air Force

that they were even employed with great effectiveness late in the war, when the government forces came up against well-trained rebel formations employing Soviet tanks and armored personnel carriers.

The paratroopers were also often used as regular light infantry, coordinating with the Army in direct search-and-destroy missions in areas containing probable enemy command centers. As the war in Angola progressed, a very successful tactic was developed by which the paratroopers were either dropped or ferried into a staging area very close to an objective, which they would then capture. The Airborne would then hold the former enemy strongpoint against counterattacks until the Army could arrive, often in the form of actual mounted cavalry, as infantry were too slow for such actions, and mechanized transport was unable to negotiate the mined and destroyed roads and bridges.

315. A parade review by high command allows a glimpse of the rare instance of the Portuguese Model AR-10 in full dress regalia, with bayonets fixed (left). Beginning in 1963, parachute drops were largely replaced by helicopter insertions, aided by the excellent French Alouette III.
courtesy Portuguese Air Force

## Stalemate in Guinea

Portugal had a considerably harder time hanging onto territory in its province of Guinea, which is today Guinea Bissau. This was the second of the Portuguese African territories to experience conflict, with coordinated and effective guerilla operations beginning in January, 1963. Just as in Angola, the

rebels here, the African Party for the Independence of Guinea and Cape Verde (PAIGC) were overtly communist and sponsored by the Soviet Union. The small size of Guinea and its proximity to both Senegal and the Republic of Guinea—both countries with governments friendly to the rebels—combined with

316. Although the terrain usually prevented it, mechanized infantry operations were sometimes possible, and when they were, the men of the Air Force Parachute Hunter Regiment were always ready to help out.

Note that everyone is armed with an AR-10 except the man at far right, who is holding a FAL.

courtesy Portuguese Air Force

its particularly uncomfortable climate and dense jungle, were all major disadvantages to the Portuguese in their attempts to combat the insurgents.

As recounted by Portuguese veterans, the ferocity of the fighting there, combined with the sheer misery engendered by the stifling heat and high humidity, caused the Air Force to shorten tours of duty for soldiers of the 12th Battalion there to 18 months, as opposed to the standard two years. The rebels in this province were quite successful from the start, taking control of much of the territory and receiving much easier resupply of relatively advanced weapons from the Soviets, Chinese, and Cubans, among others. Battalion 12 was one of the subset of the initial four main units that were not lucky enough to receive the AR-10, being issued instead the German G3 version of the CETME. Their fighting was heroic and distinguished in Guinea, but it was much harder to make progress against such a better-led and -equipped enemy in the rougher terrain. To make matters worse, many of the regular forces stationed in Guinea were still equipped with

the Portuguese variant of the Model 1898 Mauser bolt-action rifle. Indeed at the start of the war, only the élite units of the Army and Air Force were equipped with modern battle rifles; the rest using WWI-era equipment against rebels who, particularly in Guinea, generally carried modern Soviet Kalashnikov assault rifles and light machine guns.

Also particularly important in taking the fight to the rebel-held areas of the province in Guinea, were the soldiers of a particularly distinguished, storied, and ancient (by the standards of almost any modern military unit) special forces organization; the *Fuzileiros Portugueses*, or Portuguese Riflemen / Fusiliers. First commissioned in 1618, this unit of the Portuguese Navy is almost the exact equivalent of the U.S. Marine Corps. It started as naval infantry for guarding ships and boarding, but during WWI it became a landing force, being recommissioned on June 3, 1961 to perform amphibious operations against the rebels in Africa. One special company within this force; an especially élite detachment of the *Fuzileiros Especiais* (Special Riflemen); essen-

317. The utility of the airborne went far beyond vertical insertions and parachute jumps; standard infantry sweeps and patrols also made excellent use of their terrific combat abilities.                    courtesy Portuguese Air Force

318. On a combat mission through rough terrain, a paratrooper keeps two of the most important parts of his kit from being submerged or washed away: his AR-10 and his canteen.                    courtesy Portuguese Air Force

tially a special forces unit within the already-crack organization of the Portuguese Marines, received the remaining AR-10s from the final Artillerie-Inrichtingen order.

Although Guinea would end up being the most difficult and least successful of the counter-insurgency campaigns fought by the Portuguese armed forces, later in the war they did gain the initiative, going on the offensive and taking back territory that had been held by rebels for sometimes as long as nine

years. The sad fate that would eventually befall Guinea would be even bloodier than in the other unfortunate areas exposed to this conflict.

## Hard Fighting and Bravery in Mozambique

The last of Portugal's African provinces to fall victim to violent insurrection was the sprawling territory of Mozambique, known previously as Portuguese East Africa. A Soviet-sponsored, racist, nationalist organization called the Liberation Front of Mozambique (FRELIMO) began its campaign against the government forces in August and September of 1964. The nature of Mozambique was unique, in that Portugal had only really controlled a thin strip along its coastline until just a few decades before, and as such there was not much in the way of settlement, government administration, infrastructure, or military presence in most of the territory.

The relatively lawless interior and north of Mozambique helped the rebels there make solid gains early, but the quickly-organized Battalion 31 of the Portuguese Air Force led the counterattack, infiltrating rebel positions in the forests and hills and wiping them out, taking a particularly large number of prisoners in the process. The vast size of Mozambique, combined with its sharing a border with hostile (to the Portuguese) Tanzania, meant that a large increase in the number of troops in the area was crucial to effectively combatting the insurrection.

Within a few years, a second and third airborne battalion were formed—the 11th and 32nd—both armed with German G3s. The cooperation with the already-extant 31st helped the paratroopers achieve particularly remarkable successes in this theater.

Since operations did not begin until later in Mozambique, the lessons of Angola had been learned well and incorporated, and troops were usually inserted by the new French helicopters, doing a better job of surrounding rebel-controlled areas before an offensive so as not to allow as many to escape, as had occurred in Angola, where many thousands of fighters had fled unscathed into sanctuary in Zaire. For much of the conflict, overall command for the defense of Mozambique was in the hands of by-then Brigadier General Kaúlza de Arriaga; the reformer and founder of the Airborne.

Portugal also had some foreign help in their combatting of rebels in Mozambique, just as they began to enjoy towards the end of the war in Angola. The newly-independent Republic of Rhodesia was fighting its own nativist rebellion at the time, with a top-flight fighting force that coordinated with the Portuguese, and even launched strikes and raids on their own initiative against FRELIMO bases, both within their borders and over in southwestern Mozambique. The Republic of South Africa also assisted the Portuguese considerably against the communist insurgent forces that would pop up at different times along Angola's southern border.

## The Endgame for Portugal: Military Victory and Political Defeat

The Portuguese made significant gains in re-securing their overseas provinces throughout the decade of the 1960s, with Angola almost completely pacified by 1970. The FRELIMO rebels in Mozambique were not particularly more successful than those of the UPA in Angola, but were certainly more stubborn, hanging on and keeping the low-intensity conflict running into the next decade. Guinea, the toughest of the military areas, while still presenting the hardest challenge, did begin to give way in the late 1960s, as strategic reforms enacted by General António de Spínola began to give the offensive spirit and initiative back to Portugal.

The armament situation continued to improve for the Portuguese as well. Unable to equip their forces with their first choice, the AR-10, they ended up negotiating a license agreement with the German firm of Heckler & Koch, and began producing their own domestic variant of their third-choice G3 rifle. The troops issued them generally preferred the FAL to the German variant of the CETME, but by then Belgium had enacted similar arms export controls to those of the Netherlands. These domestically-manufactured G3s became the standard issue of the regular infantry, and while not comparable to the AR-10 in effectiveness, they did greatly increase the capabilities of the army regulars compared with their previous weapon, the 19th-century bolt-action Mauser.

Public works and welfare initiatives helped to sway public opinion in favor of the government, while large and successful recruitment of native troops swelled the ranks in Guinea. Until 1960, Guinea had been one of the least progressive of the Portuguese provinces, with native Africans not having de facto or de jure equal rights with whites, and with the military still segregated there. The integration policy not only engendered public sentiment for the military's mission, but it increased the effectiveness of the armed forces, when the most able men, regardless of their skin color, were able to work together.

In 1970, at the start of the new decade, and seeing the rebels severely weakened from years of successful actions against them, the Portuguese command began a series of special initiatives and operations, intended to deal a death blow to the rebel movements in the different theaters of conflict. The first of these was a coordinated air-and-ground assault named Operation Gordian Knot, the brainchild of General Arriaga. Its objective was to seal off the border infiltration routes from Tanzania, and prevent any more cross-border incursions into the north of the province. Beginning early in 1970, and ending on the 6th of August of that year, the operation employed excellent coordination between artillery, air strikes, cavalry and infantry movements, and helicopter-borne airborne attacks. This offensive incorporated some of the first units integrated with defected rebel fighters, who were usually quite reliable allies and of particular help in understanding enemy tactics and tracking their whereabouts. This experiment with reformed insurgents proved to be quite successful.

Although not completely victorious, this offensive destroyed the remaining bases of resistance in the country, relegating the rebels to unsupported hit-and-run actions in the jungle. The relatively large number of casualties sustained, however, added fuel to the already-mounting public opposition to the war, which was about to enter its second decade. Critical to the successful wiping out of rebel forces in this operation, instead of simply capturing their bases and weapons stockpiles, was the insertion of paratroopers from the 31st, 32nd, and 11th Battalions.

The next major push by the Portuguese involved both military and policy initiatives, and was aimed at finally extinguishing resistance in Angola; already the most successful front. The year 1970 saw the beginning of *Frente Leste* (Eastern Front), in which an increased troop presence against the infiltration routes out of Zaire protected a massive public works project. Education, trade, and healthcare infrastructure were constructed, and a wide-scale vaccination program saw the attitude of the populace become thoroughly supportive of the Portuguese mission. This move was entirely successful, essentially completely pacifying the eastern portion of Angola, and concluding the model counterinsurgency campaign fought by Portugal in that country.

The third of these offensives, codenamed "Operation Green Sea," was the most daring, and was intended to seize the initiative in troubled Guinea. Throughout the conflict, the rebels had received open support, backing, bases, and even logistical support from the neighboring Republic of Guinea. The government there had gone so far as to allow Portuguese prisoners of war to be held in detention centers on their soil. Their leader, Ahmed Sékou Touré, had been instrumental in orchestrating his country's independence struggle against the French, and since 1960 had ruled the country as an autocrat, making personal enemies of the Portuguese, whom he opposed fiercely. The rebel leader and the majority of their military equipment, including air and naval assets, were based in that country, and in 1970, General Spínola decided that he had had enough. His staff planned a raid to both strike at the heart of the rebel encampment, and to cripple the forces of the nation state helping them.

In the wee hours of November 22, 1970, a combined air-and-naval force began a bombardment of the capital city of Conakry, followed immediately by a landing force of Marines and native troops, led by Portuguese officers. Several amphibious assaults came ashore at the marine terminal, where ships belonging to the PAIGC insurgents and loaded with supplies were sunk with explosive charges. Another contingent made landfall near the summer home of President Touré, and proceeded to burn the compound. The leader was not at home at the time, however, and so was not killed or captured. The individual units linked up, assaulting the rebel camps and rescuing all of their prisoners of war unharmed. The continuing bombardment also accounted for several Guinean naval ships and military aircraft at bases around the capital which were struck.

When the attacking force realized that they would not be able to capture or kill President Touré, and that rebel leader Amílcar Cabral was not in the camps they raided, they pulled back and were extracted by the Navy. This operation seriously weakened the Guinean rebels, and put the Portuguese in a better position militarily. It also exposed the true colors of President Touré, who launched a Stalinist purge against his own government, military, and any foreigners in the country, imprisoning, torturing, and executing thousands; including very high officials in his own cabinet and political party (the only one permitted in the Republic of Guinea under his rule). Nevertheless the international community, especially the Eastern Bloc, reacted strongly to Portugal's invasion of the neighboring country, and these actions were condemned by the United Nations. The Soviet Union even sent naval forces to the area to protect its ally Touré against further incursions by the Portuguese.

Even as the military situation was continuously improving and the end appeared in sight, all was not well back in Lisbon. The need for skilled officers had led the Salazar government to promote to very high positions persons who were not politically friendly to the *Estado Novo* regime, and increasing privations on the civilian populace (although the economy grew considerably during the war years) to help finance the war were increasing public resentment of the prolonged conflict.

The first rumblings of political thunder were heard in Mozambique, concerning commander-in-chief Arriaga. He not only led the armed forces of that military region, but was very involved in politics, creating for himself something of a personality cult within the military and provincial government. An incident occurring on December 16, 1972 finally ended Arriaga's career. Known as the Wiriyamu Massacre, it was alleged that an Army detachment had slaughtered the inhabitants of a small town in the northwest of the province. Mystery still surrounds this event, with credible sources claiming that it was carried out by FRELIMO fighters and blamed on the Portuguese in order to hurt their standing abroad. It has also been asserted that the massacre never took place, and was concocted as a media campaign by the rebels or sympathetic African states like Tanzania.

Whatever actually happened on that tragic day, Arriaga took the blame, and by early 1974, leftists back in Portugal forced the hand of Marcelo Caetano,

Salazar's successor (Dr. Salazar had stepped down in 1968 and passed away in 1970), and Arriaga was relieved of his command and recalled to Lisbon. This hardly mattered, however, for on April 25, 1974, the full fury of the brewing storm struck, as the Armed Forces Movement, a cabal of leftist junior army officers, mounted an almost-bloodless coup, ousting Caetano from power, and overthrowing the *Estado Novo* government. Known as the Carnation Revolution after the flowers placed in the muzzles of soldiers' rifles, this event is still celebrated in modern Portugal, and spelled the end of their overseas provinces in Africa.

Almost immediately the National Salvation Junta, the political organization set up to govern the country until elections could be held, offered the rebel groups in Guinea, Angola, and Mozambique independence, ordering an immediate withdrawal of all Portuguese forces. Even as the military had essentially restored order and peace in Angola and Mozambique and were close to doing so in Guinea, the leftist sentiments of Portugal's new rulers could not countenance further rule of those areas. They therefore instituted a precipitous and sometimes-chaotic recall of forces, with almost all of the heavier military equipment being left behind to fall into rebel hands.

Of particular suddenness was the evacuation from Mozambique, where many of Battalion 31's AR-10s were left in crates in the abandoned operating bases.

The total and complete severing of military and government institutions in the former provinces caused a massive power vacuum, and in almost all cases led to horrendous atrocities. The rebel groups, already extremist and racist, sought bloody revenge against any person remaining who was of mixed or European ancestry, and any native African who was perceived as having been supportive of Portuguese rule. Prime targets of these vendettas were black soldiers who had fought for Portugal, most of whom were simply left behind by the new regime in Lisbon. Immediately upon taking power in Bissau, the PAIGC rebels brutally massacred all 7,447 natives who had served in the Portuguese military, leaving none alive and dumping their bodies in mass graves.

The moral correctness of Portugal in defending the status of its African provinces rather than letting them become independent states is an issue this author will leave for others to debate. However, it is an incontrovertible fact that the departure of Portuguese forces from those territories resulted in the formation of some of the most brutal, oppressive, corrupt, and backward regimes in history; with attendant civil wars, chaos, economic disaster, and inter-ethnic violence claiming the lives of between 1.5 and 3 million people in the following years. The standard of living for the residents of Angola, Guinea, and particularly Mozambique, dropped precipitously after independence, with Soviet-backed planned economies and collectivization programs causing horrendous famines. Even Portugal itself narrowly avoided civil war in the year after the coup, with a power struggle between extreme-left communists in the ruling junta, who were intent on collectivization and alignment with the Soviets, finally being marginalized and forced from power without violence by moderates who wanted to establish republican institutions and develop Portugal along the lines of other advanced free-market democracies.

The hasty departure of the Marines and Airborne troops from the African provinces did lead to some very interesting developments in the history of the AR-10, however. Most or all of Battalion 21's rifles were brought back to Europe by their users, but almost all of their bayonets were left in Angola. Over a decade later, these were returned in their original condition by the Angolan government, in a rare show of civility between the new African states and the "mother country." The 31st's rapid flight from Lourenço Marques (the Mozambican capital, now called Maputo) saw most of their AR-10s being abandoned. The new revolutionary government of Zamora Machel, called a butcher by some and a freedom fighter by others, immediately put these fantastic rifles into the hands of another Marxist insurgent group they were supporting against the government of one of their neighbors. This would not be the last time that these formidable weapons would see battle, and almost all of the fine rifles that were carried by the Portuguese in their 13-year adventure in Africa would eventually find new and loving owners.

# The Indelible Words of a True Hero

Of all the numerous important people involved in the AR-10's story and the fantastic resources uncovered during the writing of this book, one man stands out above all the rest. In the fall of 2014, this author had the distinct honor and privilege to conduct a series of interviews with a decorated veteran of the Portuguese Overseas War.

Sergeant Major Alfredo Serrano Rosa served as a noncommissioned officer in the *Caçadores Paraquedistas* in Africa from 1966 through 1973, being involved in an extraordinary number of counterinsurgency operations. He served three tours of duty during that time, seeing combat in all three theaters of operations. Speaking with him was a true high point of this author's research, and one of his first impressions when discussion of the ArmaLite rifle came up was, "Wow, I thought I had a high opinion of the AR-10!"

For his first and last tour of duty, 1966 – 1967 and 1971 – 1973, Sergeant Major Rosa was attached to the 31st Battalion of the Airborne, based in Mozambique, while he served one more deployment to Guinea from 1968 to 1969. He recounted how the less intense combat and better weather conditions in Angola and Mozambique meant that those areas required two years for one tour to be complete, while a standard deployment to Guinea was over in 18 months.

Having been attached to Battalion 12 in Guinea meant that Sergeant Major Rosa gained a unique perspective of having extensive combat experience with both the G3 and the AR-10, which allowed him to make a fair and informed evaluation of the two weapons. In his peerlessly-qualified opinion, there was no comparison whatsoever between the two: he found the AR-10 to be superior in every way. He commanded squads of paratroopers who also got the chance to be issued both weapons, and reports that their opinions were the same.

According to Sergeant Major Rosa, the troops were trained on all of the different variants of NATO AR-10s (standard, bipod-mounted, semi-automatic, scoped) as well as the launching of grenades, at the Airborne School located at Tancos in Portugal. However, he and his squads were only issued and used the basic and bipod-equipped models, lending credence to the evidence that very few completed rifles fitted with telescopic sights were delivered to Portugal before the Dutch arms export restrictions went into effect. Along the same lines, he remembered never having used the rifle as a platform to launch

319. An official portrait of Sergeant Major Alfredo Serrano Rosa, taken in 1996.          courtesy Portuguese Air Force

grenades, noting how little use there was for such weapons in Angola and Mozambique.

Throughout our interviews, the Sergeant Major recounted the multifarious positive attributes of his AR-10 which he well remembered. He recounted how even in the fully-automatic mode the rifle was quite controllable, making it useful in close-quarters jungle fighting and in the clearing of enemy-held structures. As a high-ranking enlisted man, he was able to keep his AR-10 in 1973 when the rest of his squad was forced to revert to G3s due to the lack of spare parts for their AR-10s. He recounted how impressive the accuracy was, and how it was so reliable that it never let him down in any situation in all of his combat experiences.

Field-stripping the AR-10 and maintaining it was reportedly very easy, and the weapon needed less attention than the other types in the arsenal of his units. Of particular satisfaction to Sergeant Major

Rosa was how little time he had to spend servicing his weapon after returning from a mission, tired and in need of rest and a bath, compared to those unfortunate souls with their G3s (of which he was one in Guinea), who had apparently had a dickens of a time scrubbing the dirt and carbon fouling from the roller-locking mechanism of their battle rifles.

This author found himself identifying deeply with the final sentiments expressed by Sergeant Major Rosa; that the Overseas War and its veterans have been largely forgotten and ignored in today's Portugal, and that their exploits deserve to be remembered by future generations. The extraordinary heroism displayed by these men and women (women were permitted to serve as combat medics even in these élite units, the duty of which often necessitated the carrying of rifles) is deeply moving to even such a humble student of military history like this author, and it is therefore his fervent opinion that their gallantry and devotion to their nation should not be forgotten.

In pursuit of the history of the Portuguese Airborne in Africa, this author also came into contact with other very kind and considerate members of the Portuguese defense community, most notably Army Chief Sergeant Carlos Alberto De Sà Canas, assistant to the commandant of the Portuguese Paratrooper Training School, and a historian involved with the base's museum of the Portuguese Airborne. In addition to useful information, Chief Sergeant Canas, on behalf of his unit and the museum, made available for the publication of this book a vast photographic archive from Portuguese operations in Africa.

One other interesting, not to say bizarre, occurrence during the Overseas Wars was related to this author by Chief Sergeant Canas, in which a supply cache, discovered near the Zaire border in a successful raid, contained not only the usual Soviet equipment, but an AR-10. What made this so peculiar was that the AR-10 discovered was not a Portuguese model, but rather one of the 2,508 delivered to the Republic of the Sudan in 1958. The rifle had almost certainly made its way from the conflict in the south of the Sudan, through Zaire, and into the hands of the rebels. Whether this sprung from an intentional move by the Sudanese government or an individual soldier or official, a rebel action to supply the Angolan insurgents, or simply the black-market sale of a

320. In a photo taken during Sergeant Major Rosa's tour in Guinea in 1968-1969, Rosa and his squadmates in the 12th Battalion re-board their Alouette III helicopter after the conclusion of a successful mission. Here he can be seen in the foreground at right, carrying his distinctly less-favored rifle, the G3.                    courtesy Portuguese Air Force

rifle that had found its way into the Congo, can never be known.

The international standing and correctness of Portugal's mission to keep control over its far-flung territories might be open to debate, but the exemplary service of the *Caçadores Paraquedistas* and their AR-10 rifles is not. The praise that the men who carried them in some of the fiercest combat ever encountered by any soldiers speaks more forcefully of the prowess and peerless capability of the AR-10 than any technical analysis or third-party opinion can. It is in large part based on these words from those who served that this author so often lavishes hyperbolic praise on this masterful battle rifle.

# Island Clashes: the AR-10 Once More in Portuguese Service

In 1975, as the Portuguese African territories were offered independence, the new military junta helped another of Portugal's foreign possessions to set up its own statehood. The peculiar province of East Timor had been part of Portugal since 1769, and had suffered terribly under Japanese occupation during WWII, with more than one tenth of the territory's population being killed by the invaders.

Unlike the western half of the island of Timor, the East was Portuguese when the Dutch colonized the rest of the Indonesian Archipelago, and so the independence of that country did not change the situation in Portuguese Timor. Instead, a brief civil war between opposing domestic camps followed Portuguese government withdrawal from the islands, but Portuguese troops stayed to help safeguard the sovereignty of the new nation from an aggressive Indonesia, which made no secret of its desire to annex the territory. Just as the Revolutionary Front for an Independent East Timor (Fretilin) government took power, the Indonesian Army invaded from the west on December 7, 1975.

Despite the Indonesians framing their mission as preventing the formation of a communist state in the Pacific, the international community recognized that they were simply making a grab for the resources and land of the former Portuguese province. This full-scale invasion was preceded by several months of raids by the Indonesian *Kopassus* special forces units into border towns, in one of which they infamously murdered five Australian journalists who were attempting to report on the action.

When the push began in December, naval landings at the former provincial capital of Dili took place, while airborne infantry from *Kopassus* dropped in, surrounding the new Portuguese-trained government troops. On hand in the territory, conducting training for the new East Timorese forces, were none other than paratroopers of the 31st Battalion, still carrying their AR-10s. Bravely engaging the invaders, the Portuguese performed a fighting retreat, to be withdrawn from the island by the Air Force and Navy.

Indonesia's invasion of the ethnically-distinct new nation of East Timor (the native inhabitants were more closely related to the aboriginal peoples of Australia and New Guinea than to the ancestrally-Malay Indonesians) looked to those observing it more like an extermination campaign than merely a bid to take territory. In storming the capital the army and special forces soldiers killed every man, woman, and child they encountered, reserving special brutality for the Chinese residents of the area. The Timorese appealed to the international community for help, and their Portuguese-trained military fought a very successful campaign against the aggressors in the mountainous interior for several years. However, the fear of a communist state trumped up by Indonesia kept the United States and other NATO members on the sidelines. Even in the freely-admitted estimate of Indonesia's own government at the time, their forces killed between fifty and eighty thousand Timorese civilians in just the first two years of the invasion.

Although there is no documentary or photographic evidence to support it (which is not surprising, since the Indonesians massacred all the journalists they encountered), it is theoretically possible that the *Kopassus* were armed with AR-10s when they engaged Battalion 31 of the Portuguese Airborne. If this were actually the case, then it would be the only such instance in history of two forces armed with AR-10s fighting one another. It is exceedingly rare for this possibility to exist even with very ubiquitous weapons. Other than both the German and British forces using Maxim machine guns during WWI, the only confirmed instance of this phenomenon of exact weapon parity occurring elsewhere in the 20th century involved the FAL, when Britain and Argentina fought for the Falkland Islands, with the infantry of both sides being equipped with versions of that rifle.

# After Many Changes of Ownership: the AR-10 in Rhodesia

As the Portuguese withdrew from Mozambique, leaving behind their AR-10s for the nefarious use of Samora Machel's band, the Republic of Rhodesia lost a staunch ally to its east. Having supported the Portuguese in many operations along their shared border, the Rhodesians were very sorry indeed to see their friends go, as almost immediately the rebels they were fighting received support in setting up camps within the frontier of newly-independent Mozambique. Receiving an overwhelming amount of Russian weaponry and standardizing their military around Soviet-issue equipment, the new government quickly presented its cache of Portuguese AR-10s to the Rhodesian insurgents. Led by the now-infamous Robert Mugabe, the Zimbabwe African National Union (ZANU) rebel group had been fighting a bloody war against the Rhodesian authorities since 1962, sometimes in cooperation and sometimes in conflict with its rival, the Zimbabwe People's Revolutionary Army (ZIPRA). Both of these organizations were Stalinist and racist in ideology, advocating extreme Marxist doctrine and the primacy of their particular tribal group.

Rhodesia (Southern Rhodesia as it was then known) had been a self-governing colony of Great Britain since 1923, and from 1953 to 1963 was part of the Federation of Rhodesia and Nyasaland. This subnational group was set up partly to prepare the three political entities within it for full independence, and by 1963, Northern Rhodesia had established itself as Zambia, and Nyasaland had declared independence as Malawi. Only Southern Rhodesia under Prime Minister Ian Smith held out, not accepting the demands the English placed on it for recognizing it as an independent country. Its two former partners in the Federation had been permitted by the Crown to declare independence, because they adopted policies of universal suffrage *ab initio*.

Southern Rhodesia, however, insisted on sticking to the voting system established under its earlier constitution. Its law was very similar to the suffrage requirements of the United States before the 14th Amendment was passed at the end of the Civil War, except that it made no distinction between race and gender. To be a member of the voter rolls in Southern Rhodesia, one had to possess a requisite level of education, income, and property. Although this had the effect of making white and Asian people disproportionately represented among those with the right to vote, the progressive policies of Ian Smith and his Rhodesian Front Party of providing universal educa-

tion and healthcare to all citizens regardless of race saw a steady increase in the percentages of black Rhodesians on the voter rolls and in government.

This was not enough for the British, however, and after years of stalled talks, Ian Smith declared Southern Rhodesia independent in 1965, beginning a period of isolation unparalleled in world history. Receiving only tacit support from South Africa, which was later cut off, Rhodesia was never recognized as independent by any other country, and was forced to make do with near-complete self-sufficiency. While fighting a guerilla war throughout its entire brief existence, Rhodesia nevertheless managed to grow its economy at a staggering rate, increase the standard of living and income for all races of its people, and by 1978, had reached a settlement that would allow black majority representation in the government.

The constant state of warfare present in Rhodesia had given rise to a military organization that was considered by contemporary observers and historians to be, man-for-man, the finest fighting force in the world. Military service was compulsory for white, Asian, and mixed-race men, and was on a volunteer basis for black men. Recruitment of black men of different ethnicities (several culturally and linguistically-distinct groups made up the black population of Rhodesia) was never a problem, however, as there was always a plethora of patriotic volunteers seeking entry to the vaunted Rhodesian African Rifles (RAR), which comprised the backbone of the regular infantry. The military wing of the British South Africa Police (the government's internal security force) was also swelled by large numbers of black volunteers, augmenting the white, Asian, and mixed-race conscripts.

In addition to the regulars, a very high percentage of the members of the Rhodesian Army belonged to especially selective élite units. The first and largest of these was the all-white Rhodesian Light Infantry (RLI). These air-mobile light-fighters operated both in vertical envelopment missions, and as regular infantry in coordination with the Rhodesian African Rifles. Similar in function, but smaller and more specialized, was the Rhodesian Special Air Service (SAS), another all-white unit that made use of air power to conduct their operations. Originally the C Squadron of the British SAS, this unit changed its allegiance to that of its new country after independence, but kept the traditions of its parent service.

Finally, and most formidable of all, were the Selous Scouts, constituted in 1973 as an all-purpose special infiltration force that carried out unconventional warfare against terrorist targets both within and outside of Rhodesia's borders. The training and selection process for the Scouts was extreme to the extent that examinees would go their first few days of training with no food, and only rotten or scavenged scraps after that. This unit was made up of volunteers from every Rhodesian ethnic community, and even welcomed former ZANU and ZIPRA terrorists who had changed affiliation after being disillusioned by the extremism and brutality of their cause. The unit would relocate the families of enemy fighters who joined it, and provide protection, medical care, and food for them, making the prospect of fighting for the government an enticing one for former rebels.

The Selous Scouts made one of their specialties cross-border raids into neighboring countries where rebels had established bases. One of the most successful of these was the Nyadzona Raid into Mozambique, where 84 Scouts killed over one thousand uniformed enemy fighters and 30 Mozambiquan Army advisers, and returned safely with only four of their own lightly wounded.

Modeling their strategy on the effective counterinsurgency operations developed by the Portuguese airborne, the Rhodesian security services refined these methods to an art, building on the concept of vertical envelopment to develop Fireforce. This tactical doctrine involved coordinated helicopter and parachute insertion, combined with true aerial support from both fixed and rotary-wing aircraft from the air force, and mechanized or traditional infantry sweeps. The Rhodesians used the same French Alouette III helicopters as their Portuguese predecessors to land members of the SAS, RAR, and RLI, but deployed them in four-man teams known as "sticks," instead of the five-man squads of the Portuguese airborne.

Throughout much of what came to be called the Rhodesian Bush War, the rebel fighters were extremely disorganized and poorly led, relying on sheer numbers and Soviet equipment to gain victory. As Mozambique became their main base of operations, the Rhodesians began to concentrate their counterinsurgency efforts in that frontier region, honing their tactics to a sharp edge in the process. The use of aerial surveillance and tracking by the Selous Scouts would give forward command advanced warning of an incoming wave of fighters about to cross the border. The next step was a deployment by airborne infantry of the SAS, RLI, or RAR to drop in behind the enemy

force. These were often joined on the ground by hidden Selous detachments who were already camped out on the border performing reconnaissance.

These units would locate the enemy routes of infiltration and exfiltration, and form a defensive line, entrenching themselves in a wide formation along the border and waiting. Next, the regular infantry of the RLI, RAR, or other units would move forward and meet the rebels head-on with mechanized infantry operating in a fire-and-maneuver pattern, with sudden aerial bombardments and strafing attacks aiding them in their forward progress. What would usually happen at this point was that the rebel force would begin to sustain casualties, waver, and then retreat.

These poorly-led, untrained, and undisciplined contingents were not militarily capable of performing an orderly withdrawal, and a mad rush of every-man-for-himself would almost invariably ensue. As they would near the border, however, they would encounter the camouflaged and dug-in line of airborne infantry, who would in turn mow down any who did not immediately surrender. This clever use of a blocking force in conjunction with a mobile one in a hammer-and-anvil type of operation was one that the Portuguese had learned to employ throughout their 13-year conflicts, which they had developed in co-operation with the French, who used similar tactics in Algeria and Indochina. This novel solution of having an entrenched line set up along the routes of retreat solved the problem essential to warfare against a mobile guerilla force; of taking a rebel position only to find that it had been mostly deserted.

Relying almost exclusively on South African equipment secretly provided to them, the regulars, RLI, and most other Rhodesian infantry were equipped with British, Belgian, or South African variants of the FAL. Opinions of this rifle were generally quite good, but when some forward units began noticing strange new weapons on the bodies of enemy fighters, members of the various special forces units decided to give them a try. Individual members of the Rhodesian SAS and Selous Scouts picked up and began to carry these third-hand Portuguese AR-10s as their main weapons. These operators, who made use of the captured ArmaLite rifles, by then somewhat familiar at least from the international press with the look of the now-famous AR-15, were very impressed with the light weight, controllability, ease of maintenance, and firepower of their new weapons. Interestingly, a disproportionate number of standard-variant rifles built on the semi-automatic-only lower receivers intended for the fabrication of sniper rifles ended up in Rhodesia, with the veteran sources of this book having observed and used several such examples.

Another very intriguing and secretive type of operator on the Rhodesian side was also known to appreciate the battlefield-pickup AR-10s that started to become available in 1974. These were brave American volunteers who traveled halfway around the world to sign up and help the Rhodesian government reinforce the ramparts against the spread of communism. Although the U.S. government was staunchly opposed to Rhodesian independence, and would only accept it as a country that offered universal and unqualified suffrage, many private U.S. citizens appreciated the progress towards racial equality the Smith government was making, and saw the potential for the horrors that would be visited upon the Rhodesian people of every color, should the fanatical Stalinist Mugabe come to power.

The Rhodesian government became increasingly aware of, and began to take advantage of, this groundswell in support among Americans for their stand against Soviet imperialism in Africa, and advertised in defense-industry publications for foreign recruits. These overseas volunteers were not brought over simply as a propaganda move to demonstrate international support, but were quickly integrated into regular and élite military units, in short order becoming actively involved in combat operations.

While Portuguese veterans returning from their valiant service to their post-revolution country did not receive nearly the warm and appreciative welcome they deserved, the aftermath of the Rhodesian Bush War was absolutely disgraceful, and sometimes even deadly, for those who had honorably served their country. Those who had worn the uniform of the Republic of Rhodesia, particularly those from the élite units like the SAS, Selous Scouts, and RLI, faced, and to this day still live in fear of, violent reprisals against themselves and their families by cowardly thugs in the employ of the current government of Zimbabwe, which Rhodesia became in 1980. For this reason, the sources interviewed by this author for this section on some of the most secretive use of the AR-10 uniformly wished not to have their identities revealed, and as such some details of the accounts herein have been altered to protect their anonymity. However, what can be publicized is that the Portuguese AR-10s gave excellent service and satisfied all who carried them there.

One complaint did come out concerning the rifle, though, by one of the persons involved in combat operations in Rhodesia. The fighting men in the field who had grown accustomed to the FAL, with its heavy-gauge steel magazines, did observe that the lightweight corrugated aluminum magazines produced by Nederlandsche Metaalfabrik for Artillerie-Inrichtingen were somewhat flimsy by comparison. Various reports cite issues where soldiers who had to either go prone quickly or dig in and crawl long distances on their belly finding that some of their magazines were damaged from this abuse. However, this problem was not the fault of either Eugene Stoner's original design, or of the excellent work of the Dutch subcontractors who had manufactured the magazines. As discussed earlier, the AR-10 magazine was not originally envisioned as a permanent fixture of the weapon, but as an expendable accessory that could be thrown away when emptied under combat conditions.

These criticisms were in sharp contrast to the reviews of the magazine design this author received from veterans of Angola and Mozambique. The Portuguese servicemen, in fact, singled out the Stoner-patented magazine as being an excellent accessory, relating how it provided sufficient strength and ease of loading, while simultaneously cutting the weight of their combat load and of the loaded rifle considerably. In fact, in response to the praise for this novel design, the Portuguese armed forces began issuing their standardized second-choice G3 rifles with aluminum magazines, instead of the original steel type.

The unique nature of the Rhodesian experience with the AR-10, in that they captured these rifles often with only a few magazines and had to make continuous use of them for many years, extended the design's use far beyond its originally-planned utility. While the Portuguese and Sudanese militaries bought huge quantities of extra magazines, from which a soldier could simply get a replacement when one was starting to become worn, the Rhodesians were restricted to using only what accessories they found on surrendered enemies or on the bodies of the dead, and so encountered this problem which was never reported by the Portuguese.

This particularly shadowed chapter in the AR-10's history has all of the excitement and intrigue of a real-life spy story, but in the end simply speaks volumes about the incredible robustness, utility, and peerless combat effectiveness of the AR-10 design. That these weapons, used for over a decade and fired tens of thousands of times, abandoned uncleaned, carried by rebels and then captured or picked from their bodies, all in some of the most corrosive and damaging climates on the planet, would be selected

321. Ian Smith, the last Prime Minister of Rhodesia, while in service with the 237th Squadron of the RAF during WWII, wherein he flew Hawker Hurricane fighter aircraft.
public domain - the Internet

and used in their very-worn condition by members of the most élite units of one of the most capable militaries of their time is a singularity in military history, as absolutely no material support was obtainable by the Rhodesian adopters of the AR-10.

Just as with the Portuguese mission in Africa, some still debate the morality of Rhodesia's resistance to immediate and complete majority rule. However, those familiar with the history of what has unfolded in that most unlucky of countries, including the oppressed people who continue to live there, almost unanimously recognize that all Rhodesians, regardless of ethnicity, were far better off under the Republic of Rhodesia. In 1979, Ian Smith and newly-

elected Prime Minister Abel Muzorewa attempted to form a black majority government that would keep the country a democracy and reject the Stalinist ZANU extremists, but this was rejected by the outside world, and in elections of 1980, in which his party made wide use of intimidation, violence, and voter suppression, Robert Mugabe became Prime Minister of the new Republic of Zimbabwe.

Within the first few years of his rule, he invited North Korean advisors to train a special military unit; the 5th Brigade, which was made up completely of members of his own tribe and answerable only to him, and not to the military chain of command. He then used this unit to massacre between 20,000 and 30,000 civilians of the Matabele ethnicity, in order to quell peaceful resistance to his plans of making the country a one-party Stalinist state controlled by his Shona ethnic group, which he has ruled unopposed through rigged elections, intimidation and violence ever since. What was once the most economically and agriculturally productive country in all of Africa is now a wasteland where starvation, disease, and poverty run rampant, and the standard of living is among the lowest in the world.

Ian Smith, who stayed in Zimbabwe as leader of the political opposition, was widely respected by both his white minority (most of whom Mugabe either slaughtered or forced out of the country) and the black majority until his death in 2007. Mugabe, on the other hand, still clinging to power with a death grip at age 91, continues to use starvation and mass killings against native Africans of other ethnic groups, cementing his reputation forever as one of the worst racists in human history. The Matabele people, the ethnic group who have been on the receiving end of most of this villain's worst brutality, are as culturally distinct from his Shona people as the white Rhodesians were, no matter how much they may resemble to outsiders the other black native groups of southern Africa. When Ian Smith passed away he was remembered fondly in his native land. Mugabe, however, for whom death will certainly come soon, will be vilified for all time as a modern-day Hitler.

*Chapter Twelve*

# The Journey Home

In the whole history of the rifle's development, no AR-10s were ever turned out on an assembly line basis in the United States, and total production reached only 47 serially numbered examples, plus Richard Boutelle's unnumbered early prototype (fig. 108), and the final one-off AR-10A, serial no. 1048 (fig. 304).

As discussed above, the intrepid but ill-fated program of production and marketing in the Netherlands had met with an untimely end by 1961, and with ArmaLite on its own and no longer having the legal right to manufacture or further develop the design, the AR-10 fell into slumber so deep that none could envision its ever waking again.

## An Occult Collector's Item: the Surplus AR-10s

The very first stirrings from the deep and dark caves into which this most precocious of battle rifles had retreated were felt as the tiniest blip on the sonogram in the late 1970s. Major purchasers like the Sudan and Portugal had finally run out of spare parts, and began to liquidate their stocks of AR-10s. With the great popularity of the AR-15 in the civilian arms market, they saw an opportunity to realize some revenue from their extremely rare copies of the weapon on which this new sensation of the sporting arms world was based.

Both countries then standardized their armament on the German G3, with Portugal manufacturing licensed versions of the wood-stocked G3, known as the M961, and later the plastic-stocked *Espingarda Automatica* M63 version, produced by Fàbrica Militar de Prata (FMP).

It says a great deal for the innate sturdiness of the AR-10 design that the these surplus rifles were even in operable condition following the arduous military service described in Chapter Eleven.

### Surplus AR-10s from Canada

Nevertheless a market was indeed very present for these surplus AR-10s, and civilian arms merchants eagerly bought up all that were offered. One particularly enthusiastic Canadian importer, E. A. Wilke, bought crates of the weapons wholesale, then marketed and shipped them to various nations where

governments allowed their citizens to collect such fine pieces of firearms history. Most of the rifles were first converted by modifications to the lower receiver, rendering them capable of only semi-automatic fire, before being sold in Canada, Australia, and New Zealand.

### Legal Restrictions in the U.S. Result in AR-10 "Parts Kits"

In its firearms regulation legislation, the United States had enacted a peculiar provision by which a firearm that had originally been capable of fully automatic fire was always considered a machine gun, making it subject to especially stringent federal regulation whether or not it was still physically capable of fully-automatic fire. To abide by this law, which

only treated the receiver of a formerly-fully automatic firearm (the lower receiver only on AR-10 and AR-15 rifles) as the prohibited part, many of the rifles had their lower receivers removed before being imported into the U.S., either directly from their original users or through Canada.

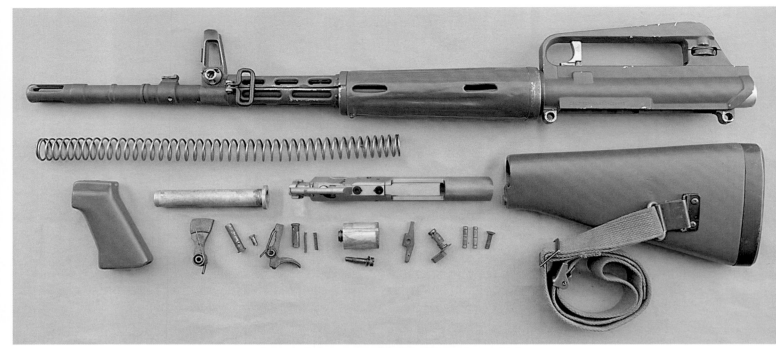

322. A typical Portuguese/NATO AR-10 "parts kit", containing every component except the lower receiver, which by itself was classed as a machine gun under U.S. law.
Charles Kramer collection

Taking another route, customers at this time wishing to buy an AR-10 from a foreign wholesaler like Mr. Wilke in Canada could have the seller cut the lower receiver in half before sending the entire weapon. This cut rendered the lower receiver no longer a firearm according to U.S. federal law (the law has since been updated to require five cuts to render such a full-automatic-capable receiver legal), and the buyer could then have the receiver halves rewelded and the hole for the auto sear pin filled, making the weapon finally an unrestricted, civilian-legal semi-automatic rifle. The very best of these reweldings were performed by Boyd Hahn, who engraved his name in a hard-to-find place on the lowers he reworked; usually under the pistol grip where it was only visible when this component was removed, giving the rifles a completely authentic original look.

Customers who either elected to have their AR-10s sent to them by Mr. Wilke, or bought them from another source without the lower receiver, took delivery of what was known as a "parts kit"—a partially or completely disassembled AR-10 with only the shell of the lower receiver missing.

# Aftermarket (Semi-Auto-Only) Lower Receivers

Although it took several years, a number of small U.S. manufacturing concerns began fabricating replacement semi-automatic-only lower receivers out of aluminum or steel that would accept the AR-10's trigger group, buffer tube, bolt-holdopen and magazine release assemblies, thus allowing those who had purchased parts kits to finally assemble a working, legal AR-10.

Although collectors petitioned them persistently, no major U.S. firearms manufacturing company ever produced AR-10 lower receivers, and thus the manufacturers' markings on the ones that were fabricated provide a nostalgic glimpse into a bygone world of small-scale firearms innovation.

Every single small American company that rose to the challenge and stepped up to meet this niche demand has since gone out of business, and thus this history of diligent cottage industries has almost been lost.

## Telko of Miamisburg, Ohio

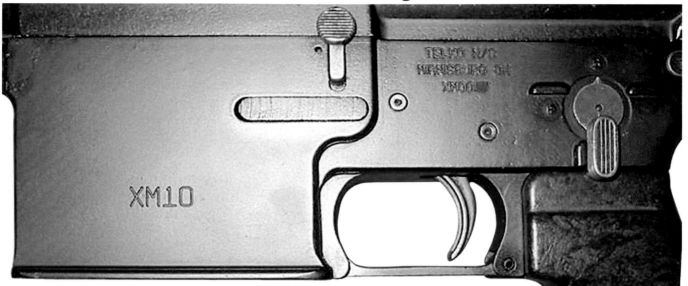

323. Left side closeup of an aftermarket "XM10" AR-10 lower receiver, machined from 7075 aluminum by Telko of Miamisburg, Ohio.   photo from Marc Miller collection

Some of the very best receivers available were produced from 7075 aluminum by Telko of Miamisburg, Ohio, which marketed their "XM10" receivers and assembled rifles they built from parts kits sourced through Paragon Sales and Service of Joliet, Illinois.

This firm produced dedicated receivers for both the Portuguese rifles (having a closed rear for the captive charging handle) and all other models, which they collectively called their "Sudanese" lowers.

A few years after Telko ceased fabricating AR-10 lowers, it was estimated by collectors that this endeavor had only resulted in around 100 receivers.

## Specialty Arms of Springfield, Ohio

Also making top-quality lowers from the same material and for both types of rifles was Specialty Arms of Springfield, Ohio. Although the production quantity was never ascertained from this company, it is believed that their total output was somewhat more than that of Telko.

324 (right). Left side closeup of an aftermarket semi-auto-only AR-10 lower receiver, serial no. S80007, showing markings indicating manufacture by Specialty Arms of Springfield, Ohio.                    Charles Kramer collection

# Central Kentucky Arms of Cynthiana, Kentucky

325. Left side closeup of a Portuguese/NATO AR-10 "parts kit," serial no. 006xxx, assembled into a complete rifle on an aftermarket milled steel lower receiver manufactured by Central Kentucky Arms of Cynthiana, Kentucky.

Charles Kramer collection

Attempting to make a cheaper and stronger product, Central Kentucky Arms of Cynthiana, Kentucky milled lower receivers for Portuguese and Sudanese AR-10s out of steel, resulting in a heavier component, but one which was quite durable. Rifles built on these receivers were marketed by Bumble Bee Wholesale, and were some of the earliest semi-auto versions offered to the American public.

CKA, as it was usually known, turned out a fairly large number of units, compared to the output of its contemporaries Telko and Specialty Arms in the early generation of receiver makers, fabricating an estimated 250 pieces.

326. Left side closeup of an even later Portuguese/NATO parts kit, serial no. 007197, built into a complete rifle on another milled steel aftermarket lower receiver from Central Kentucky Arms of Cynthiana, Kentucky.

Charles Kramer collection

## Cast Lowers from H&H Enterprises, Redwood City, California

Finally, another manufacturer, Vic Holbrook of H&H Enterprises in Redwood City, California, took a third path in producing lowers. He did most of his work in the mid-1980s, after most of the other receiver manufacturers had ceased production. Instead of milling from billets or forgings of 7075 T6 aluminum or ordnance steel, he opted to cast receivers in two parts from 6061 T6 aluminum. He would produce the two halves, and then dip-braze them together. Although the casting process, production in two joined halves, and the 6061 material are all weaker than the other methods and materials used in AR-10 lowers, Holbrook's great attention to detail, including allowing customers with parts kits to order lower receivers inscribed with their own upper receiver's serial number, and the fine fit and finish of his receivers, were particularly well-renowned.

After putting together a run of 50+ dedicated Sudanese lowers, he did what no other entity ever thought to try, and actually added a slight change to the lower receiver design perfected by Artillerie-Inrichtingen. Being singularly responsive to his customers, Holbrook had heard many horror stories about owners of parts kits being able to find a lower, only to realize that it was for the wrong type of rifle (Sudanese when they needed a Portuguese, or vice-versa). As a solution, he built the rest of his lowers with the extension found on the Portuguese receivers that closed off the channel for the charging handle, but milled a small slot in it, which would allow the Sudanese-type charging handle to protrude. Therefore, these later lowers could be used with both Portuguese and Sudanese uppers, with the extension keeping the Portuguese charging handles captive, while allowing the one-piece Sudanese handles to slide through.

The fact that he began making his products at the pleading of owners of parts kits who could not locate lower receivers after most of the other manu-

327. Left side closeup of a cast steel aftermarket AR-10 lower receiver, serial no. 00033, made by H&H Enterprises of Redwood City, California.    Charles Kramer collection

facturers had shut down, his innovative universal lower design, as well as the close contact he kept with AR-10 devotees across the country, made Vic Holbrook a particularly beloved figure in those early, formative years of scholarly and collector interest in the ArmaLite AR-10. Furthermore, although enthusiasts today who build their own AR-15s and AR-10s (this author included) would never think of using a cast lower receiver or one made from 6061 aluminum when constructing a new rifle, AR-10s put together on H&H receivers are still going strong today, with many having had thousands of rounds fired through them without issue. In fact, no reports have surfaced of one of these lowers ever failing.

Given the great care with which these pieces were hand-crafted, Holbrook's hallmark product is one of the rarest, and also one the most desirable, of the AR-10 aftermarket lower receivers, with total production numbering somewhere over 100 units.

## Pennsylvania Arms Company, Duryea, Pennsylvania

Not every entity that entered the business of fabricating AR-10 lower receivers was as responsive as H&H Enterprises's Vic Holbrook, however. John Tanis's Pennsylvania Arms Company of Duryea, Pennsylvania (affiliated with Pennsylvania Stovepipe Works) produced at least a few lower receivers for Portuguese rifles, but quickly began to amass a backlog of orders they could not fill. The receivers they did furnish

reportedly exhibited very poor workmanship, and required extensive machining and fitting before they could be assembled. Eventually the customers who were charged in advance and did not receive their lowers had their money returned to them, but this firm got a bad reputation in the collector community as a result.

## The Pennsylvania Arms Company's Short-Lived T-12 Shotgun

328. An advertisement for the short-lived Pennsylvania Arms Company's "T-12" autoloading shotgun, based on the Portuguese AR-10. This ad originally ran in *Shotgun News* in May, 1982.                    author's collection

The reason for which PAC is best remembered, however, was their ambitious and ultimately-unfruitful attempt to produce a shotgun version of the AR-10. In 1982, Tanis experimented with a Portuguese AR-10 by replacing the barrel with one with a 12-gauge bore, and having a new bolt and carrier designed that would feed 12-gauge shotgun shells. Using a Dutch AR-10 magazine with a modified follower, he succeeded in producing a single prototype that would fire very low-pressure loads, making use of a simple blowback action.

Advertised in hyperbolic and unrealistic terms in *Shotgun News* in May, 1982, the "T-12" autoloading shotgun proved a commercial and mechanical failure.

Despite advertising this weapon to dealers with the offer that they could reserve copies that would be delivered once a production version was reached, this bizarre attempt to produce an AR-10 shotgun failed, as the prototype did not function reliably, and no modifications proved successful.

## Sendra Corporation, Barrington, Illinois

Riding to the rescue in response to impassioned pleas from subscribers to *The AR-10'er* (discussed below) and other interested collectors, Gerald "Jerry" Drasen, founder of Sendra (an anagram of his last name) Corporation of Barrington, Illinois, announced in July, 1985 that he would be following the lead of the pioneering Vic Holbrook of H&H Enterprises and commencing production of AR-10 lower receivers. Taking the more traditional route of milling his receivers from fine 7075 aluminum forgings to produce the strongest and most authentic product, Jerry stated, somewhat ungrammatically, that his goal was to turn out "the lower to which all others will be judged."

At this point as few as 450 semi-auto lowers had been produced by the combined efforts of the by-then defunct small firms making them for enthusiasts, and despite the beginning of supply (never more than a trickle) from H&H, a great many owners of AR-10 parts kits were resigned to the bitter fate of never restoring their rifles to their former glory.

Sendra, unlike its predecessors, was a somewhat larger manufacturer, already having produced and marketed parts, including upper and lower receivers, for AR-15s. The actual machining work was done at its subsidiary Drasen Machine Tool. Building on its experience in the market, this enthusiastic company started shipping its products in an astounding six weeks from its initial announcement that it would be tooling up to make AR-10 lower receivers.

The first and most numerous type of Sendra lowers were for the Portuguese AR-10, with smaller installments for the Sudanese/Guatemalan pattern parts kits following.

In addition to the standard 7075 aluminum lowers, Sendra produced at least a few experimental models from milled ordnance steel. These units followed the same techniques as the CKA receivers, but

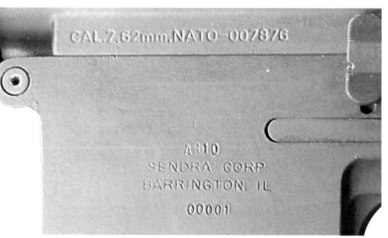

329. Left side closeup of an aftermarket AR-10 lower receiver, serial no. 00001, milled from a 7075 aluminum forging by Gerald "Jerry" Drasen of Sendra Corporation of Barrington, Illinois.

Sendra became the largest single manufacturer of aftermarket semi-auto AR-10 lower receivers, and also produced at least a few experimental models from milled ordnance steel.                         *Charles Kramer collection*

with none of the rough look, poor machining quality and substandard finish of those earliest of parts.

On the contrary, Drasen's lowers were universally considered to be the most professionally done, at least the equal of the products of Telko and Specialty Arms. They surpassed their competition in aesthetics and finish, thanks to an industrial-scale brass tumbler (a device used by ammunition handloaders to mechanically clean fired shell casings) filled with ceramic medium which polished the receivers to a bright shine before they were treated with the then-current military-specification hard anodized finish.

Sendra became the largest single manufacturer of aftermarket semi-auto AR-10 lower receivers, dwarfing the numbers of all previous makers combined. The exact quantity produced is unknown, but it is estimated that between 500 and 2,000 of these units were manufactured, enough to complete between fifty and eighty percent of all the AR-10s

imported as parts kits. It was and continues to be a boon for enthusiasts that the highest-quality aftermarket lower receivers would also become the most numerous.

## The Surplus Guatemalan AR-10s Sourced and Sold by Armex

A firm called Armex became particularly well-known for getting the drop on the whereabouts of the Guatemalan AR-10s, the existence of which was completely unknown in the U.S. collector community at the time. Armex negotiated directly with the Polytechnic School in Guatemala to receive their

entire stock of AR-10s, and imported most with their full-automatic lower receives still intact, selling them under the special regulations for machine guns in the United States. They sold a few others as parts kits, although they never tooled up to produce lower receivers for them.

## Summing Up the Surplus AR-10 Story

These barely-remembered footnotes to the history of the AR-10, replete with their own triumphs and follies, may pale in comparison to the grander endeavors of ArmaLite and Artillerie-Inrichtingen, but they nonetheless give the reader a look back at the unique figures that stepped up to feed the interest of a dedicated group of devotees of this fantastic rifle.

In that bygone age before the Internet, those who had a strong interest in an obscure piece of historical weaponry like the AR-10 had to go to amazing lengths to indulge their passion, and this author is glad to have been able to record the above short history of the key players during this period.

## The Australian AR-10 Treasure Trove is Destroyed

On a very sad note indeed, in 1997 the Australian government perpetrated against its citizens a truly heinous and unforgiveable act. Following the cowardly use of a semi-automatic rifle by a criminal to murder a large number of people, the leadership of that unfortunate country enacted a policy that essentially banned civilian ownership of almost all guns. They then went about confiscating the firearms that many Australians had spent a lifetime collecting and had used for hunting, sport shooting, and protection, and proceeded to actually destroy them. It is estimated that as many as one thousand original AR-10s, legally imported as surplus over the years, were reduced to scrap metal under this draconian program.

The reaction of punishing peaceful and law-abiding gun owners for the rare actions of madmen and criminals may be seen as lunacy, moral poverty, or at least illogical; but the fact that so many of the vanishingly few, priceless examples of this historic weapon were wantonly expropriated from their blameless owners and destroyed is particularly distressing to this author. These actions appear no different than if the United States had burned every copy of the Quran in America and outlawed the practice of Islam in retaliation for the September 11 attacks. Destroying beautiful and priceless artifacts—living pieces of mankind's history—because of how they can be misused by the evil or insane is folly in the extreme, and the Australian government of John Howard is guilty of this crime against history.

# A Group of AR-10 Scholars Coalesce

Meanwhile, in other more fortunate lands, the trickle of these strange and sublime rifles, about which almost no information was available, into the United States, Canada, and New Zealand sparked intense interest among the historical and collecting communities. By 1983, a critical mass of academic inquisi-

tiveness had coalesced around the newly-available (although exceedingly rare and prohibitively expensive, normally costing today's equivalent of over $3,000) surplus AR-10s to such an extent that an organic group of enthusiasts sprang up and started an unprecedented effort to collect and share the scattered knowledge of this singularity of battle rifles.

## Enter *The AR-10'er*

Led by, and on the individual initiative of, Louis T. Carabillo, Jr. of Wethersfield, Connecticut, a monthly newsletter, the circulation of which would eventually grow to over 100, was started. This circular, titled the *AR-10'er*, disseminated to a small group of devoted enthusiasts all that could be learned, speculated, or conjectured about the AR-10, as well as listing commercial sources for rifles, parts, and accessories.

In the days before the Internet, such an undertaking was costly and cumbersome, and was done at great personal expense in both time and money by Mr. Carabillo (LTC as he was and forever will be affectionately known in the AR-10 community). Sharing news about their beloved new acquisitions, trading parts, and reveling in what could be uncovered about the AR-10 was a passion that kept *The AR-10'er* alive and well for over two years.

Although sometimes drawing on original source documentation, a large proportion of the historical "facts" recounted in *The AR-10'er* were based merely on conjecture or rumor, and as often as not turned out to be false or misleading. For its time, however, it was a true historical masterwork, and without the intrepid efforts of LTC and his fellow scholars, much of what is known today about the AR-10 would have been irretrievably lost. This author is eternally grateful to the Herculean travails of those wise and driven men (there were apparently never any women subscribers), whose work serves as both an inspiration and a road map for a serious, primary-source inquiry into the history of the AR-10.

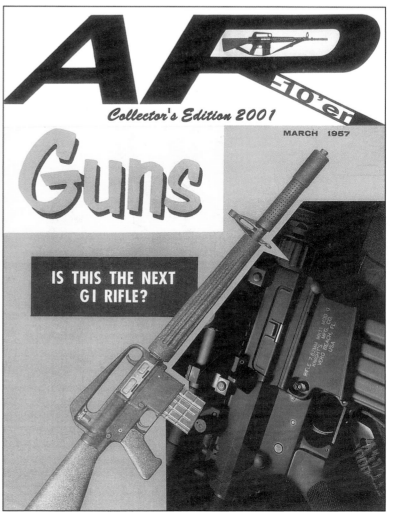

330. The cover of the 2001 Collectors' Edition of *The AR-10'er*, featuring a reprint of the famous *Guns* magazine cover from March, 1957 introducing the ArmaLite AR-10 under the heading "Is This the Next GI Rifle?".

Approximately 1¼ inches thick, the Cerlox-bound Collector's Edition was a compilation of material from previous newsletters, divided into sections containing reprints of numerous articles and features, tips, historical reports, and ArmaLite brochures.

Charles Kramer collection

# The Faucet is Turned Off - a Scorecard by Model

By 1985, the trickle of surplus AR-10s that had steadily made their way into and through Canada and the United States had dried up. That year the Sudanese liquidated the last of their stocks of rifles, and the Guatemalans sent the remainder of their decommissioned demonstration pieces to the U.S. through California. Although some Portuguese troops and small second-line units were seen carrying AR-10s for years afterward, no more shipments of surplus AR-10s out of those countries followed.

## The Cuban Unicorn

The first AR-10 model from Artillerie-Inrichtingen, being either the least- or second-least produced, consequently holds the title of the rarest variant in every country where the AR-10 is collected. With only roughly 235 Cuban models ever being fabricated, only a tiny handful have been identified in Holland, while at least one fully-functioning selective-fire example, serial no. 000312, was legally imported into Canada from a U.S. dealer source in 1966 by the editor of this work.

At least five remain in private hands in the U.S., with not one confirmed instance of a Cuban AR-10 being assembled from a parts kit on an aftermarket semi-auto lower receiver. Thus, other than the Guatemalan models, it is the only version of the AR-10 known to exist in private American hands in its original selective-fire configuration.

Not one of the 135 German test pieces, with their characteristic threaded flash hiders, has been seen since 1958 - 1959, even in a museum in Europe, and most of the rifles from the Cuban order may still be arsenaled in Santo Domingo (as well as the two the Castro brothers reportedly kept for themselves), although it is from this consignment that the few examples known in the U.S. and Canada were drawn.

## The Comparatively Rare Sudanese AR-10

As many of the AR-10s ordered by the Sudanese were subsequently captured by rebels, sold off or smuggled into local arms markets, or destroyed during the long and bloody Sudanese Civil Wars, these rifles, although originally purchased in the largest numbers, became a comparatively rare find (when considering how many were initially produced for the Sudan) in the developed countries that ended up receiving them for civilian purchase.

## The Guatemalan Model: a U.S. Exclusive

As discussed above, the 450 AR-10s that had been sold to Guatemala in 1958 by Sam Cummings were exclusively imported into the United States by Armex, with collectors in other countries often lamenting the lack of this distinct and interesting variant anywhere else.

## The Vanishingly-Rare Transitionals

The Transitionals of both the First and Second series are and remain in contention with the Cuban model for the title of the scarcest of all AR-10 models in civilian hands, especially in the United States, Canada, and New Zealand.

The first variant, used only by the Dutch in their domestic trials, has literally only been seen once outside of Holland, in a private collection in Belgium, and only one is known in private hands even in the country of its manufacture.

The second variant, while more numerous among civilian owners than the first, has only turned up a few times in North America, although a healthy number are maintained in The Netherlands.

The Austrian and Dutch governments, which purchased quantities of them for field trials, still hold them in warehouses and museums, and have never authorized their export.

The likely Indonesian rifles, which may even still be in use, have never been seen or described by any observer from the outside world.

## The NATO Model: the Most Common, and Most Robust

The Portuguese/NATO AR-10s, although never numbering nearly as many as the Sudanese, became somewhat more common than their lighter North African cousins, due to their more robust and refined construction, their use by an ordered and disciplined modern military that did not suffer from the prob- lems of disorganization and corruption that allowed soldiers to sell their weapons for cash, as happened sometimes in the Sudan, and the excellent and fastidious care that was taken of them by their appreciative Portuguese users.

## The AR-10 is Alive and Well in Holland

One place where the collecting of AR-10s has never stopped during the quarter century between their invention in the United States and the return of former military production models to that country, was the place of most manufacture; Holland. A comparatively large and extremely devoted group of collectors and historians has sprung up in that country, started by none other than the core group of Artillerie-Inrichtingen engineers. To this day regular collectors' meetings take place, and the widest and most spectacular array of parts, documents, accessories, and particularly prototypes of the most unusual, one-off designs are traded and admired. With the rifles tested at Hardewijk and Harskamp and then the lot sent to New Guinea having been returned to the home country, it is the case that the vast majority of Transitional AR-10s in non-military hands today are in The Netherlands.

## Some Day My Prince Will Come

It is the fervent hope of those who keep faith with the original AR-10, this author included, that some time in the future the remaining military stocks will be made available to the devoted collectors who would study and take good care of them.

Should they finally see the light of day, those Cuban models still stored in Germany and the Transitionals still crated up in Austria, plus the mysterious thousand—likely tucked away on a heavily-guarded island in the Indonesian Archipelago—would be an historical coup for the ages.

*Chapter Thirteen*

# The Synthesis of Old and New

If surplus military AR-10s being imported for civilian purchase during the late 1970s and early 1980s—and the attendant interest in them giving rise to a novel and devoted newsletter for collectors—could be seen as the AR-10's Medieval Warm Period, the events that followed would be its true Renaissance. In the intervening quarter-century from the early 1990s, almost a decade since *The AR-10'er* ceased publication, through to today, two great pioneers of the modern firearms world labored to bring the AR-10 home both to the people and the military of the United States; the two places it had belonged all along.

## Renaissance I: the Knight's Armament SR-25

### Working With Stoner Again

The seeds for one part of this reawakening were planted in 1974, when a young Florida firearms enthusiast and businessman by the name of C. Reed Knight Jr. first learned about an interesting type of modular light machine gun/rifle weapons system known as the Stoner 63, produced and marketed by the Cadillac Gage Company of Detroit, Michigan, and Costa Mesa, California. Fascinated since he had first heard of this adaptable design, Knight acquired his very first example of one, and later learned that the man whose name it bore was actually the inventor of the by-then-famous AR-15/M16 family of weapons.

That same year Knight was contacted by members of the Navy SEALS, who had used the modular Stoner 63 rifle-and-machine gun system to excellent effect in Vietnam, and now needed them repaired. Knight's expertise from long study of the weapon system enabled him to fabricate effective replacement parts, and get all of the Navy's initial cache of weapons running again.

When working on copying the original parts and furniture for these rifles, Knight "heard that this guy by the name of Eugene Stoner had a house in Florida." Looking him up in a local phone book, Knight called the number of the listed address just a little ways from his town, and the partnership that would play a large part in bringing the AR-10 story full circle began. By then it was late 1978 or early 1979, and Knight, not aware of the historical significance of the process he was starting, was simply on a quest for a better source of replacement parts for more of the Stoner 63 light machine guns that the Navy had commissioned him to rebuild and repair.

### Conserving the AR-10 Legacy

Over the next ten years Reed Knight and Gene Stoner collaborated on a number of different projects, and Knight became a deep historical devotee of Stoner's early work. During that time he was able to acquire virtually all of the original Fairchild AR-10 and AR-15 prototypes discussed in this book, along with most of the weapons and some of the parts that had gone with the sale of the two weapons to Colt's, and many of the rarest and most interesting examples of Artillerie-Inrichtingen design. He even held a reunion for the entire ArmaLite team in his home town in Florida, where all involved in the original development of the AR-10 were brought together for dinner and then a perusal of what was essentially the entire accumulation of their work from the 1950s, all finally gathered in one place for the first time ever.

# A New Military Sniper Rifle is Born

331. On the occasion of the introduction of the SR-25 at
the 1995 SHOT Show, Reed Knight Jr. (left) holds the rifle
with the handguards removed so that Eugene Stoner, right,
can explain details of the gas system to Uzi Gal.
courtesy Knight's Armament Co.

Having worked together at his new company, Knight's Armament, to fulfill design contracts for the military, including the invention of what were for their time state-of-the-art sound suppressors for handguns and rifles, Knight and Stoner were made aware at the beginning of the 1990s of the U.S. military's intention to choose a semi-automatic sniper rifle system. The same project that had seen the Transitional AR-10s tested at Aberdeen in 1960 undergo so many later modifications was now the impetus for seeking a new weapon to give designated marksmen and snipers a much better volume of fire, with the eventual goal being the replacement of the

M40A3 bolt-action sniper rifles that were then in use and had been since Vietnam.

The idea hit both Stoner and Knight at once to base their new rifle design—the very first complete weapon that their partnership was ever to attempt— on the Stoner gas system and the AR-10 design, on which the patents had at long last expired. Their goal was to replicate the AR-10, but with as much parts commonality with the current-production military M16 as feasible.

Seeking first and foremost a government contract, they reasoned that the commonality of replacement parts would be an excellent selling point. Even more attractive to the American defense estab-

332. Left side view of the Knight's Armament SR-25.

Alive and in production again after thirty years of lying dormant, the SR-25 bore the AR-10 spirit through-and-through. With Navy adoption in 2000, it finally brought standardization and recognition to Eugene Stoner's truly great invention.                Institute of Military Technology

lishment would be the almost-identical controls, manual of arms, and training regimen that would be possible by resurrecting the AR-10, the predecessor of the M16.

Working tirelessly on updating the AR-10's internal workings, and achieving reliable functioning with the combination of M16-compatible parts and early Dutch-produced AR-10 magazines, Stoner and Knight arrived at a rifle that was, when fitted with an extra-long barrel and free-floating handguard, capable of the most stunning accuracy ever achieved from a semi-automatic production weapon.

Acknowledging the combination of some of the best features of the AR-15 coupled with the basic design of the AR-10, the pair christened their new

creation the SR-25. The 'SR' stood for "Stoner Rifle", and the numeric designation '25' was reached by adding the numeric designations of the AR-10 and AR-15 together (10 + 15 = 25). The original production version of the SR-25 weighed 10 lbs. 12 oz.

Being built and sold as a dedicated sniper rifle in today's world of standard-issue weapons in assault rifle calibers necessitated a weapon chambered in 7.62mm NATO, which had always been Eugene Stoner's acknowledged favorite military cartridge. Its intentional similarities with the M16 helped the SR-25 gain widespread success and acceptance, and its rapid-fire controllability, capacity for a high volume of fire, and excellent accuracy have earned it high marks from its users.

## Adopting the SR-25 as the U.S. Navy Mk 11 Mod 0

After testing by the Army in the early 1990s, and the successful incorporation of their requirements for refining the system, the U.S. Navy accepted the new SR-25 in May, 2000, designating it the Mk 11 Mod 0, and issuing it for sniper use by the SEALS.

The Mk 11 Mod 0 system (fig. 333) is designed for Match-Grade 7.62x51mm NATO ammunition, and is fitted with an Obermeyer 20" (510mm) Match Target barrel, along with RAS (Rail Accessory System) forend made by KAC, consisting of an 11.35" (288mm) long aluminum Match forend which allows

for quick attachment/detachment of MIL-STD-1913 components. The forend makes no contact with the barrel, thus contributing to the extreme accuracy of which this system is capable.

As adopted, the Mk 11 system includes the rifle, 20-round box magazines, QD (Quick Detachable) scope rings, a Leupold Mark 4 Mil-dot riflescope, Harris swivel-base bipod on a Knight's mount, and QD sound suppressor, also manufactured by Knight's Armament, which slides over the barrel and is secured by a locking gate on the gas block.

## The Civilian SR-25

Early overruns in production for the Navy would also give American civilians a new choice in acquiring a modern-production variant of the AR-10. Introduced at the 1995 SHOT Show, both in its sniper configu-

ration and in a lightweight "sporter" version that was very similar to the original AR-10, the civilian SR-25 proved popular with the members of the shooting public who appreciated the light weight, modularity,

333. The U.S. Navy's Mk 11 Mod 0 on the range, with
scope, bipod and suppressor attached.
courtesy Christopher R. Bartocci

and accuracy of the AR-15, but who wanted such a weapon in a more powerful caliber. Despite the hefty price tags, the small quantities of SR-25s made avail-

able for commercial sale were an instant hit within the American shooting sports community.

## The U.S. Army M110 Semi-Automatic Sniper System

334. The U.S. Army's M110 version of the SR-25 on the
range, fitted with the Knight's Armament flash suppressor.
courtesy Christopher R. Bartocci

Although he had passed away in 1997 at the age of 74, Eugene Stoner would no doubt be happy to know that in 2008, exactly half a century after the first copies of his design rolled off the A-I assembly line in Holland, the U.S. Army adopted the Knight's Armament SR-25 as the M110 Semi-Automatic Sniper System. Thus the armed forces of his home country would at long last be equipped with an improved, modernized version of his AR-10.

On this version the KAC free-floated rail system has been replaced by a URX modular rail, with an integral folding 600-meter back-up rear sight. The scope is a Leupold 3.5 - 10x variable-power daytime optic, fitted in KAC's one-piece 30mm scope mount instead of two separate scope rings.

The M110 butstock is fixed, although the buttplate is adjustable for length of pull to match user preferences without tools, by means of a notched, hand-tightened knob on the rear right-hand side of the stock. This feature was added during the change in nomenclature from XM110 to M110. The fixed buttstock also features integral quick-detachable sling swivel sockets, located on each side near the rear of the lower receiver.

The M110 is currently undergoing renovation to achieve a lighter and more compact variant, and it appears likely that this new version will be carried by snipers and designated marksmen of the U.S. Army for a long time to come.

# Renaissance II: the New Armalite AR-10B

Concurrently with and following the AR-18 project, discussed below, ownership of ArmaLite Incoroprated changed hands a number of times and, independent of this, another two companies—Eagle Arms and Lewis Machine & Tool Co., both long-time producers of parts for the M16—capitalized on the expiration of the patents Colt's had bought from Fairchild covering the gas operating system used in the AR-10 and AR-15.

In 1994, the president of Eagle Arms, retired colonel Mark Westrom, decided to try and develop a .30-caliber rifle along the lines of the AR-10. His quest for engineering data and expertise got him in touch with the then-current owner of ArmaLite, John Ugarte.

In 1995, Westrom purchased ArmaLite Incorporated, and with the help of members of the original design team from the 1950s, such as Tom Tellefson, began work on what was to become a modern-day ArmaLite AR-10.

Owning the rights to both the name ArmaLite and the AR-10 model designation, Westrom took a different approach to developing his new rifle than Eugene Stoner and Reed Knight had done a few years before with the SR-25. Not being as concerned with military contracts or commonality with the AR-15, Westrom focused on incorporating original design elements as developed by ArmaLite/Fairchild and Artillerie-Inrichtingen, as well as completely novel parts configurations, all with an eye to robustness and reliability.

Testing his individual parts on a hand-crafted lower receiver and a shop-modified SR-25 upper receiver, Westrom made extensive and novel use of

335. ArmaLite past president Mark Westrom, proudly holding an ArmaLite AR-10B, intentionally configured as closely as possible to the original AR-10, right down to the location of the cocking handle within the carrying handle.
courtesy Christopher R. Bartocci

computer engineering software to both design new elements and test them. Given this relatively unprecedented reliance on virtual modeling, no prototypes of the modern generation of AR-10s were ever built, with the first example based on the new design being the very first weapon off the assembly line.

Christened the AR-10B, as the natural successor, both in name and in design to ArmaLite's original AR-10B (fig. 57), the new rifle was patterned very closely indeed after the early AR-10s, right down to the brown Sudanese-style furniture and the top-mounted charging handle located within the carrying handle. It became an instant success, receiving the acclaim of firearms authorities and garnering wide commercial sales.

# Variations of the ArmaLite AR-10

336. Right side view of the author's AR-10A4, home-modified with a paracord-wrapped handguard.

author's collection

Several variants were made, including a scoped model with flat-topped receiver rail (the AR-10T); the basic infantry model AR-10A2, with integral carrying handle and adjustable M16A2-type rear sight; and the AR-10A4, which as shown appears very similar to the standard AR-15A2 with a flat upper receiver with a rail for mounting sights and accessories.

Envisaged and realized as a combination of original design elements along with a small amount of AR-15 parts commonality (handguard, grip, and sights) together with newly-developed, extremely tough components, the AR-10B delivered on the promise of Eugene Stoner's original vision for the AR-10: a lightweight, accurate, dependable and modular battle rifle.

ArmaLite's new AR-10B was one of the very first modern rifles of its type to be chambered for the powerful 7.62mm NATO cartridge, and in a market that over the last few years has become flooded with 7.62x51mm copies of the AR-15 design, it stands as one of the finest, and the only one that can be called a true AR-10.

Not only becoming successful in the civilian shooting sports community, the resurrected ArmaLite AR-10 has also gained success in the international military and police field.

Still being built and going strong with excellent sales, the semi-automatic "Modern AR-10", weighing from 7 lbs. 12 oz. to 9 lbs. 13 oz., depending on the variant, surpassed the total production numbers of

337. An embroidered cloth brassard, featuring the new ArmaLite lion logo.         Institute of Military Technology

both ArmaLite/Fairchild and Artillerie-Inrichtingen within a very short time of its release.

Competing directly with the Knight's Armament SR-25, the new AR-10 lost out narrowly in trials to select the Army's M110 Semi-Automatic Sniper Rifle designation—due to its plethora of beefed-up and original-pattern parts that did not share commonality with the AR-15—but essentially being defeated by oneself is no shame at all.

In addition to its civilian sporting and defensive roles, the modern AR-10 has been adopted and is used, both as a sniper weapon and in its original role of lightweight infantry rifle, by a number of American and international military and police organizations, as well as special forces units. Notable among these (generally the ones that are not so secretive as to withhold information about what weapons they use) are the Brazilian, South African, and Canadian militaries, with the AR-10(T) being adopted by Canadian Army Special Operations elements in 2002.

# Renaissance III: the LMT L129A1 "Sharpshooter", from Law Enforcement International

As discussed and described in the 2004 Collector Grade title *Black Rifle II*, for some time the Lewis Machine & Tool Company (LMT) of Moline, Illinois has been in the forefront of innovation with such improvements as the "lobstertail" double-spring extractor and the Monolithic Rail Platform (MRP) for the M16/M4 series of weapons.

Following this lead, during 2007 - 2008 LMT developed a 7.62mm NATO caliber version of the AR-15, producing a modular weapon system based on the generic ArmaLite AR-10 to meet a perceived military requirement for a weapon with a greater range potential than that offered by the current 5.56mm small-caliber high-velocity assault rifle round.

## With the British Army in Afghanistan

In August, 2009 the U.K. MoD published an unexpected "Urgent Operational Requirement" (UOR) for a 7.62mm "Sharpshooter" rifle for use by the British Army in Afghanistan, to permit them to respond effectively to Taliban units armed with Soviet Dragunovs and PKMs. Excerpts from the original contract solicitation, dated August 4, 2009, read as follows:

> . . The Army has an Urgent Operational Requirement (UOR) for a Sharpshooter Rifle to be fielded for Pre-Deployment Training (PDT) for Operation Herrick 11 from January 2010, and fielded operationally from March 2010. These dates are not tradeable . .
>
> The User requires a semi-automatic 7.62x51mm NATO rifle capable of rapidly incapacitating an unprotected target with a single shot at ranges in excess of 600m [the actual maximum range was established as 900 meters].
>
> The Sharpshooter Rifle will be the primary personal weapon of designated Sharpshooters. The Sharpshooter therefore requires the ability to patrol with the Sharpshooter Rifle and to fight with it at close quarters, including the conduct of rapid tactical Fire and Movement, without impediment.
>
> The rifle must be of a weight and ergonomic design that allows its accurate use from any firing position and does not hinder rapid tactical movement between firing positions.
>
> The User requires 24-hour capability, including low and zero light conditions through the use of in-service and other accessories, all of which are compatible with Mil Std 1913 (i.e. Picatinny Rail) . .

The original requirement was for a "one-time" purchase of a total of 440 units, including spares, with delivery of the first 24 weapons for initial training by January 31, 2010, followed by 144 for initial PDT and deployment by March 31, 2010, and a further 272 weapons by June 30, 2010.

The solicitation continued, as follows:

> . . Weapons for initial trials and initial training must be full production build standard, including accessories.
>
> Some latitude in detachable accessories may be allowed for the single trials weapon.
>
> . . Contract conditions will reflect the situation that "time is of the essence of this Contract" for the 440 weapons detailed above.
>
> The weapon chosen to meet the Sharpshooter Rifle Urgent Operational Requirement may also subsequently be selected to meet other emerging requirements without further competition. This may ultimately include eventual fielding to all regular and reserve units of the Army, Royal Marines and Royal Air Force Regiment.
>
> In addition to the 440 UOR weapons, Tenders will therefore be also invited for Options to supply additional quantities of weapons . .

Submissions were received from H&K (HK417), FN (SCAR), Oberland Arms, Sabre Defence, Knight's Armament (SR-25), and Law Enforcement International (LEI), a U.K.-based manufacturer and supplier of specialized military small arms and accessories which had been fully involved with LMT in the development of their original 7.62mm weapon, designated the LM308MWS, production of which had just begun. The selective-fire version manufactured to LEI specifications was given the designation LM7.

Very shortly after competitive trials in 2010, the U.K. MoD short-listed two weapons - the HK417 and the LM7. To everyone's surprise, the contract was awarded for the LM7, on the grounds of both superior performance and lower price. The LM7 was accordingly adopted by the British Army as its new "Sharpshooter" rifle under the designation "Assault Rifle, Sharpshooter, 7.62mm L129A1." The "one-time buy" quantity was increased to 500 L129A1 rifles, and these weapons were immediately deployed for operational use. The L129A1 was issued to the British Army in a desert tan finish, and to the MoD Police in black.

The NATO stock number (NSN) assigned to the L129A1 is 1005-99-226-6708. The '99' country code (U.K.) is explained by the fact that LEI won the contract as the bidder, and subcontracted manufacture of the rifles to LMT, where the guns are U.S.-made to British specifications. LEI also supply, manufacture or source the accessories (carry bags, singlepoint slings, etc.) for the L129A1.

Once these weapons were issued in Afghanistan, their effectiveness quickly became apparent. The longest recorded confirmed hit was at a distance of over 1,000 meters.

After the British withdrew from Afghanistan, it was decided to retain this UOR weapon capability, and the LEI contract was extended and increased to U.S. $20 million. Today, the L129A1 has established itself as a reliable, accurate and well-respected weapon in the British Army. Offering enhanced capability and range, its role is to "supplement"—not replace—existing 5.56mm weapons.

The exact number of L129A1s in U.K. inventory today is classified, although significantly more examples have been purchased, bringing current MoD stocks well into the four figures.

Other countries have subsequently purchased LMT 7.62mm models for their military and police users. The New Zealand Army, for example, chose a version with a 20" barrel, and they are very happy with their guns. The U.K. MoD Police use them fitted with both 16" and short 13.5" barrels.

## Describing the LEI L129A1 Self-Loading Rifle

338. Right side view of an early L129A1 with scope removed and emergency or 'back-up' folding iron sights raised.                National Firearms Centre collection, photo by Richard Jones

339 (following page). The descriptive specification sheet on the L129A1, produced by Law Enforcement International Ltd in 2010.                courtesy Greg Felton

# L129A1™  7.62mm "Sharpshooter" Rifle

British Army L129A1 variant of the LM7 (with optical sight and front grip)

The L129A1 is a modular weapon system, representing the next stage in the evolution of the M16 rifle to meet the changing needs of the 21st Century. The 7.62mm calibre provides enhanced accuracy and extended range. The system features a unique one piece (patent pending) forged upper receiver with four fully integrated Picatinny rails. The receiver is machined from a solid billet of metal, providing greater strength, rigidity and heat dissipation than a conventional "two piece" Picatinny rail upper receiver/forend. Improved heat dissipation also allows a higher firing capacity than a conventional rifle.

The L129A1 incorporates a fully "free floating" barrel for improved accuracy; the barrel mounting system allows barrels to be easily changed for one of a different length to meet varying mission requirements. The inherent strength and rigidity of the integrated rail system allows a host of accessories to be securely fitted. The 540mm full length top rail accommodates the latest night vision, thermal and I² devices with ease.

The L129A1 can be supplied with several stock options, including fixed and retractable types. Barrels can be supplied in various lengths from 343mm (13.5") to 508mm (20"). A number of options and accessories are available, including an M203 40mm grenade launcher and a clip on suppressor (silencer)

In 2009, after competitive trials, the British Army adopted its new "Sharpshooter" rifle under the military designation L129A1  (NSN: 1005-99-226-6708).  A fully automatic version, the L129A2, has also been codified.

## Specifications:-

| | |
|---|---|
| **Calibre:** | 7.62x51mm NATO |
| **Operation:** | Gas operated, semi-automatic |
| **Barrel lengths:** | 343 mm, 406 mm, 457 mm, 508 mm |
| **Weight (L129A1):** | 4.5 Kg |
| **Overall length (L129A1):** | |
| Stock extended | 99 cm |
| Stock retracted | 90 cm |
| **Magazine capacity:** | 20 rd |

**L129A1. The new British Army "Sharpshooter" rifle.**

Exclusive worldwide distribution by:-

## Law Enforcement International Ltd.

P.O. Box 328, St. Albans
Herts AL4 0WA. England
Tel.  01727-826607
Fax. 01727-826615
Email. lei@lei.co.uk

340. An L129A1 (left) and an L7A2 GPMG, deployed with the Royal Marines, part of the International Security Assistance Force (ISAF) in Afghanistan.

Note the RMR red dot sight atop the Trijicon TA648 scope.        U.K. Ministry of Defence photo, courtesy LEI

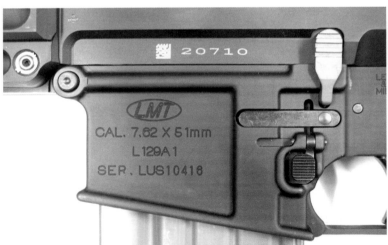

341. Left side closeup of the receiver of an L129A1, showing markings indicating manufacture by Lewis Machine & Tool Co. (LMT).

The three-letter serial number prefix "LUS" stands for LMT, U.K., and Semi-Auto.        courtesy Greg Felton

342. Left side closeup of the selective-fire LM7, as also procured by LEI from Lewis Machine & Tool.

The last letter of the three-letter serial number prefix "LUA" indicates a full-auto fire capability.

courtesy Greg Felton

343. The standard kit for the MoD for each L129A1, less the Pelican hard case and a zippered soft case (tan for the Army, black for the MoD Police). Each rifle also comes with a complement of 20 magazines.

courtesy Law Enforcement International

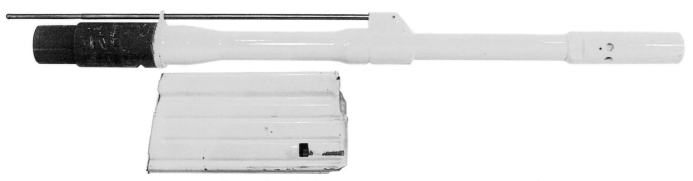

344. The "Safe" Blank Firing System (SBFS) for the L129A1. The smoothbore barrel and magazine are painted yellow for instant identification, and a bullet trap is affixed to the muzzle to stop live rounds if accidentally fired.

The magazine is restricted so that it will feed blanks only. courtesy Greg Felton

As with the LMT Monolithic Rail Platform (MRP) for the M16 series, the L129A1 is built on a one-piece forged upper receiver with four integral MIL-STD-1913 Picatinny rails. This entire unit is machined from a solid billet of metal, providing great strength, rigidity, and improved heat dissipation. The L129A1 has a fully free-floating barrel for improved accuracy, and the barrel mounting system allows barrels to be quickly and easily replaced by the user with ones of different lengths to meet varying mission requirements. A 16" (406mm) stainless steel match barrel with 1 in 11.25" rifling twist fitted with an open four-prong flash suppressor was selected as standard for the L129A1. During the selection trials the LMT 16" barrel actually outperformed 18" and 20" barrels from other bidders.

The inherent strength and rigidity of the 540mm-long top rail can accommodate a full range of optical or night vision devices, and the L129A1 comes fitted with a Trijicon TA648 6-power optic sight, sometimes enhanced with an RMR red dot sight atop it, plus emergency or 'back-up' folding iron sights. The telescoping stock of the L129A1 can be adjusted for length by the user. A detachable folding bipod and folding front pistol grip are also issued with each "Sharpshooter" rifle.

The operating system remains the Stoner direct-action gas system and multi-lugged, front-locking rotating bolt. The magazine is a detachable 20-round box. The British service L129A1 version is capable of semi-automatic fire only. It weighs 4.5kg (9.92 lbs.) and measures 900mm (35.4") overall in length with the stock retracted, and 990mm (38.9") with stock extended.

345. The "business end" of a British Royal Marine sniper team in Afghanistan. The longest recorded confirmed hit by an L129A1 during this deployment was at a distance of over 1,000 meters.

Note the free-floating barrel protruding from the rigid, monolithic receiver/rail, shown to good effect in fig. 343.

U.K. Ministry of Defence photo, courtesy LEI

# Latter Days at ArmaLite Incorporated

## Resurrecting Two Unproduced Models: the AR-12 and AR-14

Meanwhile, after Fairchild had sold ArmaLite to its original principals, who recast the company as ArmaLite Incorporated, it went through decades of uncertainty and limited success. The new source of R&D funding and primary shareholder at the time of ArmaLite's independence was the Capital Southwest Corporation of Dallas, Texas, and ArmaLite took advantage of this opportunity to chart a course for new revolutionary advancements in the field of small arms. The natural starting point for this endeavor was Stoner's final creation for the company, the design of which he had begun while ArmaLite was still a division of Fairchild.

Stoner had been taking another look at the AR-12, which had previously existed only as an artist's concept drawing (fig. 47), in the hopes of developing a weapon which would offer similar performance to the AR-10 but be much simpler to produce. It was intended to be marketed both as a finished product and as a candidate for adoption and domestic production by poorer nations which might balk at the price of an AR-10, FAL, or G3. Along with plans to incorporate the operating system of the AR-10 in the AR-12, a revived AR-14 (fig. 36), the conventionally-stocked sporting rifle counterpart which had been used to illustrate the gas system patent (fig. 37), was also in the works.

# The 7.62mm ArmaLite AR-16

346. Right side view of the wood-stocked AR-16 "rifle," caliber 7.62mm NATO, a conventional gas piston design produced after ArmaLite no longer had the right to utilize the Stoner gas system patent. This was one of only two AR-16s ever manufactured.

Institute of Military Technology

347. Left and right side views of the "carbine" version of the 7.62mm AR-16, shown with the bolt assembly locked open.                    Institute of Military Technology

With the sudden and surprising news that the gas system patent had been transferred to Colt's, it was realized that the AR-12 would need a new operating system. Stoner replaced the gas tube with a conventional short-stroke piston, and incorporated a dual recoil spring assembly held captive in the lower receiver. Gone also was the signature receiver extension tube, with the buttstock now filling a more traditional single-function role.

Leaving the reworked AR-12 prototype unfinished, Stoner nonetheless demonstrated that the design was viable, and the team got to work on a new weapon that would not infringe on the intellectual property sold to Colt's, but which would feature as much functionality in common with the AR-10 as possible.

Cosmetically resembling a shorter AR-10, the newly-dubbed AR-16 still fired the same 7.62mm cartridge from a steel magazine that was not contained in a magazine well in the lower receiver, but was removed and inserted by the more mundane "rock and lock" method, thus losing the functionality and modularity of the AR-10 and AR-15's push-button magazine release. Use of the Johnson-Stoner multi-lugged rotating bolt was carried over, with the bolt head now fitted in a large squared-off bolt carrier which was activated by direct impingement from a conventional gas piston, located above the bore.

The new rifle was purposely formulated to fill the niche for which the AR-12 had been intended, cutting manufacturing costs drastically by incorporating a receiver made not from expensive forged aluminum but from riveted and welded sheet metal stampings.

Only two finished prototypes of the AR-16 were fabricated at Costa Mesa, and while the design was basically a carbine, with both examples fitted with shorter barrels than those found on the AR-10, one was referred to as the "rifle" while the other, with a slightly shorter 16.1" barrel, was named the "carbine." Instead of the top-mounted charging handle inside an integral carrying handle, the right side of the AR-16 bolt carrier was fitted with a more conventional reciprocating cocking handle.

The top of the receiver was flat, with a rear sight assembly mounted on its back portion. The more conventionally-shaped stock and lowered sight plane meant that the completely in-line action of the AR-10, with its attendant recoil control benefits, was not carried over from the designs sold to Colt's. Although the butt of the wooden, side-folding stock was not as low as those found on more traditional designs like the M1 or M14, a certain amount of pivot and muzzle rise was introduced, like that experienced on similarly set-up weapons such as the FAL, G3, or Kalashnikov.

With the interest of the world's militaries for shoulder arm designs now moving rapidly in the direction of the assault rifle, the AR-16 was sidelined and never pursued by ArmaLite, which was experiencing a period of intense flux during these years, and only the two wood-stocked AR-16 prototypes were ever produced.

## The 5.56mm ArmaLite AR-18

348. Right side view of a 5.56mm ArmaLite AR-18, fitted with a 40-round magazine.
Designed for cheap, economical production and constructed of sheet steel and plastic, the AR-18 was the most successful design produced by ArmaLite after becoming independent from Fairchild.

courtesy Christopher R. Bartocci

With Stoner's departure in 1961, the remaining design team at ArmaLite, now capably headed by Arthur Miller, scaled down the AR-16 into a handy assault rifle chambered for the 5.56x45mm cartridge, designating it the AR-18. Also constructed largely of stamped steel, the AR-18 was, like the somewhat larger AR-16, intentionally designed to be cheap and easy to manufacture on the very simplest of machine tools.

Unlike its predecessor, the AR-18 brought back some desirable features of the AR-10, like the ejection port dust cover and the push-button magazine release, and resurrected the concept of lightweight polymer furniture, thus increasing the strength and durability of these components.

The AR-18 was extensively tested and marketed (as much as the newly-independent ArmaLite could afford) as the new answer to lightweight rapid firepower, not only for the U.S. military but for allied nations; none of which had yet adopted an assault rifle.

349. Left side closeup of the selective-fire 5.56mm AR-18,
serial no. A5413, showing markings indicating manufacture by ArmaLite, Inc. of Costa Mesa, California.
courtesy Christopher R. Bartocci

The AR-18 was evaluated by the U.S. military twice, and while gaining a good reputation, especially in its second test in 1969, it was not able to unseat the AR-15.

In November, 1969, seeing potential for the AR-18 on the international market, and having received State Department authorization to export and market it abroad, Charles Dorchester, then serving as chairman of the board, along with president Richard Klotzly, acquired the majority share in the company held by Southwest Capital.

Offered as both the selective-fire (military) AR-18 and as a semi-automatic civilian model called the AR-180, this weapon weighed 6 lbs. 11 oz. Detachable box magazines of 20-, 30-, and 40-round capacities were produced.

Manufactured in numbers far greater than the original AR-10, the AR-18 met with some commer-cial success, but never gained the military acceptance for which it was designed. No major military institutions adopted the AR-18, and only small quantities were sold, although it would end up becoming the basis for, and inspiration of, some of the most successful alternative assault rifle designs of today—like the AR-10, achieving its greatest successes through its copies and clones—but that is a subject for another book.

The vast majority of the AR-18's purchasers preferred the semi-automatic, civilian-legal version, the AR-180. Produced first at ArmaLite's Costa Mesa headquarters, where an assembly line was at last set up, then licensed out first to Howa in Japan and then to Sterling Armament in the U.K., the AR-180 sold relatively well among the sporting arms markets of countries where it was legal.

## The End of the Road for ArmaLite, Inc.

Following the ultimate failure of the U.S-built AR-18 to gain widespread military adoption, ArmaLite entered a very dark period, kept afloat only by commercial sales of the AR-180; production of which continued until 1983. That year the board of the company sold it to the Elisco Tool Company of Manila, a Philippine firm that had produced licensed copies of the M16 as the Model 613-P rifles and 653-P carbines for that country's armed forces. However, the ousting of the military dictator of that island nation in 1987 saw the end of the Pacific attempt at a revival of ArmaLite and the AR-18.

350 (right). Charles Dorchester, in a photograph taken in ArmaLite's Costa Mesa headquarters in 1984.
photo courtesy Marc Miller

# In Memoriam

## Eugene Morrison Stoner, November 22, 1922 - April 24, 1997

Eugene Stoner died of cancer on April 24, 1997, too soon to witness the successful modern-day renaissance of his patented ideas which he had first drawn up in the machine shop of the Whittaker Aircraft Controls Company in the early 1950s, and first tested as the steel-and-wood Stoner M-8 (X-01) prototype in 1954, later termed "the first AR-10" (fig. 32).

351 (right). Eugene Stoner, holding a brand-new SR-25 sniper rifle, in a photograph taken near his home in Palm City, Florida, in the spring of 1996.
courtesy Knight's Armament Co.

# The AR-10 Is Here to Stay

After a tumultuous existence that few firearms designs can claim, especially those that have ended up as successful weapons in modern production, the AR-10 is finally widely appreciated in our time for the genius of its design, with its direct gas operating system forming the basis of several successful modern variants. More than sixty years after Eugene Stoner built his first M-8, the spirit of the ArmaLite AR-10 is not only an established part of the world of modern battle rifles and the progenitor of many of them, but its popularity continues to rise.

It is likely that for generations to come, the AR-10's unique combination of heavy firepower, light weight, modularity, extreme controllability, and peerless accuracy, envisioned generations ago by a few idealistic pioneers, will continue to receive the recognition that it has deserved since many of today's grandparents were small children.

# *Bibliography*

## Books

*The Black Rifle: M16 Retrospective* (2nd ed.) by R. Blake Stevens and Edward C. Ezell. Collector Grade Publications Inc., Cobourg, ON, 1992

*Black Rifle II: The M16 Into the 21st Century* by Christopher R. Bartocci. Collector Grade Publications Inc., Cobourg, ON, 2004

*The Browning Machine Gun Volume I: Rifle Caliber Brownings in U.S. Service* by Dolf L. Goldsmith. Collector Grade Publications Inc., Cobourg, ON, 2005

*Deadly Business - Sam Cummings, Interarms & the Arms Trade* by Patrick Brogan and Albert Zarca. W. W. Norton & Company, New York and London, 1983

*The Devil's Paintbrush: Sir Hiram Maxim's Gun* by Dolf L. Goldsmith. Collector Grade Publications Inc., Cobourg, ON, Canada, 2002

*Death from Above: The German FG42 Paratroop Rifle* by Thomas B. Dugelby and R. Blake Stevens. Collector Grade Publications Inc., Cobourg, ON, 2007

*The FAL Rifle: Classic Edition* by R. Blake Stevens and Jean E. Van Rutten. Collector Grade Publications Inc., Cobourg, ON, 1993

*Famous Rifles and Machine Guns* by A. J. R. Cormack. Barrie & Jenkins, London, 1977

*The First Sudanese Civil War: Africans, Arabs, and Israelis in the Southern Sudan, 1955-1972* by S. S. Poggo. Palgrave Macmillan, New York, 2011

*The FN49: The Rifle that Ran out of Time* by R. Blake Stevens. Collector Grade Publications Inc., Cobourg, ON, 2011

*Full Circle: A Treatise on Roller Locking* by R. Blake Stevens. Collector Grade Publications Inc., Cobourg. ON, 2006

*Hitler's Garands: German Self-Loading Rifles of World War II* by Darrin D. Weaver. Collector Grade Publications Inc., Cobourg, ON, 2001

*Kalashnikov: The Arms and the Man* - A Revised and Expanded Edition of *The AK-47 Story* by Edward Clinton Ezell. Collector Grade Publications Inc., Cobourg, ON, 2001

*La Guerra de la Restauración y la Revolución de Abril* [The Restoration War and the April Revolution] by J. E. Bosch. Corripio, Santo Domingo, Dominican Republic, 1984

*The Last Adventurer* by R. Steiner. Little, Brown, and Company, New York, 1978

*The Mauser Archive* by Jon Speed. Collector Grade Publications, Inc., Cobourg, ON, 2007

*The M1 Garand Rifle* by Bruce N. Canfield. Mowbray Publishing, Woonsocket, R.I., 2013

*Proud Promise: French Autoloading Rifles, 1898-1979* by Jean Huon. Collector Grade Publications Inc., Cobourg, ON, 1995

*Rock in a Hard Place: The Browning Automatic Rifle* by James L. Ballou, Collector Grade Publications Inc., Cobourg, ON, 2000

*The Secret War in the Sudan: 1955-1972* by E. O'Ballance. Faber & Faber, London, 1977

*U.S. Rifle M14: from John Garand to the M21* (2nd ed.) by R. Blake Stevens. Collector Grade Publications Inc., Cobourg, ON, 1991

*War Baby Comes Home: The U.S. Caliber .30 Carbine Volume II* by Larry L. Ruth. Collector Grade Publications Inc., Cobourg, ON, 1993

*The World's Assault Rifles & Automatic Carbines* by Daniel D. Musgrave and Thomas B. Nelson. TBN Enterprises, Sun Valley, CA, 1967

# Handbooks and Manuals

"The ArmaLite AR-10 Infantry Rifle Caliber 7.62 (NATO)." Staadtsbedrijf Artillerie-Inrichtingen, Hembrug - Zaandam, the Netherlands, September, 1959

*Descripcion del ArmaLite AR-10 Fusil Basico de Infanteria: Calibre 7.62 NATO* ["Instructions for the ArmaLite AR-10 Basic Infantry Rifle: Caliber 7.62mm NATO."] Staatsbedrijf Artillerie-Inrichtingen, Hembrug - Zaandam, the Netherlands, 1959

"Description of the ARMALITE AR-10 Basic Infantry Weapon Caliber 7,62 (NATO)." Interarmco, Ltd., March, 1958

"Description of the ARMALITE AR-10 Basic Infantry Weapon Caliber 7,62 (NATO)." Interarmco, Ltd., May, 1958

"Description of the ARMALITE AR-10 Basic Infantry Weapon Caliber 7,62 (NATO)." Interarmco, Ltd., January, 1959

"Description of the ArmaLite AR-10 Basic Infantry Weapon Caliber 7,62 (NATO)." Staadtsbedrijf Artillerie Inrichtingen, Hembrug - Zaandam, the Netherlands (undated)

"Description of Sniper Scope 3 x 25 for ArmaLite AR-10 Rifle." Staadtsbedrijf Artillerie-Inrichtingen, Hembrug - Zaandam, the Netherlands, August, 1958

"Handbook on the ArmaLite AR-10 Infantry Rifle: Caliber 7.62mm NATO." Staatsbedrijf Artillerie-Inrichtingen, Hembrug-Zaandam, the Netherlands, 1960

"Handbook on the ArmaLite AR-10 Infantry Rifle: Caliber 7.62mm NATO." Staadtsbedrijf Artillerie-Inrichtingen, Hembrug - Zaandam, the Netherlands, June, 1961

"Illustrated Parts List of the ArmaLite Infantry Rifle Caliber 7,62 NATO AR-10." Staadtsbedrijf Artillerie-Inrichtingen, Hembrug - Zaandam, the Netherlands, March, 1960

"Instruction for Fitting the Barrel of the AR-10 Rifle and for the Use of Inspection Gauges." Staadtsbedrijf Artillerie-Inrichtingen, Hembrug - Zaandam, the Netherlands, January, 1960

"Instruction Manual for Operation of the ArmaLite light Machine Gun, Caliber 7.62 NATO." Staatsbedrijf Artillerie-Inrichtingen, Hembrug-Zaandam, the Netherlands, 1960

# Reports and Pamphlets

"The ArmaLite AR-10 Automatic Rifle." ArmaLite informational pamphlet. Costa Mesa, CA, December, 1955

"ArmaLite AR-10 Rifle Tests: Springfield Armory, January 8 through 18th, 1957" by C. H. Dorchester, January 18, 1957. Fairchild Industries collection, Smithsonian Institution Archives

"ArmaLite: Status Report & Conclusions and Agreements." ArmaLite internal report, December 19, 1957. Fairchild Industries collection, Smithsonian Institution Archives

"AR-10 ArmaLite: Lightweight Basic Infantry Rifle." ArmaLite promotional literature, Costa Mesa, CA, 1957

"AR-10 Rifle: Recommended Changes." Consultant report by Melvin M. Johnson, Jr., December 8, 1956. Fairchild Industries collection, Smithsonian Institution Archives

"Brief Description of AR-10 Automatic Rifle." ArmaLite internal report, April 23, 1955. Fairchild Industries collection, Smithsonian Institution Archives

"Brief Report: Operations." ArmaLite internal report, January 5, 1955. Fairchild Industries collection, Smithsonian Institution Archives

"Comments on the Fairchild AR-10 Rifle." Internal Report by Melvin M. Johnson, Jr., September 14, 1957. Fairchild Industries collection, Smithsonian Institution Archives

"Evaluation Test, ArmaLite Rifle, 7.62mm, AR-10," February 5, 1957. (Ordnance Corps Publication No. SA-TR11-1091). U.S. Army Ordnance Corps, Springfield Armory, Research and Development, Springfield, MA

"Fairchild Rifle in Production." Press release from Fairchild Engine and Airplane Corporation, March 10, 1958. Fairchild Industries collection, Smithsonian Institution Archives

"History." ArmaLite corporate history document. Costa Mesa, CA, 1971

"Information on Development, Production, and Sales of ArmaLite: Rifles." Confidential Intercorporate bulletin, Staatsbedrijf Artillerie-Inrichtingen, Hembrug - Zaandam, the Netherlands, July, 1958

"Loaded for Bear." Press release from Fairchild Engine and Airplane Corporation, August, 1958. Fairchild Industries collection, Smithsonian Institution Archives

"The Long-Coming Rifle." Press release by M. A. McCulloch, November 10, 1956. Fairchild Industries collection, Smithsonian Institution

"1957 Report and 1958 Prospectus." ArmaLite internal report, December 15, 1957. Fairchild Industries collection, Smithsonian Institution Archives

"Observations Prepared by Personnel of the ArmaLite Division of the Fairchild Engine & Airplane Corporation with Respect to the Second, but not Final, Report of the Springfield Armory Evaluation Test on the ArmaLite AR-10 Rifle." ArmaLite internal report, January, 1957. Fairchild Industries collection, Smithsonian Institution Archives

"Outline of Proposed Activity for 1957." ArmaLite internal report, December, 1956. Fairchild Industries collection, Smithsonian Institution Archives

"Preliminary Tests of FE&A Corp. AR-10 ArmaLite Light Automatic Rifle at Springfield Armory, 3-7 December 1956." Consultant report by Melvin M. Johnson, Jr., December 8, 1956. Fairchild Industries collection, Smithsonian Institution Archives

"Report on AR-10 in Lightweight Rifle Project." Associated Press media bulletin, November 26, 1956. Fairchild Industries collection, Smithsonian Institution Archives

"Report on A Test of Rifle, Caliber 7.62-MM, AR-10." Report No. DPS-101 (OMS Code No. 5530.11.553). Infantry and Aircraft Weapons Division, Development and Proof Services, Aberdeen Proving Ground, November, 1960

"Report on European Trip: Fairchild Aircraft Division." Fairchild Engine and Airplane Corporation, September 11, 1956. Fairchild Industries collection, Smithsonian Institution Archives

"Report on Marine Corps Tests of AR-10 Rifle" by Charles H. Dorchester. ArmaLite internal report, August 16, 1957. Fairchild Industries collection, Smithsonian Institution Archives

"Report on Pentagon Meetings, 11 December 1956, regarding German Scientific Group, regarding AR-10 Rifle." Consultant report by William Johnson, December 11, 1956. Fairchild Industries collection, Smithsonian Institution Archives

"Springfield Armory Evaluation." Internal report by George Sullivan, March 30, 1957. Fairchild Industries collection, Smithsonian Institution

"Springfield Armory Tests on AR-10 BIW". Internal report by George Sullivan, December 15, 1956. Fairchild Industries collection, Smithsonian Institution

"Springfield Tests and Future Policy, re: AR-10 Basic Infantry Weapon." Internal report by George Sullivan, December 22, 1956. Fairchild Industries collection, Smithsonian Institution

"Status of Marine Test Program on AR-10 Weapon." Internal report by George Sullivan, August 18, 1957. Fairchild Industries collection, Smithsonian Institution

"A Test of Rifle, Caliber 7.62mm, AR-10" (Ordnance Corps Report No. DPS-101). Infantry and Aircraft Weapons Division, Aberdeen Proving Ground, November 14, 1960. Armed Service Technical Information Agency, Arlington, VA

"Tomorrow's Rifle Today!" Interarmco, Washington, D.C., 1957

"What is New in the New Light Rifles." Consultant Report by Melvin M. Johnson, Jr., October 29, 1956. Fairchild Industries collection, Smithsonian Institution Archives

# Articles

"America's Biggest Arms Merchant" by William B. Edwards. *Guns* magazine, June, 1965

"The ArmaLite AR-10: The Most Modern Combat Rifle." Staatsbedrijf Artillerie-Inrichtingen, Hembrug - Zaandam, the Netherlands, 1959

"Armed Forces: The Aluminum Rifle." *TIME* magazine, December 3, 1956

"Arms Dealer Samuel Cummings Dies" by J. Y. Smith. *The Washington Post*, Saturday, May 2, 1998

"The AR-10" by C. McLoughlin. *Survival Weapons and Tactics*, September, 1982

"Better Late than Never: The M1C Garand Sniper Rifle" by Bruce N. Canfield. *The American Rifleman*, September, 2014

"The Celebration Continues: ArmaLite, Interarms 1984" by Mike Miller. *AR-10'er: The ArmaLite AR-10 Enthusiast's Monthly Newsletter*, July 16, 1984

"First Modern Combat Rifle: The Unknown Legend of the AR-10" by Major Sam Pikula. *Soldier of Fortune Fighting Firearms*, summer, 1996

"Golden Days at Armalite" By D. T. McElrath. *The American Rifleman*, December, 2004

"Great Expectations: AR-10" by Terry Edwards. *Soldier of Fortune*, January, 1978

"The Hemisphere: Dominican Republic: The Coup that Became a War." *TIME* magazine, May 7, 1965

"Info Bits" by A. M. Olszewski. *AR-10'er: The ArmaLite AR-10 Enthusiast's Monthly Newsletter*, June 1, 1983

"The Interview: L. James Sullivan" by Dan Shea. Part I, *Small Arms Review*, March, 2008

"The Interview: L. James Sullivan" by Dan Shea. Part II, *Small Arms Review*, April, 2008

"The Interview: L. James Sullivan" by Dan Shea. Part III, *Small Arms Review*, May, 2008

"Interview with C. Reed Knight, Jr., Part I" by Dan Shea. *Small Arms Review*, February, 2009. Retrieved from http://www.smallarmsreview.com/display.article.cfm?idarticles=1222

"Interview with C. Reed Knight, Jr., Part II" by Dan Shea. *Small Arms Review*, March, 2009. Retrieved from http://www.smallarmsreview.com/display.article.cfm?idarticles=1211

"The M14: Boon or Blunder?" by Jack Lewis. *Gun World*, April, 1963

"A New Automatic Rifle" by Melvin M. Johnson, Jr. *Ordnance*, May-June, 1957

"New and lighter rifles signal Fairchild's entry in arms field." *The Washington Post and Times Herald*, November 27, 1956

"New Survival Weapon" by George Sullivan. *The American Rifleman*, January, 1957

"Sam Cummings, Merchant of Arms" by Charles Petty. *Guns* magazine, April, 1985

"The Story of AR-10" by William B. Edwards. *AR-10'er: The ArmaLite AR-10 Enthusiast's Monthly Newsletter*, April 16 and June 15, 1984

"The U.S. Army's Blunderbuss Bungle that Fattened Your Taxes" by John Tompkins. *True* magazine, April, 1963

# Unpublished Material

## Letters and Memoranda

Bennett, F. S. (1954, July 16) [Floyd Bennett writing to George Sullivan]. Fairchild Industries collection, Smithsonian Institution Archives

Boutelle, R. S. (1953, August 7). [Richard Boutelle writing to George Sullivan]. Fairchild Industries collection, Smithsonian Institution Archives

Boutelle, R. S. (1954, April 12). [Richard Boutelle writing to George Sullivan]. Fairchild Industries collection, Smithsonian Institution Archives

Boutelle, R. S. (1954, January 12). [Richard Boutelle writing to George Sullivan]. Fairchild Industries collection, Smithsonian Institution Archives

Boutelle, R. S. (1954, June 4). [Richard Boutelle writing to George Sullivan]. Fairchild Industries collection, Smithsonian Institution Archives

Boutelle, R. S. (1954, May 17). [Richard Boutelle writing to George Sullivan]. Fairchild Industries collection, Smithsonian Institution Archives

Boutelle, R. S. (1954, November 22). [Richard Boutelle writing to J. P. McConnell]. Fairchild Industries collection, Smithsonian Institution Archives

Boutelle, R. S. (Telegram, 1955, November 21). [Richard Boutelle writing to George Sullivan]. Fairchild Industries collection, Smithsonian Institution Archives

Boutelle, R. S. (1956, April 12). [Richard Boutelle writing to George Sullivan]. Fairchild Industries collection, Smithsonian Institution Archives

Boutelle, R. S. (1956, January 10). [Richard Boutelle writing to George Sullivan]. Fairchild Industries collection, Smithsonian Institution Archives

Cleaveland, P. S. (1954, April 6). [Paul Cleaveland writing to Richard Boutelle] Fairchild Industries collection, Smithsonian Institution Archives

Cleaveland, P. S. (1957, July 15). [Paul Cleaveland writing to Richard Boutelle]. Fairchild Industries collection, Smithsonian Institution Archives

Cleaveland, P. S. (1958, December 10). Telegram. [Paul Cleaveland writing to L. W. Davis]. Fairchild Industries collection, Smithsonian Institution Archives

Devers, J. L. (1954, July 16). [Jacob Devers writing to George Sullivan]. Fairchild Industries collection, Smithsonian Institution Archives

Devers, J. L. (1954, July 26). [Jacob Devers writing to George Sullivan]. Fairchild Industries collection, Smithsonian Institution Archives

Diepen, F. J. L. (1957, January 5). [Fritz Diepen writing to Richard Boutelle]. Fairchild Industries collection, Smithsonian Institution Archives

Dorchester, C. H. (1957, February 5). [Charles Dorchester writing to Warren Smith]. Fairchild Industries collection, Smithsonian Institution Archives

Dorchester, C. H. (1957, September 19). "Gun Stocks." [Correspondence with Subcontractor]. Fairchild Industries collection, Smithsonian Institution Archives

Dorchester, C. H. (1957, September 30). "Gun Stocks." [Correspondence with Subcontractor]. Fairchild Industries collection, Smithsonian Institution Archives

Drane, H. A. (1954, September 21). [Hugh Drane writing to Richard Boutelle]. Fairchild Industries collection, Smithsonian Institution Archives

Fairchild Engine and Airplane Corporation (1957, June 4). [Press Release]. Fairchild Industries collection, Smithsonian Institution Archives

Fairchild Engine and Airplane Corporation (1958, December 18). [Press Release]. Fairchild Industries collection, Smithsonian Institution Archives

Fairchild Engine and Airplane Corporation Press Release, January 1, 1959. Fairchild Industries collection, Smithsonian Institution Archives

Fairchild Engine and Airplane Corporation Press Release, October 5, 1960. Fairchild Industries collection, Smithsonian Institution Archives

Hollywood Gun Shop Invoice dated December 9, 1953 [Purchase of equipment for George Sullivan by Richard Boutelle]. Fairchild Industries collection, Smithsonian Institution Archives

Johnson, M. M. (1956, November 13). [Melvin Johnson writing to Richard Boutelle]. Fairchild Industries collection, Smithsonian Institution

Johnson, M. M. (1956, November 2). [Melvin Johnson writing to Richard Boutelle]. Fairchild Industries collection, Smithsonian Institution

N.V. Nederlandsche Maschinefabriek "Artillerie-Inrichtingen" (1958, July 30). Letter to Interarmco (Canada) Ltd., '-Gravenhage, Holland, re serial numbers of first 186 AR-10 rifles shipped to Sudan. Eric Kincel collection, courtesy the late R. H. G. Koster

Sullivan G. C. & Killen, L. (1957, June 8). Telegram. [George Sullivan and Leo Killen writing to Richard Boutelle]. Fairchild Industries collection, Smithsonian Institution

Sullivan G. C. (1957, September 19). Telegram. [George Sullivan writing to Richard Boutelle]. Fairchild Industries collection, Smithsonian Institution

Sullivan G. C. (1957, September 23). [George Sullivan writing to Jacob Devers]. Fairchild Industries collection, Smithsonian Institution

Sullivan, G. C. (1953, December 14). [George Sullivan writing to Richard Boutelle]. Fairchild Industries collection, Smithsonian Institution

Sullivan, G. C. (1953, November 11). [George Sullivan writing to Paul Cleaveland]. Fairchild Industries collection, Smithsonian Institution

Sullivan, G. C. (1953, October 28). [George Sullivan writing to Paul Cleaveland]. Fairchild Industries collection, Smithsonian Institution

Sullivan, G. C. (1954, April 16). [George Sullivan writing to Richard Boutelle]. Fairchild Industries collection, Smithsonian Institution

Sullivan, G. C. (1954, December 2). [George Sullivan writing to Richard Boutelle]. Fairchild Industries collection, Smithsonian Institution

Sullivan, G. C. (1954, February 12). [George Sullivan writing to Richard Boutelle]. Fairchild Industries collection, Smithsonian Institution

Sullivan, G. C. (1954, July 13). [George Sullivan writing to Jacob Devers]. Fairchild Industries collection, Smithsonian Institution

Sullivan, G. C. (1954, June 19). [George Sullivan writing to Richard Boutelle]. Fairchild Industries collection, Smithsonian Institution

Sullivan, G. C. (1954, March 31). [George Sullivan writing to Richard Boutelle]. Fairchild Industries collection, Smithsonian Institution

Sullivan, G. C. (1954, May 22). [George Sullivan writing to Richard Boutelle]. Fairchild Industries collection, Smithsonian Institution

Sullivan, G. C. (1954, November 9). [George Sullivan writing to Richard Boutelle]. Fairchild Industries collection, Smithsonian Institution

Sullivan, G. C. (1954, September 30). [George Sullivan writing to Richard Boutelle]. Fairchild Industries collection, Smithsonian Institution

Sullivan, G. C. (1955, December 18). [George Sullivan writing to Richard Boutelle]. Fairchild Industries collection, Smithsonian Institution

Sullivan, G. C. (1955, December 27). [George Sullivan writing to Richard Boutelle]. Fairchild Industries collection, Smithsonian Institution

Sullivan, G. C. (1955, February 5). [George Sullivan writing to Hugh Drane]. Fairchild Industries collection, Smithsonian Institution

Sullivan, G. C. (1955, January 8). [George Sullivan writing to Richard Boutelle]. Fairchild Industries collection, Smithsonian Institution

Sullivan, G. C. (1955, May 16). [George Sullivan writing to Richard Boutelle]. Fairchild Industries collection, Smithsonian Institution

Sullivan, G. C. (1956, December 15). [George Sullivan writing to Col. Roy Rayle]. Fairchild Industries collection, Smithsonian Institution

Sullivan, G. C. (1956, December). Telegram. [George Sullivan writing to Richard Boutelle]. Fairchild Industries collection, Smithsonian Institution

Sullivan, G. C. (1956, February 8). [George Sullivan writing to Richard Boutelle]. Fairchild Industries collection, Smithsonian Institution

Sullivan, G. C. (1956, June 23). [George Sullivan writing to Jacob Devers]. Fairchild Industries collection, Smithsonian Institution

Sullivan, G. C. (1956, June 23). [George Sullivan writing to Richard Boutelle]. Fairchild Industries collection, Smithsonian Institution

Sullivan, G. C. (1956, November 24). [George Sullivan writing to Harry Neilson]. Fairchild Industries collection, Smithsonian Institution

Sullivan, G. C. (1956, November 24). [George Sullivan writing to Col. Roy Rayle]. Fairchild Industries collection, Smithsonian Institution

Sullivan, G. C. (1956, November 26). [George Sullivan writing to Melvin Johnson]. Fairchild Industries collection, Smithsonian Institution

Winter, J. (1957, February 15). [Jan Winter writing to Richard Boutelle]. Fairchild Industries collection, Smithsonian Institution

Woodcock, F. H. (1957, July 6). [F. Woodcock writing to George Sullivan]. Fairchild Industries collection, Smithsonian Institution

## Manuscripts

*AR 10: Ons Leger Kiest een Geweer* ["The AR-10: Our Army Chooses a Rifle"] by R. H. G. Koster (n.d.)

*Het Geweer AR 10; De Weg Naar Nederland* ["The AR-10 Rifle; the Road to the Netherlands"]. R. H. G. Koster (n.d.)

*Het 'M+G Project' van de Artillerie-Inrichtingen* ["The MG Project of Artillerie-Inrichtingen"] R. H. G. Koster (n.d.)

# *Index*

## A

# B

# C

# Z